GW01337295

Handbook of Microwave Testing

中華民國七十一年　　月初版

發行所：國　興　出　版　社
發行人：楊　　國　　藩
　　　　新竹市西門街262號
　　　　電話：(035) 223129號
行政院新聞局局版台業字第0465號
總經銷：黎　明　書　店
　　　　新竹市中正路７２　號
　　　　電話：(035) 229418號
　　　　郵政劃撥七〇二七三號

實價：330 元

Thomas S. Laverghetta

Handbook of Microwave Testing

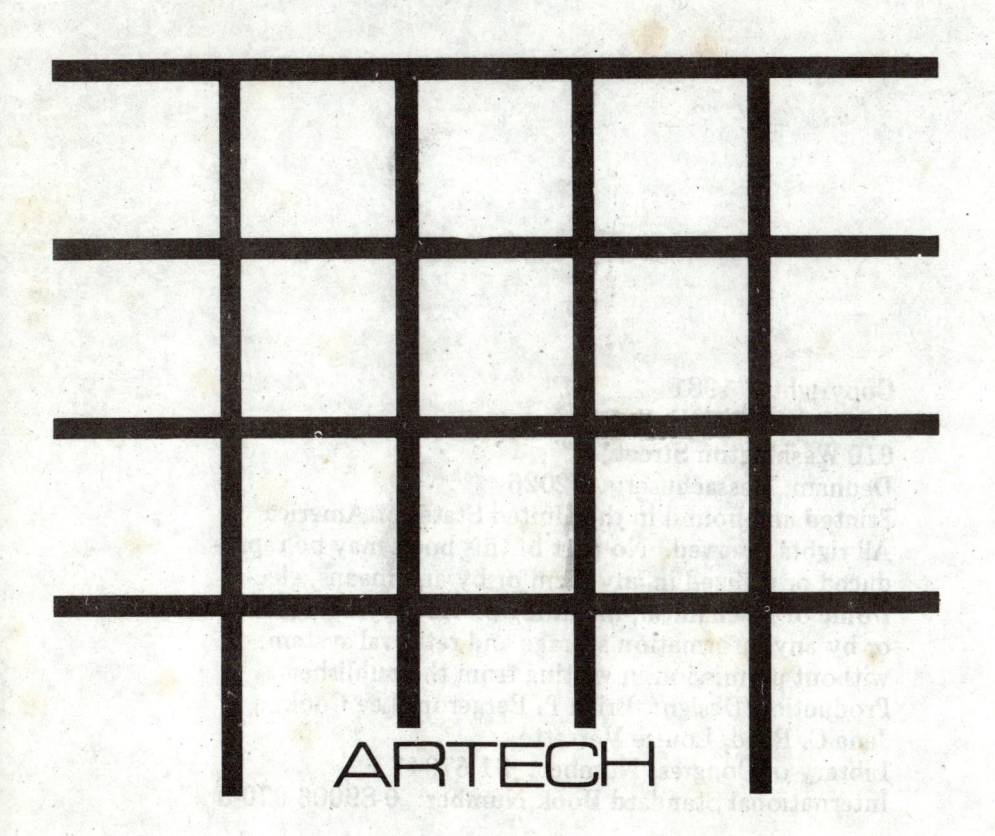

ARTECH

Copyright © 1981
ARTECH HOUSE, INC.
610 Washington Street
Dedham, Massachusetts 02026
Printed and bound in the United States of America
All rights reserved. No part of this book may be reproduced or utilized in any form or by any means, electronic or mechanical, including photocopying, recording, or by any information storage and retrieval system, without permission in writing from the publisher.
Production/Design: Brian P. Bergeron, Lea Cook, Jane C. Reed, Louise Verrette.
Library of Congress Number: 81-67941
International Standard Book Number: 0-89006-070-3

To the Memory of my Father

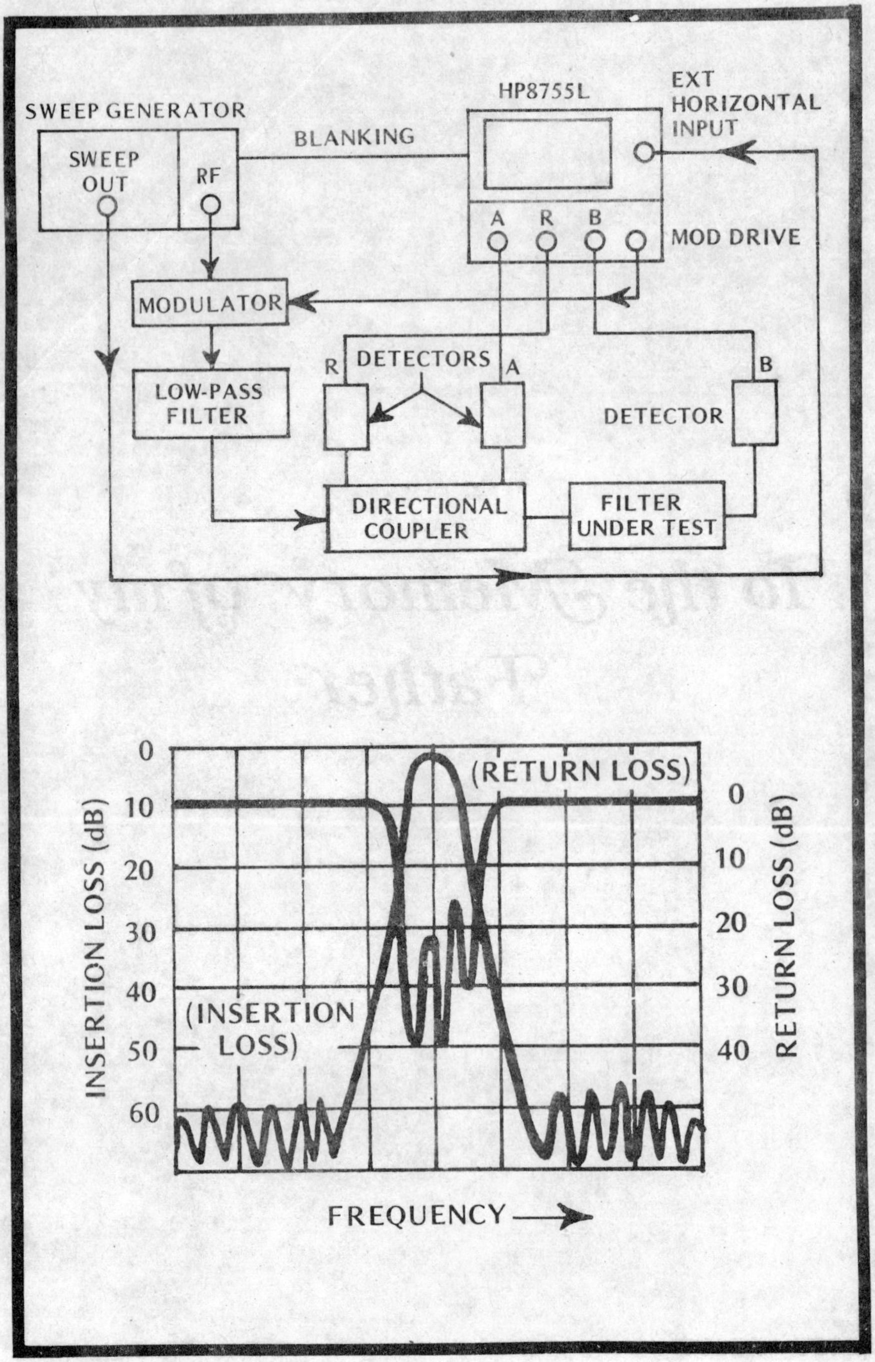

SWEEP GENERATOR
SWEEP OUT
RF
BLANKING
HP8755L
EXT HORIZONTAL INPUT
A
R
B
MOD DRIVE
MODULATOR
LOW-PASS FILTER
R
DETECTORS
A
B
DETECTOR
DIRECTIONAL COUPLER
FILTER UNDER TEST
(RETURN LOSS)
(INSERTION LOSS)
INSERTION LOSS (dB)
0
10
20
30
40
50
60
RETURN LOSS (dB)
0
10
20
30
40
FREQUENCY

Contents

PREFACE

ACKNOWLEDGMENTS

1 INTRODUCTION 1

2 TEST EQUIPMENT 13

2. Introduction 13
2.1 Signal Generators 14

    2.1.1 Signal Oscillators 14
    2.1.2 Sweep Oscillators 20
    2.1.3 Synthesized Signal Generators 24

2.2 Signal Detection/Indicating Devices 28

    2.2.1 Detectors 30
    2.2.2 Power Detectors 34
    2.2.3 Power Meters 36
    2.2.4 Spectrum Analyzer 40
    2.2.5 Noise Figure Meter 44
    2.2.6 Frequency Counters 50

2.3 Auxiliary Testing Devices 53

    2.3.1 Attenuators 53
    2.3.2 Directional Couplers 63
    2.3.3 Slotted Line 71
    2.3.4 Modulators 73
    2.3.5 Normalizer 77

2.4 Microwave Systems 78

    2.4.1 HP8755 Swept Frequency Response Test Set 78
    2.4.2 Pacific Measurements Model 1038 Measurement System 80
    2.4.3 HP8410 Network Analyzer 83

3 POWER MEASUREMENTS 93

3. Definition 93
3.1 Low Power Measurements 97
3.2 Medium Power 103
3.3 High Power 114
3.4 Peak Power 119

    3.4.1 Average Power-Duty Cycle Measurement 119
    3.4.2 Direct Pulse 122
    3.4.3 Notch Wattmeter 124
    3.4.4 DC Pulse Power Comparison 125
    3.4.5 Barretter Integration—Differentiation 128
    3.4.6 Sample and Hold 130
    3.4.7 Direct Readings 131

Chapter Summary 134

4 NOISE MEASUREMENTS 137

4. Definition 137
4.1 Noise Sources 142
4.2 Noise Meters 146
4.3 Manual Noise Measurements 151

    4.3.1 Twice-Power Noise Figure Measurement 151
    4.3.2 Y-Factor Noise Figure Measurements 154

4.4 Automatic Noise Figure Measurements 162
4.5 Errors and Accuracy 169
Chapter Summary 172

5 SPECTRUM ANALYZER MEASUREMENTS 175

5. Definition 175
5.1 Power Measurements 191
5.2 Frequency Measurements 201
5.3 Noise Measurements 206
5.4 Receiver Measurements 215

      5.4.1   Modulation 215
      5.4.2   Pulsed RF 234
      5.4.3   Distortion 245

5.5 Auxiliary Equipment 250

Chapter Summary 254

6 ACTIVE TESTING 257

6.   Definition 257
6.1 Gain 259
6.2 Gain Compression 268
6.3 Intermodulation 277
6.4 Third Order Intercept 279
6.5 Spectral Purity 282

Chapter Summary 289

7 ANTENNA MEASUREMENTS 291

7.   Definition 291
7.1 Test Range 292
7.2 Gain Measurements 301
7.3 Directivity 310
7.4 Polarization 312
7.5 Boresight Measurements 317

Chapter Summary 323

8 AUTOMATIC TESTING 325

8.   Definition 325
8.1 Section 1 (General Information) 327
8.2 Section 2 (Functional Specifications) 333
8.3 Section 3 (Electrical Specifications) 389
8.4 Section 4 (Mechanical Specifications) 394
8.5 Section 5 — Systems Applications and Guidelines for the Designer 400

8.6 Section 6 — Systems Requirements and Guidelines for the User 402
Chapter Summary 408

9 MISCELLANEOUS MEASUREMENTS 411

9. Definition 411
9.1 Phase Noise 411
9.2 Q Measurements 420
9.3 TDR (Time Domain Reflectometry) Measurement 426
9.4 Swept Impedance 433

    9.4.1 Swept Slotted Line 433
    9.4.2 Swept Reflectometer 441
    9.4.3 Swept VSWR Bridge 447
    9.4.4 Swept Systems 452

Chapter Summary 460

APPENDIX A: MICROWAVE BANDS 463

APPENDIX B: VSWR VERSUS RETURN LOSS (R) 465

APPENDIX C: MAXIMUM AND MINIMUM RESULTANT VSWR FROM TWO MISMATCHES 469

APPENDIX D: dBM TO WATTS 470

APPENDIX E: DECIBELS VERSUS VOLTAGE AND POWER 474

APPENDIX F: MICROWAVE FORMULAS 477

APPENDIX G: DECIMAL-TO-METRIC CONVERSION 479

APPENDIX H: REMOTE MESSAGE CODING 481

APPENDIX I: TIME VALUES 485

APPENDIX J: ALLOWABLE SUBSETS 486

APPENDIX K: INTERFACE MESSAGE REFERENCE LIST 496

APPENDIX L: FREQUENCY COUNTERS 501

REFERENCES 515

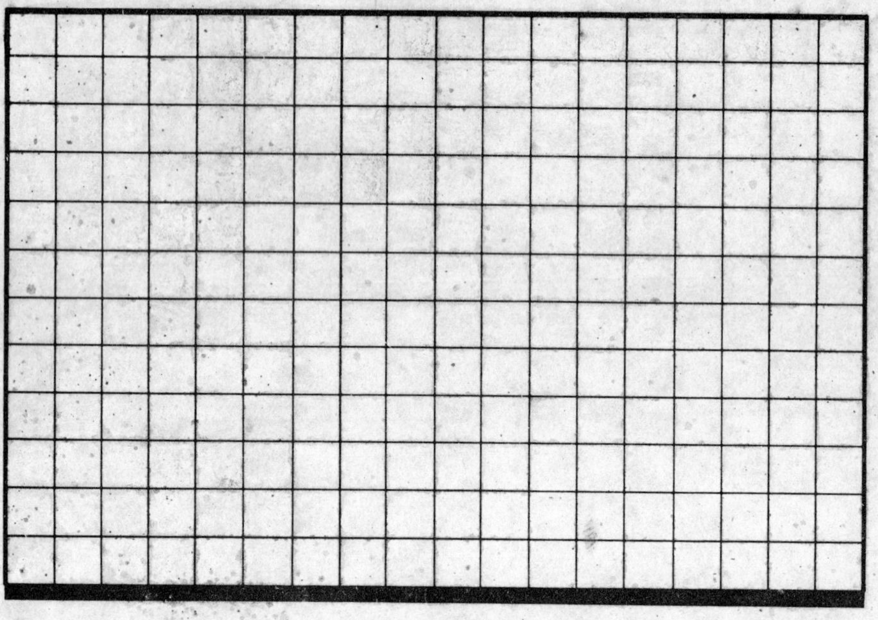

Preface

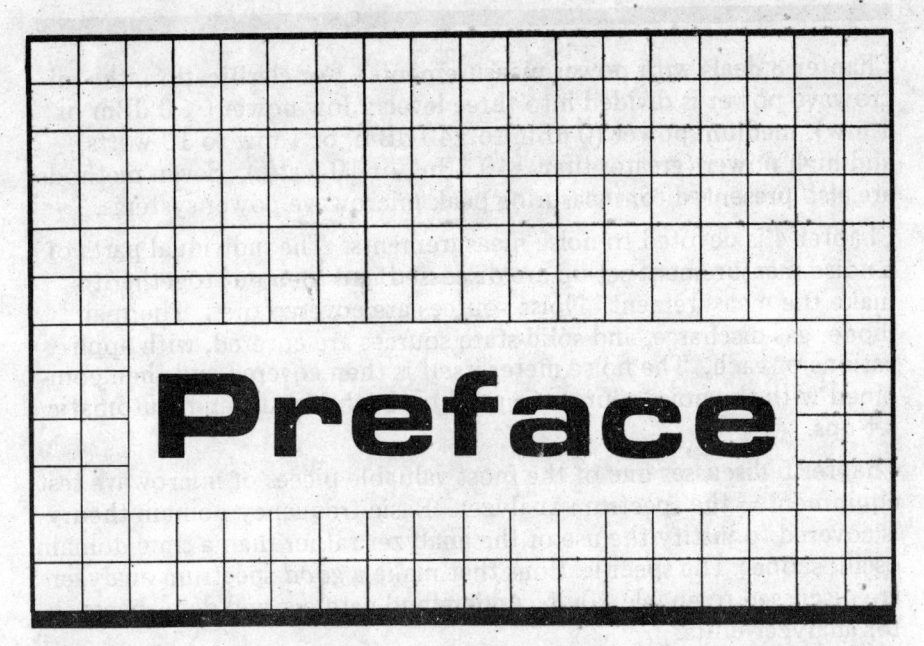

The measurement of many microwave parameters has been termed complex by many who are both in and outside of the microwave field. The intent of this book is to remove the idea of complexity and make the measurements understandable. By first defining what parameters we are going to measure, we can get a much clearer picture of what sort of procedure is needed and what criteria must be satisfied.

Chapter 1 is intended to point out areas that must be considered in order to produce the *correct* test for your particular application. Such areas as knowing the parameters you want tested, making the proper test set-up, knowing *how* to test the device, and knowing how to interpret the results once you get them, are discussed in this chapter. A summary of rules, tips, and precautions is included following these discussions.

Chapter 2 covers microwave test equipment. Five areas are covered which include: Signal Generators, Indicating Devices and Signal Detection (Power Meters, Spectrum Analyzers, Noise Figure Meters, and Counters), Auxiliary Components (Attenuators, Directional Couplers, Slotted Lines, Normalizers, and Modulators), and Systems (HP-8755 Swept Frequency Response Test Set-up, PM-1038 Measurement System and HP-8410 Network Analyzer). Individual specifications for each of these are presented so as to point out which parameters are important to obtain the proper piece of equipment for your application.

Chapter 3 deals with power measurements. For clarification, the microwave power is divided into three levels: low power ($<$ 0 dBm or 1 mw), medium power (0 dBm to +40 dBm, or 1 mw to 10 watts), and high power (greater than +40 dBm, or 10 watts). Seven methods are also presented for measuring peak microwave power systems.

Chapter 4 is devoted to noise measurements. The individual parts of a noise measurement set-up are discussed and then put together to make the measurement. Noise sources are covered first. Thermal, diode, gas-discharge, and solid-state sources are covered, with applications of each. The noise meter itself is then covered and then combined with the noise sources to produce both Manual and Automatic set-ups.

Chapter 5 discusses one of the most valuable pieces of microwave test equipment — the spectrum analyzer. Basic frequency domain theory is covered to justify the use of the analyzer rather than a time domain oscilloscope. The specifications that make a *good* spectrum analyzer are discussed to enable you to understand catalogs and data sheets for analyzer units.

Both relative and absolute power measurements are presented, as well as relative and absolute frequency methods. A procedure is also presented which enables you to use the spectrum analyzer for making noise measurements. Receiver measurements are also covered which include: Modulation (AM, FM, and combined AM and FM), Pulsed RF measurements, and Distortion measurements.

The chapter is completed by discussing auxiliary equipment that can be used with the spectrum analyzer. This equipment consists of the normalizer, tracking preselector, and tracking generator.

In Chapter 6, measurement procedures are presented for measuring active devices. Measurements such as gain (CW, swept, spectrum analyzer method, and HP-8755 swept response test set method), gain compression, third-order intermodulation, third-order intercept point, and oscillator spectral purity are all presented in this chapter.

Chapter 7 covers the world of antennas, beginning with the test ranges used for antenna measurements. Such ranges as elevated ranges, slant ranges, compact ranges, and anechoic chambers are covered. Measurements conducted on these ranges are Gain, Directivity, Polarization, and Boresight.

Chapter 8 provides an in-depth look at the IEEE-488 Interface Standard and how it can be used for automatic microwave testing. Every page of the standard is presented in down-to-earth, understandable language to enable you to adapt your particular test setup to the interface system.

Chapter 9 is called **MISCELLANEOUS MEASUREMENTS** because the tests covered do not fall into any of the previously covered chapters. The set-ups made in this chapter are Phase Noise, Q Measurements, TDR (Time Domain Reflectometry), and Swept Impedance.

Each of the procedures and test set-ups presented in the preceding chapters is written in plain, simple language so that you may analyze it and adapt it to your particular application. The intent of this book is not to provide you with a large group of measurement set-ups that you can memorize and use in every instance. It is, rather, our intention to present representative examples, along with sufficient explanation and guidance, which you can use to fabricate and create that one special set-up that will give you the information you need.

Thomas S. Laverghetta
Auburn, Indiana
August, 1980

Acknowledg

ments

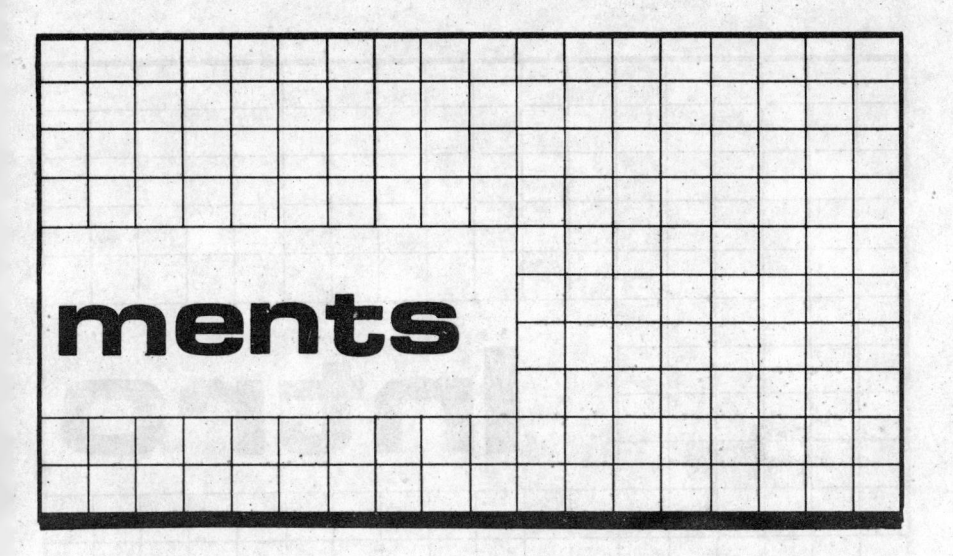

To list all of the people who contributed to this book and really made it possible would probably take another complete book in itself. They may not all be listed but I want them to know that I thank them from the bottom of my heart, because when I needed them they were there to help.

There are, however, people who deserve a special thank you: to Debbie Bair and Rita Betley who typed on the preliminary manuscript; to Vicki Bone who did an absolutely superb job on the drawings and charts; and to a gentleman named Meredith Esterline who dug out an uncounted number of manuals for me so that I could find the information I needed to finish a section or chapter — thank you.

An extra special thank you has to go to Connie Holden, who took so very many hard-to-read pages and transformed them into a manuscript that I was proud to submit.

I would like to thank Scientific Atlanta, and especially Richard Schnable, for all the material and help they supplied to aid in assembling the chapter on Antenna Measurements. Also, I would like to thank the Hewlett-Packard Company who proved, once again, how much they care about people by supplying an unlimited amount of material throughout the entire task.

The ITT Aerospace/Optical Division deserves a thank you for all of the help and encouragement they supplied to me.

Finally, I thank my family for their patience, understanding, and love and for showing me how lucky I am to have them.

Intro

1 duction

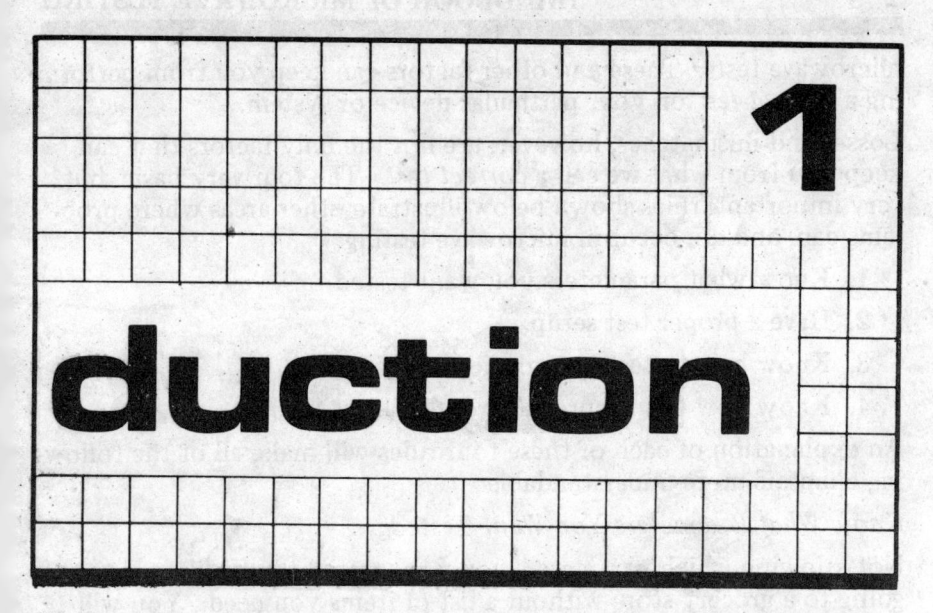

Most people, when creating a particular design, put their heart and soul into it and make it a labor of love. They calculate, re-calculate, optimize, and finally arrive at what is considered to be the masterpiece they set out to create. It may, in fact, truly be a masterpiece, but at this point it may also be almost totally useless. This is said, not to discourage creative designing, but to emphasize the fact that if a design is to be truly useful, and the masterpiece it was intended to be, it must be *tested*. Not only must it be tested, it must be *tested correctly!* The design at this stage can be likened to starting a car to run a race. It is the first step; but the race still must be run in order to prove out the car. Merely starting the engine is not a true test of the performance of the car. Similarly, designing a circuit or system says only that the design is possible. Successful testing is what determines its true value.

The most important words in the paragraph above are: "tested correctly". There are many factors involved in the proper testing of a device or system. It is much more than a signal source, the device under test, and an indicating instrument. It is, basically, realizing that no piece of test equipment is perfect and that, because of this, certain errors are going to be present. Errors such as those in couplers and meters must be considered; compensation factors for thermistor mounts come into play; and losses due to connector mismatches and cable lengths become factors which can change your results. These are but a few of the factors that can affect the final outcome of your microwave tests. These and other factors can keep you from performing a *correct test* on your particular device or system.

Losses and mismatches, however, are not the only factors that can keep you from what we call a *correct test*. The four very basic, but very important, rules shown below illustrate other areas where problems can, and do, occur in microwave testing.

1. Know what parameters you want tested.
2. Have a proper test setup.
3. Know how to test your device or system.
4. Know how to interpret the results.

An explanation of each of these four rules will make all of the following chapters more understandable.

Know What Parameters You Want Tested

Not knowing which parameters you want tested can be likened to going to a grocery store without a list of items you need. You will probably end up buying much more than you need, spending far more money than you should, and will have wasted a lot of time in the process. There is a better than even chance that you will buy everything you need within this multitude of merchandise, but what do you do with the items you don't need? It is the same situation when testing a microwave device or system. You must know which parameters will best characterize your particular device or system, and then test for them. To characterize a directional coupler, for example, you would most likely test for insertion loss, coupling, directivity, and possibly VSWR. It would make absolutely no sense at all to run a noise figure test on the coupler. First, it would do nothing to aid in characterizing the coupler and, second, it would waste time and effort that could be spent on more useful testing. So do yourself a very large favor and make a list of all the parameters you think you might need before you begin testing. Then, look closely at the device or system you are going to test and determine which of these parameters will give you the most useful information and will characterize your unit. A little bit of time spent in the beginning to select the right parameters will save time, money, and tempers later on.

Have a Proper Test Setup

When you have determined which parameters you will need to test for, the next logical step is to come up with a setup that will allow you to test them. This involves, among other things, choosing the right equipment to perform the tests you need. The right equipment

Introduction 3

means, for example, choosing the proper signal source; or being sure the auxiliary components (attenuators, couplers, detectors, etc.) have the proper power handling capability; or the output indicating device will give you the data you will be able to use.

The proper signal source for your application may be a sweep generator, a single frequency generator, a power oscillator, or even a klystron. Any one of these sources will do a particular job of microwave testing. Any one of these sources can be used as a CW source, although, it is not a good idea to use a sweep generator for CW use since its stability leaves something to be desired. Also, the fact that the instrument was not designed for CW use should keep you from using it. It is, however, used as a CW source to a much greater extent than it should be. (Many times because it is the only source available.) Choose your source wisely to suit your application.

The power handling capability of all components used in a test setup should be checked very carefully. This should be done when you are gathering them together to make a test setup, *before* the RF power is turned on. It is a little late to check power handling capability when you detect a strange odor coming from your new $100 pad or from your one and only crystal detector, or when you notice that your spectrum analyzer does not want to read right. This should always be one of your primary concerns when making a test setup, and particularly when it involves high power. *Always check the power handling capability of your microwave components and test equipment.*

The indicating device you use in your test setup should be one that will give you the information you need. You may want to simply display your output on an oscilloscope or record it on an X-Y recorder. You may be satisfied with a picture of a scope display or you may require a sheet with the data printed on it. Your requirements may even be as basic as taking readings off a power meter and recording them in a lab book. Whatever your requirements may be, make sure they are spelled out and decided before you begin testing, so that you will obtain the information that will be most useful to you.

Another requirement for having the proper test setup is to place the components you have chosen in the right position. This may sound rather basic, but it is surprising how many people get into trouble by putting components in the wrong place. To illustrate the point, we will look at an example that actually happened. The objective of the test was to sweep a low level amplifier from 2-4 GHz and use the HP8410A network analyzer as a test instrument. The display used was an oscilloscope as shown in Figure 1.1a. The particular area of interest was from 2.5 GHz to 3.7 GHz, so a wavemeter was inserted to find these points accurately. With the setup shown in Figure 1.1a it was found that the network analyzer would not remain locked to the sweep generator across the entire band. The level would be fine over a certain portion of the band, but it would drop out of lock and the display would break up on the high end of the band. All of the connectors and cables were checked to be sure they were tight; the analyzer settings were checked; and the sweeper was checked. Everything checked out. It was then decided that the setup should be looked at very carefully. At that point it was found that the problem was one component out of place in the setup: the wavemeter, which is a very high-Q device and has a deep notch when tuned to your chosen frequency. The sweeper was trying to level this notch, and was not succeeding. This could have been temporarily remedied by detuning the wavemeter, but it was needed for future verification of frequencies in our data. The solution was to put it beyond the leveling loop as shown in Figure 1.1b. This way a leveled source would be unhampered by the wavemeter within the leveling loop. There was still some effect on the analyzer, but an increase in the sweep speed of the sweep generator caused the analyzer to maintain its lock completely across the band. So you see that the position of a single component in a test setup can mean the difference between good and useful data and many wasteful hours of searching for setup problems.

A final requirement for a proper test setup, and thus a proper test, is to know how to operate the equipment you have available for the test. It certainly is of very little importance to have a lab full of the latest in microwave equipment if you do not know how to use it. That is sort of like having the most beautiful and elaborate swimming pool in the neighborhood and not knowing how to swim. It is very impressive to look at, but not very functional or useful. So, learn how to use the equipment you have available. Read the instruction manual and any application notes that you can get. Experiment with your equipment whenever there is time, so that you can learn all of the ins and outs of it and thus be more at ease with it and utilize it to its fullest capability. Most of the microwave equipment available today is highly sophisticated and has a price tag to go with this sophistication. It would be a great waste if it was only utilized to half of its intended capability simply because someone could not take the time to learn its proper operation.

To summarize the second rule, we could say: in order to have a proper setup, we must have the proper equipment; put them in the right order, and know how to operate the test equipment properly.

Introduction

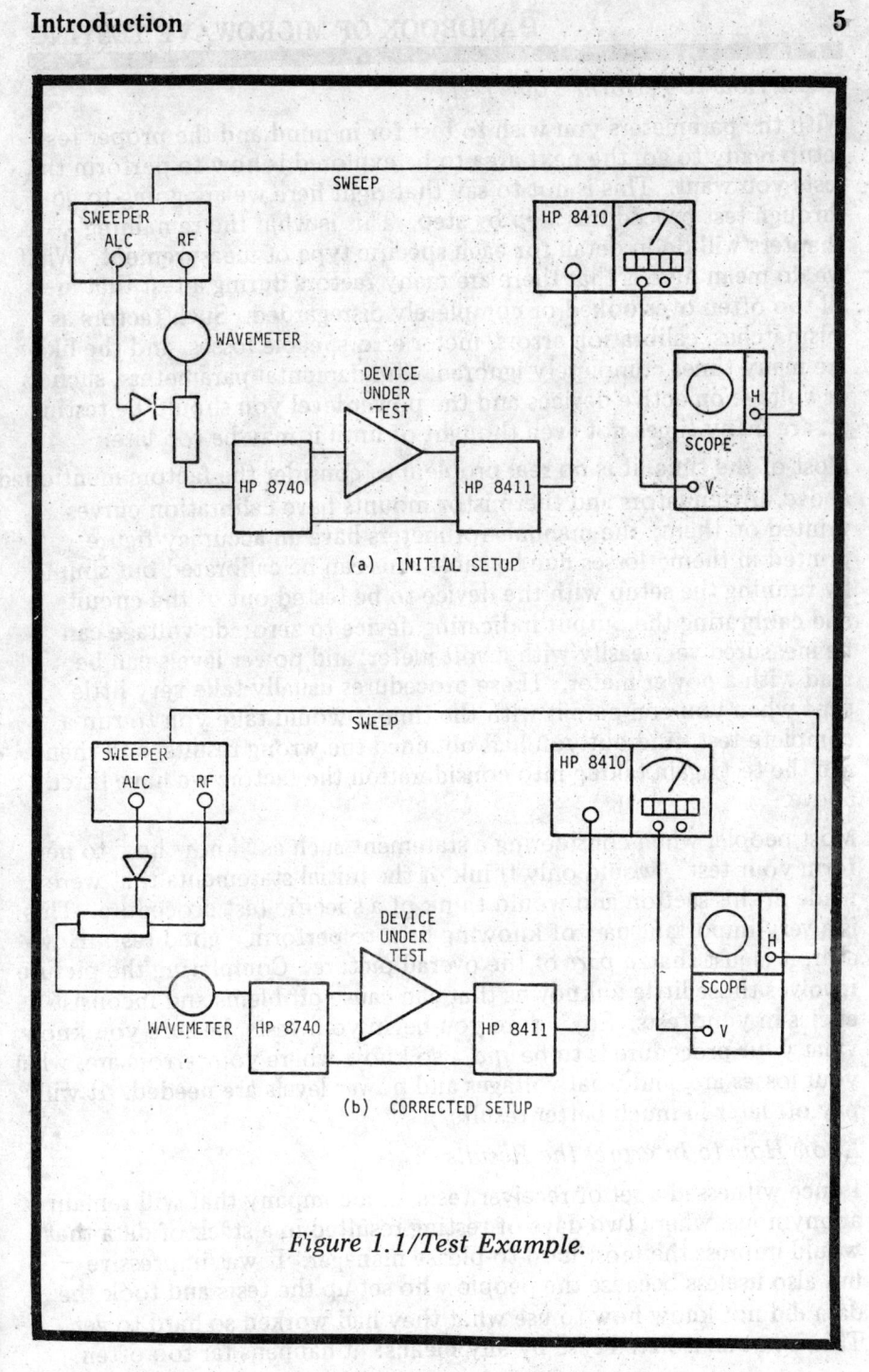

Figure 1.1/Test Example.

(a) INITIAL SETUP

(b) CORRECTED SETUP

Know How to Perform Your Test

With the parameters you wish to test for in mind and the proper test setup ready to go, the next area to be explored is how to perform the tests you want. This is not to say that right here we are going to go through test procedures step by step. This is what the remaining chapters will do in detail for each specific type of measurement. What we do mean here is that there are many factors during a test that are all too often overlooked or completely disregarded. Such factors as mismatches, calibration errors, meter errors, cable losses, and the like are many times completely ignored. Fundamental parameters, such as dc voltage on active devices and the power level you should be testing at, are many times not even thought of until it may be too late.

Most of the time, it is no real problem to consider the factors mentioned above. Attenuators and thermistor mounts have calibration curves printed on them; the manuals for meters have an accuracy figure printed in them; losses due to cable runs can be calibrated out simply by running the setup with the device to be tested out of the circuit and calibrating the output indicating device to zero; dc voltage can be measured very easily with a volt meter; and power levels can be read with a power meter. These procedures usually take very little time when you compare it with the time it would take you to run a complete test, find out you had obtained the wrong results, and then run the test again taking into consideration the factors we have listed above.

Most people, when considering a statement such as "know how to perform your test", would only think of the initial statements that were made in this section and would think of a specific test procedure. This is a very important part of knowing how to perform a good test. However, it is just that, *a part* of the overall picture. Completing the picture involves those little unknowns that can cause problems and inconsistencies in your tests. So, before you begin your tests, be sure you know what your procedure is to be *and* also know where your errors are, what your losses are, and what voltages and power levels are needed. It will pay off later in much better results.

Know How to Interpret the Results

I once witnessed a set of receiver tests, at a company that will remain anonymous, where two days of testing resulted in a stack of data that would impress the most hard-to-please manager. It was impressive — but also useless because the people who set up the tests and took the data did not know how to use what they had worked so hard to get. This is not an isolated case by any means: it happens far too often.

Introduction

It all goes back to what we have been talking about throughout all of our four basic rules for proper testing — plan your tests to obtain the maximum benefit from them. The old saying still holds: *plan ahead*. The best way that you can plan ahead is to have your final results in a form that is the most easily understandable and useful to you. If you need VSWR readings for a device, for example, you have not planned out your best setup if your printed data is in return loss or reflection coefficient. Obviously you can obtain VSWR from either or both of these parameters, but the object is to get your final data in a form you can read directly, if possible. You can consider that you have accomplished the best setup if, when you are finished testing, there is not need for calculations. What you are after is not just an answer, but the *right* answer.

There are times, however, when the *right* answer can only be obtained by taking measured data and performing calculation with it. If this is the only way that you can arrive at a useful answer, then it is the right way and you have accomplished your task of putting together the proper test setup.

So, to conclude our discussion of rule four, we can say that *an answer* should not be your objective when setting up a microwave test setup. A little time and effort used to think out the proper setup will yield the *right answer*.

The four basic rules just covered all have one area in common. They all stress the importance of planning your test setup ahead of time. Plan it all the way, from the parameters you need to the type of output you will get. We can best compare preparing the right test setup with an example: that of diving into a pool of water. If you investigate before you dive, you will find out the best way to dive, find out how deep the water is and what is on the bottom, and enjoy your swim afterwards. If you do not investigate and dive blindly into the water, you may encounter many unknown factors on the bottom, possibly crack your head open, and maybe never dive or swim again. This example illustrates the great importance of planning ahead and investigating everything you do before you do it; in swimming and in testing.

One area that follows very closely with this idea of planning ahead is that of calibration. Every instrument and component is made to function a specific way. When it was built it was set up (calibrated) á certain way to achieve its proper operation. However, just like your automobile, with time and use the performance changes and adjustments are necessary to once again obtain proper operation. This is why companies set up definite calibration schedules for their equipment. At specific time intervals throughout the year each piece of equipment, and component if necessary, is completely checked to be sure it meets all of the original specifications it had when it was new. This is very important to producing the "correct test" we have referred to throughout this chapter.

Calibration has to be one of the most important points you have to consider when preparing a test setup and when making the actual tests. With improperly calibrated equipment your results most likely will not be what you think they are. The errors, mismatches, or losses you have initially accounted for will not be accurate readings and thus will not reflect your true results.

Probably the two most important pieces of microwave equipment to keep in calibration are the power meter and the spectrum analyzer. These two are the most common pieces of equipment that are used as references in your initial setups. The power meter, for example, is used to check the level of the signal source that is used; used to check the losses due to cables in the setup; determine the coupling of a directional coupler used for input and/or output power indications; and has numerous other functions that all require it to be an accurate standard for our entire setup.

The power meter, as shown above, is used in many applications. It does, however, have one drawback, which sometimes is of great importance. That is that it cannot distinguish between the level of the fundamental and that of any harmonics that are present. It merely groups all of them together and displays a total power. Many times the level of the harmonics is very small and is of no consequence to us. However, many other times it is important to know just what level each of these harmonics is at. This is where the spectrum analyzer comes in. The spectrum analyzer will display each of the harmonics at the proper frequency and level. However, it will only do this properly if it is calibrated. By proper calibration we mean both frequency and power level. If only one or the other is calibrated, the analyzer is of no use. Both parameters must be known to have the instrument be the great value it can be.

When we say that the spectrum analyzer can be of great value, it is really a large understatement. As will be seen in a later chapter, the spectrum analyzer is a part of practically every microwave setup. One of the reasons it is so popular is that we always look for an instrument that has a scope and will give us a picture, which the spectrum analyzer does. This always seems a lot handier than interpolating

Introduction

a meter reading. I guess the spectrum analyzer has spoiled all of us a little. The main reason, though, is that so much good information can be obtained by using a spectrum analyzer — probably more than any other single instrument available today — good information, that is, if it is properly calibrated.

Properly calibrated equipment is nice to have and work with, but it is also a necessity in industry. When you have to deliver a finished product to a customer, he wants to know that he can duplicate your data once he has it delivered to him. Many people will look for calibration labels on equipment before they will allow any testing to be performed. It is not good for you or your company if a customer sees a label on a piece of equipment in a qualifying test setup that reads: "NOT CALIBRATED, DO NOT USE FOR QUANTITATIVE DATA". Actually, there is really no reason for such a piece of equipment to be in a lab. If no one knows how to calibrate it, it should be stored on a shelf until someone learns how to work with it. If it is too old or worn out and cannot be calibrated, it should be discarded. There should never be any reason for using such a label on a piece of microwave equipment or a component. Conscientious companies will set up a complete calibration facility to ensure that only the best and most accurate data is turned out.

As mentioned above, you can get some improper readings from uncalibrated equipment. You can also get improper readings by not paying attention to what you are doing. By this we mean that sometimes people will record data and not really see what they are recording. They could just be recording numbers that have no meaning. In other words, the numbers may be *too* good. A conversion loss of a mixer may be lower than its theoretical value; the directivity of a coupler may be much higher than is possibly attainable; or a system noise figure may read a fantastic 1.0 dB using Silicon Bi-Polar transistors in the front end. These things may sound far-fetched, but it is amazing how many times, in many different places, that things like this occur. It all goes back to the main theme of this chapter: plan ahead. Know what range your data should fall into. It is not necessary to know exact values, a range from minimum to maximum should be known before you begin your testing. If your results fall outside of this range, you should first check to be sure your range is correct, then check your setup, and finally check to see if there is a problem with the device or system you are testing. So, plan ahead and, as emphasized earlier, be sure you have the right numbers (answers) from your setup.

Before we get into the actual descriptions of test setups, there are a few ground rules, tips, definitions, and precautions that should be covered. Many of these will be re-stated throughout the following chapters and some will be emphasized many times before we are finished. However, they should be introduced here so that a good foundation for the testing that follows can be set down.

If we group the ground rules, tips, and precautions together we would get a list like the one below:

- Never force connectors.
- Check connectors for proper tightness and alignment.
- Use the right connectors to avoid adapters.
- Use the proper size wrench on SMA connectors.
- Do *not* use pliers on type N and TNC connectors.
- Learn how to use APC-7 connectors properly.
- Never use BNC connectors above 500 MHz.
- Keep your connectors clean.
- Be sure you have the right type of cable for your particular application.
- Always check dc voltages on active devices before applying RF power.
- Set the current limiter on your dc power supply to avoid damage to active devices under test.
- Take noise figure readings in a screen room if possible.
- Place all meters on their highest scale before beginning your test.
- Check the input RF power specification on test equipment (power meter, spectrum analyzer, etc.) *before* you turn the power on.
- Be sure all waveguide is aligned properly.
- Be sure all waveguide flanges are clean.
- Provide support for coaxial couplers, attenuators, detectors, or any other component that must be suspended from a generator, scope, or other equipment.
- Use the right equipment for the right test.
- Remember that all microwave test instruments are very precise devices; treat them with care.

Introduction

To clarify all of our references to low, medium, and high power, we will make the following definition:

Low Power: below 1 mW (0 dBm)
Medium Power: 1 mW (0 dBm) to 10 watts (+40 dBm)
High Power: above 10 watts (+ 40 dBm)

Any time that low, medium, or high power is mentioned, these are the power values we are referring to.

So the stage has now been set to proceed with each of the important microwave measurements that follow. The success of all of these measurements depends very heavily on your ability and willingness to plan ahead. None of the setups are designed to be memorized. Each of them will be presented in such a way that you can understand why each piece is used, so that you can reason out other setups that may suit your requirements better. With each of the measurements we will recommend equipment, go through a setup explaining why each piece is used, cover the measurement procedure, point out errors, and give alternate methods when they are applicable. This should give you a good understanding and background of all of the necessary microwave measurements you may need.

Throughout the following chapters, and when you need to design your own setups, remember the four basic rules covered earlier:

1. Know what parameters you want tested.
2. Have a proper test setup.
3. Know how to test your device or system.
4. Know how to interpret the results.

If you can master these four rules, your microwave testing will be trouble-free.

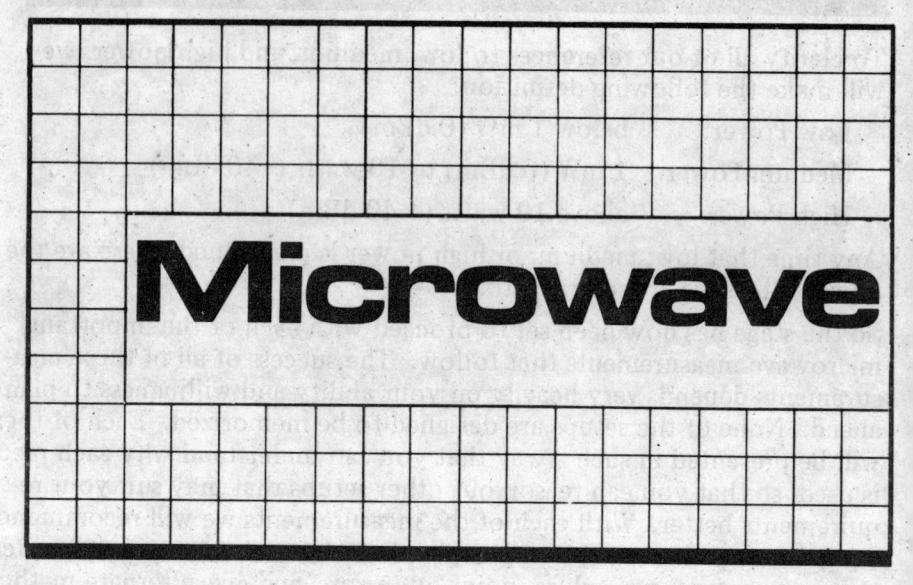

Microwave

Test Equipment 2

2. Introduction

One of the four basic rules for "correct" testing covered in Chapter 1 was: "have a proper set-up." When we defined what we meant by this rule, one of the areas covered was entitled *Choosing the Proper Equipment*. This is very important rule, but how is one to know what the proper equipment is unless one knows what equipment is available? How is one to decide which piece of equipment is best for a particular application unless one knows which parameters are important? The answers to these and other questions are the objective of this chapter. A concise explanation of major categories of equipment will aid in understanding both what is available and when is the proper time to use it. For purposes of this explanation, the equipment will be divided into four major categories:

1. Signal Generators
2. Indicating Devices and Signal Detection
3. Auxiliary Components
4. Systems

For each of these categories a representative type of equipment will be presented, pictures shown, a data sheet presented, and the terms of the data sheet explained. By learning what terms are on a data sheet, what they mean and how they apply to your particular application, you should be able to choose the proper equipment to use in your testing.

2.1 Signal Generators

The generation of a microwave signal can be from basically three types of equipment: the signal oscillator, the sweep oscillator, and the synthesized signal generator. These generators may take many different forms, from a single reflex klystron, to a multi-band sweeper, all the way to an ultrastable frequency synthesizer. No matter how simple or complex the equipment may be, however, there still are the three basic types. The particular generator you use will, as always, depend on your application. You would not, for example, need a highly stable source to check an attenuator at a single frequency; or if you need a component characterized at many frequencies you would use a sweep generator; you certainly could not use a synthesized generator to test a 60 GHz filter (they do not exist). A careful consideration of what you are required to test and what parameters are needed will simplify your choice of which generator to use.

2.1.1 Signal Oscillators

The most basic type of signal source designed to produce a single frequency output is the *signal oscillator*. Sources in this category are available up to 21 GHz for microwave applications. By using external doublers, signals to 40 GHz are obtainable.

To better understand what makes up a single frequency microwave oscillator and how it works, we will put together a block diagram of a typical unit. In other words, we will build a signal oscillator.

The most basic form of signal oscillator is simply a tunable oscillator with a dc voltage applied and an RF output. This oscillator could be a klystron, a YIG tuned oscillator, a voltage tuned oscillator, or even a cavity oscillator that is mechanically tuned. If a source with very loose requirements on output level and frequency stability will meet your requirements, this will work very nicely (Figure 2.1(a)).

The next step toward the ultimate usable general purpose signal oscillator is a means of controlling the output level. Figure 2.1(b) shows our initial tunable oscillator with a variable attenuator at the output. This will allow a setting of the output level over the range of the particular attenuator you select. This method of output level adjustment, although adequate in most cases, does have one drawback: it will cause the oscillator frequency to change as the attenuator is changed (the frequency will *pull*). Figure 2.1(c) shows the oscillator with a buffer amplifier at its output to isolate it from the variable attenuator and thus virtually eliminate the frequency pulling effect.

Test Equipment

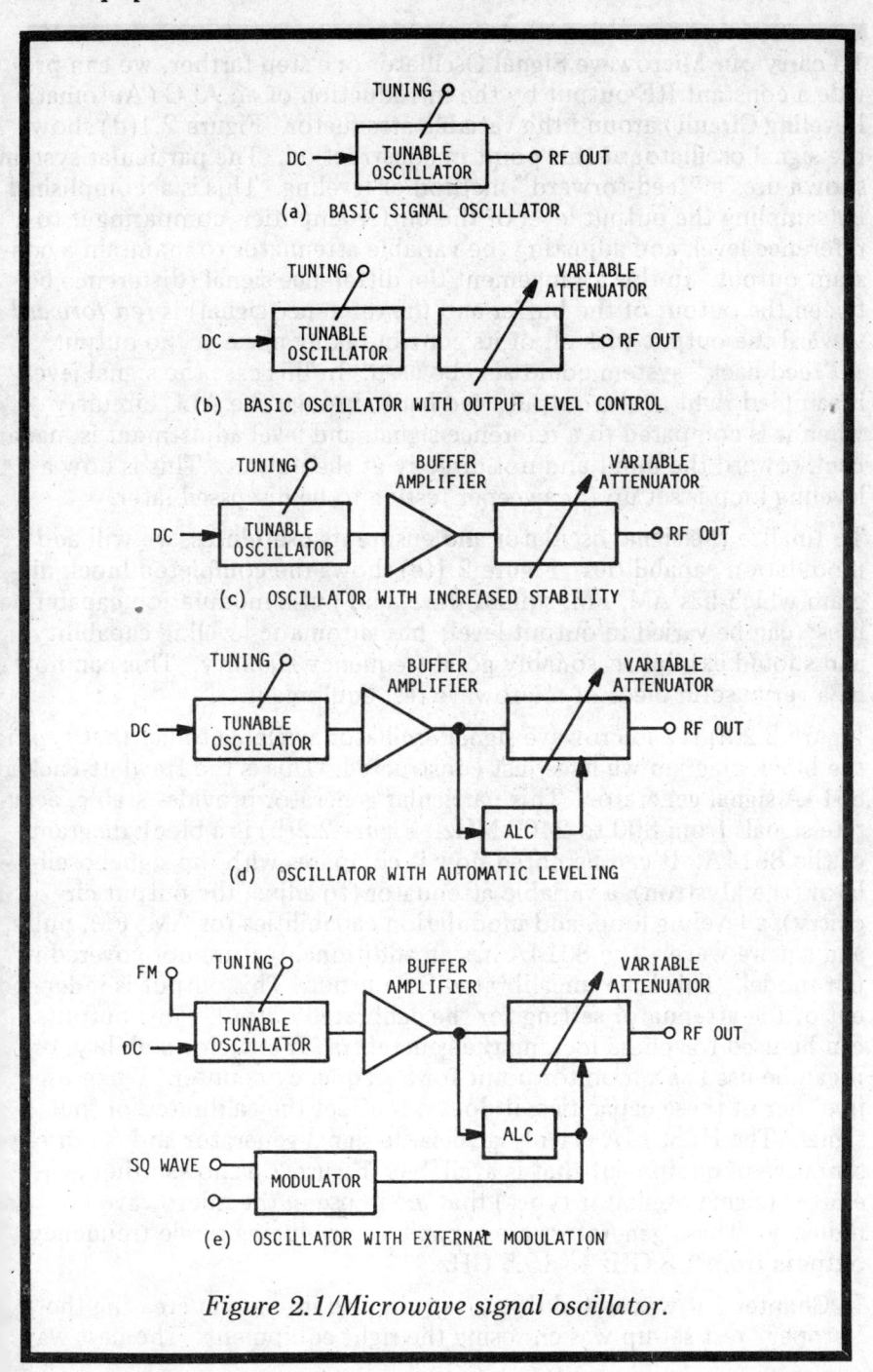

(a) BASIC SIGNAL OSCILLATOR

(b) BASIC OSCILLATOR WITH OUTPUT LEVEL CONTROL

(c) OSCILLATOR WITH INCREASED STABILITY

(d) OSCILLATOR WITH AUTOMATIC LEVELING

(e) OSCILLATOR WITH EXTERNAL MODULATION

Figure 2.1/Microwave signal oscillator.

To carry our Microwave Signal Oscillator one step farther, we can provide a constant RF output by the introduction of an ALC (Automatic Leveling Circuit) around the variable attenuator. Figure 2.1(d) shows the signal oscillator at this point in construction. The particular system shown uses a "feed-forward" method of leveling. This is accomplished by sampling the output level of the buffer amplifier, comparing it to a reference level, and adjusting the variable attenuator to maintain a constant output. In this arrangement the difference signal (difference between the output of the buffer and the reference signal) is *fed forward* toward the output with all of its control taking place at the output. A "feed-back" system could also be used. In this case the signal level is sampled right at the output; fed *back* through the ALC circuitry when it is compared to a reference signal, and level adjustment is made back toward the input and not directly at the output. This is how a leveling loop is set up for sweeper testing to be discussed later.

To finalize the signal oscillator and ensure its usefulness, we will add modulation capabilities. Figure 2.1(e) shows the completed block diagram which has AM, FM, square wave, and pulse modulation capabilities; can be varied in output level; has automatic leveling capability; and should exhibit reasonably good frequency stability. This can now be a very useful piece of microwave test equipment.

Figure 2.2(a) is a microwave signal oscillator available today that typifie the block diagram we have just constructed. This is the Hewlett-Packar 8614A signal generator. This particular generator provides stable, accurate signals from 800 to 2400 MHz. Figure 2.2(b) is a block diagram of the 8614A. It can be noted how it compares with the signal oscillator (the klystron), a variable attenuator (to adjust the output circuitry), a leveling loop, and modulation capabilities for AM, FM, pulse, and square wave. The 8614A has an additional feature not covered in our model. This is an uncalibrated RF output. This output is independ ent of the attenuator setting for the calibrated output. This output can be used for phase locking the generators for extreme stability, or it can be used as a monitor point for a frequency counter. When used in either of these capacities, it does not effect the calibrated output signal. The HP8614A is truly a versatile signal generator and is representative of equipment that is available. Figure 2.3 shows other generators (signal oscillator types) that are in use in the microwave industry. These generators are capable of producing single frequency outputs from 1.8 GHz to 15.5 GHz.

In Chapter 1 it was stated that a very important step in creating the "proper" test set-up was choosing the right equipment. The best way

Test Equipment

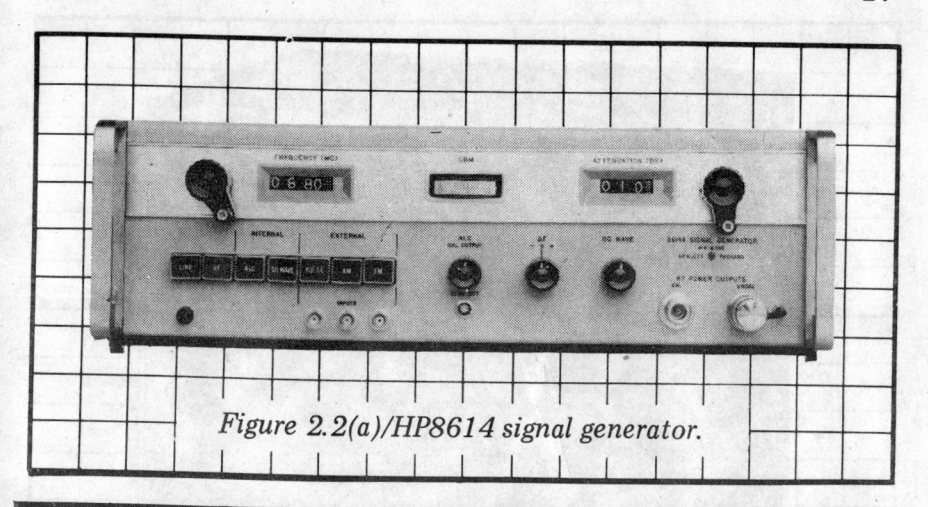

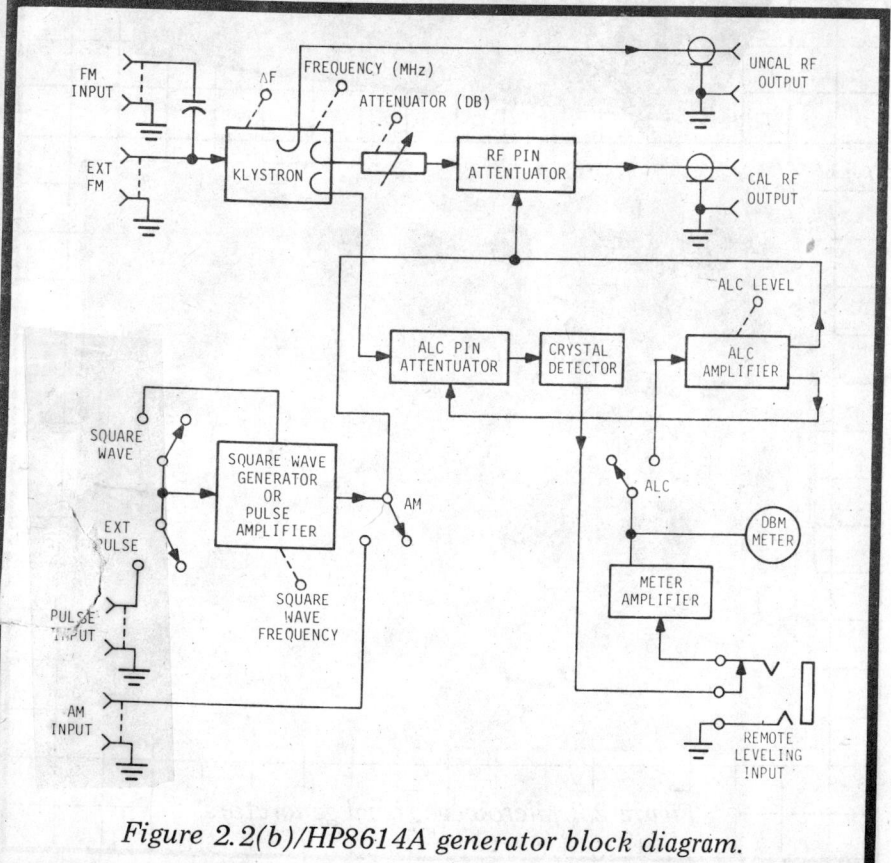

Figure 2.2(a)/HP8614 signal generator.

Figure 2.2(b)/HP8614A generator block diagram.

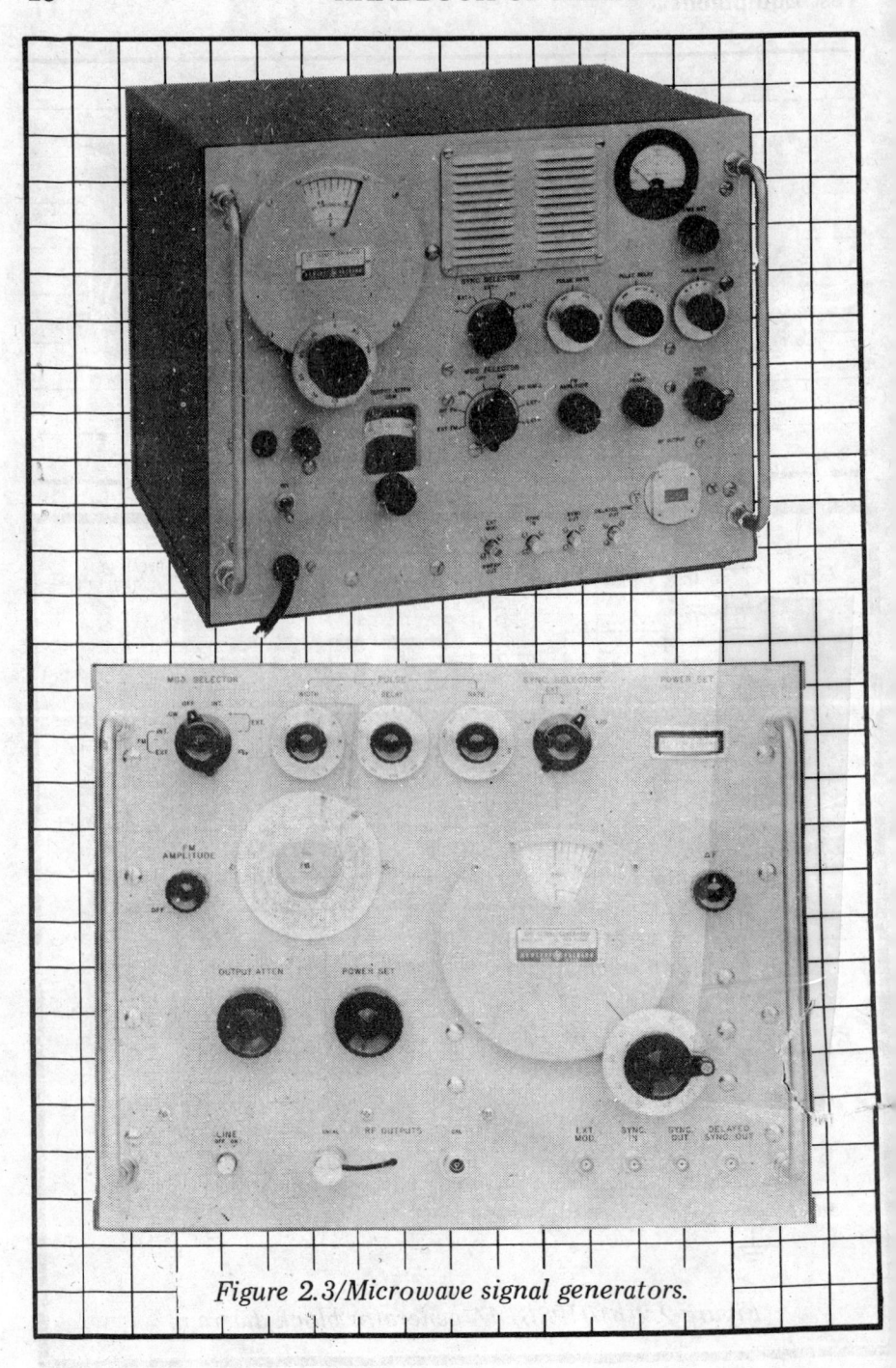

Figure 2.3/Microwave signal generators.

Test Equipment

to choose the right equipment is to examine a data sheet and know what each term means. To aid in selecting the right signal oscillator for your set-up, we will present a data sheet and explain the terms in it. The HP8614A will once again be the example used.

Frequency Range	800 to 2400 MHz
Frequency Accuracy	± 5 MHz (0 dBm and below)
Frequency Stability	50 PPM/°C, less than 2500 Hz peak residual FM
RF Output Power	+10 dBm (0.707V) in to 50 ohm
RF Output Accuracy (with respect to attenuation dial)	± 0.75 dB + Attenuator Accuracy (0 to −127 dBm)
Attenuator Accuracy	+0, −3 dB from 0 to −10 dBm; ± 0.2 dB ± 0.06 dB/10 dB from −10 to −129 dBm

The data sheet terms are defined as follows:

Frequency Range — This, of course, is the range of frequencies over which the generator will operate. You should take precautions not to operate a generator at its upper or lower frequency limits. Operation in these areas may result in instabilities or level drop-off which will make good measurement very difficult. Try to use a generator that has *your* operating range near the center of *its* range.

Frequency Accuracy — This is a measure of how close the generator actually is to the set frequency. This figure can be misleading in many cases, since some generators contain counters and digital displays and it is sometimes assumed that the display accuracy is the generator accuracy. This is not usually the case. The distinction to be made here is that what you see on the display is the *resolution* of the display. The *accuracy* of the generator is the figure on the data sheet and a measure of the instrument's capability.

Frequency Stability — A measure of how well the generator remains on frequency when subjected to outside effects such as temperature and noise. The first specification (50 PPM/°C) says that the HP8614A will vary 50 parts per million for each °C of temperature change from ambient operation. This is the normal specification for the signal oscillator type of generator we have been discussing. The *residual FM* number is a common means of specifying short-term stability (short term is defined as being within a fraction of a second). It is a measure of the small amount of FM (or short term jitter) inherent in the generator in the CW mode with all modulation turned off. The residual FM is virtually all noise related.

RF Output Power — This is the amount of power delivered to a 50 ohm load. You will note that the specification is also given as a voltage into 50 ohm. These are very common methods of specifying the RF output of a microwave signal oscillator.

RF Output Accuracy — This is how close to the theoretical output the actual output is. That is, it is the amount of deviation from the ideal. This deviation is caused by many sources: temperature on detectors, detector and meter linearity errors, and the actual accuracy of the output attenuator. The accuracy is usually specified over an attenuation (or level) range.

Attenuator Accuracy — A part of the overall output level accuracy of the signal generator. It is the variation of the attenuator from its theoretical design value. The specification is almost never given as a single number over the entire attenuation range. Rather, the range is split into segments and each segment is specified separately.

2.1.2 Sweep Oscillators

In the previous section we were concerned with a type of signal generator that would produce a single frequency output. For application when a component or system has to be checked at one frequency, or a group of set frequencies, this is more than adequate. However, there are many times when it is necessary to perform measurements at not just a few, but every frequency within a band. For these applications, an oscillator (or generator) that will provide continuous coverage of that specific band is required. This is where the sweep oscillator (or *sweeper*) is used.

The sweep oscillator is a rather unique piece of test equipment and is one of the most versatile. Much of the versatility comes from the fact that most sweep oscillators consist of a mainframe and an RF plug in. By changing plug-in you are able to operate and test in as many bands as you have RF units available.

Figure 2.4 shows a block diagram of a typical sweep oscillator. Notice the separation between the mainframe and the plug-in. Many of the operations of the mainframe are to provide control functions. Such functions as tuning voltage, square wave modulation, blanking signals, and the sweep voltage are all found in the mainframe of a microwave sweep oscillator. The mainframe must be like a universal socket which is able to accept a variety of RF plug-in heads and produce satisfactory results. The importance of a good mainframe cannot be overemphasized.

Test Equipment

Figure 2.4/Block diagram of sweep oscillator.

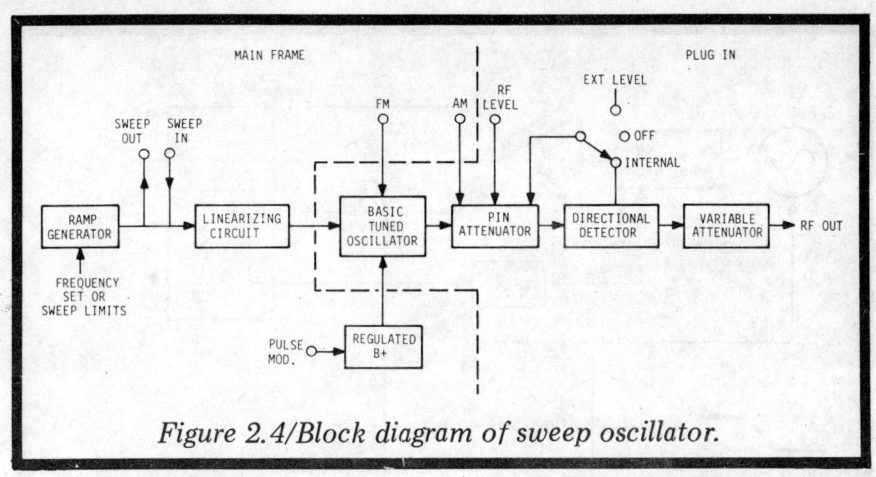

As valuable as the mainframe is to a sweep oscillator combination, it is useless without an RF plug-in. This is the heart of a good and useful sweeper. It is here that the ultimate RF output is generated, filtered, leveled, sampled, and generally massaged to result in a microwave signal at the proper frequency and swept over the proper band. It is here that a signal of the proper stability and at the desired level is produced. To repeat an earlier statement, it is the *heart* of the sweeper.

We saw in the previous section how a single frequency generator produced an output. But, what does it take to result in a generator that sweeps from one frequency to another over octave bands and wider with the stability and accuracy to rival a single frequency source? A large part of the magic that produces a sweeper is the oscillator that is employed. There are two types of oscillators widely used in RF plug-ins for sweepers. They are varactor tuned oscillators (VTO) and YIG tuned oscillators (YTO). Figure 2.4 shows a block of a sweeper that uses a varactor tuned oscillator, as well as a linearizing circuit. This is because the characteristics of a varactor exhibit a larger degree of non-linearity. In order for an oscillator to be swept efficiently over a wide frequency range, it must a linear. Therefore, a linearizing circuit is necessary.

The most popular of the tuned oscillators in use today is the YIG (Yttrium Iron Garnet) tuned oscillator. The YIG oscillator does not use a tuning voltage applied to a non-linear device, but rather has a current applied to YIG spheres to determine the frequency of operation. An RF interaction with the YIG spheres causes a resonance which produces a frequency dependent only on the magnitude of

Figure 2.5/RF plug-in block diagram.

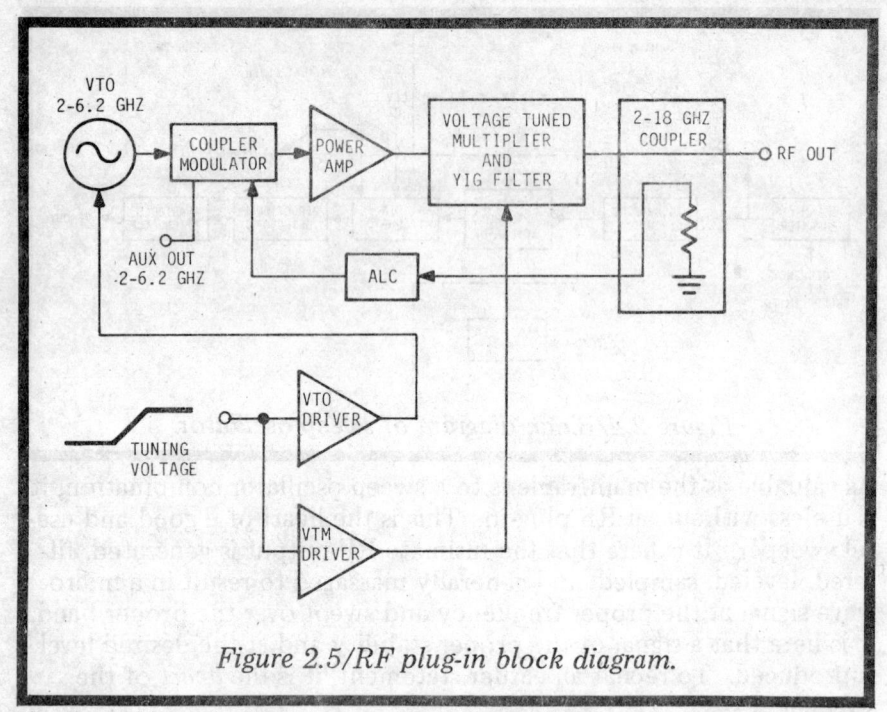

the dc field, and which is entirely independent of physical dimensions. A change in current causes a change in frequency. It is simple to see at this point that frequency is a linear function of current. Therefore, no linearizing circuitry is needed. However, driver units for the YIG circuits are needed. This is illustrated in Figure 2.5, which is a block diagram of an RF plug-in to go from 2 to 18 GHz. This is a plug-in that is available today and exhibits excellent performance.

The fundamental oscillator is a YIG-tuned transistor oscillator (YTO) which generates 2 to 6.2 GHz directly. This drives a 2 to 6.2 GHz amplifier which supplies more than 100 mW to a YIG-tuned multiplier (YTM). In the YTM a step-recovery diode produces harmonics of the 2 to 6.2 GHz fundamental signal and a tracking YIG filter selects the desired frequency. This tracking filter limits output harmonics and harmonically related signals.

Output power considerations demand exceptionally close tracking between the two YIG circuits. This is made possible through development of extremely linear magnet structures for the YIG devices which, in turn, results in excellent frequency accuracy. Even at 18 GHz, frequency can be set from the dial scale to ± 20 MHz, which is better than frequency meter accuracy. Signal drift in CW is typically less than

Test Equipment 23

0.005% per 10 minutes, which means narrow bandwidth measurements can be made reliably. Temperature compensation also minimizes frequency drift with ambient changes. Hysteresis effects in the tuning are imperceptible.

Between the oscillator and the modulator is an auxiliary output coupler. This output on the rear panel allows external monitoring of frequency and provides a sampling signal for phase locking from 2 to 18 GHz with only 6.2 GHz RF hardware. Since this signal is taken from the fundamental oscillator before the modulator, it is relatively unaffected by amplitude modulation. This is very similar to the "uncalibrated output" jack discussed for the HP8614A signal generator in the previous section.

With the basic idea of sweep oscillators in mind, let us examine a typical data sheet and its terms.

Frequency Range	2-6.2 GHz
Frequency Accuracy (25°C)	
All Modes	± 30 MHz
Frequency Linearity	± 8 MHz
Frequency Stability	
with Temperature	± 0.5 MHz/°C
Residual FM	± 100 kHz
Drift	10 kHz Peak
Maximum Leveled Power	5 dBm @ 25°C
Spurious Signals	
Harmonics	> 25 dB
Non-Harmonics	> 50 dB

Frequency Range — The band of frequencies over which the sweeper will operate and meet all of the other specifications listed. This could be one continuous sweep or may be switched in distinct bands. If the frequency range is broken into bands, the data sheet will say this.

Frequency Accuracy — This is how close to any frequency, throughout the band you are sweeping, you actually are. It is a + or − figure to allow for drift in either direction.

Frequency Linearity — This is how well the *sweep out* voltage tracks the actual frequencies being swept. It tells you how linear the relationship is between the voltage that drives an X-Y recorder or oscilloscope horizontal deflection circuitry, and the frequency that is being swept.

Frequency Stability — This is a measure of how steady the frequency of the sweeper remains when subjected to various conditions other than standard operation. Such factors as temperature, voltage changes, power changes, and VSWR at the output are some of the parameters that affect frequency stability. Residual FM, explained in the previous section, is also an important factor to consider.

Maximum Leveled Power — This is the maximum amount of power you can expect out of the sweeper and still maintain a level sweep according to listed specifications. You will note that this is specified at a temperature (25°C). This is because the maximum power will decrease at an elevated temperature and will increase at a much lower temperature. This should be kept in mind when considering what is needed in the way of *leveled* power for your particular application. One point should be stated here. That is that we have been talking about *leveled power* only. Unleveled power will be much higher at some points and will be much lower at others. So when discussing power output of a sweeper, be sure you are talking about the right type of output.

Spurious Signals — These are the harmonics and non-harmonics that are generated within the sweeper. In Figure 2.5, the YIG tracking filter has the task of removing harmonics and harmonic related signals. Usually a combination of good design and carefully placed filters will keep the spurious signals to a level where they will not interfere with the sweeper's operation.

2.1.3 Synthesized Signal Generators

We have covered single frequency microwave signal oscillators and sweep oscillators that can cover an octave or multi-octave of the microwave spectrum. We now come to one of the most sophisticated means of generating a microwave signal of today, that of the *microwave frequency synthesizer*. I am sure you have heard the term many times, but what exactly does the word *synthesizer* mean?

Technically speaking, a synthesizer is a device which creates something synthetically. This sounds like one of those typical dictionary definitions, but it serves as a base for a much more understandable definition. Anything synthetic is created by artificial means. This goes for fabrics, metals, colors, or even microwave frequencies. A synthetic item has changed in some way from the basic original substance so that it does not resemble this original substance in its final form. In a microwave synthesizer, the final output does not resemble the original oscillator signals because the output is many times higher in frequency. It creates these frequencies by using a very stable low frequency oscillator and

Test Equipment

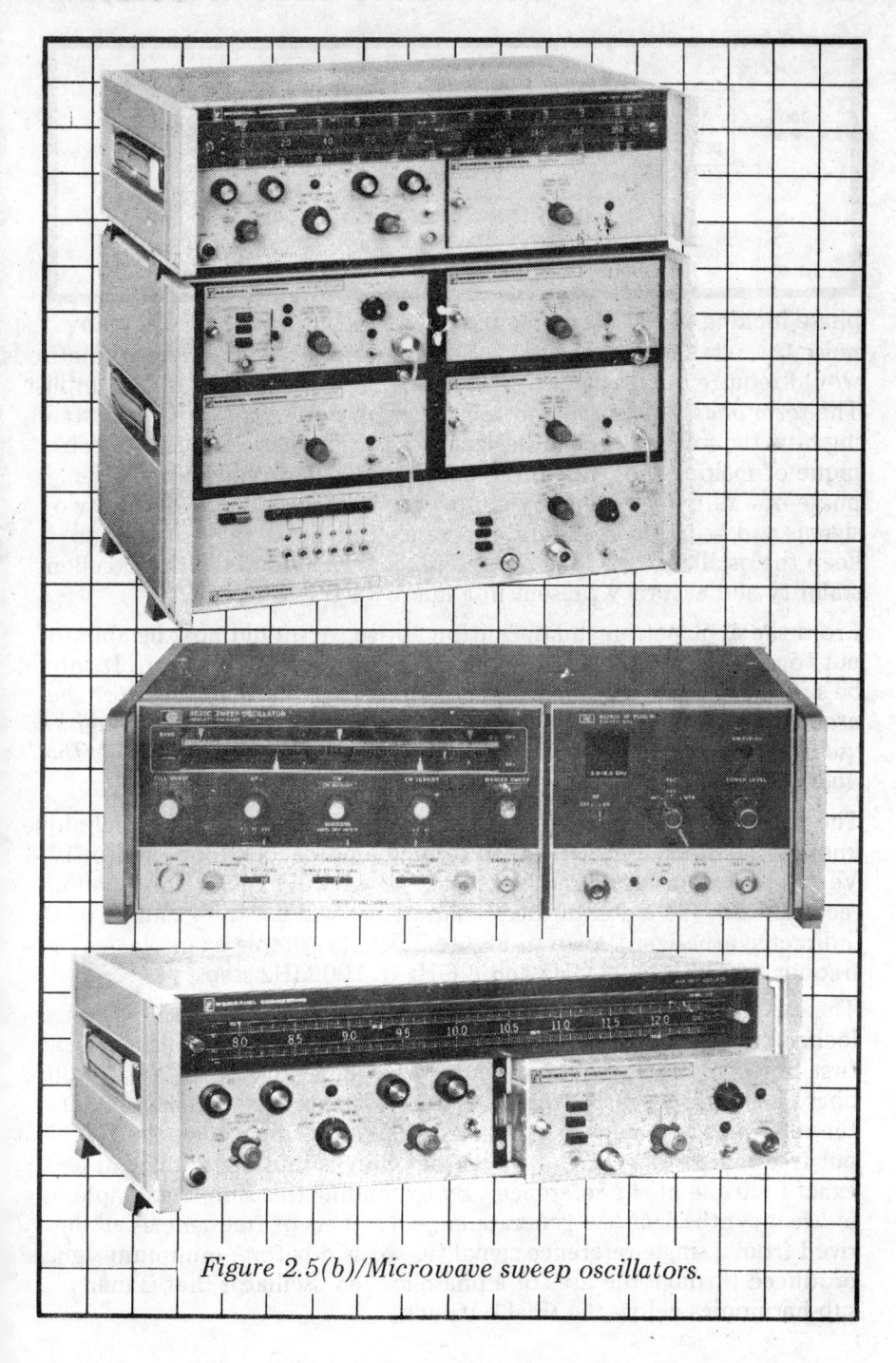

Figure 2.5(b)/Microwave sweep oscillators.

Figure 2.6/Indirect synthesizer.

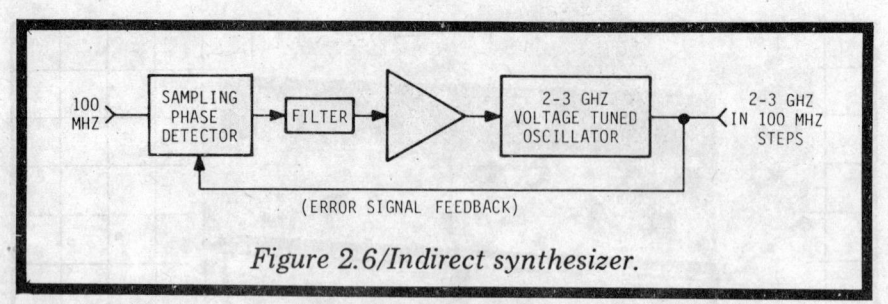

phase locking to the harmonic produced by it. This results in many noise free, stable frequencies that, using normal signal generator methods, would require hundreds or thousands of separate crystals to accomplish. The term *phase-lock* used above is a very important part of understanding how the microwave synthesizer works. Phase locking is the technique of making the phase of an oscillator signal follow exactly the phase of a reference signal by comparing the phases between the two signals and using the resultant difference to adjust the frequency to keep the oscillator tracking the reference. This results in the excellent stability and accuracy present in a microwave synthesizer.

From the explanation and definition above we should now be able to put together a definition for the term Microwave Synthesizer. It should be sufficient to say: *a microwave synthesizer is a signal generator that produces highly stable, accurate, and very low-noise signals through the use of a phase-locked oscillator that is many sub-harmonics below the final output.*

The majority of microwave synthesizers available today use a technique known as *indirect synthesis* (also termed a phase lock loop method). Very few systems use direct synthesis, so we will explain only the indirect method. To aid in our explanation, we will use the example of an indirect synthesizer shown in Figure 2.6. This simple example generates frequencies between 2 GHz and 3 GHz in 100 MHz steps.

The output is generated by a voltage controlled oscillator (VCO) phase locked to a harmonic of the reference signal. In operation, the VCO is first tuned to the approximate output frequency desired. The sampling phase detector compares the VCO with the appropriate harmonic of the reference and feeds back an error signal that fine tunes the VCO output frequency. The VCO output frequency is thus maintained at an exact multiple of the reference. By expanding this simple example, the indirect synthesizer can generate large numbers of frequencies all derived from a single reference signal (as we said before, minimum signals produced through the uses of a phase locked oscillator that is many sub-harmonics below the final output).

Test Equipment

At microwave frequencies, one of the advantages of indirect synthesizers is their low phase noise. They have better wideband phase noise performance than either direct synthesizers or lower frequency synthesizers multiplied to microwave frequencies. Indirect synthesizers accomplish this by taking advantage of the difference in noise characteristics of the crystal reference and the VCO for optimum noise performance.

With indirect synthesizers, the output phase noise is that of the reference multiplied up to microwave frequencies within the phase locked loop bandwidth. As the offset from the carrier increases, the effects of the phase locked loops decrease and the noise performance approaches that of the VCO only. The result is an overall improvement in noise performance, because close to the carrier the multiplied reference has the lower phase noise while at larger offsets, the VCO is actually cleaner. The synthesizer then includes the best regions of both signals for optimum noise performance.

With the basic operation of the microwave synthesizer explained, the next step is to define what you should look for in a commercially available instrument. Below is a typical synthesizer data sheet with the terms defined.

Frequency Range	4-6 GHz
Frequency Stability	1×10^{-9} per day
Frequency Resolution	10 kHz
Switching Time	< 20 msec to be within 10 Hz
Output Power	+10 dBm
Harmonics	−20 dBc (Maximum)
Single Sideband Phase Noise	−90 dBc (Maximum)
Non-Harmonic Spurious	−60 dBc (Maximum)

(Before explaining the above terms, it should be pointed out that the term dBc refers to the dB reading with respect to the carrier.)

Frequency Range — The range of frequencies that the synthesizer is capable of producing at the output.

Frequency Stability — This is usually the long term stability of the instrument and refers to the slow change in the average frequency with time. This is in contrast to short term stability, which refers to change in frequency over a time sufficiently short that a change in frequency due to long term effects is negligible. These fluctuations are due to noise modulating the carrier.

Frequency Resolution — This is how accurately the synthesizer frequencies can be read.

Switching Time — This is the time required for the synthesizer to *settle* to a new frequency when switched from an existing one. It is usually expressed as the time required to come within a certain frequency accuracy (for example, within 10 Hz).

Output Power — The RF power available at the output of the synthesizer. This can range all the way from +13 dBm in the frequency range below 1.0 GHz to +7 dBm at frequencies to 18 GHz.

Harmonics — These are the signals present at the output that are multiples of the desired frequency (for example, a 1.0 GHz signal would have harmonics at 2.0, 3.0, 4.0, etc.).

Single Sideband Phase Noise — This is a measure of noise energy and is usually expressed in a 1 Hz bandwidth versus frequency offset from the carrier (100 kHz for example). It is often specified for both signal generators and synthesizers. This measurement gives more information than any other about noise distribution, and all other short term stability parameters can be derived from it. Pay very careful attention to the parameter when specifying a microwave synthesizer.

Non-Harmonic Spurious — These are the signals present at the output that are *not* multiples of the desired frequency. These are the most difficult to eliminate, since they can be at any frequency.

Figure 2.7 shows some of the microwave synthesizers that are available today. The shapes, displays, and keyboards may be different, but their functions are the same: to produce stable, accurate, low noise microwave signals.

2.2 Signal Detection/Indicating Devices

The previous section covered the generation of microwave signals. This is the first step in building a usable microwave measurement setup. A second step is to detect the microwave signal and display it in a useful form. This section will cover six areas of signal detection and display (or indication). Some of them are only for detection; some only for indication; and some are capable of both detection and indication. The areas to be covered are: Detectors, Power Detectors, Power Meters, Spectrum Analyzers, Noise Figure Meters and Counters. Each will have a basic explanation followed by a data sheet with the terms explained.

Test Equipment   29

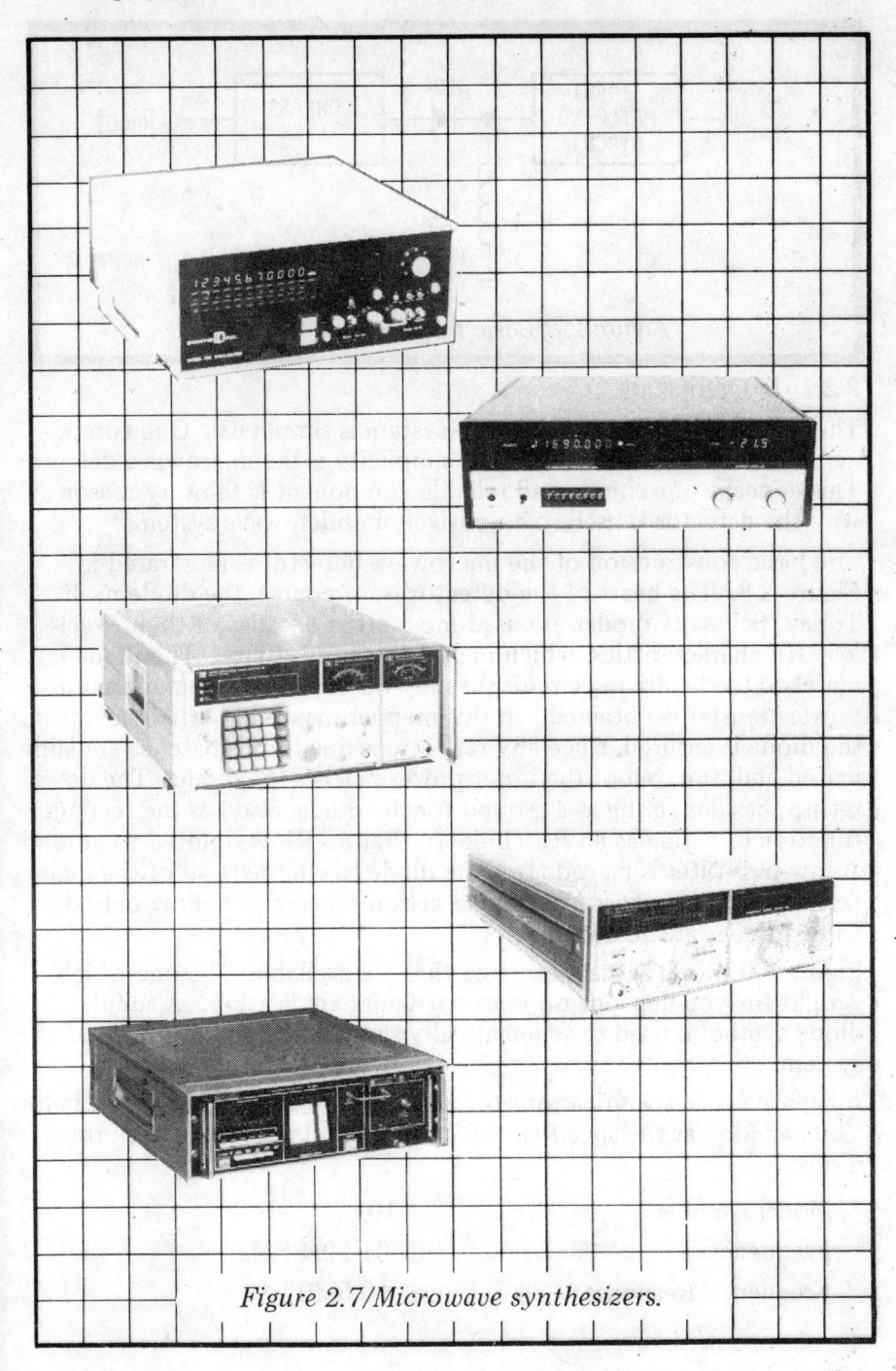

Figure 2.7/Microwave synthesizers.

Figure 2.8/Basic microwave detector.

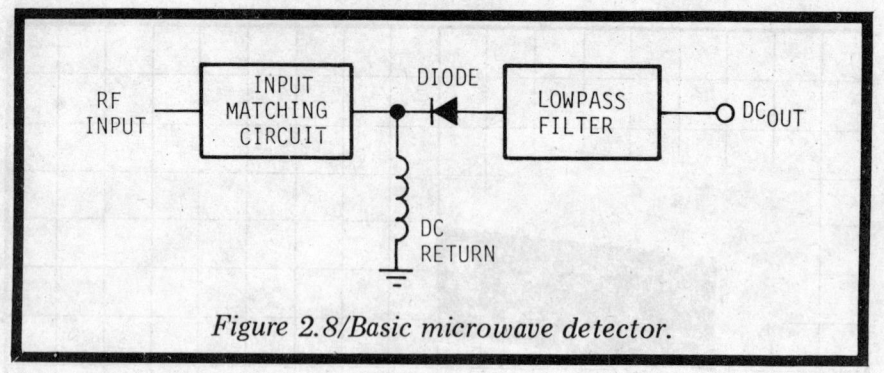

2.2.1 Detectors

The key to a reliable, trouble-free system is simplicity. One component whose operation is based on simplicity is the microwave detector. This concept of a simple and reliable component is the very reason why the detector is used so extensively in microwave systems.

The basic construction of the microwave detector is illustrated in Figure 2.8. The heart of the detector is, of course, the diode itself. Today, Schottky diodes are used most often because of their excellent RF characteristics, which include high sensitivity. The diode is matched to the driving circuit (usually 50 ohms) so that maximum power transfer is obtained. In this manner maximum efficiency of the diode is ensured, since any reflections due to mismatches are eliminated and, thus, all of the input power reaches the diode. The dc-return, besides acting as a ground for the diode, also has the second function of acting as an RF choke so that no RF is shunted to ground. A low-pass-filter is placed after the diode to eliminate all of the high frequency ripple caused by the detection process and allow only dc to be present at the output.

Figure 2.9 shows actual detectors that are available. The one which is right for you depends on your particular application. A simple diode symbol is used to schematically show that a detector is in a system.

A typical set of specifications for a microwave detector is shown below. You can look at the specification, then we will define the important terms.

Model Number	101
Frequency	.01-12.4 GHz
Frequency Response	± 0.5 dB

Test Equipment

Figure 2.9/Microwave detectors.

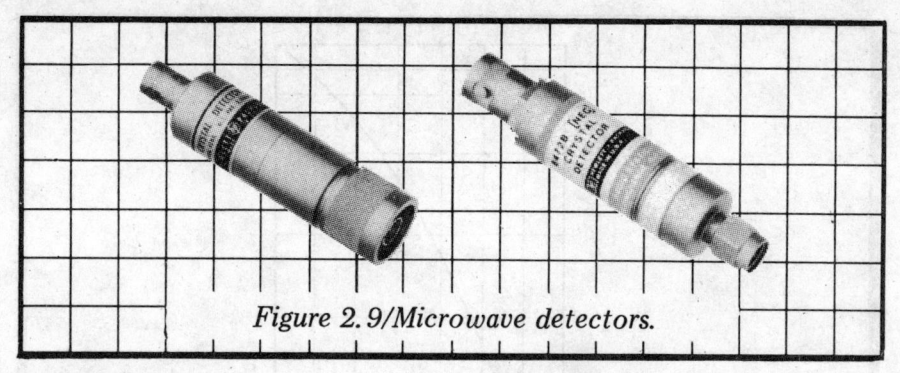

Sensitivity	0.4 mV/microwatts
Impedance	50 ohm
Maximum Input	100 mW
Polarity	Negative (see options)
VSWR	1.3:1 (Typical)
Connectors	Input — N (Male)
	Output — BNC (Female)
Weight	.25 Lbs.
Dimensions	2.75" Long/0.65" Dia.
Options	101-1; Positive Output, 50 ohm
	101-2; Negative Output, 75 ohm
	101-3; Positive Output, 75 ohm

Frequency Response — Frequency response is a term used to indicate the RF performance of the detector; it is sometimes given in decibels per octave or in one figure for overall performance. Basically, it is a measure of the variation in sensitivity of the detector, expressed in dB.

Maximum Input Power — This term refers to one of the most important considerations when using a detector — the highest level of RF power that can be applied to its input. It is always good practice to stay below this figure by a reasonable margin. The reason being that there is a diode directly in line in the circuit; wherever there is a semi-conductor there is a certain potential power level that could destroy the detector.

Polarity — Polarity is either positive or negative. It is important to know the polarity of the output of your detector, since it will have to be joined with some external circuit that requires a certain polarity to operate properly.

Figure 2.10/Detector sensitivity curve.

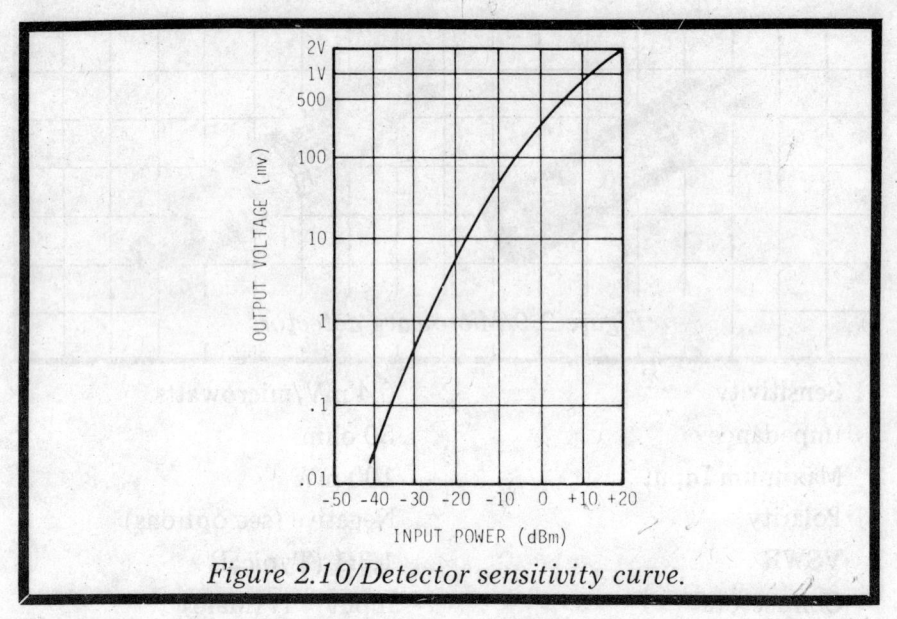

Sensitivity — Figure 2.10 is a typical sensitivity curve for a microwave detector. It can be seen that one factor affecting sensitivity is the RF power applied to the input of the detector; another is the dc output voltage. Therefore the detector sensitivity is the dc voltage produced at the output for a specific power input, usually expressed in millivolts per microwatt CW. Sensitivity is simply how much power you need to produce a certain voltage.

It is important to note, before leaving detector specifications, that you must always consider the type of connectors on the detector. It cannot be emphasized strongly enough that when the devices utilize the proper connectors, the electrical and mechanical operation of your system is improved.

The microwave detector finds applications wherever there is a need to control or convert an RF signal to a dc voltage for display purposes.

Figure 2.11 shows three such instances. Figure 2.11(a) is one of the most common applications, that of monitoring an RF line. Power is coupled off the main line; the detector provides a dc voltage proportional to this power.

The monitoring device, which has been previously calibrated, is deflected to give an instant and accurate reading of the main line power.

Figure 2.11(b) is a scheme that is very useful when a leveled power is needed, such as in swept measurements. Once again, a coupler and

Test Equipment   33

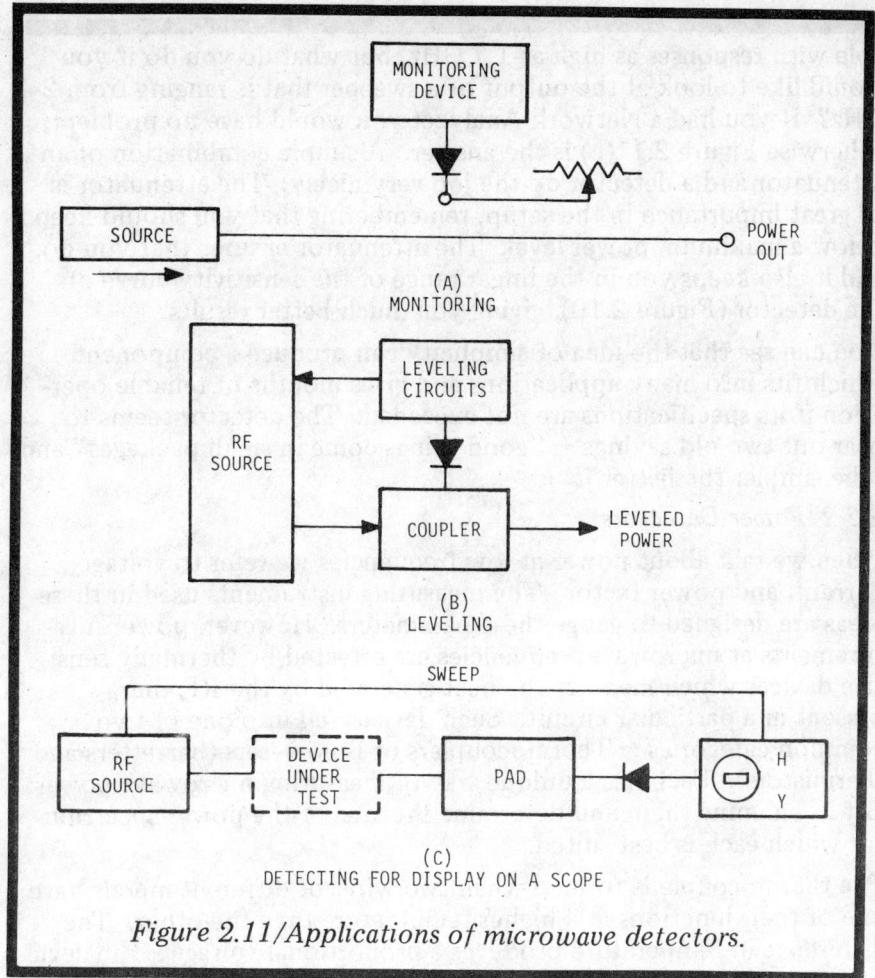

Figure 2.11/Applications of microwave detectors.

detector work together to sample the main line RF; the detector converts the RF to dc to activate the leveling circuits. This dc voltage causes either the RF power out of the source to be attenuated, or the attenuation to be removed in order to raise the power level. The result of this instantaneous pattern of action and reaction of coupler and detector is that the RF power remains level to the system output. Automatic Leveling Control (ALC) in microwave systems often employ just such a process.

Figure 2.11(c) is a method of displaying RF energy on an oscilloscope. It is impractical to attempt to view signals at frequencies above 100 MHz directly on a conventional oscilloscope. Storage scopes are available with responses as high as 1.7 GHz, but what do you do if you would like to look at the output of a sweeper that is ranging from 2-4 GHz? If you had a Network Analyzer you would have no problem; otherwise Figure 2.11(c) is the answer. A simple combination of an attenuator and a detector do the job very nicely. The attenuator is of great importance in the setup, remembering that you should keep below a maximum power level. The attenuator ensures that you do, and it also keeps you in the linear range of the sensitivity curve of the detector (Figure 2.10), giving you much better results.

You can see that the idea of simplicity can produce a component which fits into many applications and gives months of reliable operation if its specifications are not exceeded. The detector seems to bear out two old sayings — "good things come in small packages" and "the simpler the better".

2.2.2 Power Detectors

When we talk about power at low frequencies we refer to voltage, current, and power factor. The measuring instruments used in these areas are designed to gauge these parameters. However, power measurements at microwave frequencies are effected by thermally sensitive devices which measure the heat generated by the RF energy present in a particular circuit. Such devices fall into one of two common categories — Thermocouplers or Bolometers (barretters and thermistors). Each has a unique way of measuring microwave power. Let us examine them and determine the area of the power spectrum for which each is best suited.

The thermocouple is formed when two wires of different metals have one of their junctions at a higher temperature than the other. The difference in temperature produces a proportional voltage. The device is used to measure the temperature rise of a load which dissipates microwave power; by appropriate calibration, the temperature change is converted to an indication of power. Thermocouplers have excellent sensitivity and are especially useful as power monitors; their primary applications are in low and medium power areas.

Bolometers are classed as barretters or thermistors. The most common usage of the term "bolometer" refers to a barretter. However, we describe both classes to provide a familiarity with all types of microwave power measuring devices.

The barretter consists of an appropriately mounted short length of fine wire, usually platinum, with sufficient resistance to enable it to be impedance-matched as a termination for a transmission line. It

Test Equipment

Figure 2.12/Microwave barretters.

Figure 2.13/Microwave thermistor.

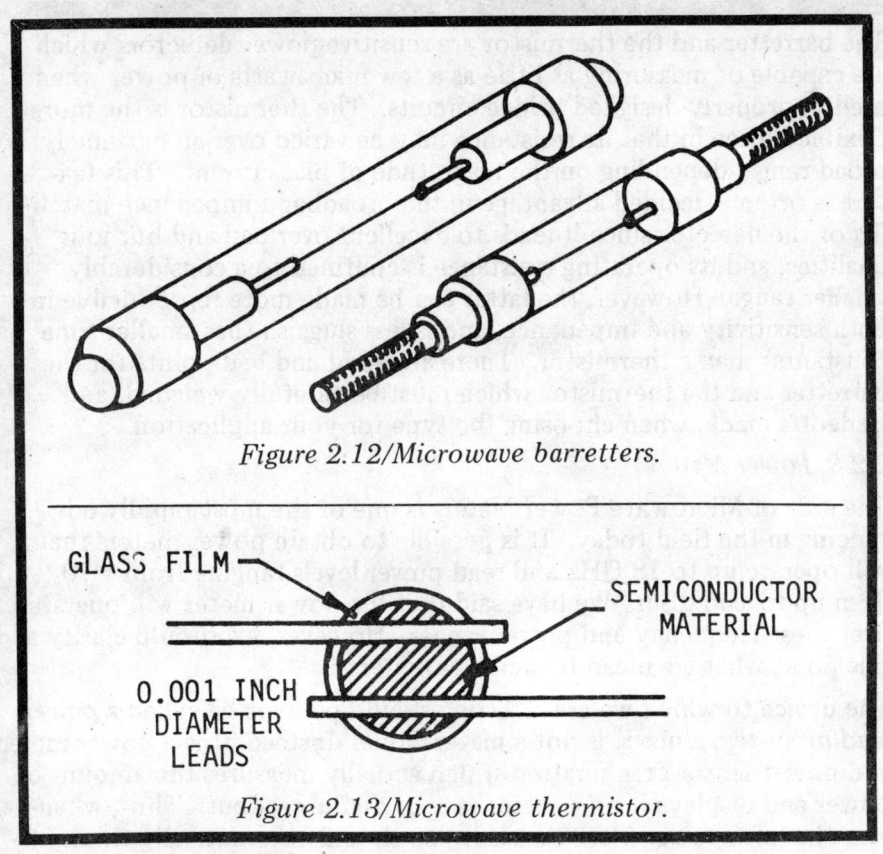

has a positive temperature coefficient; that is, the resistance of the barretter increases as the temperature rises. It is usually biased to an operating resistance of from 50 to 400 ohms; 200 ohms is hte most common. Figure 2.12 shows two common barretters.

The thermistor consists of a tiny bead of semiconductor material which bridges the gap between two fine, closely spaced, parallel supporting wires. This construction is shown in Figure 2.13. Effectively, all the resistance of the thermistor is concentrated in the bead material which has a negative temperature coefficient of resistance (an increase in temperature results in a resistive decrease). They usually are made of complex metallic-oxide components using oxides of manganese, nickel, copper, cobalt, and other metals.

The changes in resistance resulting from the dissipation of microwave power in the bolometer (barretter or thermistor) are commonly measured by using the devices as one arm of a dc or audio frequency wheatsone bridge circuit.

The barretter and the thermistor are sensitive power detectors which are capable of measuring as little as a few microwatts of power when used in properly designed bridge circuits. The thermistor is the more flexible device in that its resistance may be varied over an extremely broad range, depending on the magnitude of bias current. This feature is often a decided advantage in the broadband impedance matching of the detector since it leads to excellent overload and burnout qualities, and its operating resistance is confined to a considerably smaller range. However, the latter can be made more reproducible in both sensitivity and impedance, and is less sluggish (has smaller time constants) than a thermistor. There are good and bad points for the barretter and the thermistor which must be carefully weighed, and tradeoffs made, when choosing the type for your application.

2.2.3 Power Meters

The area of Microwave Power Meters is one of the most rapidly advancing in the field today. It is possible to obtain power meters that will operate up to 18 GHz and read power levels ranging from −70 dBm up to +35 dBm. We have said that the power meter will operate over these frequency and power ranges. However, we should clarify at this point what we mean by a *power meter*.

The device to which we are referring should perhaps be called a *power reading system* since it is not a meter alone. Instead, it is a power meter and power sensor combination which actually measures the amount of power and displays it on a meter face or digital readout. Thus, whenever the term *power meter* is used, you should automatically think of the meter/sensor combination and realize that they must work together to have an operational setup.

To realize the sensitivity of the power sensors used to measure microwave power, consider the lowest level available: −70 dBm. This is a value of 100 picowatts (100×10^{-12} watts). When you write that figure out you get: .0000000001 watts, which is a pretty small amount of power for a sensor unit to detect. Efficient and thorough thermal design of the unit and excellent noise characteristics enable readings to be made at the extremely low power level.

Now consider the upper end of the power spectrum, the +35 dBm maximum which is very readily available in power sensors today. It was not long ago when the only type of reliable power sensor available was the HP478A thermistor mounts to use with the 430 series of power meters. This mount had a line printed on it in red that warned that 50 mW (+17 dBm) maximum power was all it could

Test Equipment

Figure 2.14/Microwave power sensors.

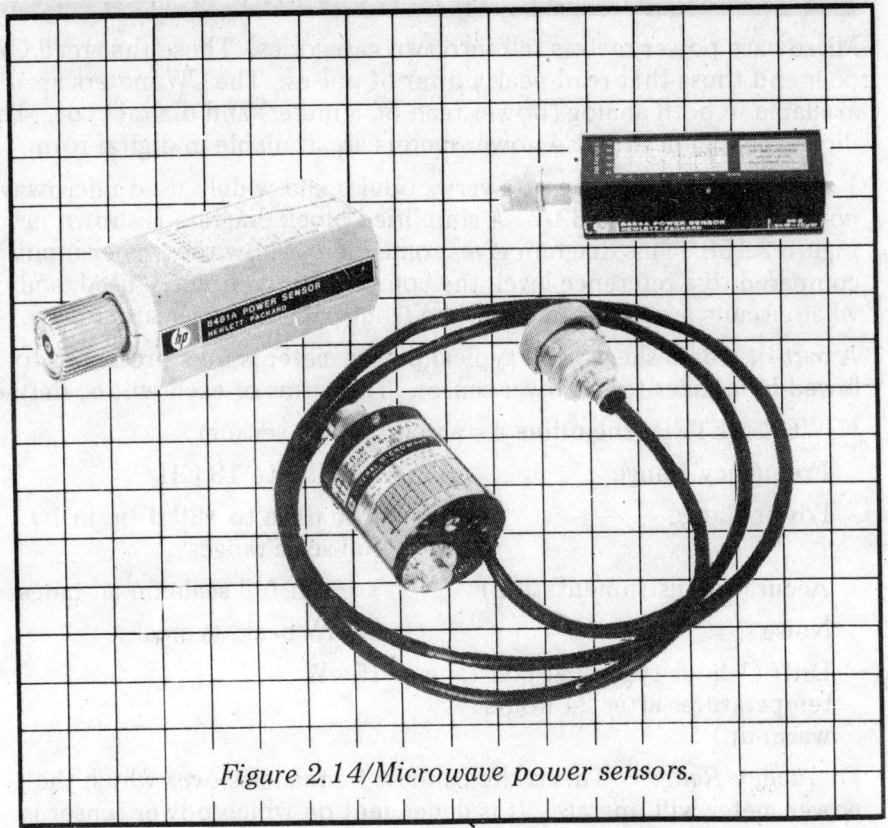

handle. If any higher power needed to be measured, it was necessary to obtain attenuator (pads) of sufficient power rating to handle the necessary power and "pad the signal down." Today, with a power sensor having a +35 dBm rating, a power of 3 watts can be read directly with no "padding" at all. Power sensors that are available are shown in Figure 2.14.

There is no one power sensor that covers the entire range mentioned above. Instead they are different sensors that cover overlapping ranges. Some of the typical ranges are:

−70 dBm to −20 dBm
−30 dBm to +20 dBm
−17 dBm to +35 dBm
−10 dBm to +35 dBm

The most popular range is the one that covers −30 dBm to +20 dBm. About 90% of all microwave power measurements will fall into this range.

Microwave power meters fall into two categories. Those that read CW poer and those that read peak power of pulses. The CW meters are available in both analog (power read on a meter) and digital type. The direct reading peak pulse power meters are available in digital form.

Figure 2.15a is a picture of a very popular and widely used microwave power meter, the HP432A. A simplified block diagram is shown in Figure 2.15b. This diagram gives some idea of how the power input is compared to a reference level, the compensation circuitry used, and what circuits are adjusted when the front panel controls are moved.

A part of a data sheet for a typical power meter is now presented followed by a sheet for a power sensor. The terms of each will be defined.

POWER METER (Including a standard power sensor)

Frequency Range	10 MHz to 18 GHz
Power Range	−25 dBm to +20 dBm in 10 full-scale ranges
Accuracy (Instrumentation)	± 1% of full scale on all ranges
Noise	30 nW peak, typical
Drift (1 hour typical at constant temperatures after 24 hours warm-up)	15 nW

Frequency Range — This is the band of frequencies over which the power meter will operate. It is dependent on which power sensor is used. Ranges such as 10 MHz to 18 GHz, 100 kHz to 4.2 GHz, and 100 kHz to 2 GHz are common.

Power Range — This parameter also depends on the power sensor used. The typical range of −30 dBm to +20 dBm has a range of 50 dB with 10 full-scale ranges as follows: −30, −25, −20, −15, −5, 0, +5, +10, +15, and +20 dBm.

Accuracy — This is how close to the actual power level the meter is reading. If, for example, you were putting in a reference 0 dBm signal (1.0 mW), the meter would have an accuracy of ± 0.01 mW. (It could read from 0.99 mW to 1.01 mW and still be within specification.)

Noise — This is a very important parameter since a meter with a high internal noise level will not be able to distinguish low level signals from the noise. It will also cut down the dynamic range of the meter since the level will have to be much higher on the low end.

Test Equipment

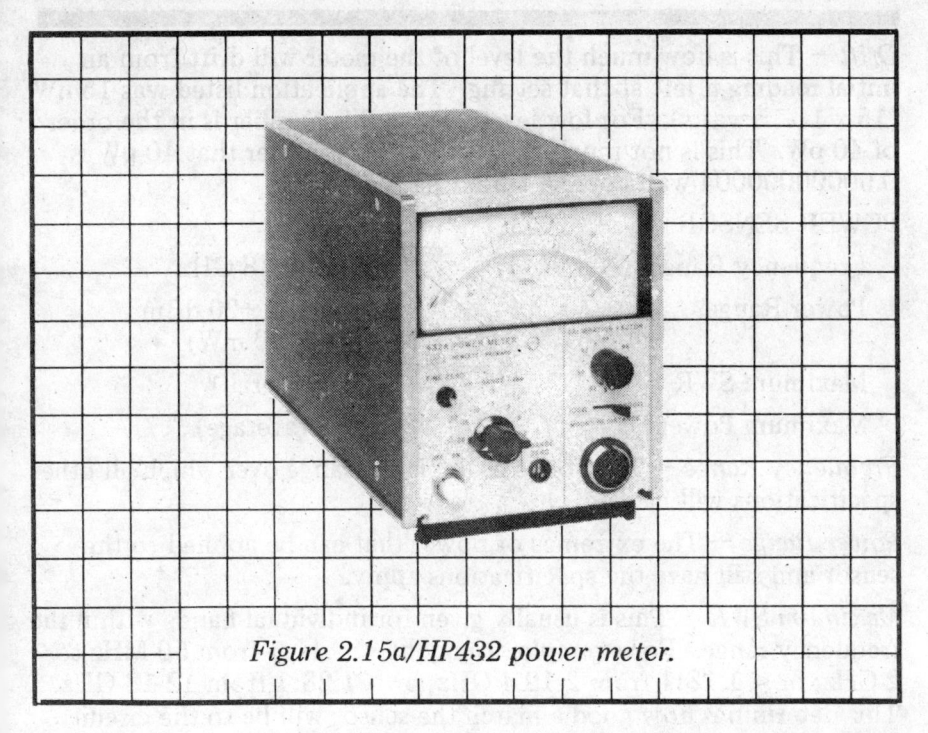

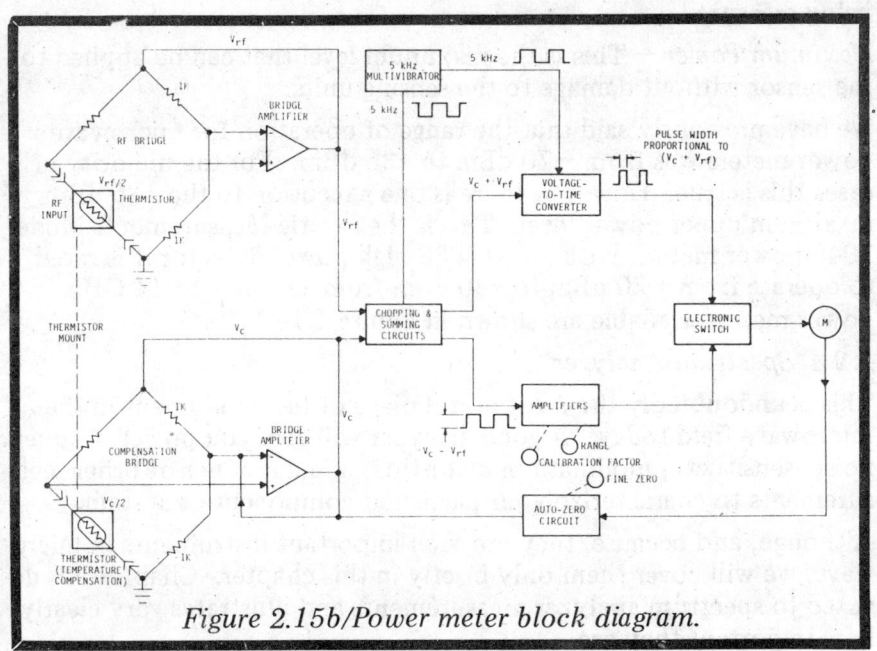

Figure 2.15a/HP432 power meter.

Figure 2.15b/Power meter block diagram.

Drift — This is how much the level of the meter will drift from an initial reading if left at that setting. The application listed was 15 nW (15×10^{-9} watts). For low level sensors this number is in the order of 40 pW. This is not much drift when you consider that 40 pW is 0.00000000004 watts over a 1-hour period.

POWER SENSOR

Frequency Range	10 MHz to 18 GHz
Power Range	−30 dBm to +20 dBm (1 μW to 100 mW)
Maximum SWR	< 1.4:1 (overall)
Maximum Power	300 mW (average)

Frequency Range — This, once again, is the range over which all other specifications will be valid.

Power Range — The extremes of power that can be applied to the sensor and still have the specifications apply.

Maximum SWR — This is usually given for individual bands within the frequency range. For example, it may be < 1.10:1 from 50 MHz to 2 GHz; or < 1.18:1 from 2-12.4 GHz; or < 1.28:1 from 12-18 GHz. This determines how good a match the sensor will be to the circuit being measured.

Maximum Power — This is the maximum level that can be applied to the sensor without damage to the sensing units.

We have previously said that the range of operation for microwave power meters was from −70 dBm to +35 dBm. For the majority of cases this is true. However, there is one exception to the +35 dBm maximum upper power level. This is the Pacific Measurements Model 1045 power meter. With its #13838 high power detector it is rated to operate from −20 dBm to +40 dBm from 1.0 MHz to 14 GHz. Power meters available are shown in Figure 2.16.

2.2.4 Spectrum Analyzer

This is undoubtedly the most useful piece of test equipment in the microwave field today. A good analyzer will give you power, frequency, noise, sensitivity, modulation, distortion, and a number of other measurements to characterize your particular component or system.

Although, and because, they are such important instruments in microwave, we will cover them only briefly in this chapter. Chapter 5 is devoted to spectrum analyzer measurements and illustrates very clearly how important they are.

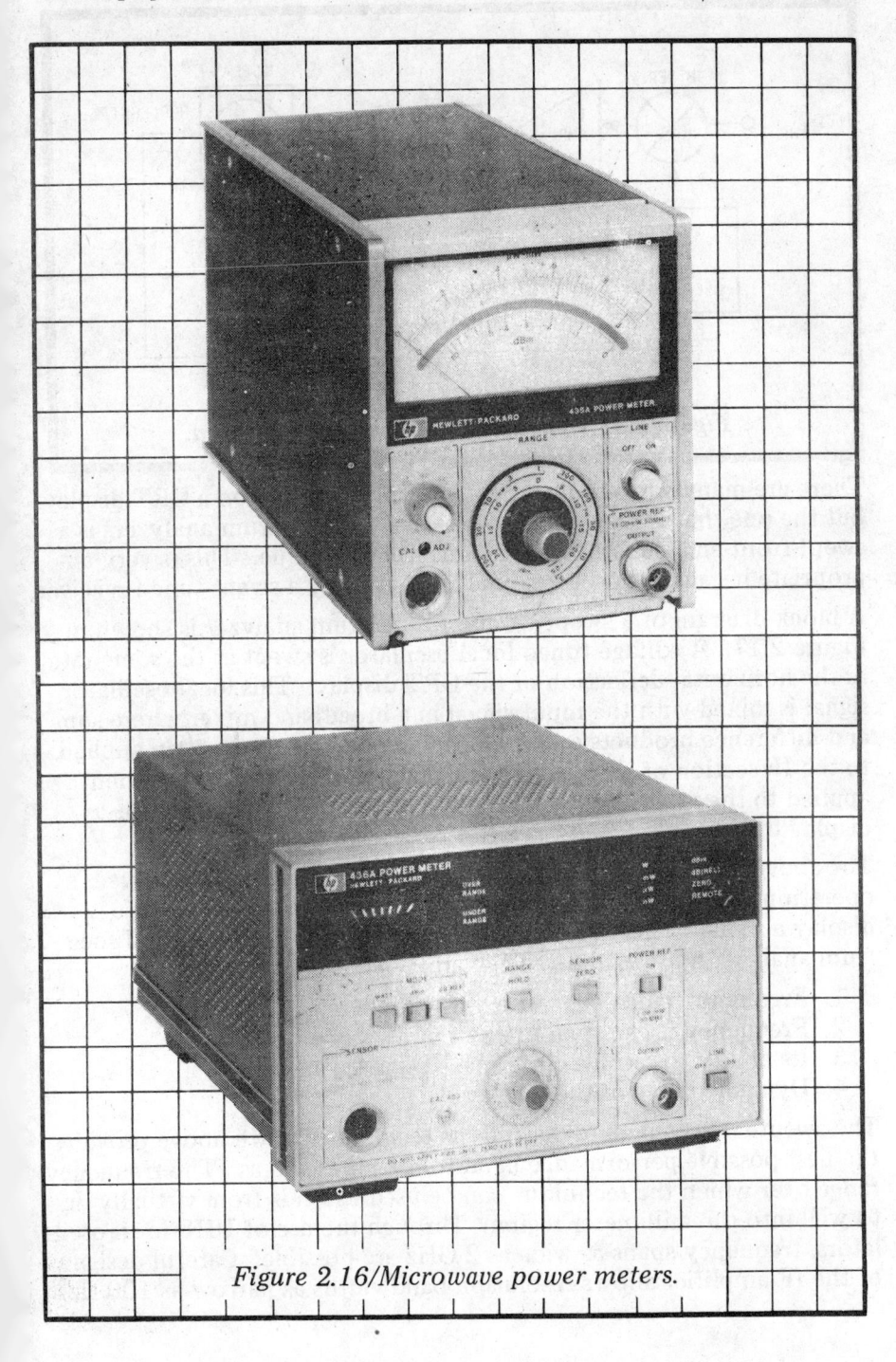

Test Equipment

Figure 2.16/Microwave power meters.

Figure 2.17/Spectrum analyzer block diagram.

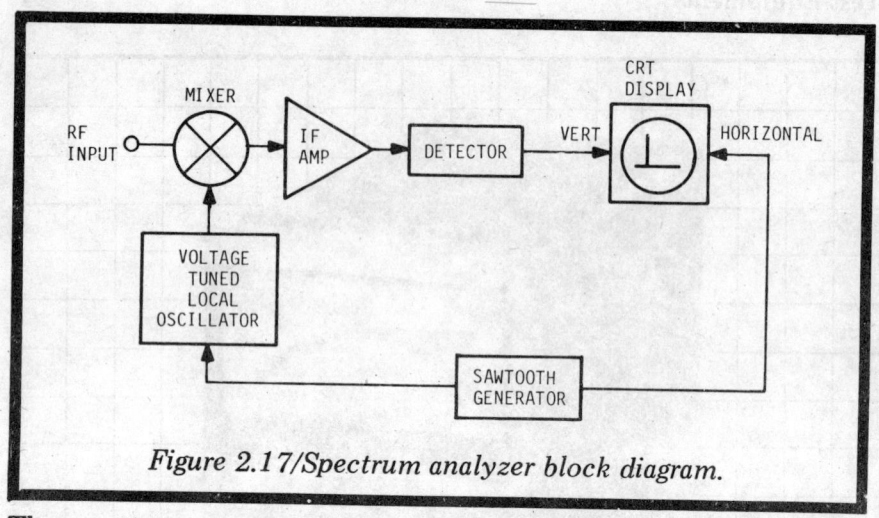

There are many ways to process and display a signal on a CRT display, but the one that is used almost exclusively in spectrum analyzers is a swept-front-end, superhetrodyne receiver technique. This is very appropriate because a spectrum analyzer is simply a swept-tuned receiver.

A block diagram of a swept-front-end spectrum analyzer is shown in Figure 2.17. A voltage tuned local oscillator is swept at the same rate as the horizontal deflection of the CRT display. This local oscillator signal is mixed with the input signal in a broadband mixer where sum and difference products are produced. These signals are then applied to the IF section of the analyzer. The detected IF signals are then applied to the vertical deflection plate of the CRT. The result is a display of signal amplitude as a function of frequency.

The above explanation is a good illustration of a spectrum analyzer's operation. However, a spectrum analyzer has to do more than just display a signal. There are four basic characteristics that a good spectrum analyzer must exhibit. They are:

1. Frequency range over which it will operate,
2. Frequency it can span in a sweep,
3. Its ability to resolve close spaced signals, and
4. Dynamic range of the display.

The swept-front-end, superhetrodyne technique shown above provides the best possible performance in all four of these areas. The frequency range over which the technique can be used extends from virtually dc to well into the millimeter region. Through the use of UHF local oscillators, frequency spans as wide as 2 GHz are possible. Careful design of the IF amplifier allows resolution bandwidths as narrow as 100 Hz.

Test Equipment 43

By including a logarithmic amplifier prior to the vertical deflection plate, a display dynamic range as wide as 70 dB can be achieved.

The swept-front-end superhetrodyne scheme is ideal for use in spectrum analyzers.

A portion of a typical data sheet for a spectrum analyzer is shown below. The terms will be explained following the listing.

Frequency Range	0.01 to 22 GHz
Frequency Spans:	
1.7 to 22 GHz span	Spanned in one sweep
Full Band	Band selected swept in one span
Per Division	1 kHz to 500 MHz/division in a 1, 2, 5 sequence
Zero Span	Analyzer is a fixed-tuned receiver
Resolution Bandwidth	1 kHz to 3 MHz in 1, 3, 10 sequence
Measurement Range	Noise level < −106 dBm to +30 dBm damage level

There are many other parameters presented in an analyzer data sheet but these are some of the more important ones you should know to select an instrument for your particular use. Explanations of these terms are now presented.

Frequency Range — This is the frequency coverage of the spectrum analyzer. Using internal circuitry, analyzers are available up to the 20 GHz area. Using external mixing processes the range can be extended up to 60 GHz.

Frequency Spans — This is the range of frequencies displayed on a single sweep of the CRT. In one case the entire range of the analyzer is displayed. In other positions a full band, selected increments of frequencies, or single fixed-tuned settings are displayed.

Resolution Bandwidth — These are the bandwidths of filters that allow an operator to expand the input signal sufficiently to characterize it. Most analyzers have selectable bandwidths in 1, 3, and 10 sequences.

Measurement Range — This is the minimum and maximum level between which you must adjust your settings to protect the analyzer and to obtain an accurate measurement. All spec sheets show the maximum level

Figure 2.18/Microwave spectrum analyzers.

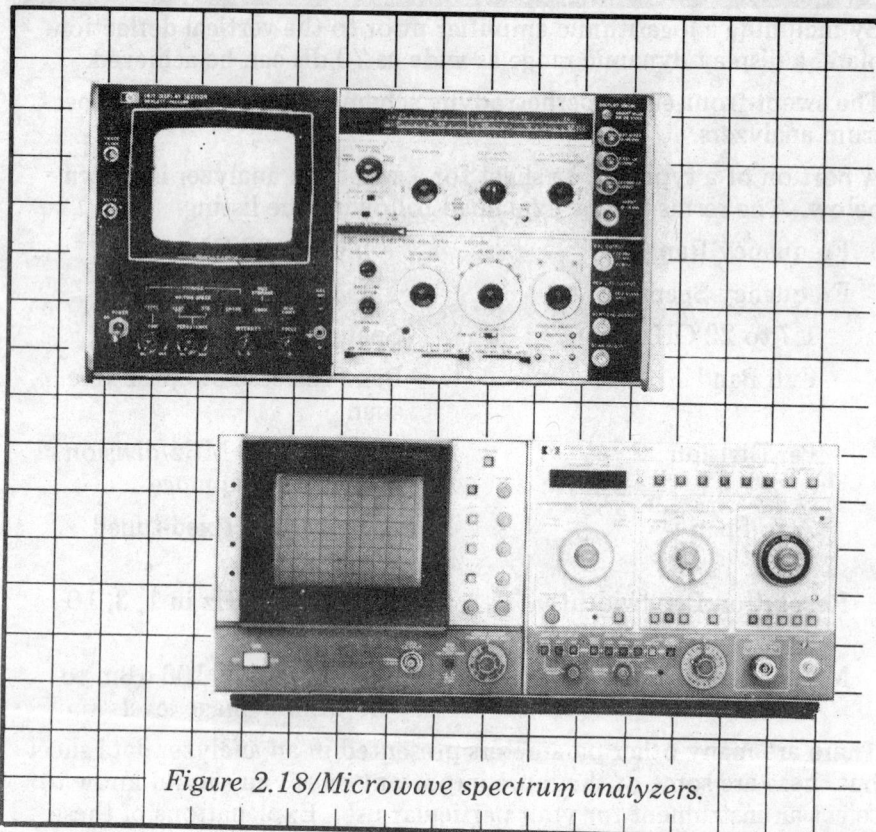

but there are some manufacturers who do not list the noise level. This is as important as the maximum level since a signal below, or at, the noise level of the instrument will not be displayed properly on the CRT.

Some of the spectrum analyzers available today are shown in Figure 2.18.

2.2.5 Noise Figure Meter

The ultimate sensitivity of a system is set by the noise presented to that system with the desired signal. To be able to characterize the noise system is one of the most important requirements in microwave system testing.

There are manual and automatic methods used for noise figure measurements. We will concentrate at this point on the *automatic noise figure meter*. A block diagram of an automatic meter is shown in Figure 2.19. You can see that the meter provides an input to the noise sources, a timing signal, and receives a noise level from the device under

Test Equipment

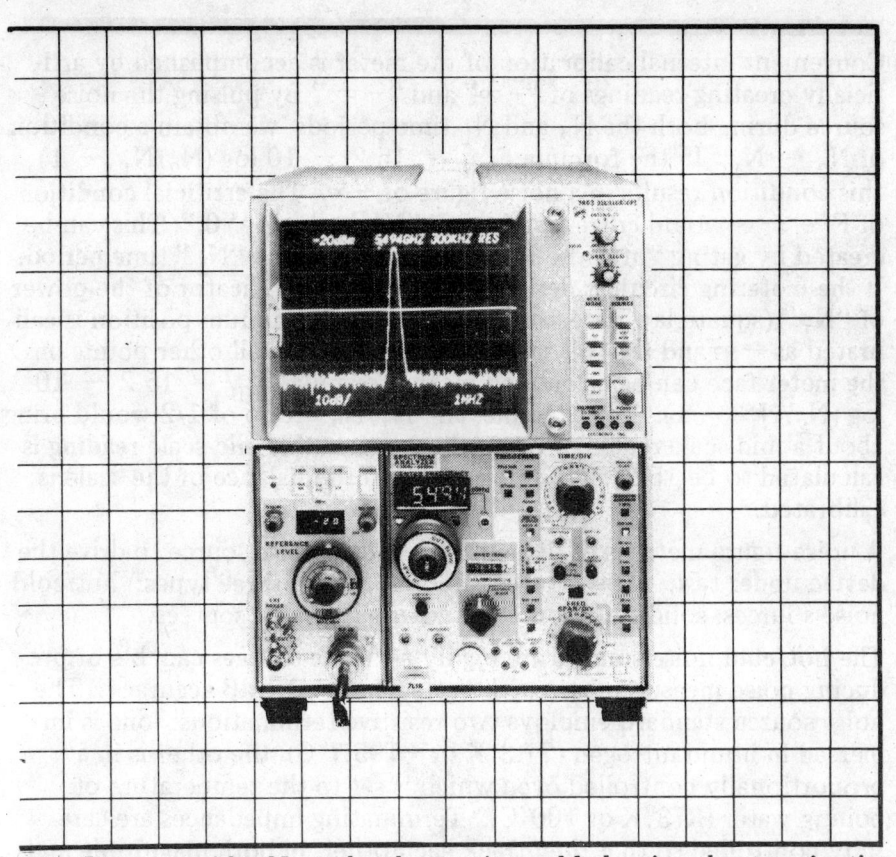

test. If we consider N_1 to be the receiver-added noise plus termination noise and N_2 to be the receiver-added noise, termination noise and the excess noise times the system gain (shown in Figure 2.19b), the operation of the meter is as follows:

The gating source pulses the noise source at a rate of 500 Hz; N_1 and N_2 pulses arrive at the IF amplifier. Noise sources have a finite noise build-up time, so the IF amplifier is gated to pass only the final amplitudes of N_1 and N_2 to the square law detector. The detected N_2 pulse is switched to an AGC integrator, where a voltage for gain control of the IF amplifier is derived. The time constant of this circuit is made long enough to control the IF amplifier gain even when the N_1 pulse is passing through it. Since the AGC action keeps the detected N_2 pulse at a constant level, a measurement of the detected N_1 pulse is, in effect a measurement of the pulse ratio. The N_1 pulse is measured by switching it to the meter integrator and meter.

Convenient internal calibration of the meter is accomplished by artificially creating readings of "$+\infty$" and "$-\infty$." By pulsing the noise source during both the N_2 and N_1 time periods, we obtain a condition of $N_2 = N_1$. In the formula $F_{dB} = 15.2 - 10 \log (N_2/N_1 - 1)$ this condition results in a noise figure of $+\infty$. The artificial condition of $F = -\infty$ would correspond to an "N_1" value of "0." This can be created by gating "off" the IF amplifier during the "N_1" time period. If the metering circuit is designed to be a linear indicator of the power of "N_1" (square law detector) and the meter minimum position is calibrated as $-\infty$ and the full scale deflection as $+\infty$, all other points on the meter face can be calculated by the formula $F_{dB} = 15.2 - 10 \log (N_2/N_1 - 1)$. For example, an "$N_1/N_2$" ratio of 1/2 would bring about a mid-scale reading. From the formula this mid-scale reading is calculated to be 15.2. In a similar fashion the balance of the scale is calibrated.

A noise figure meter is of little use without a noise source to drive the device under test. Noise sources are available in three types: hot/cold noise sources, solid-state sources, and gas-discharge sources.

The hot/cold noise sources are highly accurate sources capable of producing noise measurements with better than ± 0.1 dB accuracy. The noise source standard employs two resistive terminations: one is immersed in liquid nitrogen (77.3°K or −195.7°C), the other is in a proportionally controlled oven which is set to the temperature of boiling water (373°K or 100°C). Terminating impedances are carefully controlled so that they track each other, in both magnitude and phase, over the full frequency range of the instrument. Thus, mismatch uncertainties are virtually eliminated.

Solid-state noise sources are available to cover the range from 10 MHz to 18 GHz. These devices are small in size and high in reliability. The noise-generating diode used is isolated from the output by 12 to 20 dB; consequently, there are almost no mismatch errors as the source is switched on and off, and no destructive transients are present in the output.

The gas-discharge noise sources usually use argon discharge tubes, although some sources can be furnished in neon. This type of noise source is available from approximately 4 to 40 GHz. Needless to say, these sources are of waveguide construction. One end of the waveguide is open while the other end must be terminated in a low SWR load. The termination is placed on the source by the manufacturer and should not be removed.

Test Equipment   47

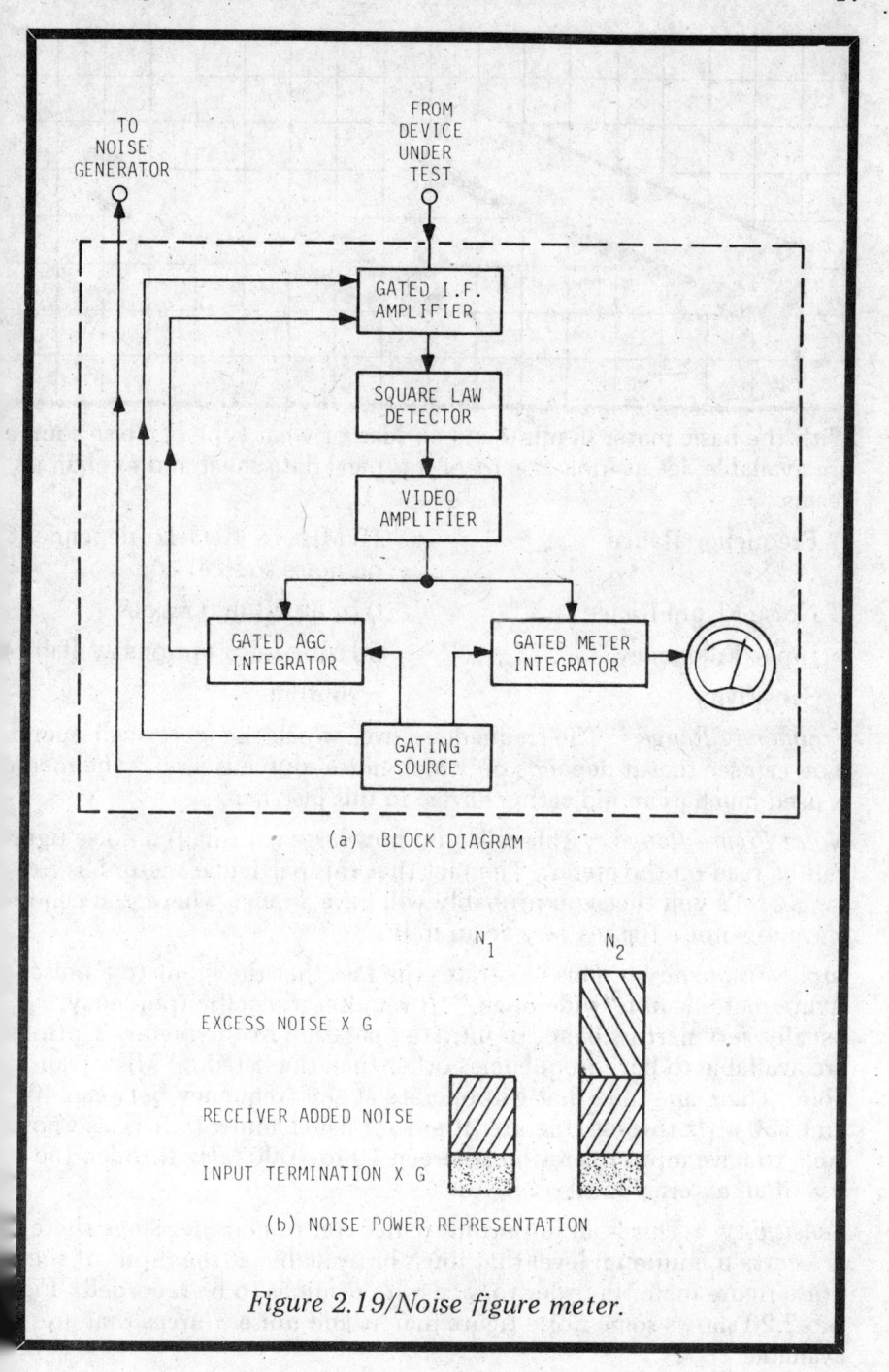

(a) BLOCK DIAGRAM

(b) NOISE POWER REPRESENTATION

Figure 2.19/Noise figure meter.

48   HANDBOOK OF MICROWAVE TESTING

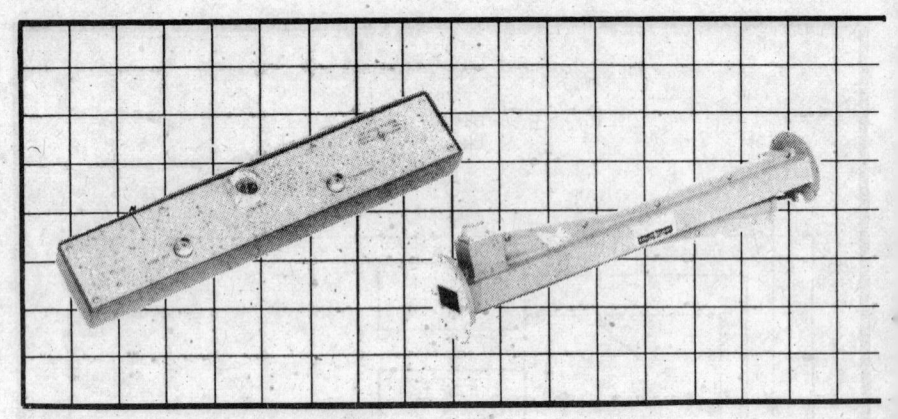

With the basic meter in mind and an idea of what type of noise source are available, let us look at part of a typical data sheet and explain its terms.

Frequency Range	10 MHz to 40 GHz (dependent on noise source)
Noise Figure Range	0 to 33 dB in 5 ranges
Input Frequency	30 MHz (also options available)
Sensitivity	−76 dBm

Frequency Range — The frequencies over which the meter will operate. You can see that it depends on which noise source is used. The meter is used much as an indicating device in this instance.

Noise Figure Range — This tells you how large (or small) a noise figure can be read on the meter. The fact that this particular meter has five ranges tells you that you probably will have a range where you can resolve low noise figures very accurately.

Input Frequency — This illustrates the fact that the input to a noise figure meter is not "wide open." It requires a specific frequency, usually very narrow band, to initiate operation of the meter. Options are available to have frequencies other than the listed 30 MHz available. There are units that will operate at any frequency between 40 and 150 MHz through the use of a front panel control. It is also possible to have input operation between 10 to 1000 MHz through the use of an external local oscillator.

Sensitivity — This is an important parameter to consider since there is a certain minimum level that must be available at the input of the noise figure meter in order for accurate readings to be recorded. Figure 2.20 shows some noise figure meters and noise sources that are available.

Test Equipment

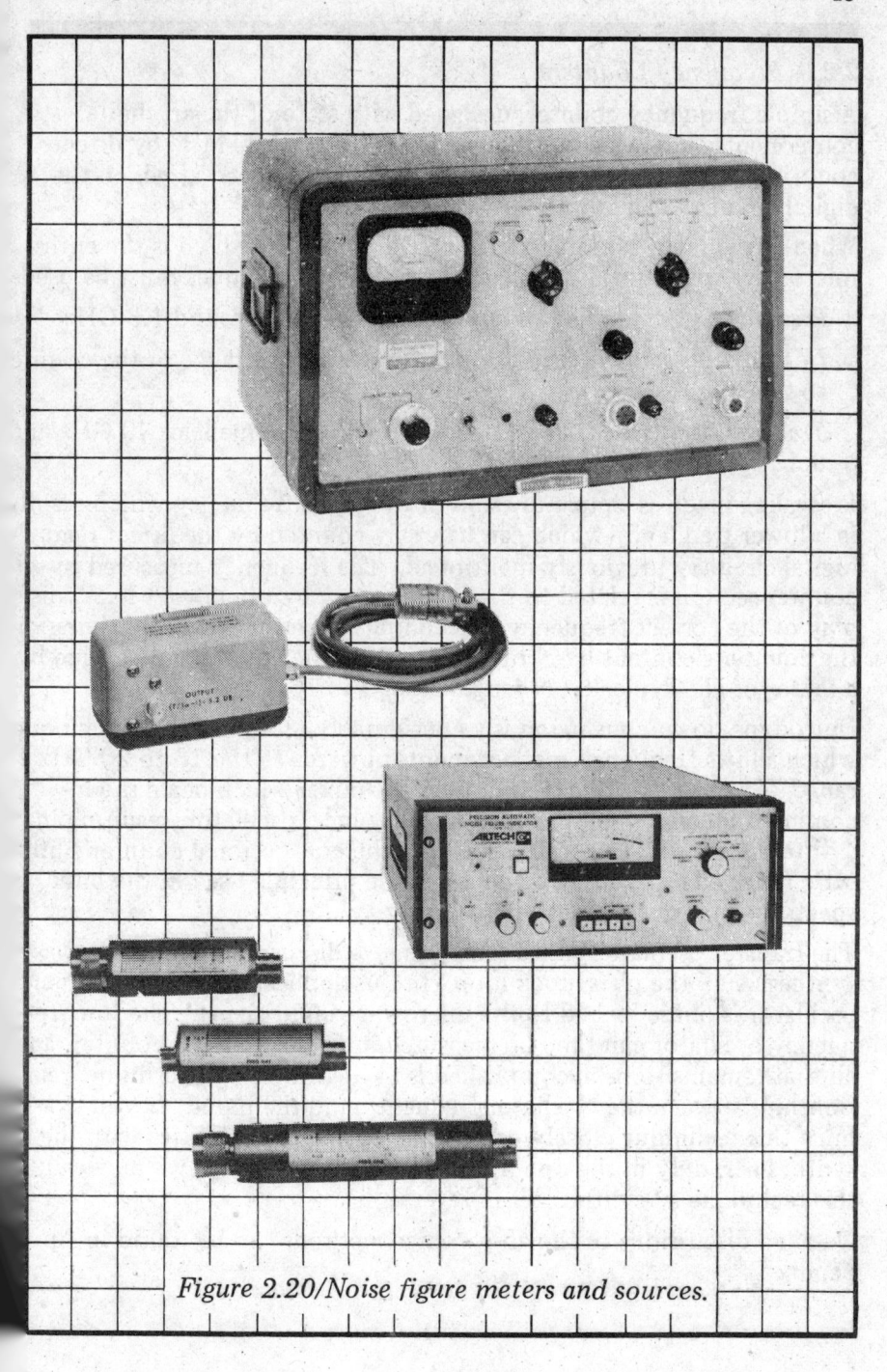

Figure 2.20/Noise figure meters and sources.

2.2.6 Frequency Counters

A simple frequency counter, designed with state-of-the-art digital components, can measure frequencies as high as 500 MHz by direct counting. This upper limit is determined by the speed at which the digital circuitry can switch between the "1" and "0" states.

When a frequency counter useful above 500 MHz (which is the entire microwave spectrum) is needed, one of three techniques must be used.

Prescaling — Extending the upper frequency to around 1.3 GHz.

Hetrodyning Down-Conversion — Increasing the range of the counter to 20 GHz.

Transfer Oscillator Down-Conversion — For frequencies 40 GHz and up.

Prescaling involves simple division of the input frequency which results in a lower frequency which can be easily counted by the direct read digital circuitry previously mentioned. The frequency measured by the counter section is related to the input simply by the integer N. A display of the correct frequency is accomplished either by multiplying the counter's content by N or by increasing the counter's gate time by a factor of N. Typically, N ranges from 2 to 16.

Hetrodyne down-conversion is a considerably more involved technique which allows frequency measurements upwards to the 18 to 20 GHz range. The heart of this technique is the mixer which beats the incoming frequency against a high-stability local oscillator, resulting in a difference frequency which is within the conventional counter's 500 MHz range. This, you will recall, was the principle used in the basic spectrum analyzer presented previously.

The transfer oscillator uses a technique we discussed under microwave synthesizers: the phase-lock loop. In this application, a low frequency oscillator is phase-locked to the microwave input signal. The low frequency oscillator can then be measured in a conventional counter, and all that remains to be accomplished is to determine the harmonic relationship between the oscillator frequency and the input. If you examine this technique closely, you can note many similarities with the synthesizer, only in the opposite direction. As a result, we may call this technique a "synthesizer-in-reverse."

Detailed discussions of the above three methods can be found in Appendix L.

Test Equipment

The techniques discussed above are all designed to measure CW signals. In many cases, this is more than sufficient. But, what about measuring frequencies of pulsed signals or measuring frequencies at a desired point in time of a time varying signal? Until recently it was impossible, but now counters are available which count and display frequencies of pulses and at a desired point in time of a time varying signal (called a dynamic frequency measurement). These types of measurements are possible by using two of the previously mentioned methods for measuring CW signals: prescaling and hetrodyne down-conversion. By using this combination to control the gate time and thus measure pulses at any point in time.

The following is a portion of a data sheet of a typical microwave counter. Its terms are explained for a better understanding of their importance.

Frequency Range	0.8 to 24 GHz
Sensitivity	−30 dBm to 10 GHz; −25 dBm to 18 GHz; −20 dBm to 24 GHz
Maximum Input	+30 dBm continuous
Recycle Rate	0.5 ms to 10 sec and manual

Frequency Range — These are the frequencies over which the instrument operates. It should be emphasized that sometimes all frequencies are covered on one connector and other times there may be two or three inputs, depending on the band you are using.

Sensitivity — This is the minimum level signal required for proper operation. Without this minimum level the counter will simply run by itself and read harmonics or spurious signals. Some counters have a light on the front panel that indicates when the proper level is applied. These lights may say "lock" or "operate" or any other term sufficient to describe the existing condition.

Maximum Input — This is the maximum signal level that can be applied to the input of the counter without damaging it. *Do not exceed this level.*

Recycle Rate — This is how often the counter cycles through its counting process. It is front panel adjustable and, in most cases, it can also be manually controlled.

Some typical microwave counters are shown in Figure 2.21.

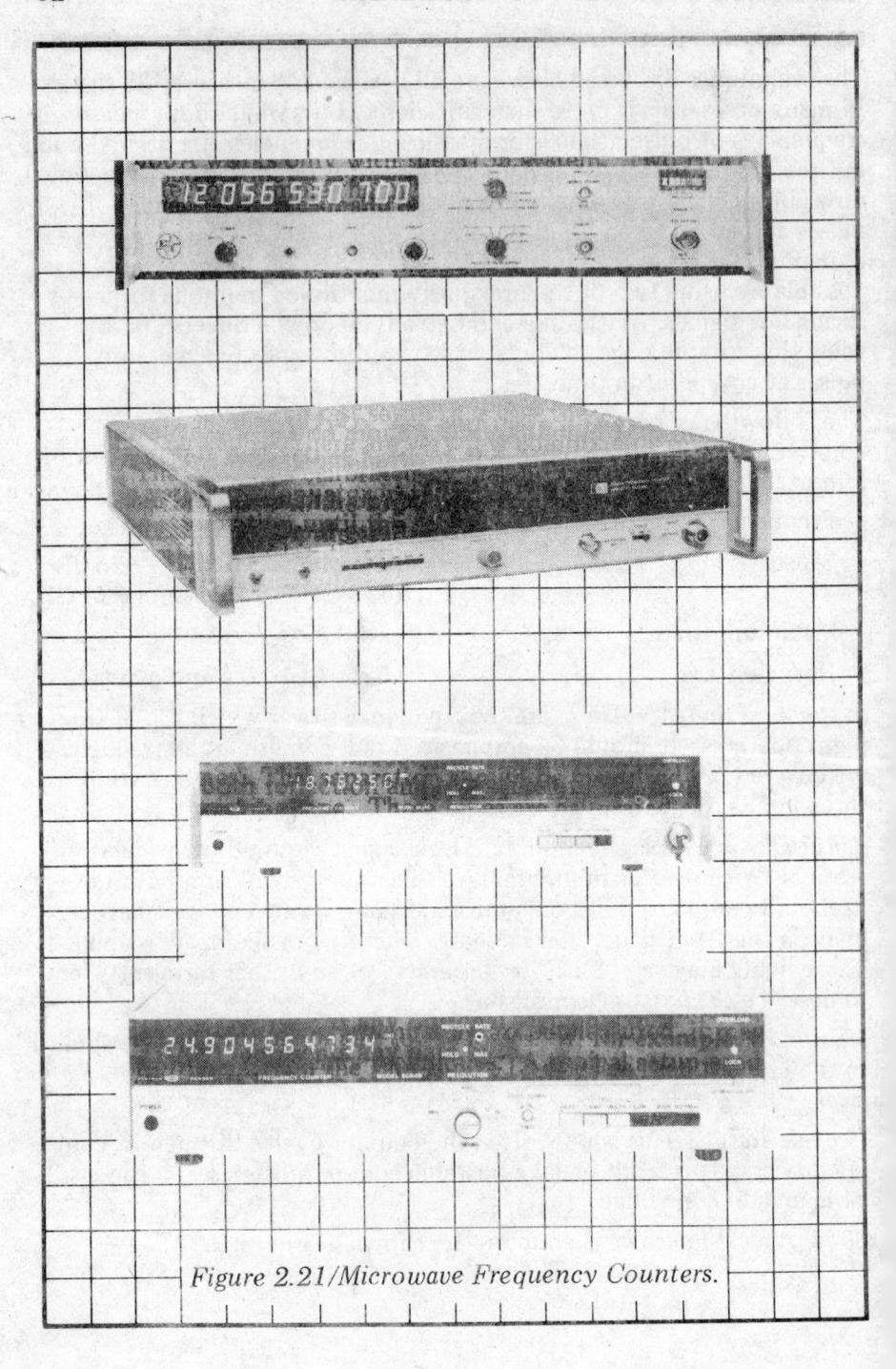

Figure 2.21/Microwave Frequency Counters.

Test Equipment 53

2.3 Auxiliary Testing Devices

Auxiliary testing devices are those extra components or instruments which are important to accurate measurement, but are not considered major instruments. Those we will cover are: attenuators, directional couplers, slotted lines, modulators, and a new device for expanding the capabilities of a network analyzer and spectrum analyzer called a normalizer.

2.3.1 Attenuators

In microwave testing there are two types of attenuators in common usage: the fixed attenuator (or pad) and the variable attenuator (continuously variable and step). Their usage depends on the particular application. If a constant attenuation is required the fixed attenuator is normally used, although a variable attenuator can be set at one point and left there. If various power levels are required, the variable attenuator is inserted and stepped or varied as required.

Probably the most familiar type of attenuator is the fixed coaxial attenuator shown in Figure 2.22. This attenuator usually is constructed in either of the two ways shown in Figure 2.23; it may be a "T" section or a "π" section. You can easily see why there are severe restrictions on the power input to this type of attenuator — each case utilizes a series film resistor whose power handling capability is limited to approximately 2 watts. Provisions are made in some models to handle higher powers, through the use of finned heat sinks to dissipate the excess heat produced by the greater amounts of power. Attenuators (or pads, as they are called sometimes), are available that are capable of handling up to 50 watts (CW) and 2 kW peak power, but they grow very rapidly in size as their power ratings increase. An attenuator that can handle 50 watts would be more than 7 inches long and nearly 1-3/4 inches in diameter — a far cry from the miniature, 2-watt attenuator which is around 1-1/4 inches long and less than 1/2 inch in diameter. Once again, there is a trade-off. If you need high power, you must have the space to put the component. True miniaturization is more easily accomplished at lower power levels.

There are also Distributed Series Resistor types of attenuators which have a single center tubular film resistor on an insulating substrate. This is the only resistor in the attenuator. There are no shunt resistors used. This type of attenuator is not one that is used where size is a concern. An attenuator operating from 1-18 GHz and attenuating 10 dB, for example, has a 6 inch attenuating element. The long attenuators with type N connectors on them are distributed series type of attenuators.

Figure 2.22/Attenuators.

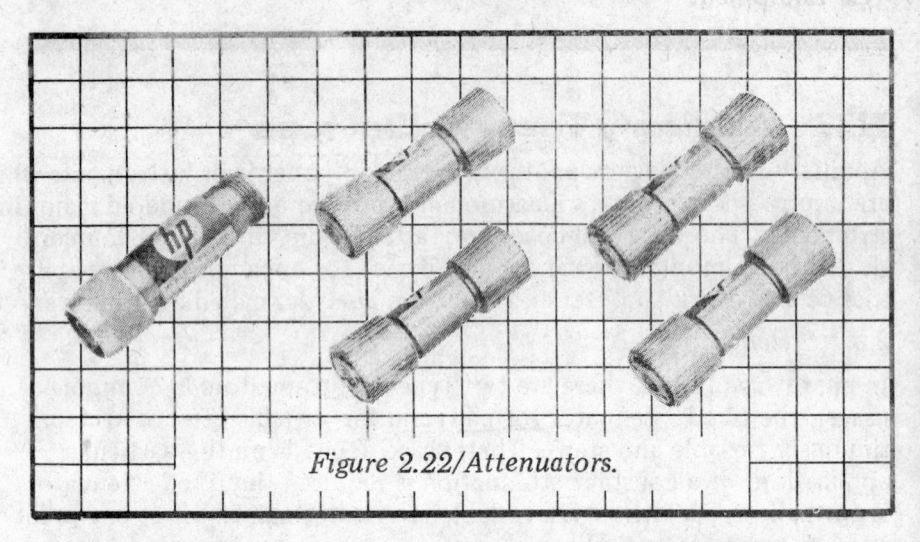

A typical data sheet for a coaxial attenuator is shown below. Most of the terms are self-explanatory, however we shall examine the most important more closely.

Model Number	ATTEN-20
Frequency	dc-12.4 GHz
Attenuation	20 dB
Accuracy	± 0.5 dB
VSWR	1.2:1 (dc-8 GHz)
	1.3:1 (8-12.4 GHz)
Input Power	2 W (CW)
Connectors	Type N

Accuracy — Accuracy measures how closely the attenuator stays to the attenuation figure (20 dB in the above example) over the specified frequency range. Most coaxial attenuators have calibration marks imprinted on the body of the component giving the attenuation as a function of the frequency, which is an actual reading of the accuracy of the component.

Attenuation — Attenuation is the figure, given in dB, which indicates how much RF power is lost between the input of the attenuator and its output. In other words, it is how much the input power is "lessened" by the time it gets to the output. Expressed mathematically, it is:

$$\text{Attenuation (dB)} = 10 \log P_{in}/P_{out}$$

Test Equipment

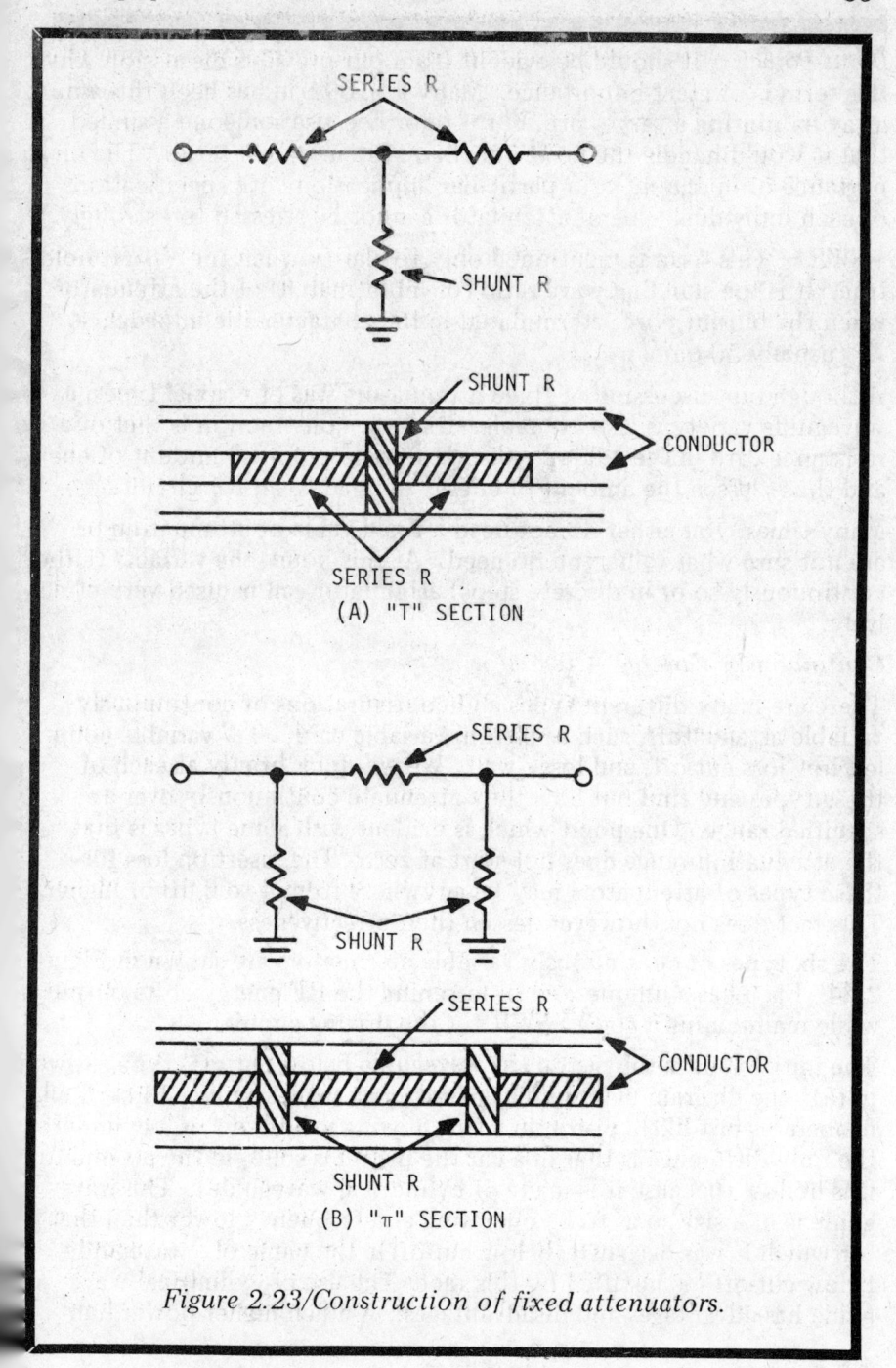

Figure 2.23/Construction of fixed attenuators.

Input Power — It should be evident from our previous discussion why this term is of great importance. Many a $100 pad has been thrown away exhibiting a very sharp, burnt odor because someone assumed that it would handle the power of their system or test setup. The importance of matching your particular application with specifications of each individual type of attenuator cannot be stressed too strongly.

VSWR — This term is mentioned only to clarify when the VSWR holds true. It is the standing wave ratio (or input match) of the attenuator when the output port is terminated in the characteristic impedance, Z_o (usually 50 ohms).

Although our discussion of fixed attenuators was of coaxial types, a waveguide variety is also available. Its basic construction is that of a resistance card in the guide to absorb a predetermined amount of energy and thus "lessen the amount of energy applied to an RF circuit."

Many times, you either do not need a fixed value of attenuation or are not sure what value you do need. At this point, the variable (either continuously so or in discrete steps) attenuator can be used very effectively.

Continuously Variable Attenuator

There are many different types and configurations of continuously variable attenuators, such as piston, variable card, "T," variable coupler, low loss cut-off, and lossy wall. We will look briefly at each of these types and find out how they attenuate continuously over a specified range. One point which is evident with some types is that the attenuation range does not start at zero. The insertion loss for these types of attenuators may be anywhere from 4 to 6 dB or higher. This fact does not, however, lessen their effectiveness.

The six types of continuously variable attenuators are shown in Figure 2.24. Each has a unique way of lessening the RF energy at its output while maintaining a good VSWR for the driving circuit.

The initial type is the piston (or waveguide below cut-off) type shown in (a); the diagram makes it clear how both names apply. First of all, it operates just like a piston in a car, moving in and out of a cylinder; the only difference is that in a car the piston is solid, in the attenuator it is hollow (because it is made of cylindrical waveguide). The waveguide is of a size so as to be operating at a frequency lower than that for which it was designed (below cut-off); the name of "waveguide below cut-off" is justified by this fact. The use of cylindrical waveguide has advantages and disadvantages. A much higher power han-

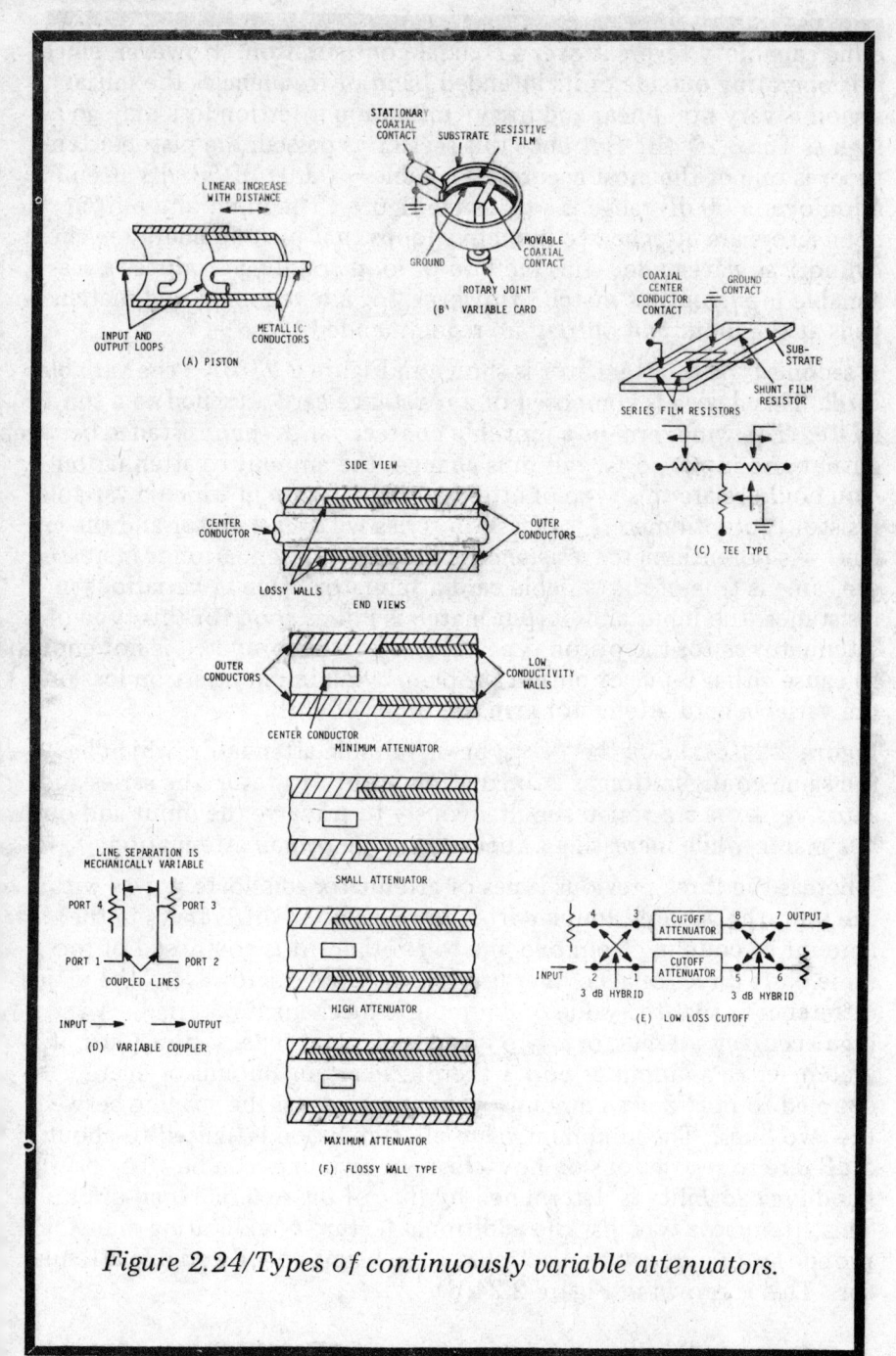

Figure 2.24/Types of continuously variable attenuators.

dling capability favors it over a coaxial configuration. However, since it is operating outside of its intended band of frequencies the initial region is very non-linear and lossy; minimum insertion loss may go as high as 15 to 20 dB. But once this section is passed, the piston attenuator is one of the most accurate available — 0.001 dB/10 dB attenuation over a 60 dB range is a plausible figure. The input and output connectors are attached to inductive loops that pick up energy in the cylindrical waveguide; this method of loop coupling produces a reasonable input/output match. However, for a much improved match, pads at the input and output are recommended.

A second type of attenuator is shown in Figure 2.23(b) — the variable card. This device is composed of a resistance card attached to a substrate. The wiper arm of a movable contact varies the resistance between output and input ports, and thus changes the amount of attenuation. You could relate this type of attenuator to a setup in which a variable resistor (potentiometer) is placed in series with a generator and the circuit. As potentiometer resistance is increased, attenuation is increased; the same is true of the variable card attenuator. Due to variations in resistance, the input and output match is not as good for this type of attenuator as for the piston type. However, the mismatch is not enough to cause either input or output problems. Minimum insertion loss on the variable card attenuator is in the order of 4 dB.

Figure 2.24(c) shows the "T" type of variable attenuator, which has the same configuration as a fixed attenuator. However, the series and shunt resistors are varied simultaneously to preserve the input and output match while increasing or decreasing the overall attenuation.

Whereas the three previous types of attenuators dissipate power within the unit, the variable coupler attenuator relies on differences in the amount of coupling from one line to another. It is composed of the same basic directional coupler used throughout microwaves; the major difference is that the value of coupling is mechanically varied. A very high accuracy attenuator is produced with an excellent input/output match. With an input at port 1 there is a certain amount of energy coupled to port 2 — an amount which depends on the spacing between the two lines. The minimum value of attenuation is limited to about 5 dB due to restrictions on how close the two lines can be. The power handling capability is determined by that of the external load at port 4. This attenuator type has the additional feature of exhibiting many properties of a directional coupler while being a good variable attenuator. This is shown in Figure 2.24(d).

Test Equipment

The low-loss cut-off (Figure 2.23e) is another type of attenuator that does not rely on dissipation within the unit itself. Instead, the device uses two quadrature hybrids (to divide and re-sum the RF energy), between which are sections of mechanically coupled variable cut-off waveguide. By mechanically varying the cut-off point of the waveguide, the attenuation at the desired frequencies can be varied accordingly. The insertion loss of such a device is very low since both legs of the system are balanced. With a zero dB setting, the insertion loss should be 1 dB or less. Due to the use of quadrature hybrids, the input and output match is excellent in this type of attenuator. The configuration shown in (e) is sometimes referred to as a "constant impedance" attenuator.

The last type of continuously variable attenuator is shown in Figure 2.24(f). Perhaps this variety is the one most easily understood. The main idea of an attenuator is to introduce a loss into the system — what better way to accomplish this function than to have a lossy wall attenuator? This attenuator is a stripline construction with a section of its outer conductor replaced by a lossy material, the position of which is varied to cause the attenuation. As illustrated in (f), with zero attenuation, there is simply a stripline conductor with a low loss dielectric above and below. As the lossy material is moved over the conductor, attenuation increases; maximum attenuation occurs when the conductor is completely covered.

At that point, it is surrounded by a lossy material rather than its original low loss dielectric.

An attenuator of this sort exhibits many interesting features. First, because of the availability of lossy material at that frequency, it is generally used above 2 GHz. The attenuator once again operates by means of dissipation within the unit, limiting the power handling capability to around 10 watts. The insertion loss is generally below 1 dB since, with zero attenuation, the attenuator is composed of nothing more than a single piece of transmission line. The VSWR is also very low because of the construction and the dissipation of properties of the transmission lines involved.

A typical data sheet for a general type of continuously variable attenuator is shown below.

Model Number	VAR-1
Frequency	1-2 GHz
Attenuation Range	6-120 dB

Accuracy	± 1.5 dB
Power	100 W (CW)
VSWR	2.0:1 (6-15 dB)
	1.5:1 (15-30 dB)
	1.2:1 (30-120 dB)
Connectors	Type N

Accuracy — Accuracy could be referred to as setability or resetability; that is, the level of precision to which an attenuator can be set.

Attenuation Range — Very simply put, this range runs from the minimum to the maximum amount of attenuation possible with the individual attenuator.

Connector — This term is listed to emphasize the need for the proper connectors which eliminate any additional loss or VSWR problem. Care must be taken to fit the correct connectors to your unit.

Frequency — The frequencies indicated by this specification are those over which the attenuator characteristics hold true. Above or below this range, the figures change. Some attenuators, such as variable couplers, have a limited frequency range (usually one octave). Others, like the variable card or "T" type, are less sensitive to frequency and thus have a much wider range of operation.

Power — Power is a key term. There is wide variation in the handling capabilities of the different types of attenuators — some are limited by external terminations while others are restricted by their internal construction. Great care should be taken when choosing an attenuator to make sure that you have not exceeded any of the power limitations.

VSWR — The true reading of the input match of the attenuator over its range of operation is the VSWR. The data sheet above lists more than one figure for VSWR; these different values are at different attenuation levels. As attenuation increases, more power is absorbed by the attenuator and, therefore, is not reflected back to the input. As a result, the VSWR is lowered.

Step Attenuator

When you think of a step you think of a distinct increment by which you move or progress. This idea also applies to step attenuators. This device is a set of fixed attenuators arranged in a turret or in a slab for switching between stationary coaxial contacts to obtain discrete steps of attenuation. Attenuators are available in 0.1, 1.0, and 10 dB steps.

Test Equipment

The 0.1 dB step attenuator usually has a total range of 10 dB, the 1.0 dB step usually has 10-12 dB of range, and the 10 dB step can have over 120 dB of total range.

The step attenuator has many outstanding features, among which are flatness (over a frequency range), repeatability (.03 dB is possible), wide range, long life, and either manual or programmable capability. These characteristics are readily apparent from the data sheet shown below.

Model Number	STP-1
Frequency	dc-12.4 GHz
Attenuation Range	0-60 dB (in 10 dB steps)
Accuracy	± 2 dB
Insertion Loss	1.5 dB
VSWR	1.5:1
Power	2 W (CW)
	100 W (peak)
Connectors	Type N

To understand the STEP attenuator, let us examine each of these terms.

Frequency — This term indicates the range of operation over which the attenuator specifications apply. An outstanding feature of the device is its very wide range (dc-12.4 GHz); it is possible because the construction of each fixed distributed attenuator is not frequency dependent. The attenuator is only specified to 12.4 GHz because the connectors begin to limit performance above that point, and the attenuators used may be lumped elements.

Attenuation Range — This specification is the total attenuation (at the maximum and minimum settings) and the increments available in between (0.1 dB, 1.0 dB, or 10 dB steps).

Accuracy — Accuracy is how close each step actually is to its specified value. For example, 10 dB steps in the attenuators above could be anywhere from 8 to 12 dB. This information is useful when making precise measurements.

Insertion Loss — This loss is that through the attenuator when its front dial is set at zero dB.

VSWR — VSWR is a measure of the match of the attenuator when it is stepped through its range and terminated in its characteristic impedance (Z_o), usually 50 ohms.

Figure 2.25/Continuously variable attenuator.

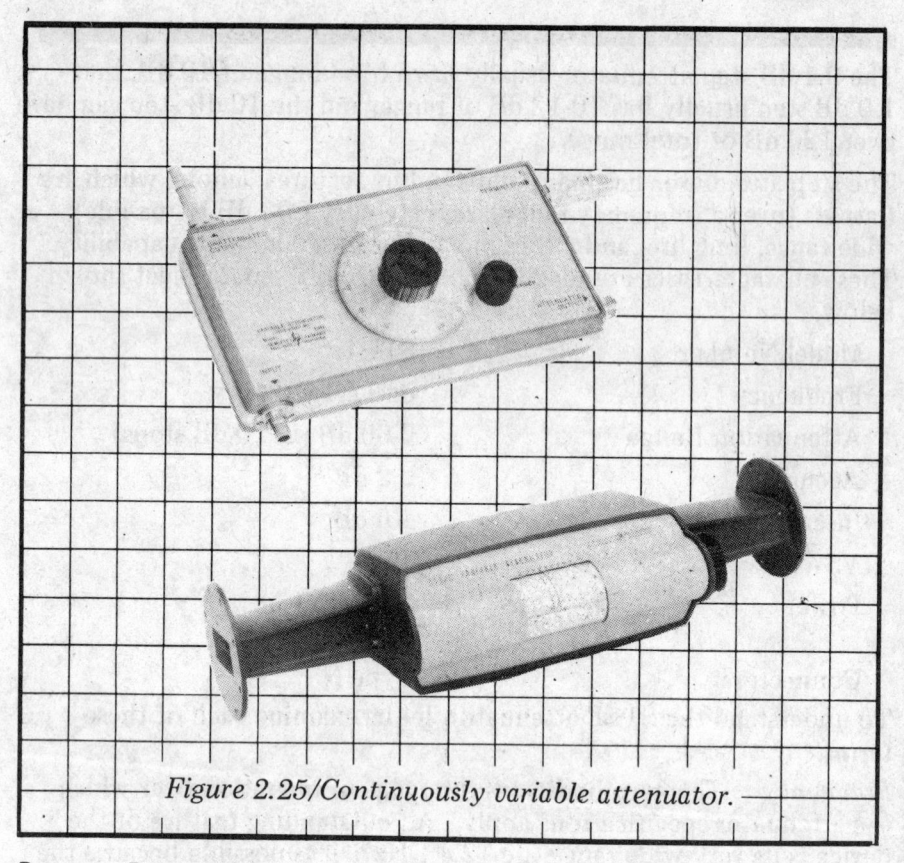

Power — Power is the maximum amount that the attenuator can handle without damage. Notice that we are back to a 2 watt CW power capability, as we were when we discussed the fixed attenuator. This similarity is understandable because the step attenuator is actually a series of fixed attenuators.

One feature not discussed above is repeatability. If you think of switching between each component in a series of fixed, and accurate, attenuators in a row, the concept of repeatability can be understood. An attenuator remains at the same value no matter how long you are switched to another value attenuator; when you return to a specific device, you can be sure that it has the same value as before. The only element that will degrade repeatedly is the switch used to switch the attenuators in and out.

Another attribute is long life, though it has one string attached — its longevity is dependent upon observation of its power handling specifications. If you observe the specifications rigorously, you have one

Test Equipment   63

Figure 2.26/Step attenuators.

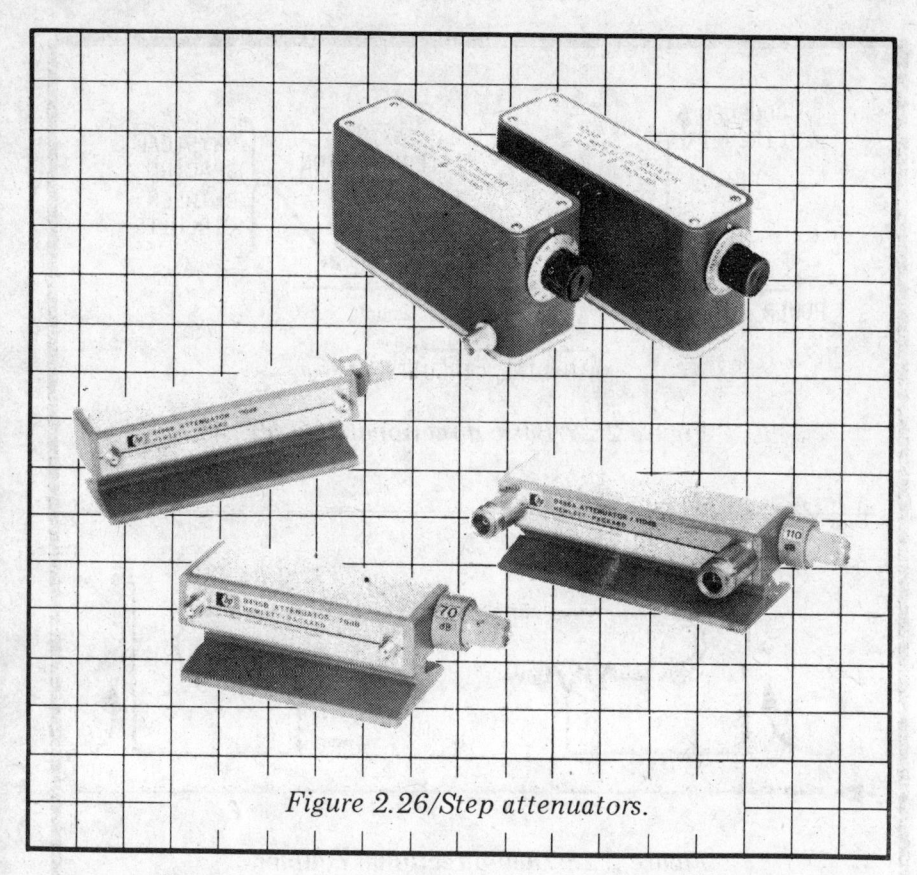

of the most useful test components available. Figure 2.25 shows some continuously variable attenuators that are available, and Figure 2.26 shows step attenuators.

2.3.2 Directional Couplers

Directional couplers are very useful in microwave measurements if you know how and where to use them.

A directional coupler is a component which allows two microwave circuits to be combined into one integrated system in one direction, while being completely isolated from each other in the opposite direction.

Figure 2.27 illustrates the basic directional coupler and how the definition applies to it. It should be pointed out that, although the majority of the time there is a physical spacing in directional couplers, it is not necessary. An example of such a coupler is the Branch Line which has complete dc continuity.

Figure 2.27/Basic directional coupler.

Figure 2.28/Dual-directional coupler.

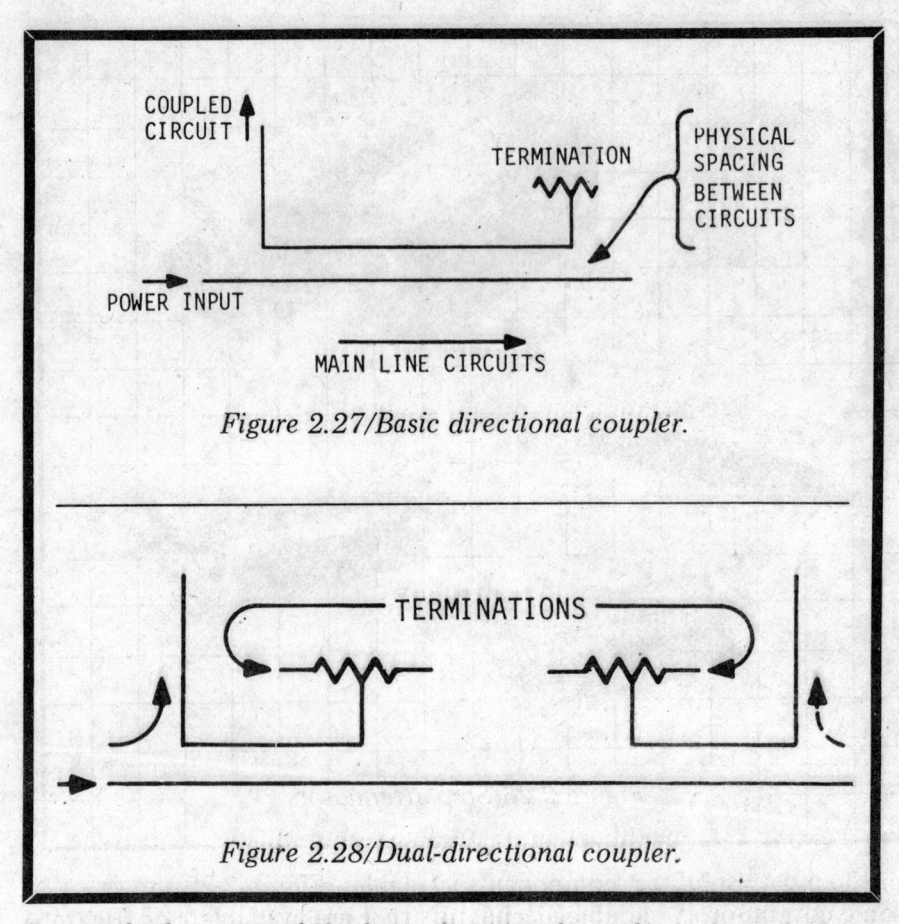

One method to expand the basic directional coupler is through the construction of a dual-directional coupler. Just as the name implies, this procedure involves two couplers put back-to-back as shown in Figure 2.28. The main advantage of this type of coupler is that it allows you to monitor both forward and reflected power at the same time. It greatly aids in obtaining data which indicates the input match and, thus, the VSWR.

Figure 2.29 is a photograph of various directional couplers as they would appear on a bench or in a catalog. The type of coupler that should be used depends on many factors. It may be necessary for it to have a large value of coupling, a very high directivity (as in accurate VSWR measurements), or a low input VSWR of its own. Any of these specifications may be of great importance when choosing the appropriate coupler for your job.

Test Equipment

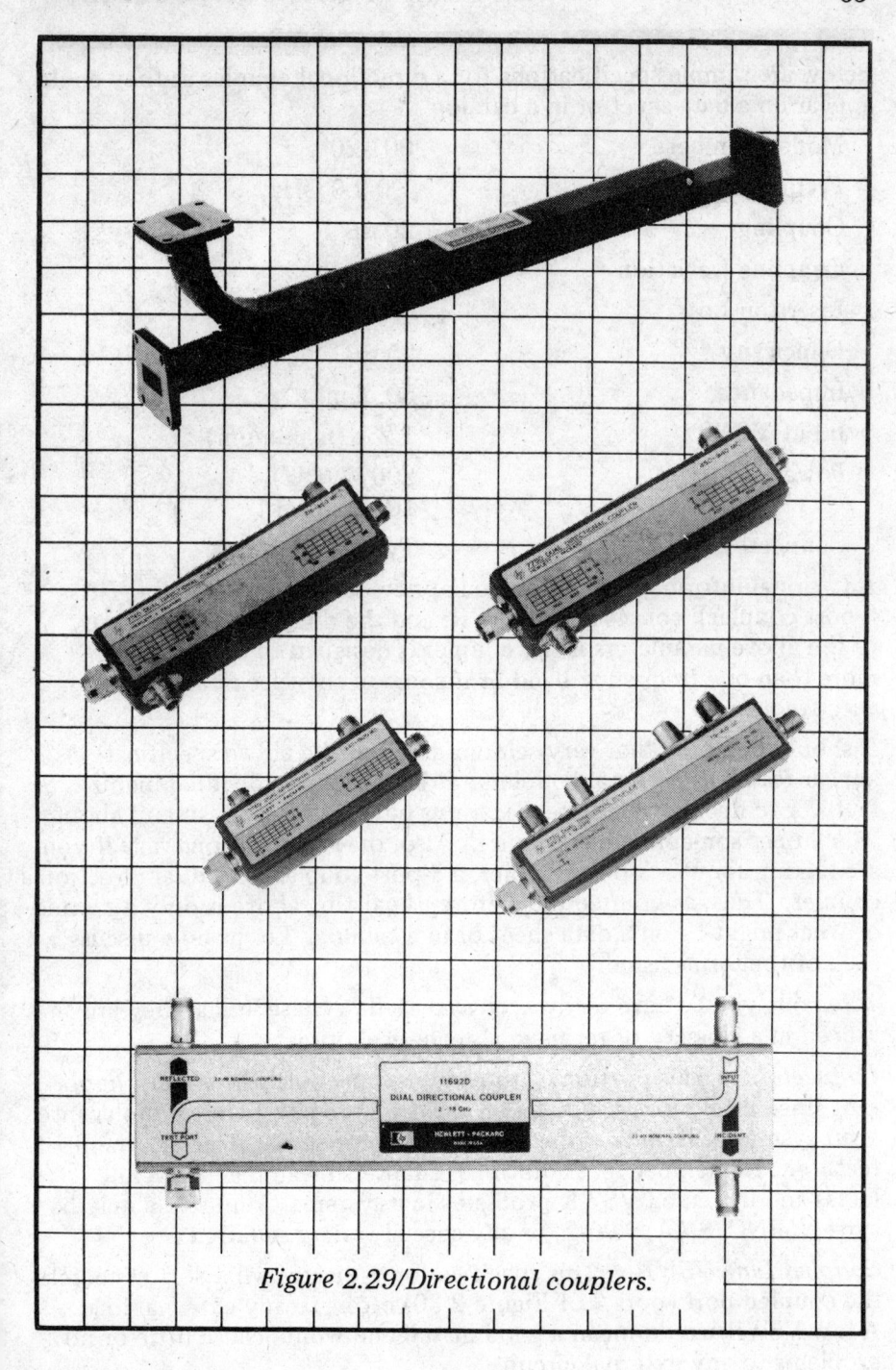

Figure 2.29/Directional couplers.

Below are sample specifications for a directional coupler as they could appear on a data sheet or in a catalog.

Model Number	001-20
Frequency	1.0-2.0 GHz
Coupling	20 dB
Coupling Deviation	± 1.0 dB
Insertion Loss	0.2 dB
Directivity	25 dB
Impedance	50 ohm
Input VSWR	1.2:1 (maximum)
Power	200 W (CW)
	10 kW (peak)
Connectors	Type N (female)

Additional information may include price, reverse power rating (if a 3-port coupler), coupled line VSWR, and the different values of any of the above parameters if the coupler is designed to be used over more than one frequency band or if some of the specifications vary with frequency.

It should be noted that very seldom do all of the above specifications appear for all directional couplers. It is up to the individual manufacturer to decide which specifications he wants to publish and about which ones someone must inquire. Also, the specifications vary if you are talking about a 3-port coupler, a 4-port coupler, or a dual-directional coupler. Thus, as pointed out before, the listing above is only a sample of what might be on a data sheet or in a catalog. Let us now discuss each of the terms listed.

To enable you to have ready access to each of these items, they are arranged in a glossary of terms in alphabetical order.

Connectors — This portion of the coupler specifications is very important since it tells you if you either need adapters to fit the coupler into your system or have to order your desired connector from the manufacturer. Remember that additional adapters mean higher overall losses and increased VSWR problems in a system. Connectors may be type N, BNC, SMA, APC-7, or any special variety required.

Coupled Line VSWR — This specification indicates what sort of match the coupled port (port 3 of Figure 2.30) offers to any external load. A low VSWR would mean a good match and would cause little or no problems to any external circuit.

Test Equipment

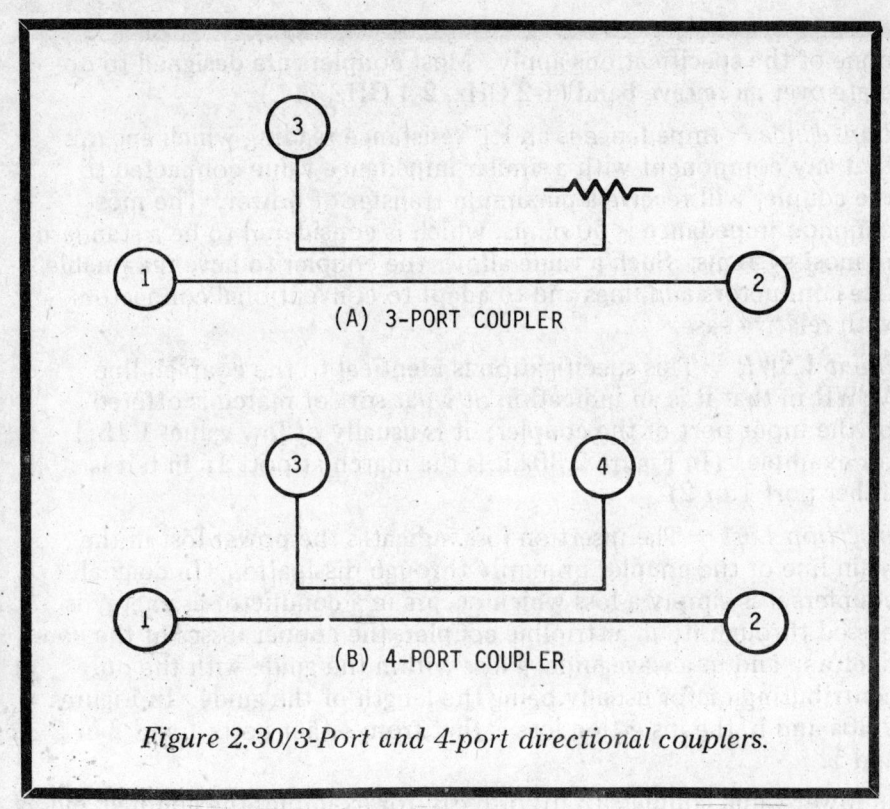

Figure 2.30/3-Port and 4-port directional couplers.

Coupling — The ratio of the power available at the coupled port to the power at the input port (port 3 to port 1 in Figure 2.30). It is the amount of attenuation of the input power as a result of the coupling structure. This term is expressed in dB and has standard values of 6, 10, 20, and 30.

Coupling Deviation — This measurement indicates how flat the coupling is over a specified band of operation; it is expressed as a (+) or (−) dB value. If you rely very heavily on a specified value of coupling over a wide range of frequencies, the term should be low.

Directivity — This figure indicates how accurately your are able to measure parameters in your system. It is the difference between the desired and undesired couplings. If a coupler has low directivity, the forward and reverse powers in the coupler interfere with one another and cause great inaccuracies to occur. Ports 1 and 4 in Figure 2.30b are the undesired coupling; 1 and 3 are the desired combination.

Frequency — This term indicates the range of frequencies for which all the other specifications are true. Above and below this range, none of the specifications apply. Most couplers are designed to operate over an octave band (1-2 GHz, 2-4 GHz, etc.).

Impedance — Impedance is an RF resistance reading which ensures that any component with a similar impedance value connected to the coupler will receive a maximum transfer of power. The most common impedance is 50 ohms, which is considered to be a standard in most systems. Such a value allows the coupler to have reasonable size conductors and lines and to adapt to conventional connectors with relative ease.

Input VSWR — This specification is identical to the coupled line VSWR in that it is an indication of what sort of match is offered by the input port of the coupler; it is usually of low value, 1.15:1 for example. (In Figure 2.30a it is the match at port 1; in b it is either port 1 or 2).

Insertion Loss — The insertion loss indicates the power lost in the main line of the coupler primarily through dissipation. In coaxial couplers it is simply a loss which occurs in a conductor as energy is passed through it; in a stripline coupler, the copper losses of the conductors; and in a waveguide, a loss within the guide with the only contributing factor usually being the length of the guide. In Figures 2.30a and b, the insertion loss is that from either ports 1 and 2 or 2 and 1.

In lower value couplers (3 dB or 6 dB, for example) the coupled energy due to the coupled port also is included in the total through-line insertion loss, thereby increasing the overall value. It is the net loss after summing all output power and comparing with the input power.

Power Rating — This specification refers to the amount of both CW and peak power that the coupler is capable of handling. Any power level greater than this value may cause arcing or a deterioration of performance. This term is one of the most important of which to take careful note if you will be operating at a high power level.

Reverse Power Rating — This term applies only to a 3-port coupler; it is usually a lower value than that of the forward power rating because the fourth port is internally terminated and so, generally, can not handle any amount of power. This specification is also quoted as a CW and peak power number; it is the power handling capability of the coupler if power is applied in a reverse direction. (At port 2 instead of port 1 in Figure 2.30(a)).

To conclude the discussion on directional couplers we show how the properties described in the preceeding paragraphs can be put into

Test Equipment

Figure 2.31/Directional coupler symbols.

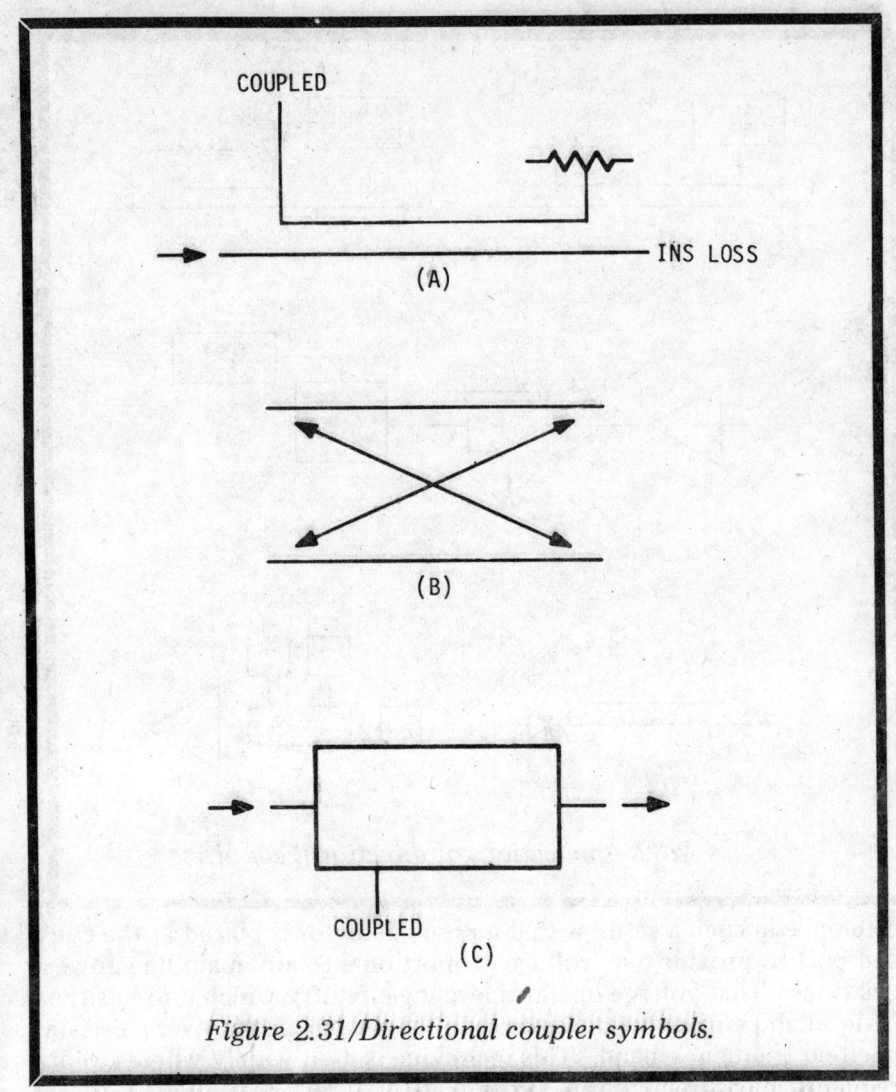

practical application. So that the individual may recognize a directional coupler in a practical application, Figure 2.31 shows three symbols that are employed.

The most obvious use of a directional coupler is to monitor an output power. Its directional properties ensure that only the forward power is coupled; its property of circuit isolation leaves the main line power undistributed. Figure 2.32(a) shows the orientation of a directional coupler when used as a power monitor. A scheme similar to the monitoring technique described above is shown in Figure 2.32(b). The big

70　　　　　　　　　HANDBOOK OF MICROWAVE TESTING

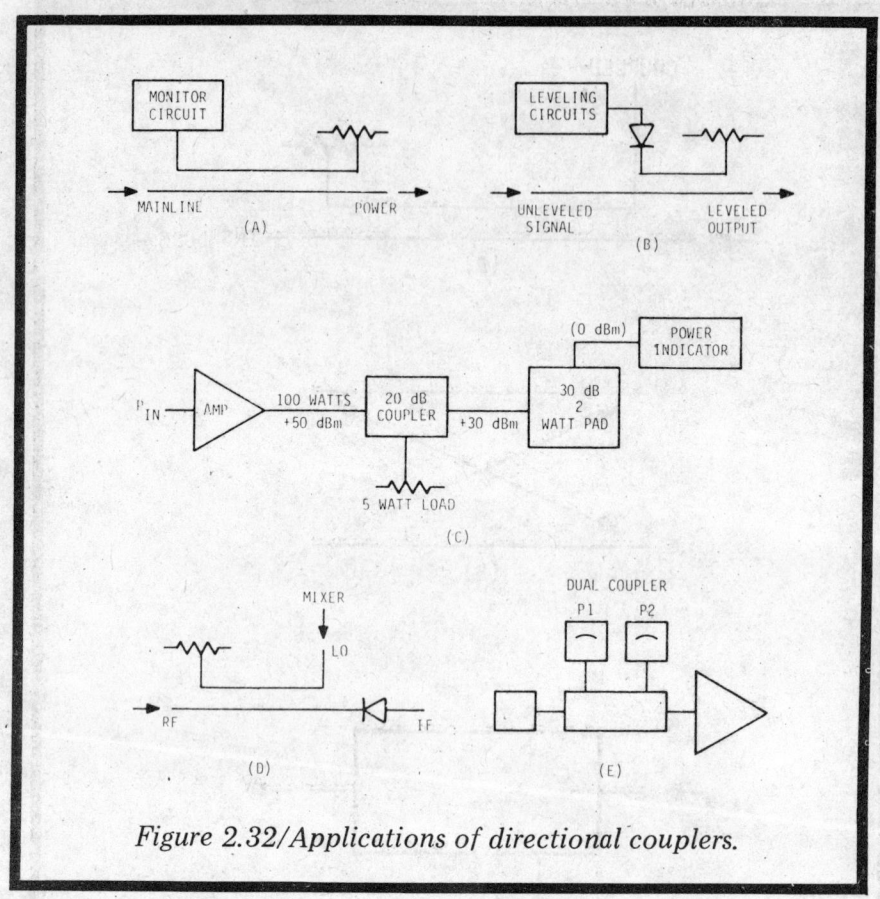

Figure 2.32/Applications of directional couplers.

difference in such a setup is that a diode detector is placed at the coupled port to provide a dc voltage proportional to any main line power variations. That voltage operates leveling circuitry which corrects the RF level and eliminates differences in the output power over a certain specified frequency band. This technique is used widely where swept frequency measurements are taken.

Figure 2.32(c) illustrates an application which takes advantage of the directional coupler's ability to handle higher CW power; it is an actual example of a system used to check a 100 watt power amplifier. The only components available were two-watt attenuators, five-watt terminations, and the directional coupler. It can be seen that a more adequate setup was created by utilizing the coupler's high power capability, directivity of 25-30 dB and reverse power handling capability achievable in some couplers. None of the components used had its

Test Equipment

Figure 2.33/Microwave slotted line.

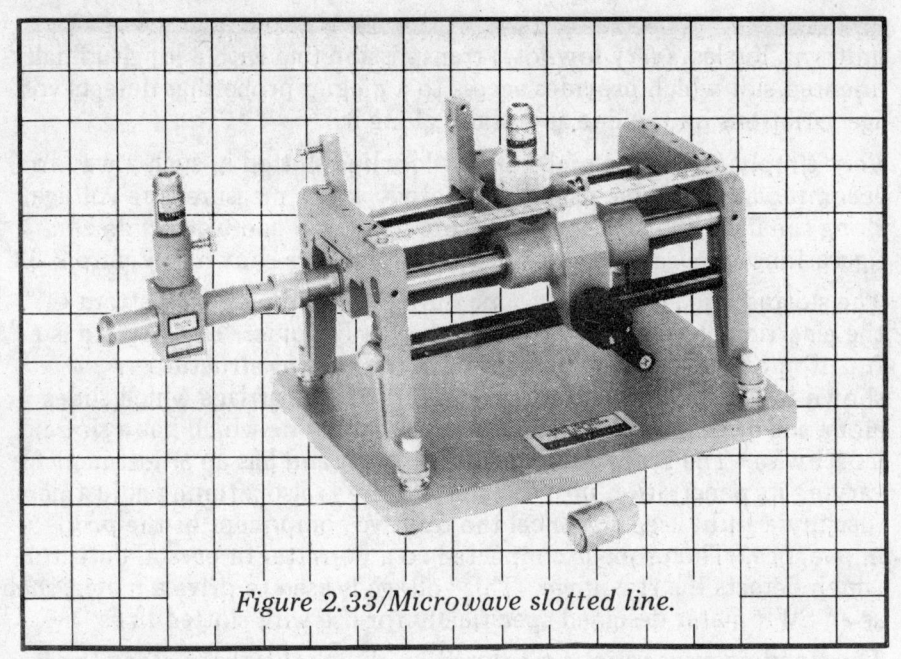

power rating exceeded; the directional coupler was doing the majority of the work.

Figure 2.32(d) shows a coupler being used as part of another component. In this case, it is a very important part of a single diode mixer. The circuit is of a very basic mixer, but it does show that by using a coupler, the RF and LO signals have a high degree of mutual isolation.

The dual-directional coupler, as stated before, is very useful in VSWR measurements (Figure 2.32(e)); to put it more accurately, it is employed to measure the return loss which then is converted to VSWR. Power meter #1 reads the forward power to the device under test; meter #2 measures any reflected power. By comparing these two figures a return loss is determined and converted to the input VSWR of the particular device under test.

The directional coupler is one of the most widely used components in microwaves. Its property of being directional, its isolation, its low input and coupled line VSWR, and its adaptability to being 3- or 4-port or dual-directional make it a vital tool in the repertoire of the microwave engineer.

2.3.3 Slotted Line

One of the most useful and most accurate instruments used in microwave measurements is the slotted line. A slotted line is a section of uniform, lossless (very low-loss) transmission line with a longitudinally-oriented slot which provides access to a pickup probe that detects voltage variations on the line as it slides along it.

Very simply put, it is a piece of low-loss line slotted in such a way to accommodate a probe, usually inductive, which measures the voltage along the line. Slotted lines are made in coaxial and waveguide configurations; typical examples of slotted lines are shown in Figure 2.33.

The slotted line is designed to measure the standing wave pattern of the electric field within a coaxial line or waveguide. This pattern is a function of the lengthwise position in the guiding structure. As is shown in Figure 2.33, a probe is mounted on a carriage which slides along the outside of the coaxial line or waveguide which has a slot cut lengthwise. The probe extends into the slot and has an adjustment for varying its penetration into the slot. There is also a tuning adjustment (usually a stub) used to cancel the reactive component of the probe impedance. The probe is connected to a Barretter or crystal detector which detects the rf voltage. This voltage is used to drive a meter, such as an SWR meter designed specifically for use with slotted lines.

The standing-wave ratio is measured by sliding the probe along the line for a maximum and minimum indication on the output meter. The standing-wave ratio is calculated from the above data or read directly from the indicating device if the indicator is calibrated in SWR.

The wavelength of the signal frequency can be measured by obtaining the distance between minima, since the distance between successive minima or maxima is one-half wavelength.

A data sheet for a slotted line is shown below:

Frequency Range	1.8-18 GHz
Residual SWR	1.06 (Maximum)
Irregularities	0.2 dB
Maximum Power	2 Watts
Probe Travel	10 cm.

Although all of the above specifications are important, the most critical are Residual SWR and irregularities. For this reason, we shall look at these terms first.

Residual SWR — This is the standing wave ratio of the line itself. This figure must be as low as possible in order to faithfully reproduce the standing wave pattern of the device under test. A high residual SWR will cause ambiguities in your readings, thus masking the actual characteristics to be measured.

Test Equipment

Irregularities — These are variations in the line (much like a reflection loss in an attenuator) that cause the slotted line to receive false information concerning the device under test. These are more of a concern in waveguide slotted sections since milling machines may cause regular, repeatable mechanical variations which interact with the standing wave pattern. By careful milling a regular lengthwise uniform surface in the guide, these irregularities can be kept very low.

Frequency Range — The range of operation over which the slotted line will perform its required measurements.

Maximum Power — This is a figure which should not be exceeded. If too much power is sent down the line, the probe will pick it up in proportion to its depth in the guide and may burn out the detector if the probe is inserted too far into the slot. Therefore, the maximum level should be observed.

Probe Travel — This is the distance that the carriage will travel, with the probe, down the slotted section of line. This length will tell you how many half wave maximum and minimum points you will see on the entire line.

2.3.4 Modulators

Many applications of microwave testing require amplitude modulation and pulse modulation of signal sources. In some cases, highly stable CW sources are amplitude modulated in a functional block external to the source itself.

Other applications require fast pulse modulation not possible with previous methods of switching on oscillators such as klystrons, BWOs, or solid state sources. These pulse characteristics usually are desired without frequency "pulling" or incidental FM.

PIN diode modulators offer an ideal way to amplitude and pulse modulate microwave signals through the wide range of frequencies. PIN diodes are electrically controlled microwave resistors. By appropriate design into microwave structures, attenuation may be electrically set. Both AM and pulse characteristics are possible. A simplified block diagram of a PIN modulator is shown in Figure 2.34.

The available PIN modulators utilize PIN devices distributed at appropriate quarter-wave spacings along a transmission line. The advantage of this technique is that low attenuation is required of any single diode so that good impedance match is preserved throughout the depth of modulation. Maintaining low SWR for both on and off conditions permits pulsing of sources without causing reflective signals which might cause frequency or phase shift.

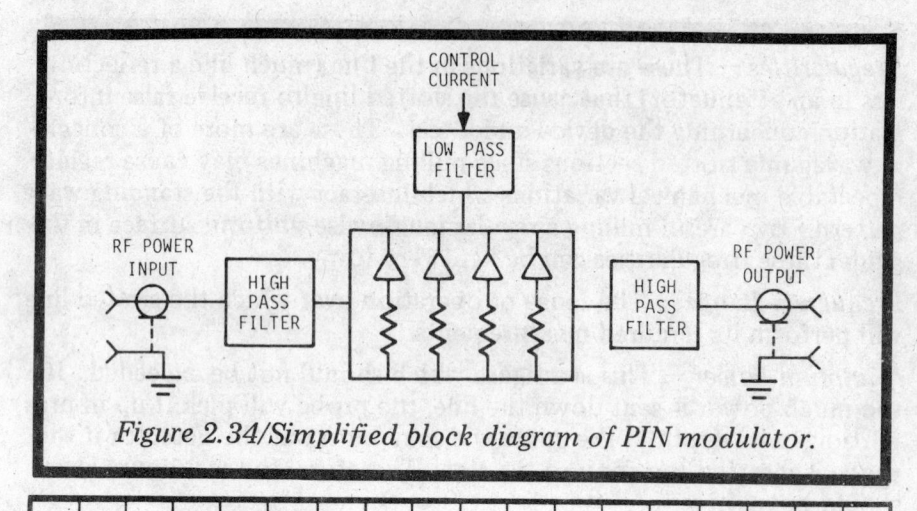

Figure 2.34/Simplified block diagram of PIN modulator.

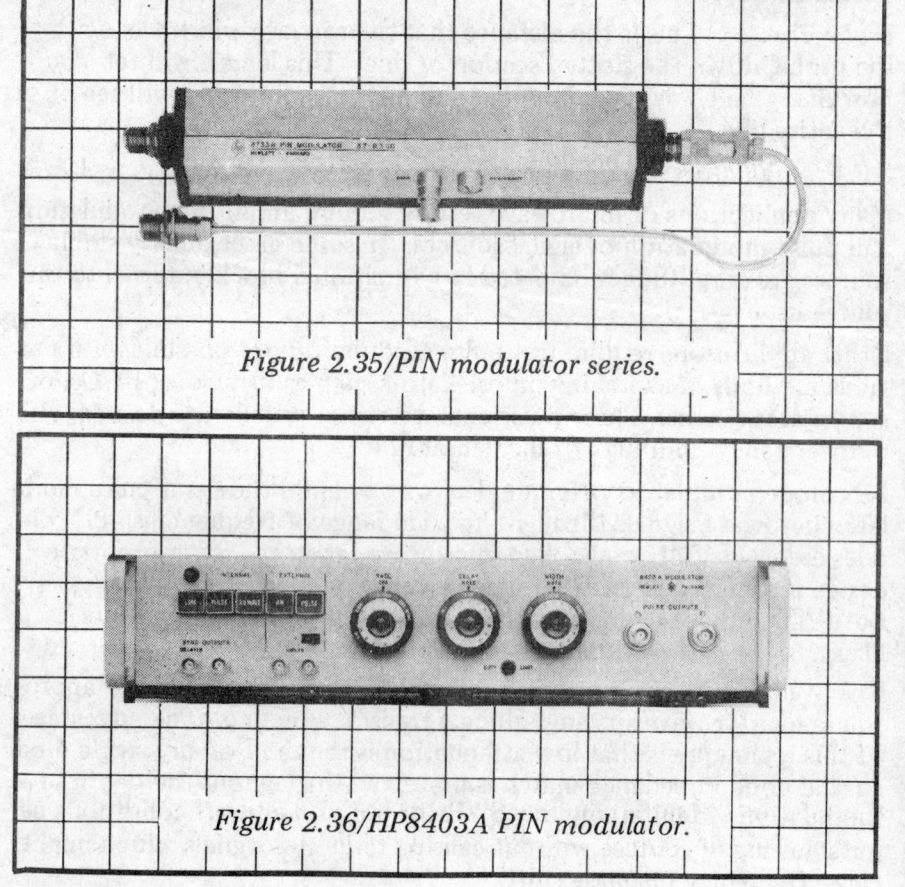

Figure 2.35/PIN modulator series.

Figure 2.36/HP8403A PIN modulator.

Test Equipment

Figure 2.37/Pulse modulator.

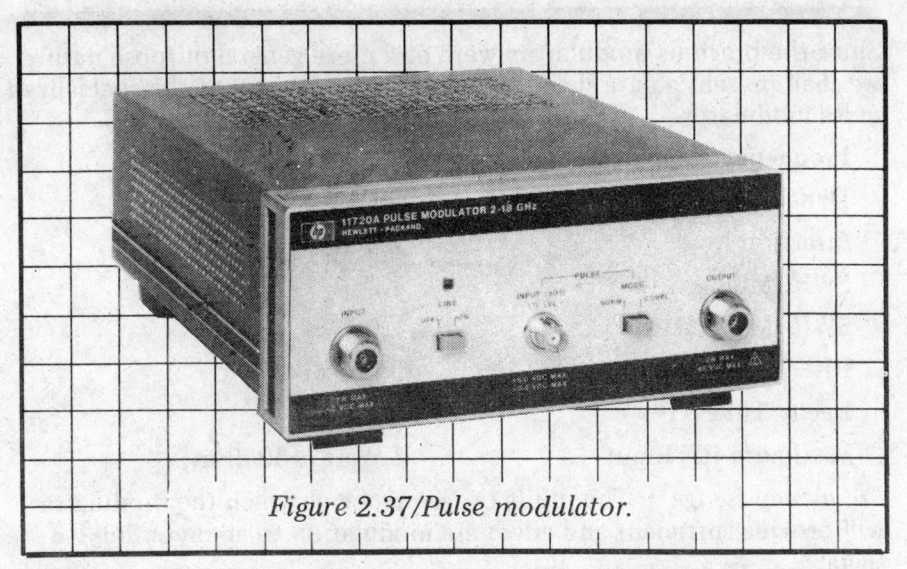

When individual PIN modulator units are encased into a modulator chassis you have an instrument that provides complete control over the individual unit supplying the appropriate modulation waveshapes and bias levels for fast rise times, rated on/off ratios, and amplitude modulation. An internal squarewave and pulse modulator with variable PRF (Pulse Repetition Frequency) and adjustable width and pulse delay also provide squarewaves and pulses for general pulse application. Figure 2.35 shows an individual PIN modulator while Figure 2.36 shows a modulator control unit which may be used to drive an external PIN modulator or may have the modulator incorporated within it.

The modulators described above are designed to handle AM, squarewave, or pulse modulation. These are, however, units that are used only as pulse modulators. Figure 2.37 shows one such unit. This is a versatile, broadband (2-18 GHz), general purpose pulse modulator with very short rise and fall times and high on/off rates. It is ideal for signal simulation in pulse applications such as radar testing, where excellent performance is required. Such figures as < 10 msec rise and fall time (very important in preserving the original shape of the input pulse) and > 80 dB on/off rates (a form of isolation reading) are typical performances.

When you couple these with a 2-18 GHz RF range, you have an excellent modulator for many applications.

Since the previous modulators were of a more general purpose nature, we shall present a data sheet for one of them instead of the specialized pulse modulator.

Frequency Range	3.7-8.3 GHz
Dynamic Range	35 dB
Insertion Loss	2.0 dB
SWR (Min. Attn)	1.8:1
SWR (Max. Attn)	2.0:1
Rise Time	30 nsec.
Decay Time	20 nsec.
Maximum RF Input	1 Watt (+30 dBm)

Frequency Range — The limits of frequency at which the modulator will provide sufficient and adequate modulation to an incoming RF signal.

Dynamic Range — The amount of attenuation that the modulator exhibits from minimum bias to maximum bias. The input will experience an attenuation equal to this number if you begin at minimum bias voltage and increase to a maximum level.

Insertion Loss — The amount of loss through the modulator when the bias voltage is set at its minimum value. For example, if the modulator bias was specified from +5 to +15 volts, the Insertion Loss would be taken at +5 volts. This is the minimum loss experienced through the modulator.

SWR — This is the standing wave ratio of the modulator. It is a measure of the input match with respect to 50 ohm. Notice the difference in SWR with minimum and maximum atttenuation. This is due to the fact that most modulators are not constant impedance devices over their entire range. (These figures do not guarantee a flat VSWR over the entire range. If a specific attenuation is needed, you should check the VSWR at that point.)

Rise and Decay Time — These are important parameters when considering pulse modulation. They are the time it takes for a pulse to increase to 90% of its final value and fall by 90% of its maximum value, respectively. They determine how faithfully the forward and trailing edges of the original pulse applied for modulation are reproduced at the modulator output. The modulation bandwidth is also a function of rise and fall time.

Test Equipment 77

Figure 2.38/Normalizer (HP8750).

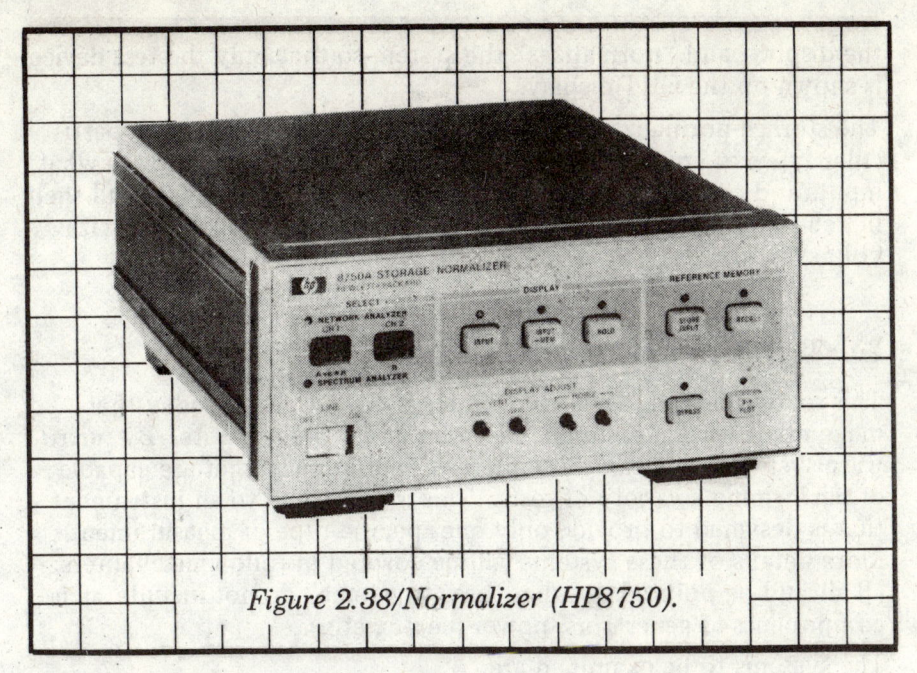

Maximum RF Input — This is the value of RF power that is just under a level that would burn out the PIN diode of the modulator. For a long life of performance from your modulator, you should pay strict attention to the specified value.

2.3.5 Normalizer

Swept measurements of network characteristics generally include the response of the test system itself and all of its components thus making it necessary for the operator to take the systems effects into account. This is often done by drawing the system's responses on the display with a grease pencil or plotted grid line with an X-Y recorder. Now a new accessory (or auxiliary test device), the storage-normalizer can store the systems' response in memory and then subtract it from the measured data for direct display of the test device alone. This greatly reduces the possibility of results being erroneously interpreted. The normalizer is compatible with most network analyzers and many spectrum analyzers.

Figure 2.38 is a picture of the HP8750A storage-normalizer. It is directly compatible with such instruments as the 8410A and 8505A network analyzer; the 8755A frequency response test set; the 8557A and 8558B spectrum analyzers; and the 140 series of spectrum analyzers to name only a few. It digitally displays system data, updates the display, and "normalizes" the system so that only the test device is shown on the CRT display.

The storage-normalizers are usually specified to operate with particular types of equipment. It is advisable that you check to see what inputs and outputs are available on your particular analyzer and then match the specification to the storage-normalizer compatible with your system.

2.4 Microwave Systems

This section introduces some of the systems available today that make microwave measurements easier and more accurate. By microwave systems we mean those pieces of equipment that are capable of performing a variety of tests. This is compared to an instrument that is designed to provide only one specific type of measurement. More details of these systems will be covered in following chapters. (It should be pointed out that these "systems" do not include such components as generators, power meters, etc.).

The systems to be examined are:
- HP8755 Swept Frequency Response Test Set
- PM1038 Measurement System
- HP8410 Network Analyzer

2.4.1 HP8755 Swept Frequency Response Test Set

Figure 2.39 is a photograph of the 8755 Swept Frequency Response Test Set. The 8755 series consists of precision detectors and display systems for making the basic microwave measurements of insertion loss/gain and return loss (VSWR) from 15 MHz to 18 GHz. Each working system (whether bench or rack mounted) includes the necessary detectors, modulators, and an appropriate display. The dual channel 8755 allows simultaneous swept frequency display of two ratio measurements or measures absolute power at the push of a button. The 8755 offers a number of advantages besides covering a wide frequency range; the 11665B modulator allows ac signal processing for virtually drift-free operation with time and temperature. Use of hot carrier diode detectors, which are completely interchangeable, enables a —50 dBm sensitivity. Therefore, a 60 dB dynamic range is available with solid state sweepers having a 10 mW output. Front panel controls are both easy to understand and operate. Each channel is separate, but identical, and all functions are push-button controlled. A direct reading digital dB offset thumbwheel allows the magnitude of any displayed signal to be easily determined.

Test Equipment

Figure 2.39/HP8755 swept frequency response set.

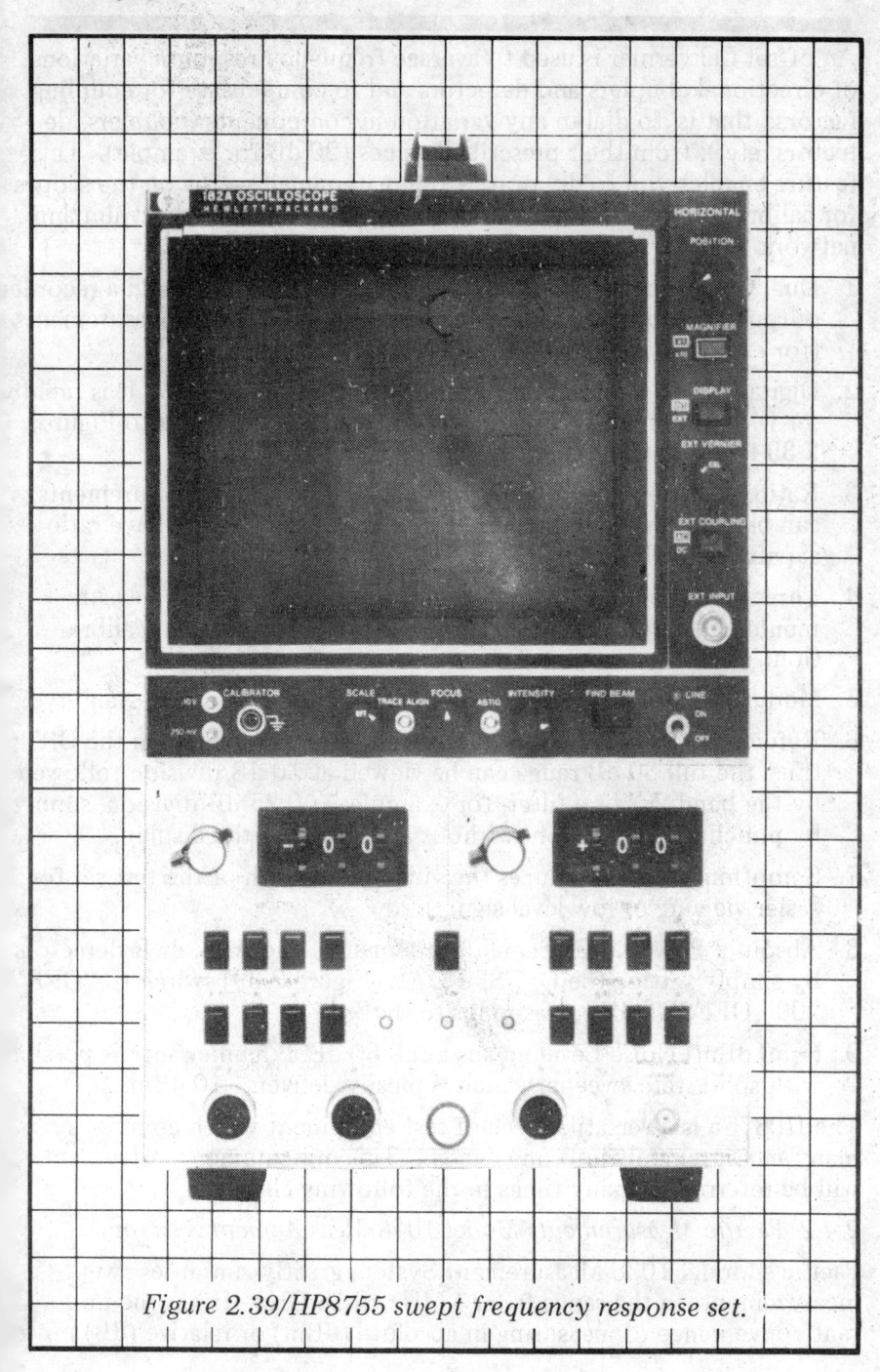

An offset Cal vernier is used to average frequency response variations of directional couplers and detectors and to compensate for coupling factors; that is, to dial in any variation in components (couplers, detectors, etc.), from their prescribed values (20 dB for example). This feature enables you to do away with grease pencil marks on the scopes for calibration much as the storage-normalizer does for spectrum and network analyzers. Important features of the 8755 are:

1. Dual Channels — Each trace has independent controls and a recorder output, providing flexibility in viewing simultaneous measurements (for example, insertion loss and return loss).
2. Digital Offset Readout in 1 and 10 dB steps of ± 59 dB. It is rapidly set with ±, ten, and units thumbwheel switches. (Refer to Figure **2.39** to see the switches marked OFFSET DB).
3. Ratio Measurement capability means that accurate measurements can be made from an unleveled source. Both channels have ratio capability (A/R and B/R).
4. Three Detectors make possible two simultaneous ratio measurements. Detectors are interchangeable without system recalibration.
5. Modulated systems gives excellent time and temperature stability.
6. Reference Position of either trace can be set anywhere on the CRT; thus, the full 60 dB range can be viewed at 10 dB/division followed by the bandpass of a filter, for example, at 0.25 dB/division, simply by punching one button, without any offset of the display.
7. Smoothing control reduces the video bandwidth of the test set for easier viewing of low level signals.
8. Absolute Power Measurements are made at the A, B, or R detectors by simply setting the OFFSET CAL (Figure 2.39) switch to "OFF" "00" OFFSET dB corresponds to 0 dBm.
9. (−50 dBm) Noise Level means a full 60 dB dynamic range is possible with solid state sweeper which typically delivers +10 dBm.

The HP8755 is a versatile piece of test equipment which combines many measurements into one chassis. This measurement instrument will be referred to many times in the following chapter.

2.4.2 Pacific Measurements Model 1038 Measurement System

Pacific's Model 1038 Measurement System greatly simplifies swept measurements in the range from 1 MHz to 18 GHz. It has the important convenience of measuring in absolute (dBm) or relative (dB) power

Test Equipment

Figure 2.40/PM1038 measurement system.

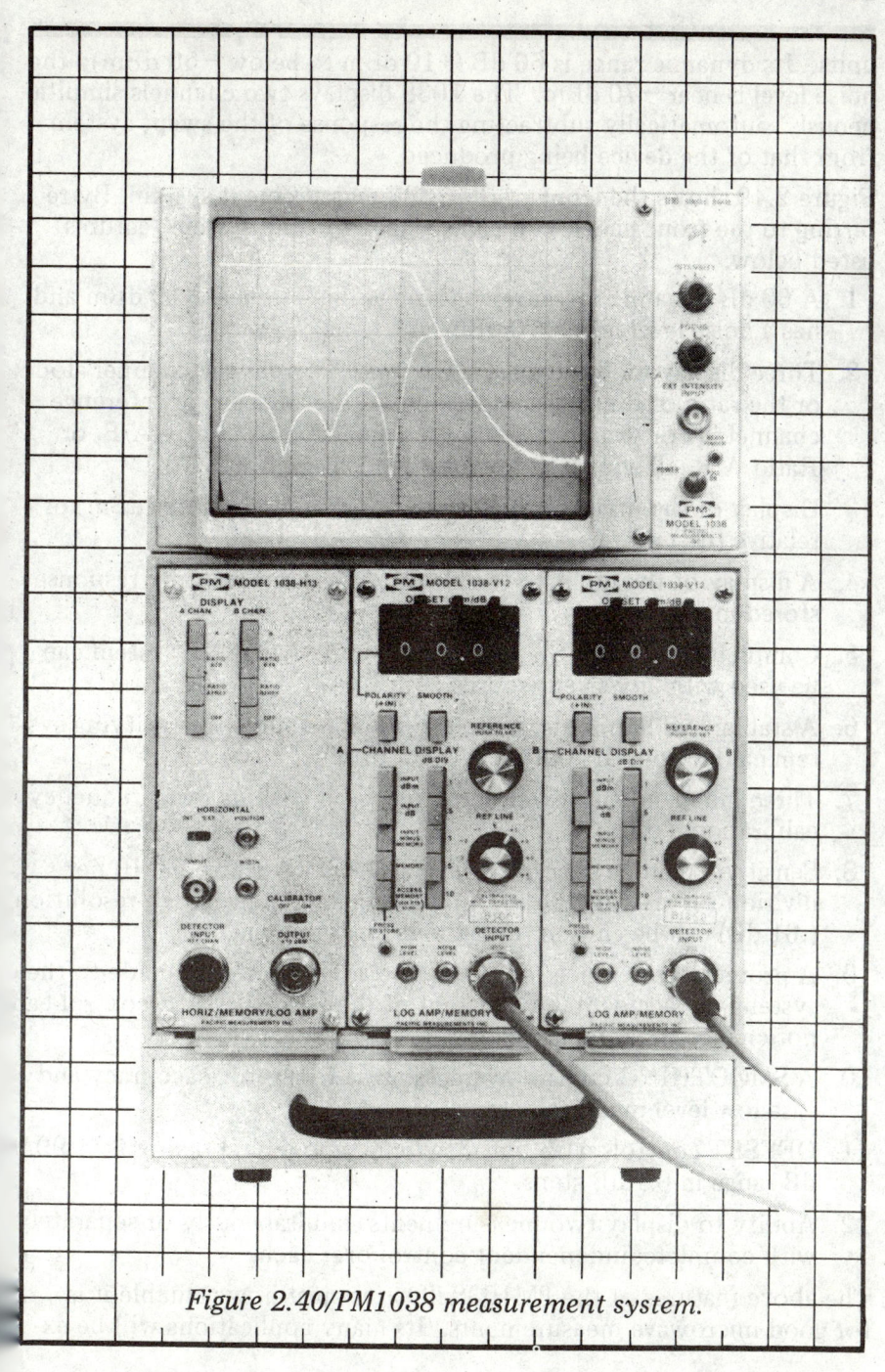

units. Its dynamic range is 60 dB (+10 dBm to below −50 dBm); the noise level is near −70 dBm. The 1038 displays two channels simultaneously, automatically subtracting the response of the swept system from that of the device being produced.

Figure 2.40 shows the front of the 1038 measurement system. By referring to the front panel, you can see many of the system features listed below.

1. A 60 dB dynamic measurement range that starts at +10 dBm and has a noise level near −70 dBm.
2. Three display modes for each channel. There is the channel alone or the ratio of the channel with either the adjacent or reference channel. (For example, there is channel A only, Ratio A/B, or Ratio A/R, all shown in the upper left of Figure 2.40).
3. Display of the input to the power detector in absolute (dBm) or relative (dB) power.
4. A display of the input to the power detector minus any response stored in the memory.
5. Complete control of the horizontal input so that the system can be used with any swept source.
6. A stable, low harmonic, matched 50 ohm source for verifying system calibration. Its output level is 10 mW (+ 10 dBm).
7. Three independent power detectors, each with its own frequency calibration data.
8. Sensitivity can be set on the display from 10 dB to 0.1 dB per division for viewing the full 60 dB dynamic range; high resolution (.01 dB) can be chosen for flatness measurement.
9. A protected reference set. Once the adjustments are made to the system any accidental movement of the REFERENCE control can not change the original setting.
10. A SMOOTHING control averages noise to improve accuracy and ease low level measurements.
11. OFFSET controls with linear switches and digital readouts, ± 99.9 dB range in 0.1 dB steps.
12. Ability to display two measurements simultaneously or separately with complete, independent control over each.

The above features of the PM1038 illustrate just how valuable it is for good microwave measurements. Its many applications will be ex-

Test Equipment 83

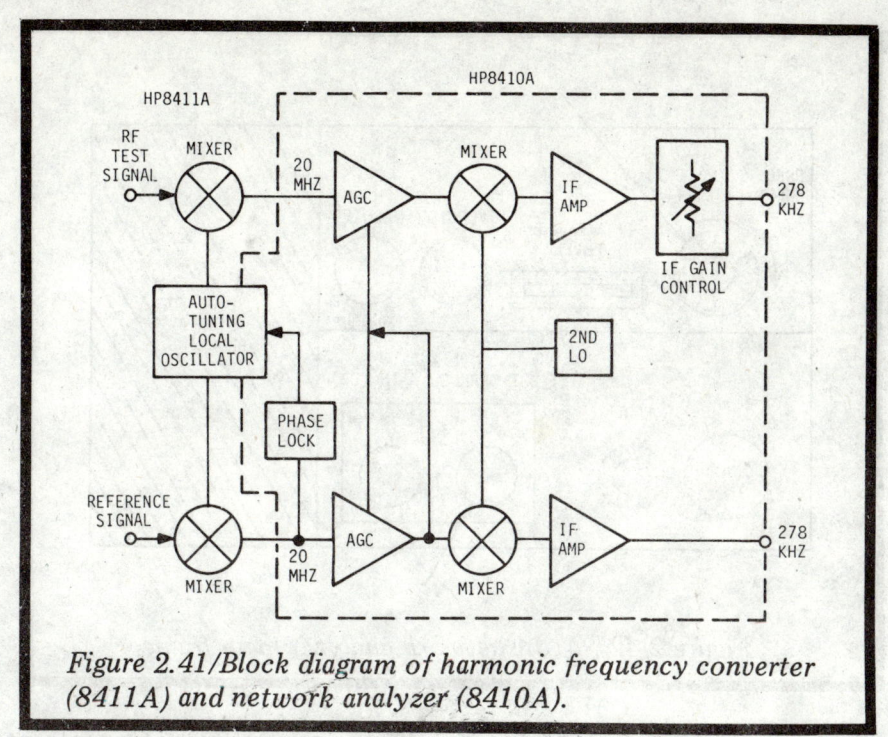

Figure 2.41/Block diagram of harmonic frequency converter (8411A) and network analyzer (8410A).

panded late in the chapters covering transmission and impedance measurements.

2.4.3 HP8410 Network Analyzer

The network analyzer is very difficult to classify as a "System". It is really many systems within one another.

The 8410A network analyzer can be broken into three separate sections: the harmonic frequency converter/mainframe section; transducers; and display units.

The HP8411A Harmonic Frequency Converter and the 8410A Mainframe are discussed together because they are the real heart of the network analyzers. Figure 2.41 shows a block diagram of these units and how they work together. Information from the test and reference channels enter the frequency converter/network analyzer. By harmonic sampling, the input signal is converted to a fixed IF frequency at which low frequency circuitry can measure amplitude and phase relationships. Sampling has the advantage that a single system can operate over an extremely wide input frequency range; in this case, from 110 MHz to 12.4 GHz. It also enables the system to be relatively insensitive to source harmonics.

Figure 2.42/8410A network analyzer main frame.

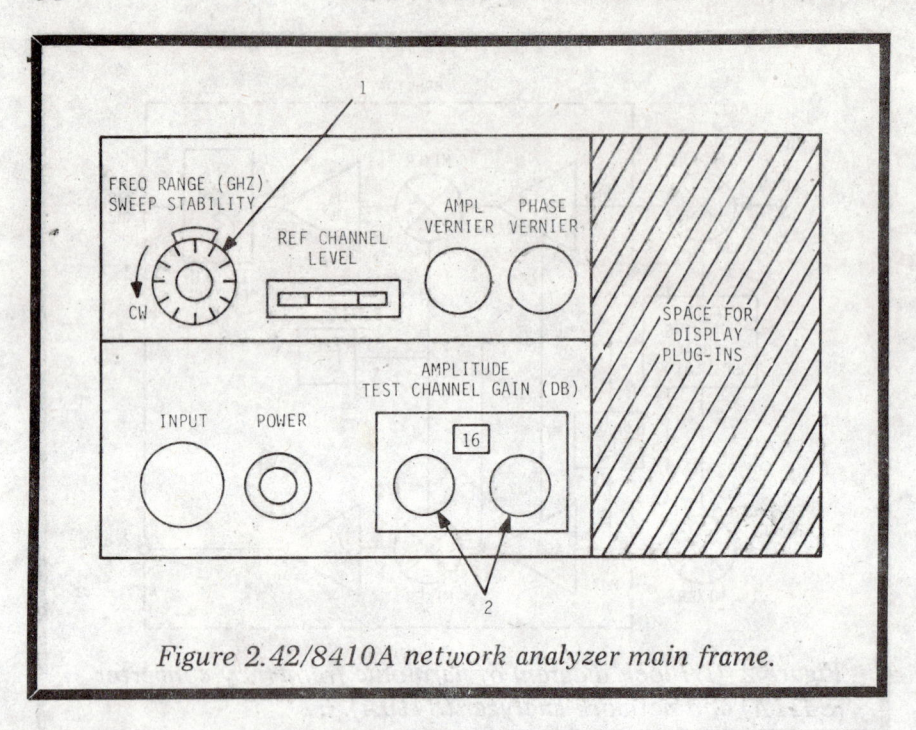

To make the system capable of swept frequency operation, the internal phase lock loop shown in Figure 2.41, keeps the reference channel tuned to the incoming signal. This loop ensures repeatability and accuracy of test measurements since it automatically tunes back and forth across the octave frequency band selected on the front of the 8410A in order to obtain a constant IF frequency. The range switch is shown as number 1 in Figure 2.42.

The IF signals going to the 8410A mainframe are both 20 MHz signals. Since frequency conversion is a linear process, these signals have the same relative amplitude and phase as do the microwave reference and test signals. Thus, gain and phase information are preserved and all signal processing and measurements take place at a constant frequency.

A leveled source is unnecessary for the network analyzer. As shown in Figure 2.41, there are two matched AGC (automatic gain control) amplifiers in the analyzer. One keeps the signal level of the reference channel constant and applies an error signal to a matched amplifier in the test channel so that the latter signal level does not change. The ratio of the test to reference channel signals can then be measured directly on one of the display units.

Test Equipment   85

Figure 2.43/8410-8411 network analyzer.

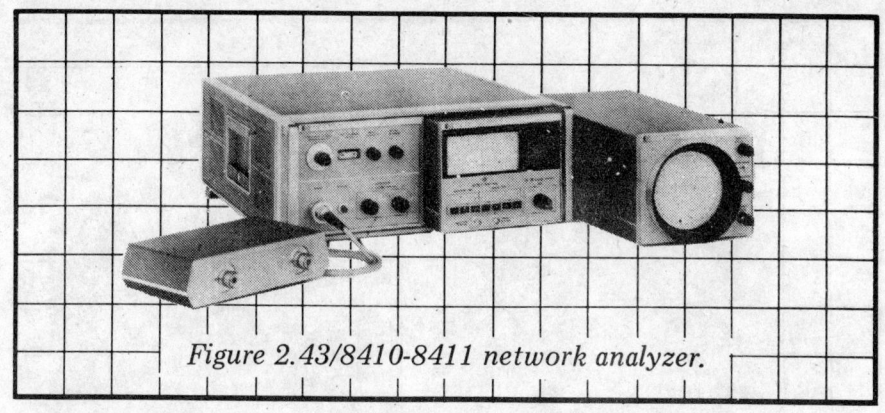

By down-conversion of the input frequency, attenuation can be added to or subtracted from the test channel at the one stable point. This very accurate technique is known as the "IF substitution method." This technique is used during calibration and permits expansion of the output display resolution to ensure greater accuracy during the actual measurement. On the front panel of the 8410A, there are two test channel gain switches (number 2 in Figure 2.42) that are calibrated by increments of one and ten dB. By using the IF substitution method, the measurement range of the analyzer is extended so that values of attenuation greater than 60 dB can be determined. By referring to Figure 2.41, once more you now should realize why the tag on the top of the 8411A says "MAXIMUM INPUT 50 mW" — any greater input level would burn out the mixers at the input to the converter. Figure 2.43 shows the HP8410A mainframe and the HP8411B Harmonic Frequency Converter.

The transducer unit is placed between the microwave signal source and the harmonic frequency converter and has three functions. First, it splits the incoming signal into reference and test signals. Secondly, it provides the capability of extending the electrical length of the reference channel so that the distances traveled by the references and test signal are equal. Finally, it connects the system correctly for transmission and reflection measurements.

To split the incoming signals, two basic methods are used — the power splitter (employed in the 8740B transmission test set) and directional and dual directional couplers (utilized by the other transmission and/or reflection units). To be effective the couplers must exhibit high directivity and the power splitter must have good isolation.

Transducers available with the HP8410A network analyzer are designed for transmission and/or reflection measurement in a variety of

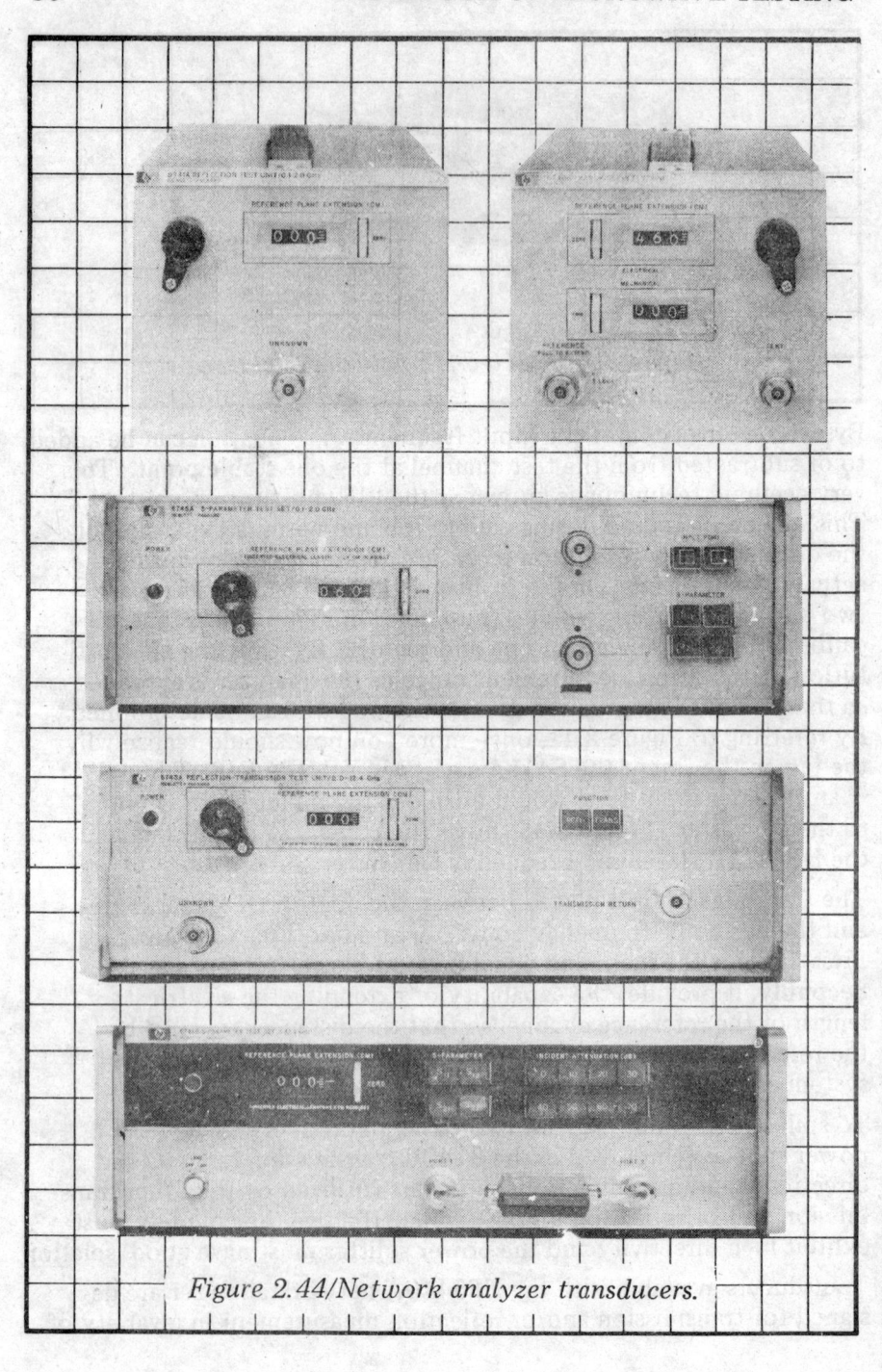

Figure 2.44/Network analyzer transducers.

Test Equipment 87

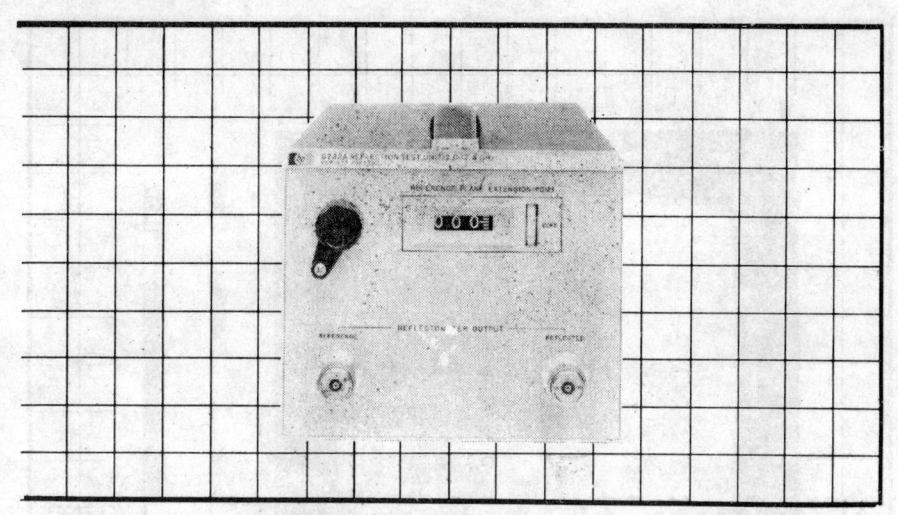

frequency ranges. The listing below shows the available transducers according to function and frequency. Figure 2.44 shows all of these transducers.

Frequency Range	Transmission Only	Transmission & Reflection	Reflection Only
0.11-2 GHz	HP8740A	HP8745A	HP8741A
2-12.4 GHz	HP8740A	HP8743A	HP8742B
0.5-12.4 GHz	(S-Parameter Test Set - HP8746B)		

Three plug-in displays are available. These units give a choice of convenient and varied readouts. In addition, equipment such as an oscilloscope or an X-Y recorder can be tied to each system for improved displays or permanent records.

The first device is the HP8412A Phase-Magnitude Display Plug-In. This unit displays amplitude and phase versus frequency; that is, insertion gain/loss or return loss can be shown directly. It has 80 dB and ± 180° range with selectably resolution up to 0.25 dB and 1° per major division (allowing measurement accuracy of 0.05 dB and 0.2° using the inter-division graticular lines). Its dual channel scope allows amplitude and phase to be displayed either simultaneously or separately. (The 8412A is shown in Figure 2.45).

The second display unit is the HP8413A Phase-Gain Indicator. It uses a meter readout for both phase and gain. The amplitude ranges are ± 30, ± 10 and ± 3 dB full scale, phase ranges are ± 180, ± 60, ± 18, and ± 6° full scale. Resolution of phase and amplitude measurements is selected at the front panel by push-button controls, thereby permit-

88   HANDBOOK OF MICROWAVE TESTING

Figure 2.45/HP8412 phase-magnitude display.

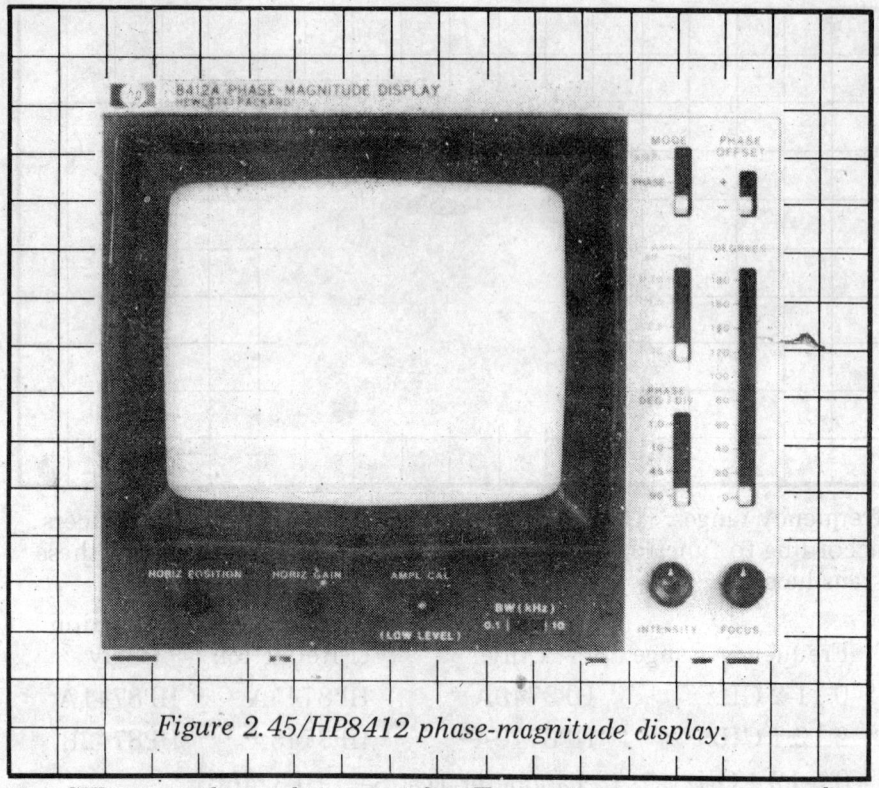

ting CW tests to be made accurately. Two output connectors on the front panel provide the phase and amplitude signals for an external recorder or a dual channel oscilloscope for simultaneous swept measurements (The 8413 is shown in Figure 2.46).

The third type of indicator is the HP8414A Polar Display Plug-In. It provides a polar plot of the magnitude and phases of transmission and reflection coefficients, and can be used with either CW or swept frequency. (The unit is shown in Figure 2.47.) If the device under test is a well matched 50 ohm device, the CRT displays a dot at the center of the screen. But, if there is a mismatch from the 50 ohm transmission line impedance, the device's reflection coefficient, magnitude, and phase can be read off the display at each frequency.

This has been an introduction to the HP8410A Network Analyzer "System." There are many different combinations of the units explained. Their combinations depends on what your particular application might be.

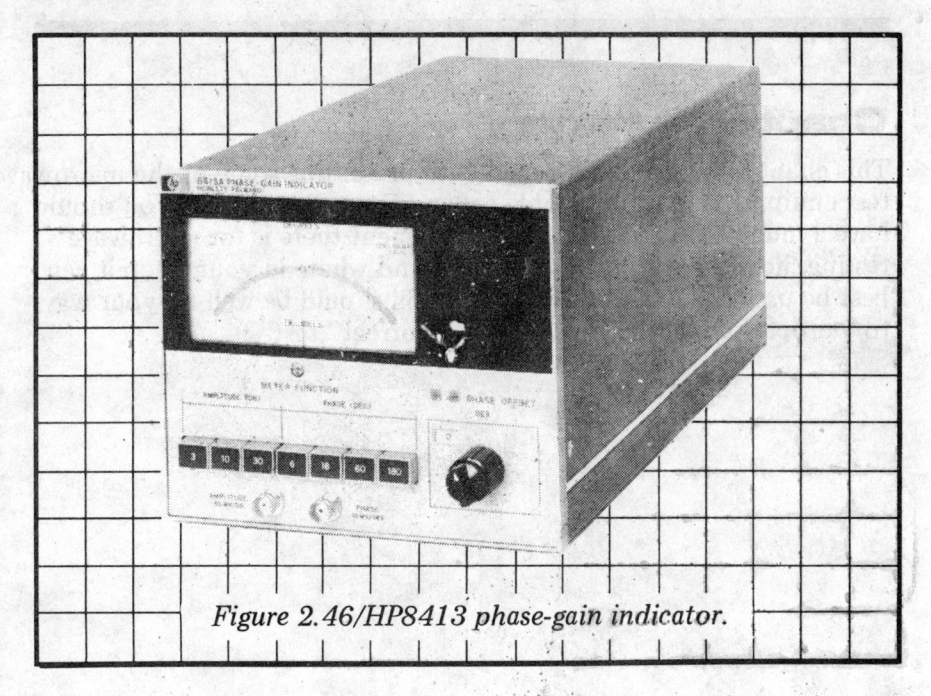

Test Equipment 89

Figure 2.46/HP8413 phase-gain indicator.

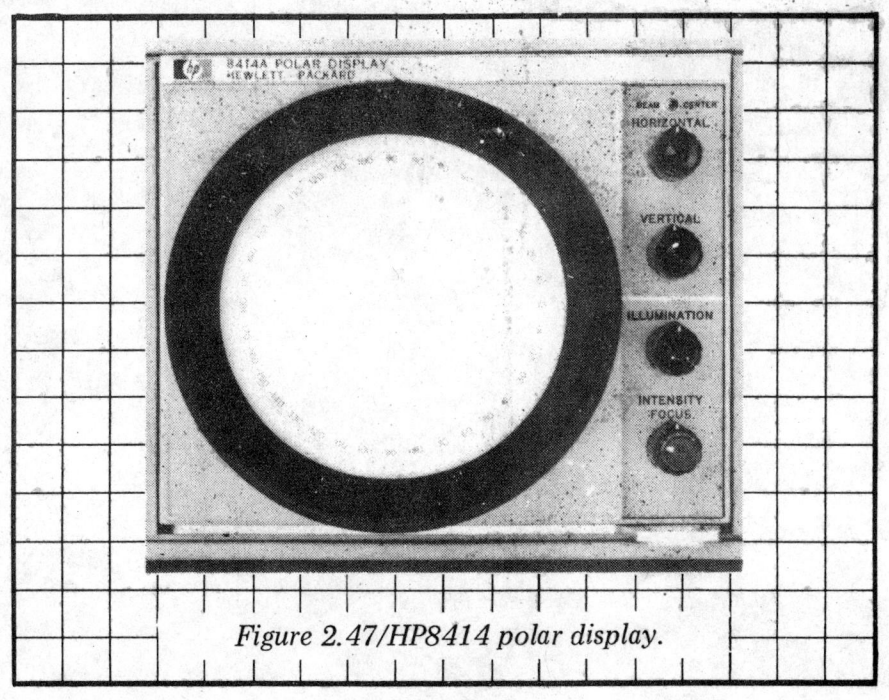

Figure 2.47/HP8414 polar display.

Chapter Summary

This chapter has been concerned with presenting some of the microwave test equipment that is available today. From this chapter you should have a much better idea of what equipment there is for microwave testing, how the equipment operates, and where in your setup it can best be used. With this information you should be well on your way to a proper setup which results in a "correct" test.

Power Measure

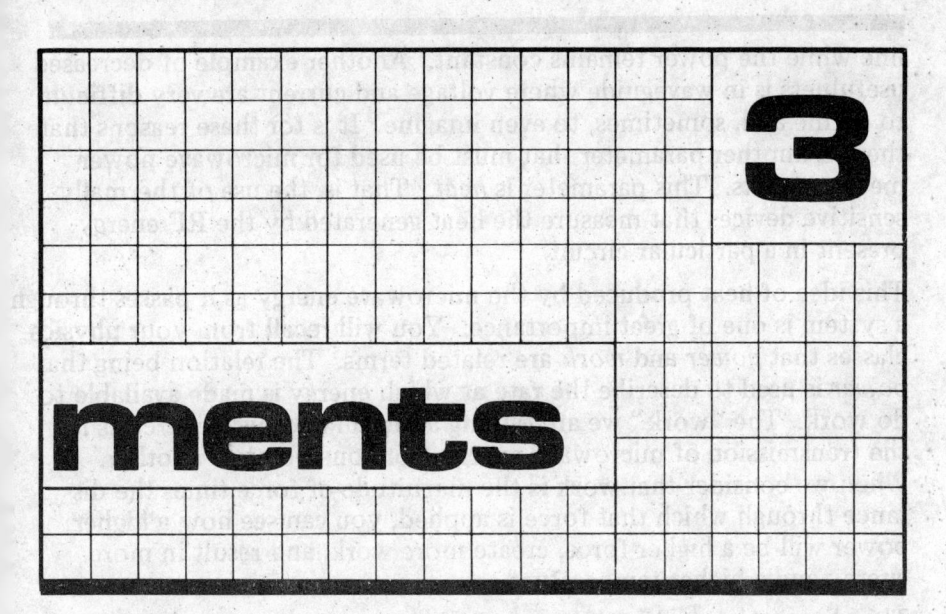

3. Definition

We have used the term power very liberally throughout the first two chapters. We have referred to power handling capability of microwave components and to maximum power input level for test equipment. Knowing what these levels were was very important to the proper operation of the test setups. The importance of knowing and heeding them was stressed many times, but a definition of microwave power and how it necessitates these set levels in the first place was not discussed. That will be our task in this section.

When power is referred to in a dc circuit it is the product of the voltage across a specific circuit and the current traveling through that circuit. In other words, the old familiar ohms law relation:

$P = E \times I$

where P = power
 E = Voltage
 I = current

This is not the case when measuring microwave power. It is true that microwave circuits have voltages and currents set up by electromagnetic fields when they are operating, but they are not of such a nature that one can put a voltmeter and ammeter in the circuit to measure them. One reason is that as the frequencies of operation approach 1.0 GHz, voltage and current readings vary with position along the transmission line while the power remains constant. Another example of decreased usefulness is in waveguide where voltage and current are very difficult to define and, sometimes, to even imagine. It is for these reasons that there is another parameter that must be used for microwave power measurements. This parameter is *heat*. That is, the use of thermally sensitive devices that measure the heat generated by the RF energy present in a particular circuit.

This idea of heat produced by the microwave energy as it passes through a system is one of great importance. You will recall from your physics classes that *power* and *work* are related terms. The relation being that *power* is used to describe the rate at which energy is made available to do work. The "work" we are talking about in microwave circuits is the transmission of microwave energy from one point to another. When we consider that work is the magnitude of force times the distance through which that force is applied, you can see how a higher power will be a higher force, create more work, and result in more energy and a higher temperature.

This direct relationship shown above between the magnitude of microwave power and temperature is valuable in arriving at our definition of power promised at the beginning of this section.

At this point we can define Microwave Power as:

> The energy in a microwave circuit which causes a change in the amount of work being done within that circuit. This change in work causes a corresponding change in temperature which can be measured with a high degree of accuracy.

In order for us to be able to determine how accurately we can measure the power we have just defined, we should know a little something about what units are used to measure power. The units used are watts, dB, and dBm. The *watt* has been established by the International System of Units as the main unit of power. Its official definition is that one watt is one joule per second (the joule being the standard unit of work and energy). Electrical quantities do not even enter into the definition of power (as you can see from our definition presented above). As a matter of fact, other electrical units are derived from the watt. A volt, for example, is one watt per second. So you can see the importance of the watt in power terminology.

The dB (decibel) is a unit used to express the ratio between two amounts of power, P_1 and P_2, existing at two points. This is a measure of relative power between the two points and is dimensionless because the units of both powers are watts. Expressed mathematically:

Power Measurements

$$\text{dB (power)} = 10 \log_{10} \frac{P_1}{P_2}$$

There are also relationships for voltage and current expression in dB. They are:

$$\text{dB (voltage)} = 20 \log_{10} \frac{E_1}{E_2}$$

$$\text{dB (current)} = 20 \log_{10} \frac{I_1}{I_2}$$

The use of dB has two advantages. First, the range of numbers commonly used is more compact; for example +63 dB to −153 dB is more concise than 2×10^6 to 0.5×10^{-15}. The second advantage is apparent when it is necessary to find the gain of several cascaded devices. Multiplication is then replaced by the addition of the power gain in dB for each device.

The dB discussed above was referred to as a relative parameter since it was the ratio of two powers, each of which could vary. The dBm does away with this relative status and allows absolute power expression. The formula for dBm looks very similar to that used for dB.

$$\text{dBm} = 10 \log_{10} \frac{P_1}{1 \text{ mw}}$$

The one difference, which makes absolute readings possible, is that P_1 is the only quantity that can vary. The denominator (formally a P_2 variable) is a constant reference of 1.0 milliwatt. An example of how the formula gives absolute readings is an oscillator that has an output of 13 dBm. By solving for P, in the above equation, the power output is found to be 20 mw (an absolute reading of power). Notice the difference from saying that an amplifier has 13 dB of gain, for example. All this is saying is that there is an increase in signal from input to output of 13 dB (a factor of 20). It says nothing about how much power is coming out of the amplifier, it only tells you that there is 20 times more (13 dB) coming out than you put in. If, for example, you specified that this same amplifier had an input applied at 0 dBm (this is 1.0 mw if you run through the formula), you would then know that the absolute output level of the 13 dB amplifier was +13 dBm (or 20 mw). So you can see that dBm actually means "dB above one milliwatt", but a negative number of dBm is to be interpreted as "dB below one milliwatt".

A case of a negative dBm number would occur if you were testing an attenuator. Suppose we have a 20 dB attenuator and apply a +7 dBm (5.0 mw) input to it. At the output we would have a signal that was 20 dB lower, or −13 dBm (.05 mw). This means that the output is 13 dB below one milliwatt or 1/20 of a milliwatt; which it is.

You can see from the examples above the advantages of dBm and how they are identical to those listed for dB. The only difference to remember is that dB results in relative readings and dBm is absolute readings.

Before proceeding to the first section of the chapter it would be wise for us to provide definitions of the power levels we will be discussing. For purposes of our discussion we will define them as follows:

Low power	− 1 milliwatt and below (< 0 dBm)
Medium power	− 1 milliwatt to 10 watts (0 dBm to +40 dBm)
High power	− 10 watts and up (> +40 dBm)

These levels are set up simply for our own reference to enable the power measurements to be covered by some designated level. There is certainly no standard set up which will govern what is called high, medium, or low power.

A last order of business would be a brief discussion of the devices which are used to sense and translate the temperature variations in a microwave circuit that we are calling *power*.

Chapter 2 covered different power sensors and pictured some that are commercially available. Basically, two types are used for power sensing: the thermocouple and the bolometer.

The thermocouple is formed when two wires of different metals have one of their junctions at a higher temperature than the other. The difference in temperature produces a proportional voltage. The device is used to measure the temperature rise of a load which dissipates microwave power; by appropriate calibration, the temperature change is converted to an indication of power. Thermocouplers have excellent sensitivity and are especially useful as power monitors; their primary applications are in low and medium power areas.

Bolometers are classed as barretters or thermistors. The most common usage of the term "bolometer" refers to a barretter; however, we describe both classes to provide a familiarity with all types of microwave power measuring devices.

The barretter consists of an appropriately mounted short length of fine wire, usually platinum, with sufficient resistance to enable it to

Power Measurements

be impedance matched as a termination for a transmission line. It has a positive temperature coefficient; that is, the resistance of the barretter increases as the temperature rises. It is usually biased to an operating resistance of from 50 to 400 ohms; 200 ohms is the most common.

The thermistor consists of a tiny bead of semi-conducting material which bridges the gap between two fine, closely spaced, parallel supporting wires. Effectively, all the resistance of the thermistor is concentrated in the bead material which has a negative temperature coefficient of resistance (an increase in temperature results in a resistive decrease). They usually are made of complex metallic-oxide components using oxides of manganese, nickel, copper, cobalt, and other metals.

The changes in resistance resulting from the dissipation of microwave power in the bolometer (barretter or thermistor) are commonly measured by using the devices as one arm of a dc or audio frequency wheatstone bridge circuit.

The barretter and the thermistor are sensitive power detectors, and are capable of measuring as little as a few microwatts of power when used in properly designed bridge circuits. The thermistor is the more flexible device in that its resistance may be varied over an extremely broad range, depending on the magnitude of bias current. This feature is often a decided advantage in the broadband impedance matching of the detector, it leads to excellent overload and burnout qualities, and its operating resistance is confined to a considerably smaller range. However, the latter can be made more reproducible in both sensitivity and impedance, and is less sluggish (has smaller time constants) than a thermistor. There are good and bad points for the barretter and the thermistor which must be carefully weighed, and tradeoffs made, when choosing the type for your application.

With a definition of power, units to be used, and a basic discussion of power sensors behind us we are now ready to investigate the actual measurement setups and procedures. They will be covered in the three levels we have set up; low, medium, and high power. Each will cover CW measurements with swept setups discussed where applicable.

3.1 Low Power Measurements

The measurement of low power (1.0 mW and below) requires great care in the selection of the equipment and components you use. Many instruments are noise limited if you operate at too low a level. That is, the internal noise of the instrument is higher than the signal you would be measuring. Also, the match of the instruments that you use should be much better than when you are testing at higher power. This is because you can not use attenuators so freely since they will attenuate the signal too much while they are matching the devices. We will be emphasizing points such as these as we move through our discussion.

We shall cover two methods of measuring low power circuits. The first setup will use a power meter as an indicator. Until recently this was not a very good means of measuring very low powers since power sensors were not available to do an adequate job. There is now a good selection of sensors that read to −70 dBm with good accuracies. The second method we shall cover is by far the most widely used for low power readings. This method uses the spectrum analyzer as an indicator. The accuracies obtained and wide dynamic range of the analyzer make it very suitable for low power work.

Figure 3.1 shows a low power test setup using a power meter. The generator used depends upon your application in regards to frequency range and power level. The particular setup shown uses a CW generator, but it can be adapted with relative ease to a swept setup using a sweep generator. The swept setup would probably have a leveling loop (ALC) at the RF output and would have a *sweep out* voltage going to an indicating device (scope or X-Y recorder, for example). These are unnecessary for CW measurements since only a single frequency is measured. The generator, however, should have reasonable amplitude and frequency stability so that once the system is calibrated you could leave it for a while and have it still maintain that calibration when you return for a measurement. It must also have enough range to allow you to make the necessary low level tests that you want. In other words, it must be able to be adjusted low enough in power to give you the data you need and still maintain its original specifications.

The second block in Figure 3.1 is a low pass filter (it could also be a bandpass filter if you so desired). This component is used to ensure that the readings seen on the power meter are the signals that you really want. As was discussed in Chapter 2, a power meter's sensor unit is completely independent of frequency. It does not care if it sees a 1 GHz, 2 GHz, or 3 GHz signal; it only knows that there is microwave power present which produces heat that it can pick up. This results in a meter reading. The low pass (or bandpass) filter is thus inserted to ensure that only the desired fundamental frequency is being indicated. This may seem to be a rather trivial and unneces-

Power Measurements

Figure 3.1/Low power test setup.

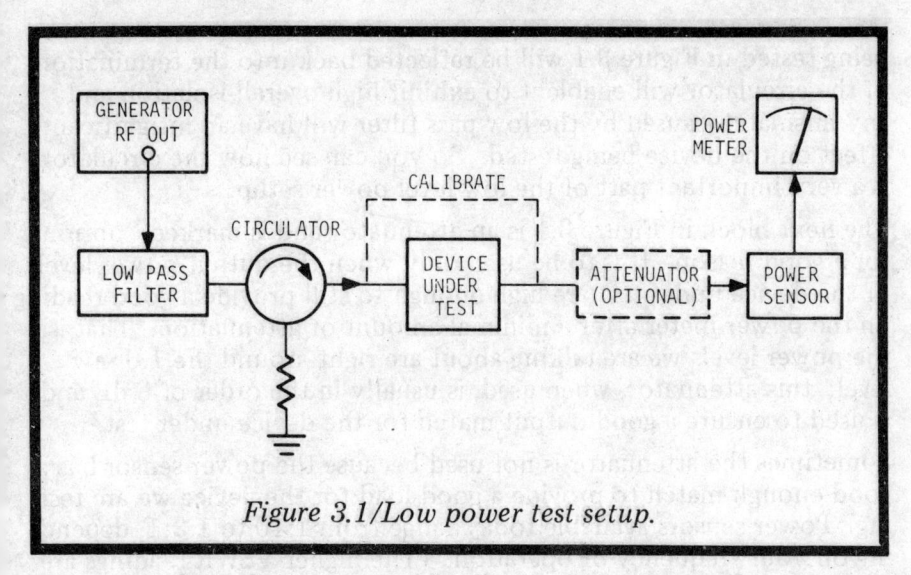

sary worry, but when you consider that some generators have their outputs produced by a comb generator and have very little in the way of output filtering, a filter in the setup is well worth the trouble. Some harmonics are only a few dB different from the desired frequency in the generator as discussed above.

The next component in the setup is the one that replaces the input attenuator in many setups. The device is a circulator and its main purpose is to provide a good match for the device under test with very little loss to the input signal. As was mentioned earlier, great care must be taken not to decrease the levels throughout the low level setups. When an attenuator is being used, a minimum value is usually 3 dB. When a circulator is used, the loss is in the order of 0.3 to 0.5 dB. The major item to be concerned with when using a circulator is that it is broad enough to cover your frequency range. There are octave band devices available, but many circulators are narrow band devices. It pays to check your particular component for its coverage.

The reason that a circulator is able to provide a good match for the system with low loss is its non-reciprocal properties. That is, it only allows microwave energy to move in one direction. There is a high isolation in the reverse direction. This is due to a gyromagnetic (rotating) effect set up within the device when the microwave energy interacts with the ferrite material which is a major part of the construction of the circulator. This rotation is either clockwise or counterclockwise and forces the energy to follow the field and keep it from being reflected back to the input. Any mismatch caused by the device being tested in Figure 3.1 will be reflected back into the termination of the circulator will enable it to exhibit high overall isolation and any mismatch caused by the low pass filter will have an insignificant effect on the device being tested. So you can see how the circulator is a very important part of the low level power setup.

The next block in Figure 3.1 is an attenuator and is marked "optional" for a good reason. It is to be used only when the output power levels of the device under test are high enough to still provide a good reading on the power meter after a minimal amount of attenuation. That is, the power levels we are talking about are right around the 1.0 mw level; this attenuator, when used, is usually in the order of 3 dB and is used to ensure a good output match for the device under test.

Sometimes the attenuator is not used because the power sensor has a good enough match to provide a good load for the device we are testing. Power sensors available today range from 1.10 to 1.3:1, depending on your frequency of operation. (The higher VSWR readings are for sensors operating in the 12-18 GHz range. These VSWR readings will also vary according to the power range the sensor has to operate over.) Some representative power ranges for sensors are:

 —70 to —20 dBm
 —30 to +20 dBm
 —17 to +35 dBm
 —10 to +35 dBm

We have, of course, been referring to the next block in the test setup, the power sensor. As mentioned, the power range and VSWR are prime concerns when choosing a sensor for your test setup. Any of the sensors listed above would do very nicely for the low level setup shown in Figure 3.1. The power meter shown works hand in hand with the power sensor. The meter is the device which processes and displays the power detected by the sensor. They are a team which should be chosen wisely.

With an idea of what makes up the setup in Figure 3.1, let us go through a test to see how we use them.

Calibration

    (1)   Connect the setup as shown in Figure 3.1, with the device to be tested removed and a calibration cable inserted.

    (2)   Turn on the AC power to the generator and power meter. Be sure that the generator output is turned to its minimum output.

Power Measurements

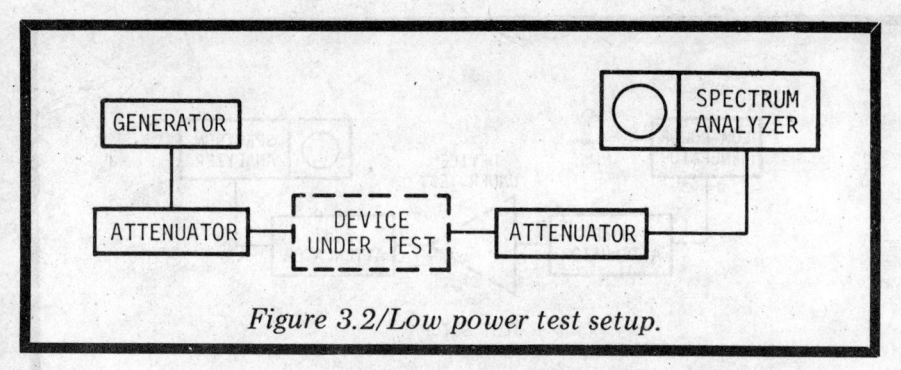

```
┌─────────────────────────────────────────────────────┐
│                                                     │
│  ┌──────────┐                    ┌───────────────┐  │
│  │GENERATOR │                  ○ │    SPECTRUM   │  │
│  └──────────┘                    │    ANALYZER   │  │
│        │                         └───────────────┘  │
│  ┌──────────┐   ┌─ ─ ─ ─ ┐    ┌──────────┐          │
│  │ATTENUATOR├───┤ DEVICE ├────┤ATTENUATOR│          │
│  └──────────┘   │ UNDER  │    └──────────┘          │
│                 │ TEST   │                          │
│                 └ ─ ─ ─ ─┘                          │
│                                                     │
│           Figure 3.2/Low power test setup.          │
└─────────────────────────────────────────────────────┘
```

(3)   If a relative power reading is desired, turn the RF generator output until a convenient level is indicated on the power meter. Record this level as a reference.

(3a)   If an absolute reading is desired, turn the RF generator output until the power meter reads 0 dBm. (If an active device such as an amplifier is to be tested, check to see what the maximum input level is to the device. The 0 dbm signal may overdrive it. A —5 dBm or a —10 dBm signal may have to be used as a reference.)

Measurement

(1)   Remove the calibration cable and insert the device to be tested. If an active device is to be tested, be sure it has the proper dc voltages applied before inserting it into the test setup.

(2)   If a relative reading is desired, record the reading on the power meter and subtract it from the previous reference reading above.

(2a)   If an absolute reading is desired, record the reading on the power meter directly as your results.

Although the power meter method of measuring low power is used more than it used to be, the old standard way of using a spectrum analyzer is still one that many people prefer.

Figure 3.2 is a typical setup that can employ a spectrum analyzer for a low level power reading. The analyzer, placed directly at the output of the device under test, can read either a relative or absolute power. Notice the simplicity of the setup. This is because the spectrum analyzer input is a microwave signal that does not need to be converted. The attenuators are shown in the figure as impedance matching devices. Once again, the circulator could also be used as described above,

102   HANDBOOK OF MICROWAVE TESTING

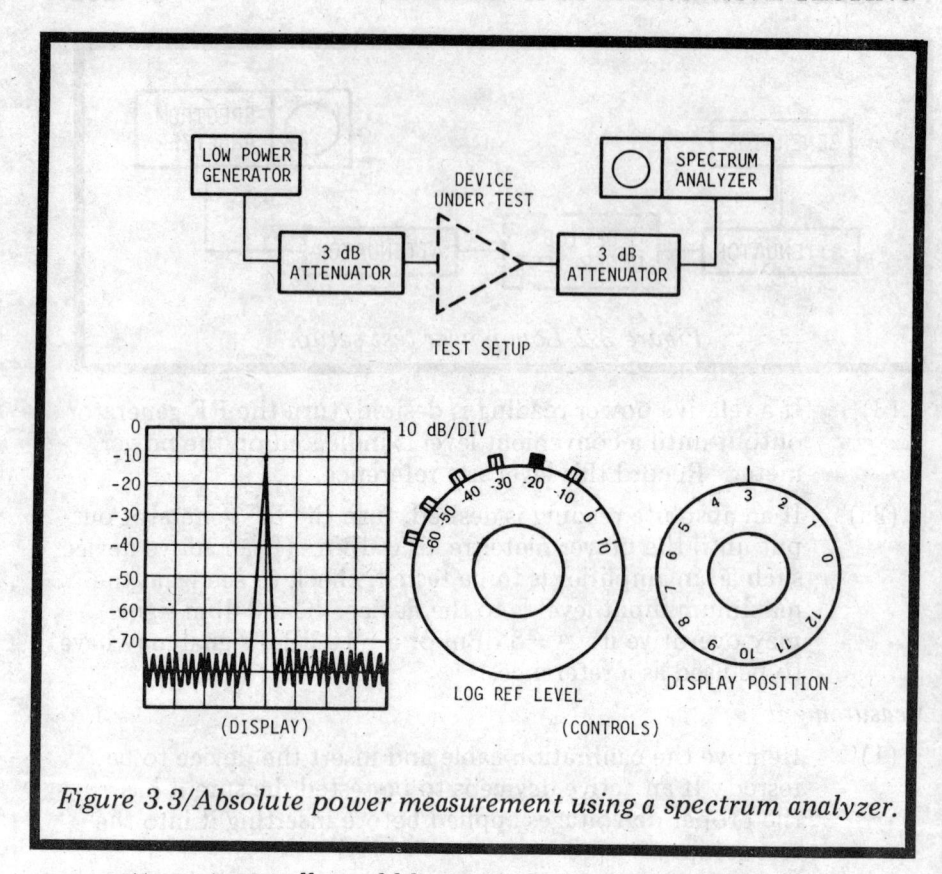

Figure 3.3/Absolute power measurement using a spectrum analyzer.

or no attenuator at all would be needed if the matches between components were good enough. A lowpass filter could also be inserted at the input if there is any question as to the spectral purity of the generator output.

We mentioned that this setup can read either relative or absolute power. With our coverage in this chapter thus far, we should be reasonably familiar with relative reading procedures. Let us take a moment and see what an absolute reading is like.

Figure 3.3 illustrates an analyzer display and how the absolute power reading is made; the test setup is shown above the display. The controls are set at −20 dBm on the log reference level and −3 dB on the display position. Therefore, the top of the display, at the 0 line, is −23 dBm. Our signal is on the −10 line; thus, the level of the signal at the input to the analyzer is −33 dBm. If there were no padding at the output of the test device we would have a −33 dBm signal and that would be it. However, as you can see from the setup, we have a

Power Measurements

3 dB pad at the output — effectively making our readings 3 dB worse than they actually are. And so, the real output power from this particular device is —30 dBm. If you were to connect a direct reading power meter, with adequate filtering to remove harmonics, to the output of this device you would find a value of —30 dBm. (The term "adequate filtering" once again refers to the need to eliminate harmonics and other spurious signals when using a wide open frequency independent power sensor.) The point of filtering is one great advantage of using a spectrum analyzer. You are able to see the fundamental frequency, the second harmonic, and any other signals generated, along with their associated absolute level. Each can be very clearly distinguished from one another.

We have seen how some special precautions must be taken when measuring low level powers. It is possible to use a power meter setup or the more popular spectrum analyzer. In each case proper matching and isolation of components, when necessary, will greatly enhance the accuracy of your readings. You must remember that when measuring power levels such as —20 or —30 dBm, you do not have much margin for error because you do not have a lot of power to work with. We can emphasize this more when we consider that —20 dBm is 10 μwatts and —30 dBm is only 1 μwatt. That is only 0.000001 watts of power to be measuring. So take extra care when making a test set-up to be used for low power measurements. It will pay off in the end.

3.2 Medium Power

Of all the power measurements made in microwaves, probably 95% of them fall into the category of *medium power*. We have defined medium power as that range between 0 dBm (1.0 mw) and +40 dBm (10 watts). If you stop and think of the last power measurement you made, it probably fell somewhere into this range. In this range we must take into consideration the precautions (covered in the previous section) on low power and we must also take into consideration some precautions to be covered in the following section on high power. It is for this reason that medium power will be covered more extensively than either the low or high power case.

A typical medium power set-up is shown in Figure 3.4(a). This first set-up is for the low end of the medium power range. It involves a power level of 100 mW (+20 dBm). This level is chosen because it is a fairly common power level in microwave circuits and it also involves component power handling capability that now must be taken into consideration. You will note that Figure 3.4(a) is very similar to the low level set-ups discussed in the previous section. The generator

Figure 3.4/Medium power setup.

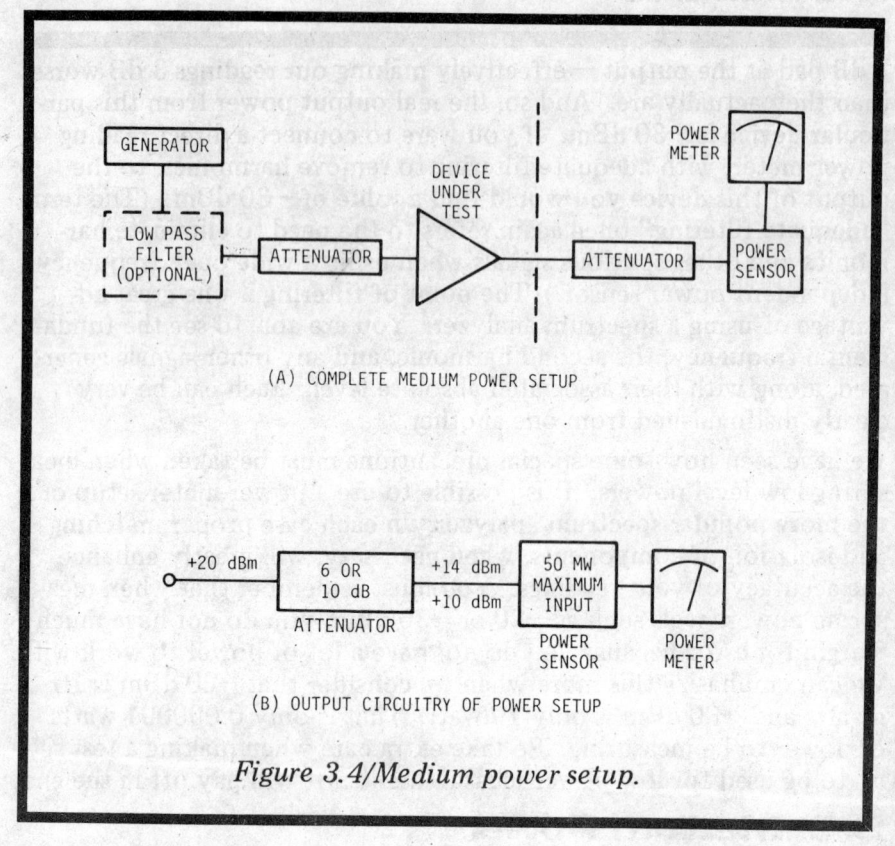

can be either a CW oscillator, a sweep oscillator in a CW position, or a sweep oscillator sweeping a band of frequencies. The low-pass filter in this set-up is listed as "optional". This is usually a matter of preference depending on the individual and the spectral purity of the generator used. Some generators are notorious for their harmonic and spurious contents and a filter is an absolute necessity regardless of the preference of an individual.

A circulator can also be used in this set-up, much the same as the low power setup. In this set-up it would be placed at the output of the generator. It can be placed there because the power levels are not as critical as in the low level case and thus an attenuator can be used for matching. It is not a common practice to use a circulator followed by an attenuator unless the generator output is too high and you want to decrease the level going to the device under test. The attenuator in the set-up in Figure 3.4(a) is used to aid in matching rather than level reduction. This could be a 3 or 6 dB attenuator and will do a good job

Power Measurements

of matching the set-up to about any device you may wish to test. The power rating on this attenuator is not a critical item since we are still talking about a low level signal coming from the generator.

The major differences are in the set-up beyond the device under test (to the right of the dashed line). The differences arise in the considerations to be given to component selection. In the low level power set-up you could use practically any component you may have available and not really worry about its power handling capability or input power rating. To some extent this is still the case. However, we are now approaching the power levels where these ratings become the difference between a good reading and a burned out power sensor. Consider, for example, the fact that some power sensors have a 30 mW (+15 dBm) maximum power input rating. Figure 3.4(b) can be used to show how to protect such an input rating. This is the portion of Figure 3.4(a) to the right of the dashed line. The input to Figure 3.4(b) is the +20 dBm output of the device under test. At this point we have to skip one block and move to the power sensor. You can see that the sensor has a 30 mW (+15 dBm) maximum input. With this in mind, we must attenuate the +20 dBm input signal at least 6 dB to protect the sensor and keep it from being burned out. A 6 dB attenuator, however, offers little margin for the set-up. If the device under test, for one reason or another, increase its power output we will have to put in a requisition for a new power sensor. I do not know about you, but I sure would hate to ask my boss to spend $300 for a new component because of a "slight miscalculation". The best protection for you and the sensor is a 10 dB attenuator, as Figure 3.4(b) also indicates. If you really want to be sure, you would use the 10 dB and a 3 dB attenuator. This will give you a +7 dBm (5 mW) input to the power sensor and also put 26 dB more return loss between the device under test and the power sensor. This would ensure a good match between them. This is not to say that the 6 dB attenuator would be unsatisfactory, if that was all that was available or if that was what you really had your heart set on using. The attenuator would result in +14 dBm (25 mW) being applied to the sensor, which is still 1.0 dB below its maximum rating.

The attenuator in Figure 3.4(a) and (b) is being dwelt upon because it is an important component in the overall setup. We have determined a value for this attenuator, but what type of power handling capability should it have? Suppose, for example, we have a miniature coaxial 10 dB pad (attenuator) available, which we would like to use in a power setup similar to Figure 3.4. The pad has a power rating of 0.5 watts and operates from dc to 4 GHz. The power input to the pad is 100 mW (or 0.1 watts). Therefore, it has more than enough power handling capability for the job that it must do. If you use a 50% safety factor as a rule of thumb, you should have little problem with components being under-rated in your test setups. This case obviously exceeds this 50% safety factor by a considerable margin.

The presentation above of a 100 mW power setup was intended to get you accustomed to considering power input levels and the power handling capability of components and equipment. We will now put the exercise to the test by creating and analyzing a setup which measures microwave power of 1 watt and 5 watts.

Figure 3.5 shows two methods of making a 1 watt test setup. The two methods are very similar, except for the means by which the energy is moved from the device under test to the power sensor (thermistor mount). In Figure 3.5(a), a 20 dB coupler and a 1 watt attenuator are used to reduce the +30 dBm signal to a safe level for the thermistor mount. In Figure 3.5(b), a 30 dB attenuator with a 2 watt power handling capability is used to accomplish the same purpose. Let us look at each of these setups, see how they come about, and see how to use them.

In Figure 3.5(a), we will assume that the only devices available for use in the setup are 1 watt attenuators, a 20 dB directional coupler, and a 2 watt 50 ohm termination. With these limitations, there are certain areas that require some consideration. To examine these areas more closely, we will set up the equivalent circuits in Figure 3.6. In Figure 3.6(a) we have the +30 dBm (1 watt) signal from the device under test as an input to the directional coupler. At the coupled port, there is a +10 dBm (10 mW) signal which could be used safely for the input to the thermistor mount (power sensor). This is a choice that must be made based on how much margin you personally feel that you must have. The straight-through port of the coupler is assumed to have 0.5 dB insertion loss, so there is a +29.5 dBm (0.9 watts) signal at the 50 ohm termination. With this level present at this point, the 2 watt termination is more than adequate to do the job. There should be no concern as to the power handling capability of the directional coupler, since most coaxial couplers are rated at 200 watts CW.

We mentioned above that the coupler output of +10 dBm is a safe level for the power sensor. The addition of an extra attenuator between the coupler and the power sensor, however, adds two important features to the set-up. First, there is an additional 10 dB safety margin in the input power to the power sensor. Second, the reduction

Power Measurements

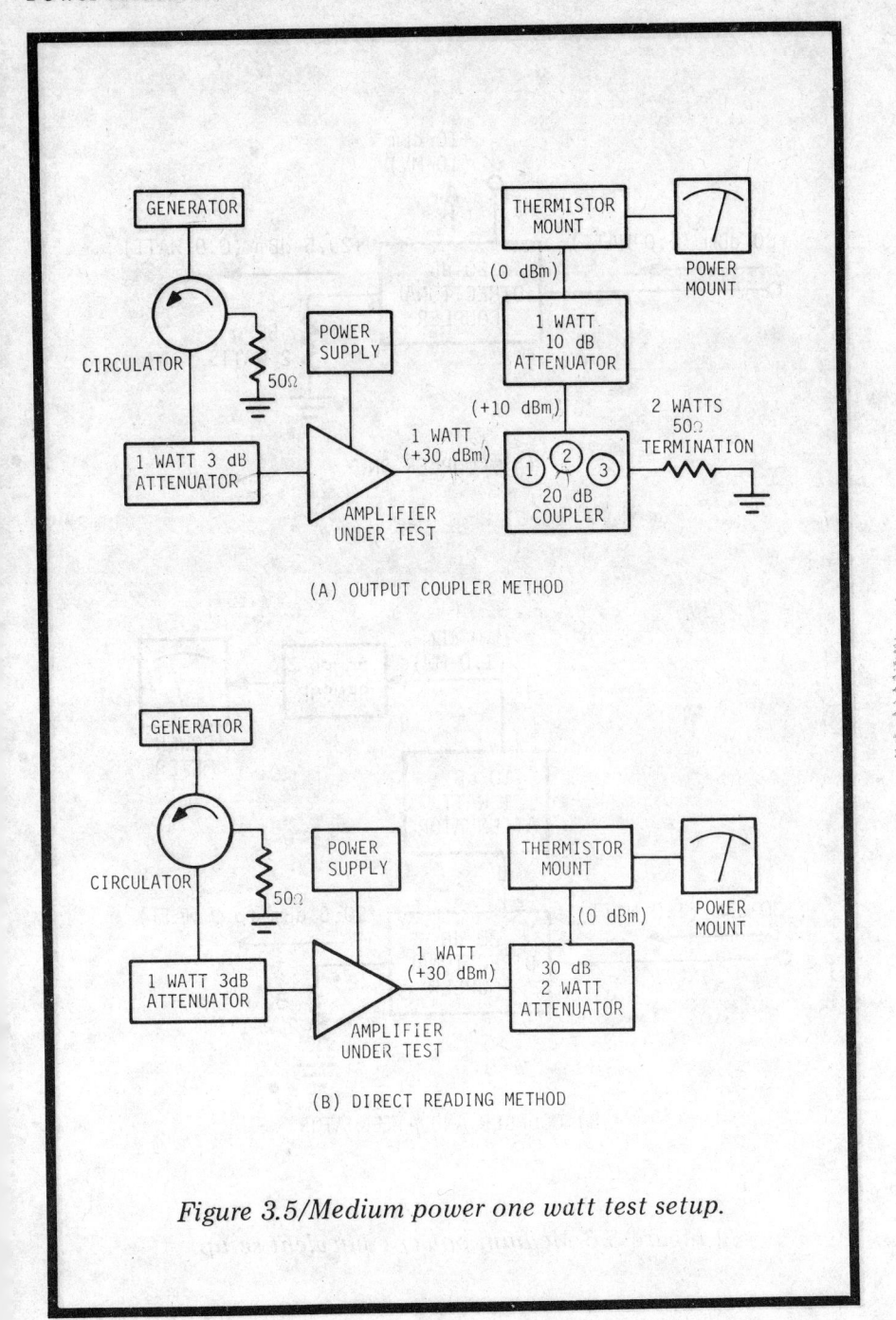

(A) OUTPUT COUPLER METHOD

(B) DIRECT READING METHOD

Figure 3.5/Medium power one watt test setup.

Figure 3.6/Medium power equivalent setup.

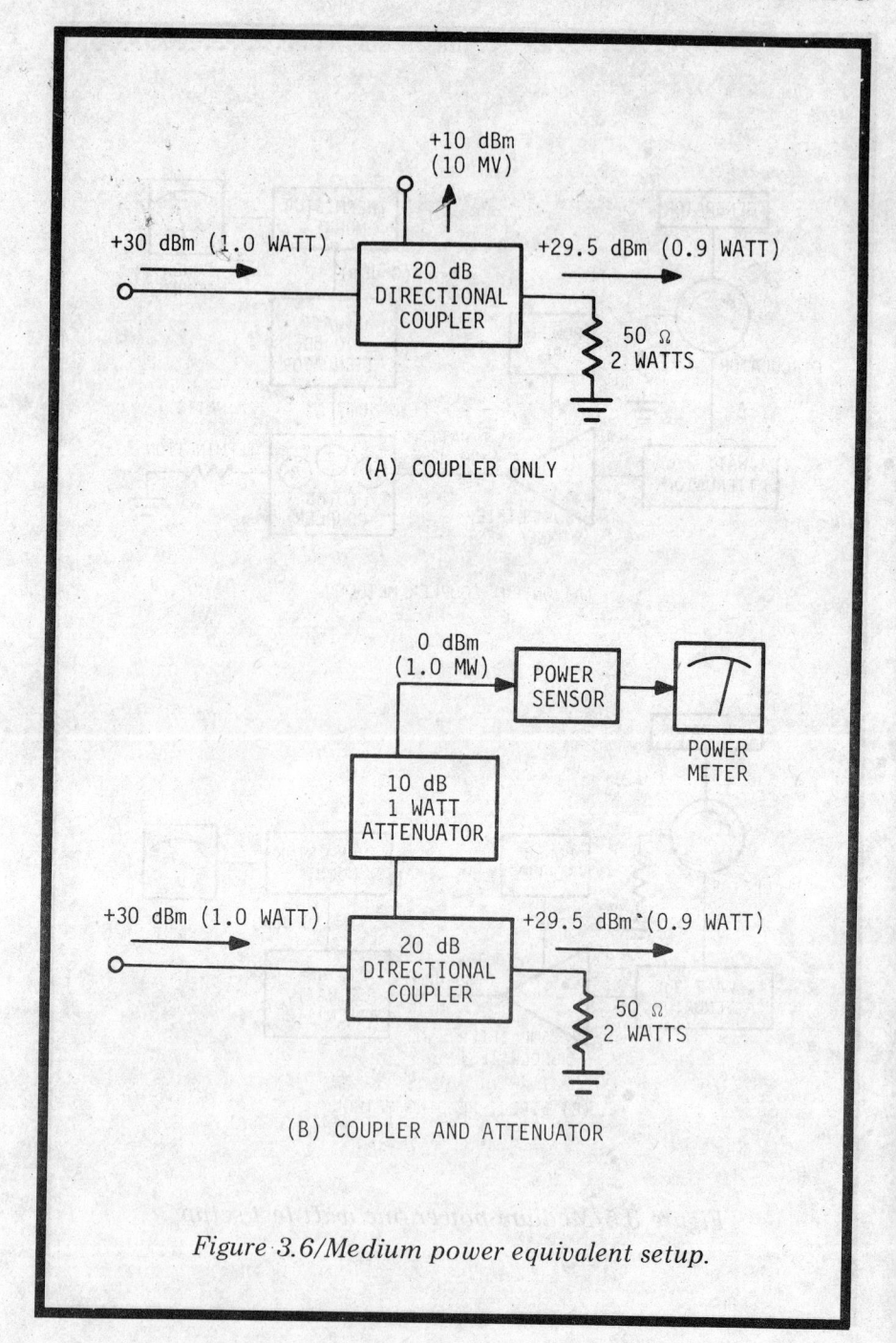

Power Measurements

in level allows readings to be taken at a scale that is in the mid-range of most power meters. This improves accuracy since you are not taking readings at either extreme (low end or high end of the meter scales). This arrangement is shown in Figure 3.6(b). We are now applying a 0 dBm signal to the thermistor mount, which is well below the maximum allowable input to most power sensors (thermistor mount). The lowest maximum power rating of commercial power sensors is +10 dBm (10 mW), so the 0 dBm is a very safe level. You should, however, check your particular power sensor to see what its maximum input level is and adjust your test set-up accordingly.

Figure 3.5(b) shows a much simpler, and often much preferred, set-up for 1 watt power testing. Instead of a 20 dB coupler and a 1 watt 10 dB attenuator, the set-up uses a single 2 watt 30 dB attenuator to transform the 1 watt output of the test device to a 0 dBm signal to be applied to the power sensor. The value of attenuation of the pad (30 dB) is good, since we want to attain the 0 dBm for the power sensor, as mentioned above. Its 2 watt power rating allows it to be placed directly in line with the 1 watt output of the device under test with no effect at all on the pad. The power rating provides a very substantial safety margin for the device.

As mentioned above, the method of using a single series attenuator is much preferred when making a power set-up of this sort. There are many reasons for this. One is very obvious when looking at the set-up shown in Figures 3.5(a) and (b); this is the simplicity of the set-up. There are less components in the set-up and thus there are less chances for error.

Another reason for this arrangement being preferred is that calibration is simpler and thus the results are more accurate. Each component used in a test set-up has its own characteristics as a function of frequency and power. Very seldom will you find two that are the same. Therefore, when you have a directional coupler (with its characteristics) and an attenuator (with its characteristics) in the same set-up, you must consider both characteristics and make adjustments for them. However, when only a single attenuator is involved, you only need consider its characteristics and can adjust your set-up for them rather easily.

This idea of simplicity in set-up and in calibration can best be seen in a swept set-up using such an arrangement. Figure 3.7 is such a test set-up. The sweep generator used is part of an ALC loop identical to those covered earlier for previous chapters. You will recall that the unleveled RF output is sampled by the directional coupler and any

Figure 3.7/Swept medium power test setup.

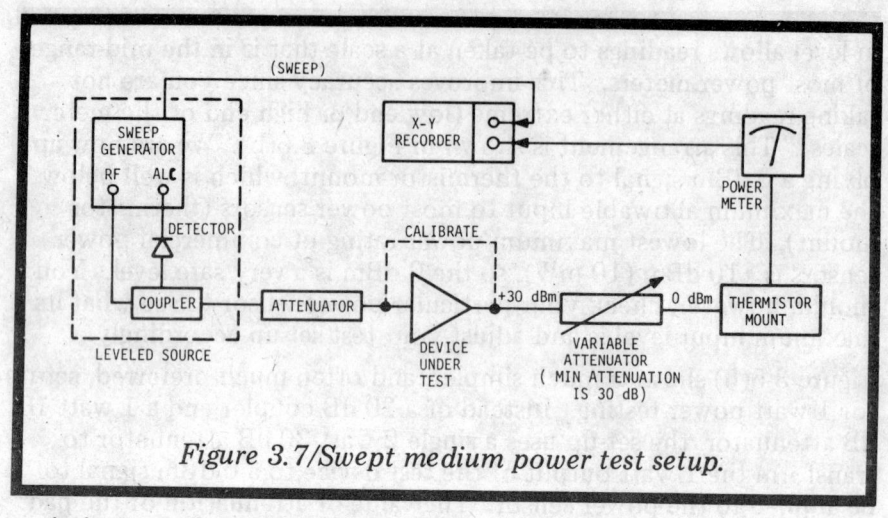

variations are detected and used to control a variable attenuator within the generator itself. This loop increases the output of the generator when a decrease in power is sensed, and decreases the output when an increase is sensed. The result is an output that is of a constant level over the entire frequency range of operation.

You may ask why it is always necessary to use an ALC loop when swept measurements are made. The answer is that it is *not* always necessary. For many narrowband applications, or when a generator output that is level within ± 0.5 dB or less is not of prime importance, there is no need for the extra components that make up an ALC loop. However, when wider bandwidths are swept or when you need a narrowband leveled input for a circuit or system, the ALC loop is a must. We use it in Figure 3.7 so that the power meter does not have to be switched to different ranges to accommodate a wide variation in power due to an unleveled sweep.

The attenuator prior to the device under test is there to improve the match between the device and the generator. The attenuator following the device under test is the same one we referred to in Figure 3.5(b) to decrease the level of the device output and protect the thermistor mount. It is variable in this case to allow the operator to set in a group of power levels on the X-Y recorder. You will note, however, that there is a requirement placed on this attenuator. That requirement is that the minimum amount of attenuation you should set in is 30 dB. This will ensure that the thermistor mount will see a level no higher than 0 dBm. In order to meet the power handling requirements, you may have to put a fixed value attenuator before the variable attenuator. A 2 watt 10 or 20 dB attenuator should do the job.

Power Measurements

A brief description of a calibration and measurement procedure will show how the set-up operates.

Calibration

(1) Connect the test set-up as shown in Figure 3.7 with at least 30 dB set into the variable attenuator. (Connect a cable in the "calibrate" portion of the set-up in place of the device to be tested.)

(2) Turn on the ac power to the sweep generator, power meter, and X-Y recorder. (Leave the generator in the *standby* or *RF off* position.) Place the generator sweep control to *manual sweep*.

(3) When the equipment has gone through any warm-up delays, turn the *RF on* and set the X-Y recorder pen to the top left hand corner of the paper. Manually sweep the generator to set up the right hand edge and also to be sure the sweeper is leveled.

(4) Put the sweeper in the *trigger* mode, set the sweep speed for approximately a 5 second sweep time and record the first power level. Repeat this procedure for a variety of power levels.

(5) Set the attenuator to its original setting (as in step 1) and put the sweeper in a *standby* position. Remove the "calibration" cable.

The set-up is now calibrated and we can proceed with the measurement.

Measurement

(1) Place the device to be tested as indicated in Figure 3.7 and apply any external voltages it may need to operate.

(2) Turn the RF on the sweep generator; change color pens from the calibration run; turn the *servo* on the X-Y recorder; press the *trigger* button on the sweeper; and record the power level of the device under test.

You can see from the previous discussion how the attenuator can be used for 1 watt (and below) power readings for both CW and swept applications. The main precaution that was stressed was to check the power rating of the attenuator so that they would not be destroyed. The versatility of this type of set-up was also shown when we indicated that a fixed 10 or 20 dB attenuator with a 2 watt rating in conjunction

112 HANDBOOK OF MICROWAVE TESTING

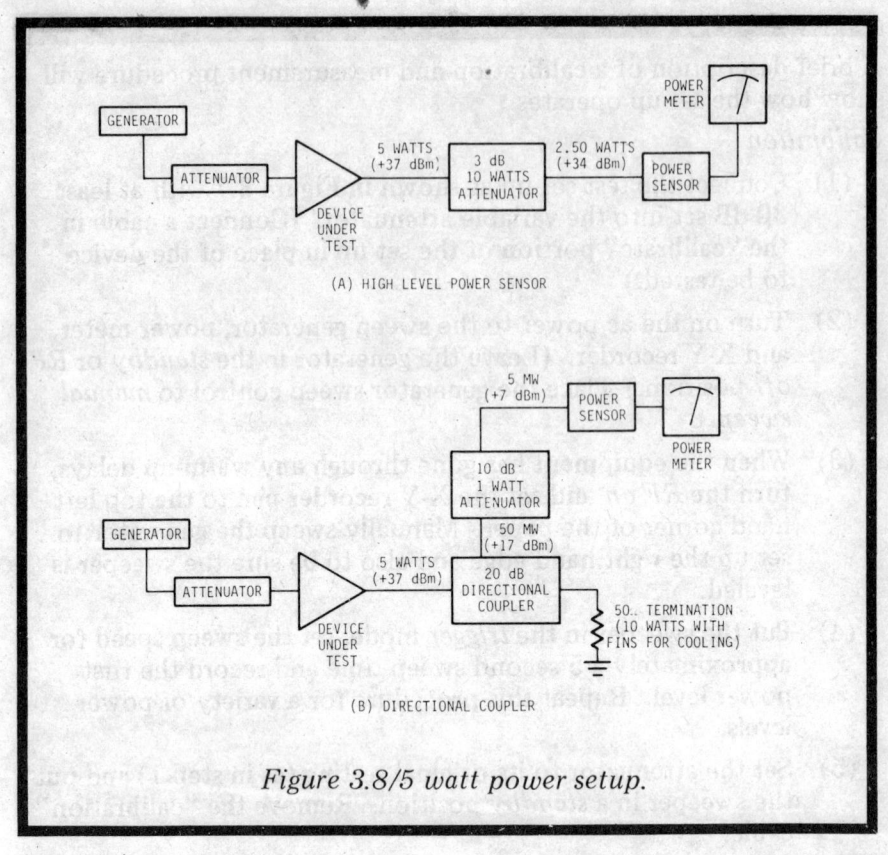

Figure 3.8/5 watt power setup.

with a lower power (less expensive) variable attenuator could be used to obtain accurate and fast swept power readings at the 1 watt level.

We have thus far covered the low end (100 mW) and center (1 watt) portions of what we have referred to as the *medium power* measurements. Now let us proceed one step farther and investigate a 5 watt set-up. The 5 watt level is chosen instead of the upper limit of 10 watts to illustrate how your thinking must change when you move only a short way from the 1 watt level. These same principles will apply between 5 and 10 watts also, only they become more critical. These higher powers will be covered in the next section.

Figure 3.8 shows two methods that can be used to measure 5 watts of microwave power. The first method is a result of modern day technologies which make direct power readings to 3 watts possible. Figure 3.8(a) shows a 5 watt device attenuated 3 dB by a 10 watt attenuator. This type of attenuator can vary greatly in size, depending on the connectors you need and the manufacturer. If you are able to

Power Measurements 113

use SMA connectors, the attenuator can be as small as 2 to 3 inches in length and approximately 0.5 inches in diameter. Restrictions such as type "N" connectors make the attenuator grow in size very rapidly.

With the signal attenuated by 3 dB, it can now be read directly by the power sensor/power meter combination. Until recently, the maximum level that could be placed directly on a power sensor was the 50 mW level printed in large red letters on the sensor. But, as the industry progresses and finds new methods, the levels go up and so do the accuracies. Accuracy increases primarily due to the fact that less complex test set-ups are needed, resulting in less components used. The lower the number of components used, the less chance there is for error in the system, since there are less areas where errors must be considered.

A set-up using a high level power sensor may not be possible for you to assemble, for the simple reason that you probably do not have such a power sensor in your lab. That is why Figure 3.8(b) presents an alternate method of measuring your 5 watt device (and higher).

This figure looks very similar to some we have discussed earlier. The main difference in this set-up is the power handling capability of the components involved. You will note, in particular, the 10 watt rating on the coupler's 50 ohm termination. This is necessary because almost the entire 5 watts of the device under test is fed directly to this termination (approximately 0.5 dB is lost in the coupler as insertion loss). Therefore, there must be an adequate safety factor in order to preserve the component.

The power handling requirement can be eased on this termination if it is too much of a problem for your particular set-up. The 3 dB 10 watt attenuator shown in Figure 3.8(a) can be inserted between the coupler and the termination. With this attenuator in the circuit, the power level at the termination is now +34 dBm (2.5 watts) and a more reasonable 5 watt termination can be used.

One point should be brought up before we leave the 5 watt set-up. That point is that we have been pretty much ignoring the power rating of the attenuator which is placed between the generator and the device under test. We have been concentrating, and rightly so, on the attenuator after the device we are testing. We are, however, fast approaching power levels where it will be necessary to supply a substantial amount of input power to obtain the required output power. There is a good chance that we should probably look closely at the attenuator as it appears in Figure 3.8. If you consider that 5 watts generated by a transistor stage or a complete amplifier will only have a gain of 5 or 6 dB, the input power (worst case) will be in the order of 1.5 watts.

This will necessitate the use of at least a 2 watt attenuator and maybe higher if you will be making a lot of tests that will require the set-up to be operating for long periods of time. So, you see it does not take much to place additional restrictions on the components used in power set-ups.

We made a statement at the beginning of this section that 95 percent of all microwave power measurements fall into the category of *medium power* (0 dBm to +40 dBm). You can see from our discussions that this is a very true statement. Your measurement may not be at 100 mw or 1 watt or 5 watts, but it undoubtedly falls between 1 mw and 10 watts. Thus, the setups outlined and the procedures discussed can be used with the same degree of success and accuracy.

3.3 High Power

There are very few areas in microwave testing that require more care in choosing components, and actually making measurements, than that of high power testing. We could probably expand that to include not only microwaves, but any area of electronic testing. There are special precautions to be taken in making high power measurements, but they should not cause you to fear the high microwave power. You should rather develop a respect for it. When you have eliminated the fear and attained the respect, you will be able to obtain accurate and reliable information from your test setup.

Two areas of high power testing are of prime importance and should be mentioned before any test setup is presented. They are component power ratings and test equipment maximum input ratings. We have been emphasizing this factor more and more through this chapter as the power levels being tested have gotten higher. They now are to a point where this factor is not only mentioned, but *shouted*. It is not a very pleasant experience to pump 50 watts of power into a power sensor rated at a maximum level of 300 mw and watch $500 go up in a puff of smoke. Your boss would not find it a very pleasant experience, either. So very careful consideration should be given to the power handling capability of the component used and the maximum input level for the equipment.

A high power setup which considers power handling capability and maximum input levels is shown in Figure 3.9. It also shows what can be accomplished when high power must be measured with limited components to make the setup. This particular setup is one that was actually constructed to check out a 100 watt amplifier. At the time it was to b

Power Measurements

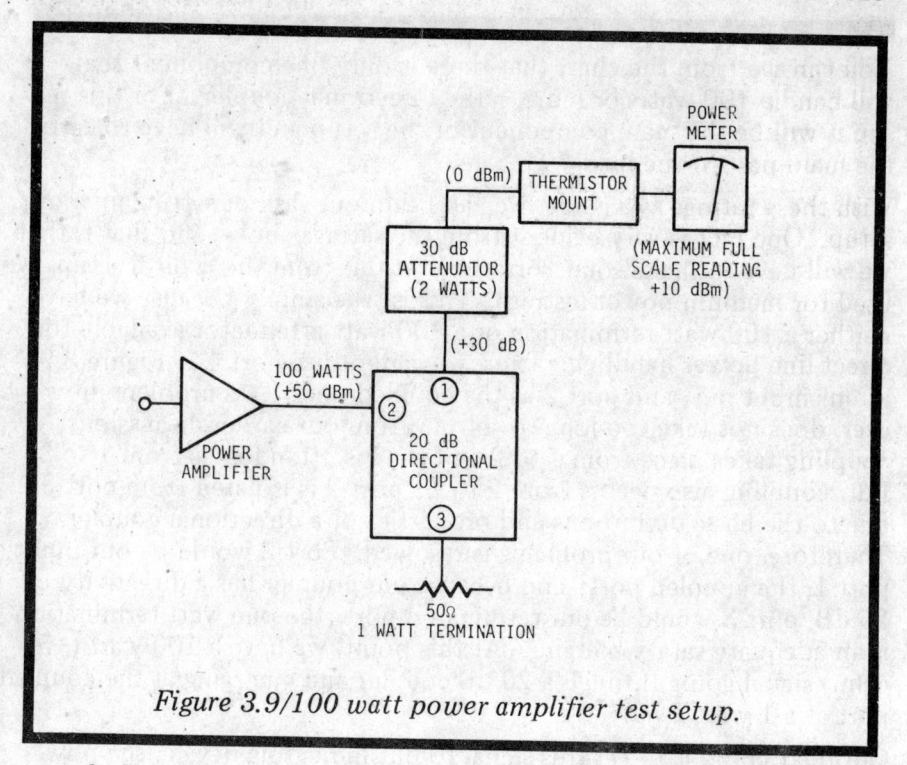

Figure 3.9/100 watt power amplifier test setup.

tested, there were limited components available, including an HP432A power meter with a 478A thermistor mount; a series of 2 watt attenuators with values of 6, 10, 20, and 30 dB; a type N 1 watt 50 ohm termination; and a 20 dB coaxial coupler in the frequency range of interest. Of course, there was a generator available which would drive the 100 watt amplifier; however, for the purposes of the setup we are only concerned with how to measure the resulting 100 watt level.

The very first area to be considered, of course, is the power handling capability of the available components. The following table lists each component and either its power rating or the maximum input it can handle.

Component/Device	Power Rating
Attenuators	2 Watts
Termination	1 Watt
Directional Coupler	2000 Watts, Forward; 500 Watts, Reverse
Thermistor Mount	30 mw (+15 dBm) Maximum Input Level

You can see from the chart that there is only one component that will handle 100 watts or more — the directional coupler. For this reason it will be the main component of the setup and will have to carry the main part of the load.

With these ratings as a guide, we can begin our task of setting up a test setup. One fact is very evident from the ratings above, and that is that we will have to make some sort of variations from the typical setups we used for medium power testings. This is true simply because we have neither a 100 watt termination or a 100 watt attenuator available for direct line power handling. Thus, we cannot use port 1 in Figure 3.9 as our input port and port 2 as the coupled one. This problem, however, does not take too long to solve. From our earlier discussions, coupling takes place from 1 to 2 and the insertion loss is from 1 to 3. But, coupling also occurs from 2 to 1; port 2 is isolated from port 3 due to the basic definitions and properties of a directional coupler. Therefore, one of our problems is resolved. Port 2 would be our input; port 1, the coupled port; and because our coupler has a directivity of 25 dB, port 3 would be our terminated port (the one watt termination is an adequate safety margin). At this point, we have a 100 watt (+50 dBm) signal going through a 20 dB coupler and emerging at the coupled port at a 1 watt level.

Our next chore is to get this signal to our indicating device, the power meter. The thermistor mount can take only 30 mW (+15 dBm) before it burns up; therefore, we need a minimum of 15 dB of attenuation between the coupler and the thermistor mount. As an extra safety margin, 30 dB of attenuation is added, so that our thermistor mount has an input level of 0 dBm, well below its maximum rating. The power meter, when put on the 0 dBm scale, has a full scale reading of 100 watts — just what we needed to read. However, a much more accurate reading is obtained close to center scale. Therefore, we move the meter one scale higher to the +5 dBm scale — full scale is now 300 watts. Our desired reading is in the first 1/3 of the scale and our reading accuracy is increased.

We should note that since the 20 dB coupler is used, as opposed to a 10 dB coupler, the output level of 1 watt enables us to use the 2 watt attenuator with a large safety factor. This fact is emphasized below in the table which lists each component again, gives its power handling capability, and shows how much power it is handling in the setup.

Component/Device	Power Rating	Power Being Handled
Attenuator (30 dB)	2 Watts	1 Watt

Power Measurements

Figure 3.10/Alternate setup for 100 watt test.

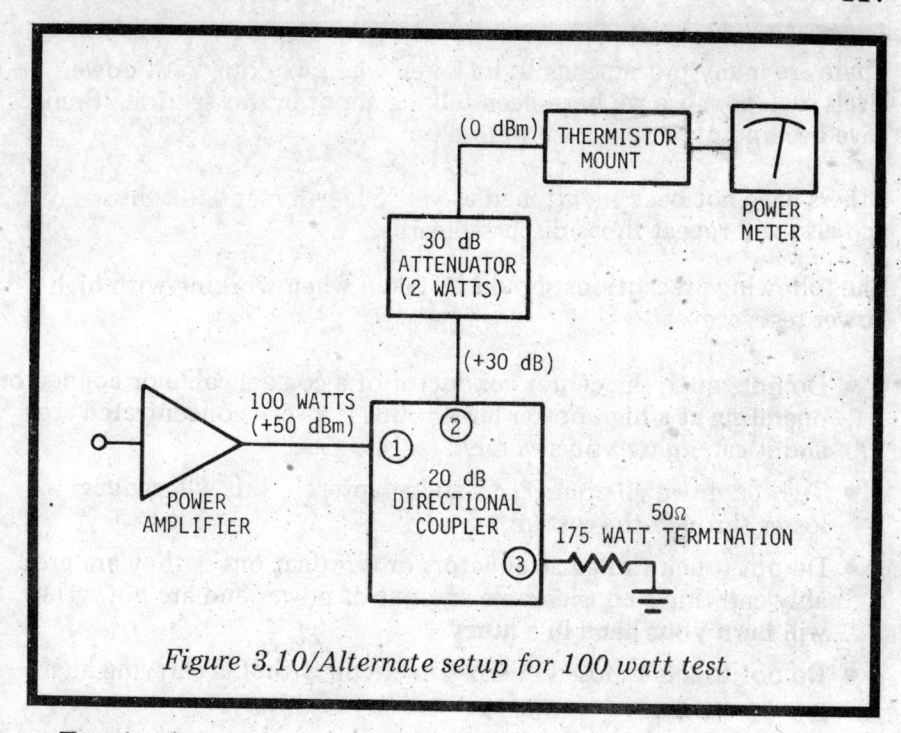

Termination	1 Watt	300 mw
Directional Coupler	2000 Watts, Forward	100 Watts
Thermistor Mount	30 mw (+15 dBm) Maximum Input Level	1 mw (0 dBm) Maximum Input Level

It is very obvious from the chart that nothing in the setup is being stressed or is having its ratings exceeded. As a matter of fact, when this particular setup was first put together and run, the 100 watt amplifier was on for 45 minutes and it was the only thing that got warm.

An alternate approach to measuring the 100 watts would be to have a high power termination. For example, you can buy a coaxial termination off the shelf that can handle 175 watts of power. With this termination your test setup would be as shown in Figure 3.10. You will notice the similarity of the setup to those used in medium power test setups. This is because it is identical to those setups with the exception of the power handling capability of the termination. Utilizing this high power load, the coupler now assumes a more familiar orientation — the input at port 1, the coupled port at port 2, and the straight through (or insertion) port at port 3.

There are many precautions to be taken when working with power levels such as those we have been talking about in this section. Some have been pointed out as we went along.

Others have not been mentioned as yet. We will mentaion those now and also will repeat those discussed earlier.

The following precautions should be taken when working with high power test setups:

- Do not touch the center conductor of a coaxial cable or connector operating at a high power level — this is a very concentrated area and it can injure you severely.
- Tighten down all connectors and adapters — this will reduce losses through the system.
- Do not touch finned attenuators or terminations — they are probably carrying a considerable amount of power and are hot. They will burn your hand in a hurry.
- Do not touch exposed microstrip circuitry that is carrying high power energy — once again, it will burn severely.
- Operate in a shielded area if possible — high power circuits have a tendency to radiate and may interfere with other circuits in the area.
- Always check the power handling capability of your components before applying RF power.
- Always check the maximum input level of your test instruments *before* applying RF power.
- Check the losses of all components (couplers, attenuators, etc.) when making up a test setup — the importance of accounting for all losses can be realized when you consider that a 0.5 dB loss at a 100 watt level is a loss of 11 watts.

These precautions should *always* be considered when making high power test set-ups. The fact that RF power can burn must be realized and respected if you are going to work at these power levels. If there is any doubt that RF can burn, you only need to consider a very useful and increasingly popular appliance found in many homes today — a microwave oven.

Power Measurements

3.4 Peak Power

Up to this point, all of the power measurements discussed have been for CW power; that is, levels of power that are continuously applied to the indicating device or power sensor. There was no on-off time, no rise or fall time of the power, and no specific length of time the power level was to be read. There are instances, however, where these factors are important to an efficient and accurate measurement. These instances are for *peak power* measurements.

Measurement of pulse (peak) power has been a frequent requirement in microwave work since the early development of pulse radar. There are a variety of ways of measuring this peak power which, of course, are different from those used in the CW measurement. We will cover seven of these methods in this section. They are:

- Average Power-Duty Cycle
- Direct Pulse
- Notch Wattmeter
- DC Pulse Power Comparison
- Barretter Integration - Differentiation
- Sample and Hold
- Direct Reading

3.4.1 Average Power-Duty Cycle Measurement

The first type of set-up is the Average Power-Duty Cycle method, as shown in Figure 3.11. It uses the idea that the peak power of a rectangular RF pulse may be determined by measuring average power, pulse width, and repetition rate. The product of pulse width and repetition rate is called the duty cycle of the pulsed source; it relates the duration of the pulse to the period of the pulse rate. Thus, the name "Average Power-Duty Cycle" specifies what actually is measured to obtain the peak power reading. Referring to Figure 3.11, it is very important that the maximum peak power and energy-per-pulse ratings of the particular thermistor being used are not exceeded — they are printed on the mount as watt-microsecond figures. If for any reason they are not obvious, you should consult a manual. Another point that should be made is that, when measuring peak power in a waveguide system, you should be sure that all flanges are carefully aligned and that the waveguide is clean and dry to avoid arc-over. This particular set-up is shown in waveguide.

Figure 3.11/Average power-duty cycle method of measuring peak power.

$$P_{PK} = \frac{P_{AVERAGE}}{(PRF \times \tau)(LOG_{10}^{-1} \frac{dB}{10})}$$

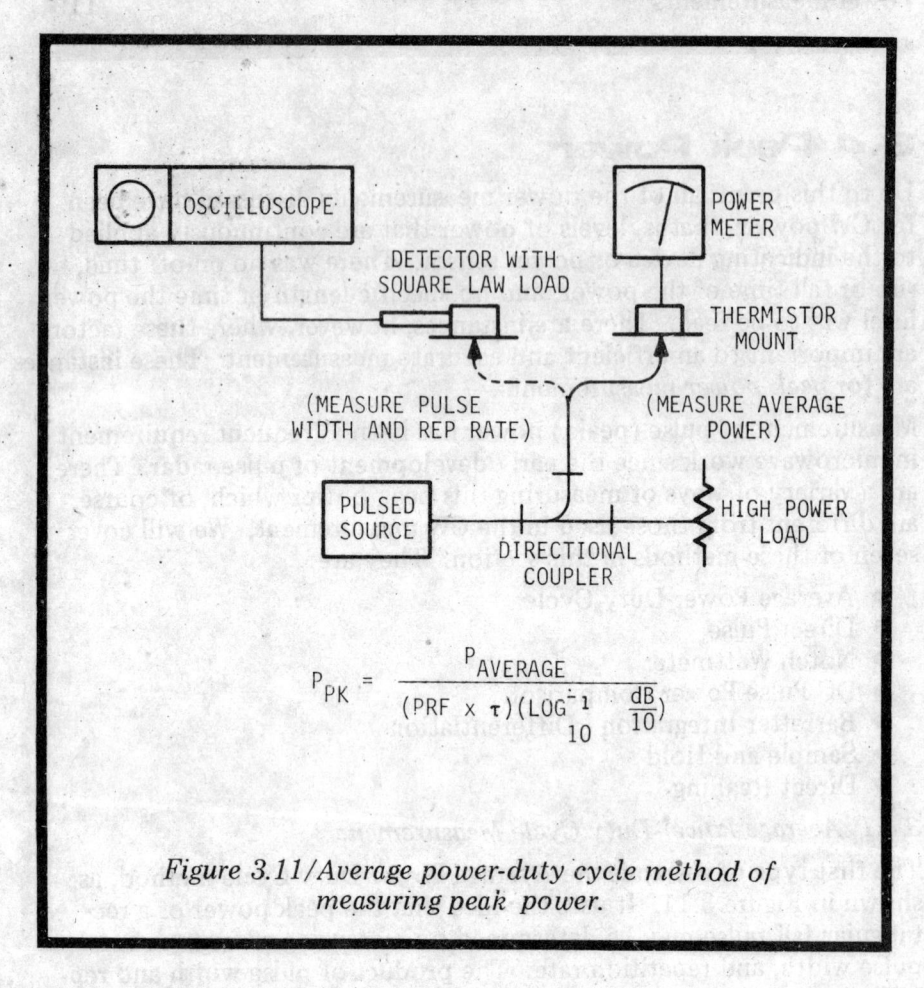

The procedure is as follows:

(1) Connect the crystal detector and its square law load to the coupled arm of the directional coupler. The square law load is used to improve the rise and fall times — which are very important in pulse measurements — of the detector.

(2) Measure with the oscilloscope and record the pulse repetition frequency (PRF) and the pulse width (τ).

(3) Replace the detector with the thermistor mount and the power meter.

(4) Note the average power in the meter.

Power Measurements

(5) Calculate the peak pulse power as follows:

$$P_{peak} = \frac{P_{average}}{(PRF \times \tau) \; \log_{10}^{-1} - \frac{dB}{10}}$$

when:

P_{peak} = peak power of the source

$P_{average}$ = average power read on the power meter

dB = coupling factor of the directional coupler

The term, $\log_{10}^{-1}$, indicates the anti-log of a number. To illustrate, let us take the log of a number and then reverse the procedure. If we want to determine $\log_{10} 200$, we can review a log table and find a value of 2.303 (because 200 is 2×10^2, which gives the first significant number, and the log of 2 is 0.303).

Now, if we were given a ratio of two numbers equal to 2.303 and we want to find its anti-log, we would reverse the order. We could look in tables or use a slide rule. For the latter, first find 0.303 on the log scale; looking on the C or D scale would give us 2, meaning that it is some number times 20. That number we have already determined to be 2. Therefore, the anti-log of 2.303 is 2×10^2, or 200. Written mathematically:

$\log_{10} 200 = 2.303$

$\log_{10}^{-1} 2.303 = 2 \times 10^2 = 200$

When using a calculator this can be done directly.
In the case above there is another (—) sign in front of the dB/10 term, meaning that the power of 10 is negative. Thus, $\log_{10}^{-1} (-2.303) = 2 \times 10^{-2} = 1/200$.

To illustrate this method let us show an example. Suppose we took the following readings:

PRF = 1 kHz

τ = 1.5 μs (1.5×10^{-6})

$P_{average}$ = 7.5 mW (7.5×10^{-3})

Coupling factor = 40 dB

Figure 3.12/Direct pulse power method.

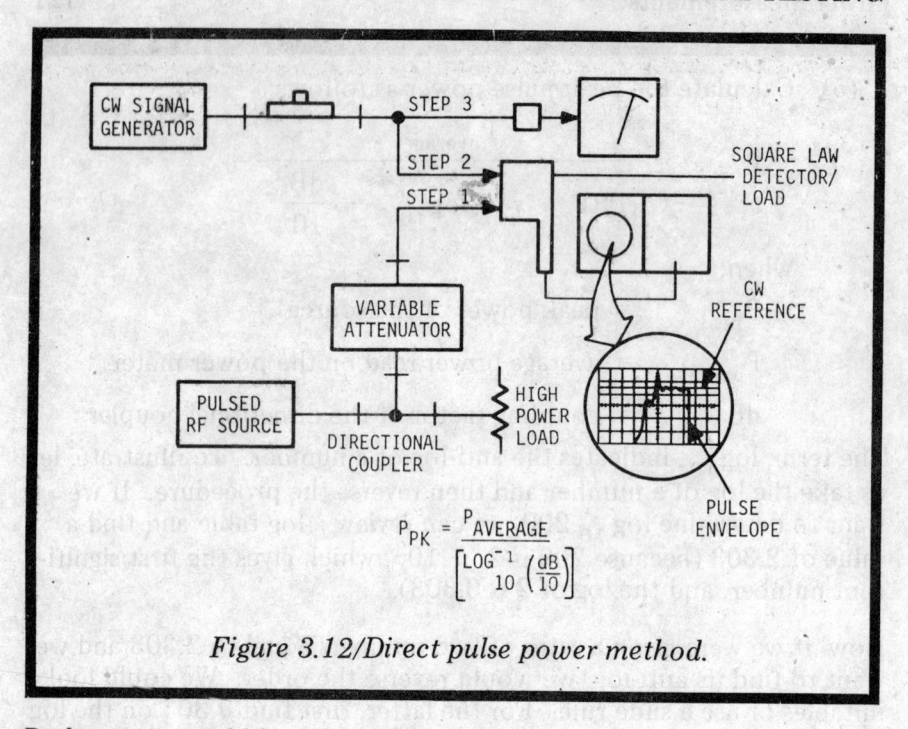

$$P_{PK} = \frac{P_{AVERAGE}}{\left[\text{LOG}^{-1}_{10}\left(\frac{dB}{10}\right)\right]}$$

Peak power would be calculated as:

$$P_{peak} = \frac{7.5 \times 10^{-3} \text{ watts}}{(1 \times 10^3)(1.5 \times 10^{-6}) \quad \log^{-1}_{10} - \frac{40}{10}}$$

$$= 50 \text{ kW}$$

This seems like an awfully lot of power for a system to read, but you must remember that this is not continuous power. The *duty cycle* is very short. This is the product of the pulse width (1.5×10^{-6} seconds) and the pulse repetition frequency (1×10^3). The total duty cycle is 1.5×10^{-3}, which is 0.0015. This corresponds to 0.15 percent of the time the power is on. So you see, the 50 kW does not seem as large when viewed this way.

3.4.2 Direct Pulse

The peak power range of the system shown in Figure 3.11 is around 100 kW at X-band, limited by the maximum power ratings of the termination, thermistor mount, detector, and coupler.

An improved waveguide set-up for peak power measurement of non-rectangular pulses or multipulse systems is shown in Figure 3.12. In

Power Measurements

this arrangement, the pulse amplitude is first detected by a fast response crystal detector with a square law load. The detected pulse amplitude is noted on the scope and then accurately duplicated with a known CW power.

The procedure is as follows:

(1) Connect the equipment as shown in Figure 3.12, Step 1. Set the attenuator at a convenient reference to limit the expected peak power to about 1 mW.

(2) Adjust the oscilloscope trigger and vertical sensitivity controls for a stable pulse display. Use dc coupling at the oscilloscope vertical input, and make no further adjustments to the oscilloscope sensitivity once the reference pulse amplitude is set.

(3) Move the detector/square law load combination to the output of the slide screw tuner and signal generator, as shown for Step 2.

(4) Adjust the signal generator for a CW output frequency equal to the pulse source's RF frequency.

(5) Adjust the slide screw tuner for maximum deflection of the oscilloscope trace. Do not adjust the vertical sensitivity. Use the output attenuator on the generator to keep the trace on the screen during these adjustments.

(6) Set the generator for a trace deflection on the oscilloscope equal to the pulse amplitude noted in Step 1. Make no further adjustments to the output attenuator.

(7) Connect the thermistor mount and power meter to the generator as shown for Step 3.

(8) Re-adjust the slide screw tuner for a maximum reading on the power meter and note the average power reading.

(9) Calculate the peak power from the equation:

$$P_{peak\ (source)} = \frac{P_{average}}{\log_{10}^{-1} - \frac{dB}{10}}$$

where:

dB = coupling factor of the directional coupler (including the variable attenuator setting) in dB

$P_{average}$ = average power measured on the power meter

An example would be given:

$$P_{average} = 0.5 \text{ mW}$$
$$dB = 80 \text{ dB}$$

$$P_{peak} = \frac{5 \times 10^{-4} \text{ watts}}{\log_{10}^{-1}\left(-\frac{80}{10}\right)} = 50 \text{ kW}$$

So we have two set-ups using different methods to arrive at the same figure. The set-up you use will depend on your particular requirements.

3.4.3 Notch Wattmeter

The notch wattmeter system is used in applications where the average power in a pulsed RF wave is too low to be measured by a power meter. Various null, or notch, techniques are possible; one is shown in Figure 3.13. The RF pulse to be measured is compared with a CW RF signal which is pulsed off to form a notch. Waveform (a) in Figure 3.13 represents the detected CW signal from the signal generator. Waveform (b) shows the detected RF pulse and the detected notch when they are not coincident. At waveform (c), the two waves are coincident and the pulse to be measured is placed in the notch. Coincidence is obtained by adjusting the pulse width, delay, and repetition frequency at the pulse generator. The signal generator has been adjusted to be equal to the pulse power of the pulse under consideration. The peak power can be determined from the calibrated output of the signal generator or from the reading of the power meter to which the two signals are simultaneously applied. When the latter method is used, the following formula applies:

$$P_{peak} = \frac{P_{power\ meter}}{\log_{10}^{-1}\left(-\frac{dB}{10}\right)}$$

If the power read on the meter were 1 mW and the total attenuation between the source and the thermistor mount were found to be 73 dB, the peak power would be:

$$P_{peak} = \frac{1 \times 10^{-3}}{\log_{10}^{-1}\left(-\frac{73}{10}\right)} = 20 \text{ kW}$$

This set-up is one that is commonly used for systems that have a low level of power available to be read.

Power Measurements

Figure 3.13/Notch wattmeter setup.

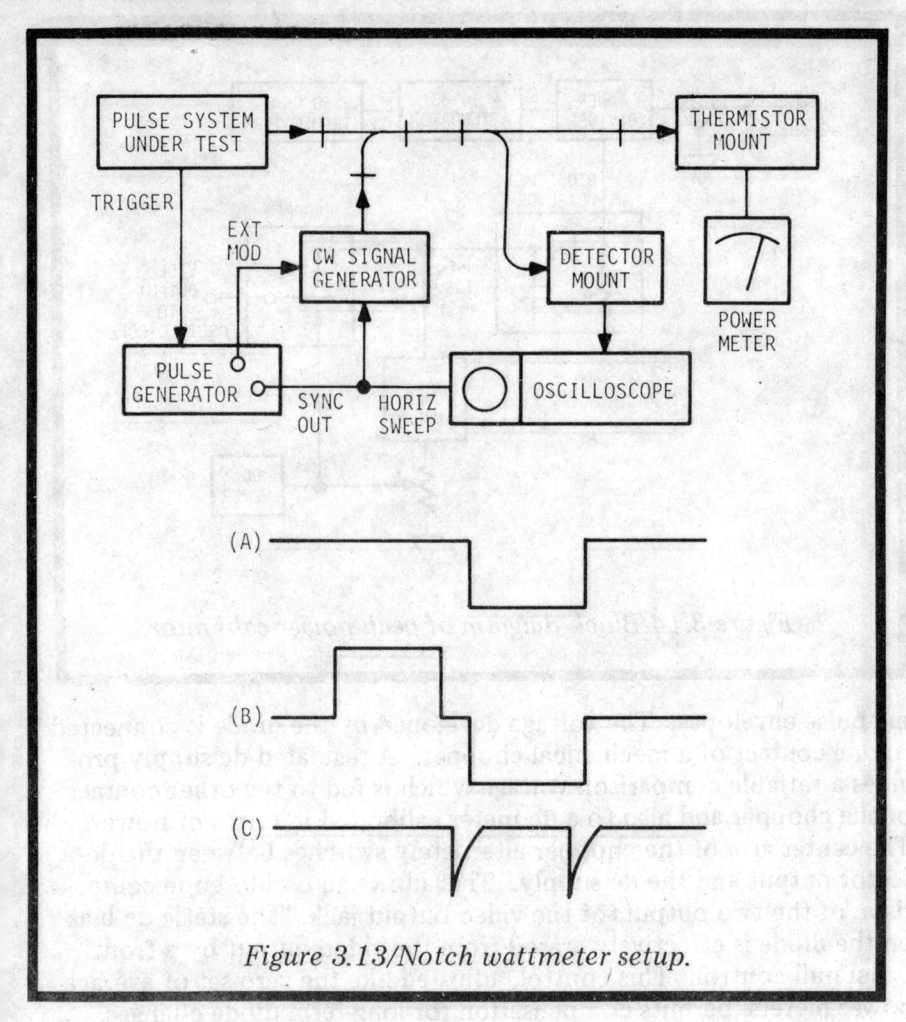

3.4.4 DC Pulse Power Comparison

An example of the DC Pulse Power Comparison technique is the use of the HP8900B Peak Power Calibrator. Unlike some systems, the 8900B does not rely on pulse width or on repetition rate measurements for its accuracy. The 8900B Peak Power Calibrator provides the opportunity to measure either pulse or peak envelope power.

The block diagram of the HP8900B, Figure 3.14, shows that a power divider splits the input pulse and feeds part of it to a diode peak detector. The peak detector develops a dc level proportional to the peak voltage of the input RF pulse. The diode is forward-biased to bring its operating point to a satisfactory impedance level for follow-

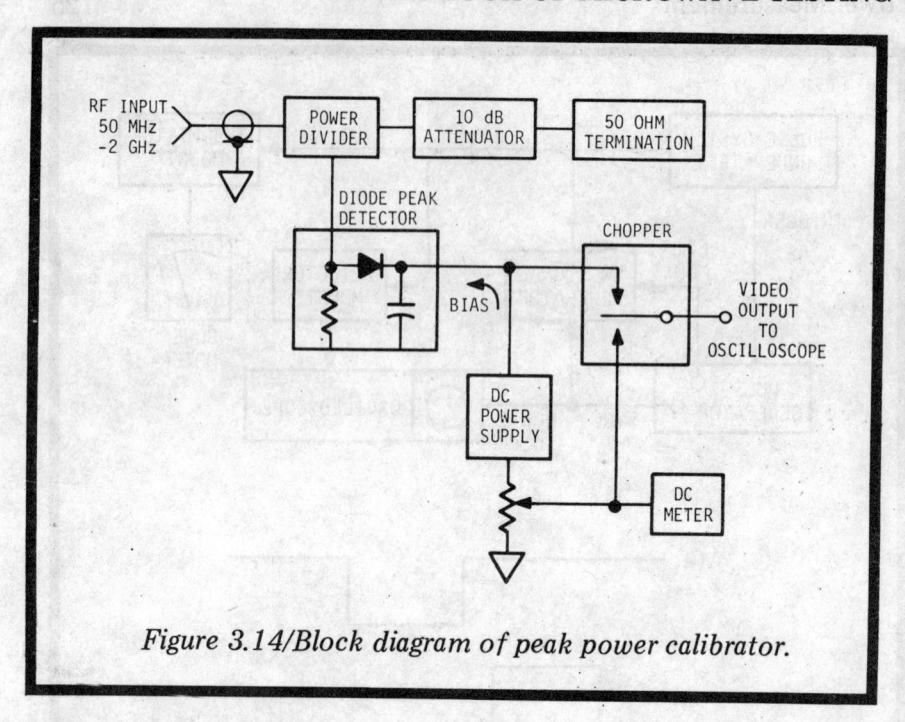

Figure 3.14/Block diagram of peak power calibrator.

ing pulse envelopes. The voltage developed by the diode is connected to one contact of a mechanical chopper. A regulated dc supply provides a variable comparison voltage which is fed to the other contact of the chopper and also to a dc meter calibrated in terms of power. The center arm of the chopper alternately switches between the detector output and the dc supply. This allows an oscilloscope comparison of the two outputs at the video output jack. The static dc bias on the diode is effectively erased from the video output by a front panel null control. This control, adjusted like the zero-set of average power meters, permits compensation for long-term diode changes.

In operation, unknown pulsed power is applied to the 8900B input, the envelope is detected and displayed on an oscilloscope. The dc voltage supplied by the 8900B is adjusted for coincidence with the detected peak pulse amplitude. At this point the dc voltage is equal to the peak pulse voltage input, and the meter indicates the equivalent peak envelope power.

Although some of the power is directed to the peak detector, the remainder of the power is fed to a 10 dB attenuator and 50 ohm termination. This power is convenient for calibrating the instrument. If the 50 ohm termination is replaced by an accurate average power

Power Measurements

measurement system and a CW power source is applied to the 8900B input connector, the effect of applied power can be monitored on the average-reading CW power standard and the peak-reading diode detector simultaneously. Only the attenuation between the input connector and the CW standard needs to be known to determine the effect of a known power level on the peak detector. The 10 dB pad reduces the input power so average-reading power meters may be used for calibration.

The 8900B enables convenient and rapid power measurements of pulses greater than 0.25 μs in duration at RF frequencies of 50 MHz to 2 GHz. Pulse repetition rates (PRFs) up to 1.5 MHz may be measured because of the wideband detector. The 0.25 μs specification gives the peak detector time to charge to the true peak value of an input pulse; 0.1 μs is a more realistic limit with normal cable lengths and oscilloscopes connected to the video output. Overall accuracy is ± 1.5 dB, but an optional correction chart for frequency effects reduces this error to ± 0.6 dB maximum. This accuracy is based on an absolute worst-case error analysis which includes the following sources of error: (1) attenuation measurement between the input connector and 10 dB pad output; (2) the average power measurement; (3) mismatch; (4) meter tracking and repeatability; (5) aging effects on detector diode; and (6) readout resolution.

A test set-up for a peak pulse power is shown in Figure 3.15. The directional coupler is a 50 dB coupler to ensure that the calibrator input is no greater than 200 mW (+23 dBm) peak. The test procedure used is as follows:

(1) Calibrate the peak power calibration as outlined in its operating manual. Leave the pulsed source power off.

(2) Set the oscilloscope Vertical Sensitivity to about a 50 mV/cm range and use a dc input setting.

(3) Set the oscilloscope sweep time and trigger for a free running trace. The display should appear as two horizontal lines on the CRT.

(4) Adjust the Null on the calibrator until the two horizontal lines are on top of one another (balancing out the detector diode bias from the video output waveform, which might otherwise be read as RF power).

(5) Turn on the pulsed RF source.

(6) Synchronize the oscilloscope for a stable display.

(7) Using the Coarse and Fine controls on the calibrator, adjust the dc reference voltage (which now appears as a straight line on the screen) until it coincides with the peak of the displayed pulse. The scope display should be as in Figure 3.15.

(8) Read the peak power on the calibrator and use the formula below to find the actual power:

$$P_{peak\,(source)} = \frac{P_{peak\,(calibrator)}}{\log_{10}^{-1}\left(-\frac{dB}{10}\right)}$$

where

$P_{peak(calibrator)}$ = reading of peak power on the peak-power calibrator

dB = coupling factor of the directional coupler

As an example of the above procedure, suppose we read 135 mW on the calibrator and the coupling factor is 50 dB. Then,

$$P_{peak\,(source)} = \frac{135 \times 10^{-3}}{\log_{10}^{-1}\left(-\frac{50}{10}\right)} = \frac{135 \times 10^{-3} \text{ watts}}{1 \times 10^{-5}}$$

$$= 13.5 \text{ kW}$$

3.4.5 Barretter Integration-Differentiation

Peak power meters are available which operate in conjunction with barretter mounts and special barretter elements that integrate the pulsed power input. The integration comes from the barretter's comparatively long thermal time constant (greater than 120 μs). The barretter forms one leg of a Wheatstone bridge that receives excitation and bias current from a constant current supply. Input pulses to the barretter change the element resistance, resulting in a bridge output signal which is the integral of the pulses. By amplifying and differentiating the bridge signal, the input pulse shape is reconstructed. The reconstructed pulses are peak detected in a voltmeter circuit calibrated in peak power. Pulse sensitivity characteristics of the barretter must be known in order to translate the relative pulse amplitude to an absolute power level. Calibration of typical peak power meters using the Barretter Integration technique is based on: (1) previous measurements of representative barretter output-per-milliwatt input-per-microsecond (pulse sensitivity); and (2) a calibrating signal generator used in adjusting amplifier gain so that a predetermined level is indicated on the meter before measuring pulse power.

Power Measurements 129

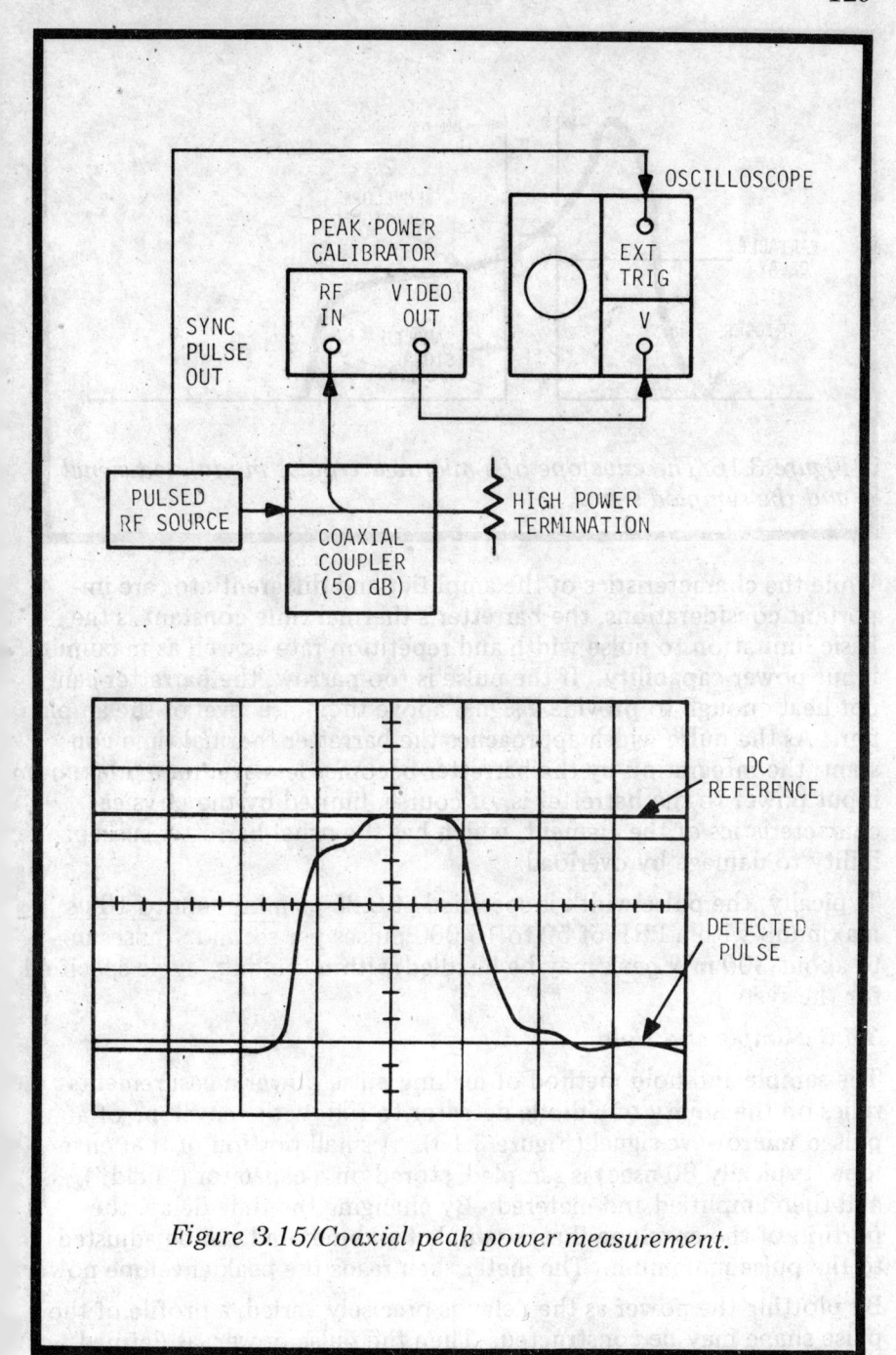

Figure 3.15/Coaxial peak power measurement.

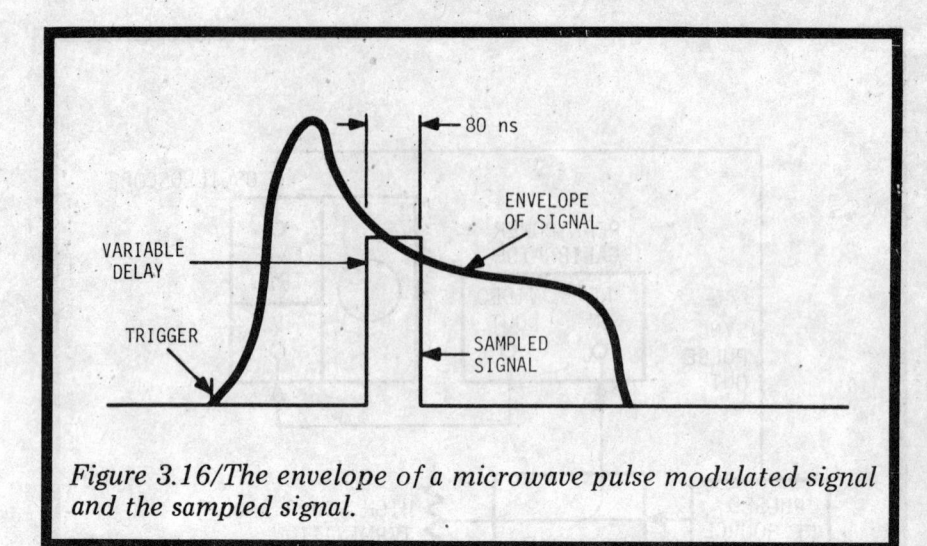

Figure 3.16/The envelope of a microwave pulse modulated signal and the sampled signal.

While the characteristics of the amplifier and differentiator are important considerations, the barretter's thermal time constant is the basic limitation to pulse width and repetition rate as well as maximum input power capability. If the pulse is too narrow, the barretter cannot heat enough to provide a signal above the noise level of the amplifier. As the pulse width approaches the barretter thermal time constant, the integration by the barretter becomes less accurate. Maximum input power to the barretter is, of course, limited by the physical characteristics of the element, which has the usual barretter susceptibility to damage by overload.

Typically, the pulse width is specified at 0.25 μs minimum to 10 μs maximum with a PRF of 50 to 10,000 pulses per second. Pulses up to about 300 mW peak may be handled within the duty cycle specified for the system.

3.4.6 Sample and Hold

The sample and hold method of making pulse power measurements relies on the ability of a diode detector to follow the envelope of a pulsed microwave signal (Figure 3.16). A small portion of that envelope (typically 80 nsec) is sampled, stored on a capacitor ("held"), and then amplified and metered. By changing the time delay, the portion of the envelope that is sampled and metered can be adjusted to the pulse maximum. The meter then reads the peak envelope power.

By plotting the power as the delay is precisely varied, a profile of the pulse shape may be constructed. Then the pulse power, as defined earlier, could be calculated.

Power Measurements

The small percentage of the total time that the signal is being sampled results in an effectively large noise level and therefore a large minimum detectable signal. The CW diode detectors, with the techniques discussed in previous chapters, could measure levels from −70 dBm to −20 dBm, but this sampling technique covers from −20 dBm to +10 dBm. At these levels, the diode detector is not very square-law. But the amplifiers include shaping (gain designed to depend on signal level) to make the diode appear square-law and indicate the output power. Such square-law correction of diode output must be made before the averaging is done; that is, the correcting circuits must be quick enough to follow the modulation waveform. With this sample and hold technique, however, once the signal is "held" it is basically a CW signal and proper correction for diode square-law variation is possible. Still other shaping circuits can be used so the indicated value is in dBm or relative dB.

Measurements using this technique are not limited to pulsed signals. Sampling a CW signal is not only valid, but is also a convenient means of calibrating the meter. Pulses as narrow as 0.35 μs may be measured without correction and pulses as narrow as 0.2 μs may be measured with correction.

3.4.7 Direct Readings

The coaxial measurement referred to in Section 3.4.4 still has the inconvenience of necessitating the taking of a reading and the plugging of variables into a formula. What is really needed is a method by which you can obtain a direct reading of the peak power in a coaxial system; that is, a specific device which indicates peak power. One such unit is the Pacific Measurements Model 1018A peak power meter.

The Model 1018A measures RF pulses with widths as short as 0.25 μs. The instrument measures signals of amplitude between 10 μW (−20 dBm) and 10 mW (+10 dBm). The unit employs a 50 ohm matched coaxial RF detector to make the measurement, and its frequency range extends from 100 MHz to 12.4 GHz. The instrument responds to each pulse individually, and so is not sensitive to pulse width, repetition rate, duty factor, or pulse shape. The pulse is detected, sampled, and held and the data is stored digitally each time the unit is triggered. Display is in the form of a 3-digit readout.

These qualifications make the PM1018A very similar to the power meter we were talking about when covering CW power measurements. This instrument is even more convenient, since it has a digital readout instead of a meter that would have to be interpreted.

Figure 3.17/Direct coaxial peak power setup.

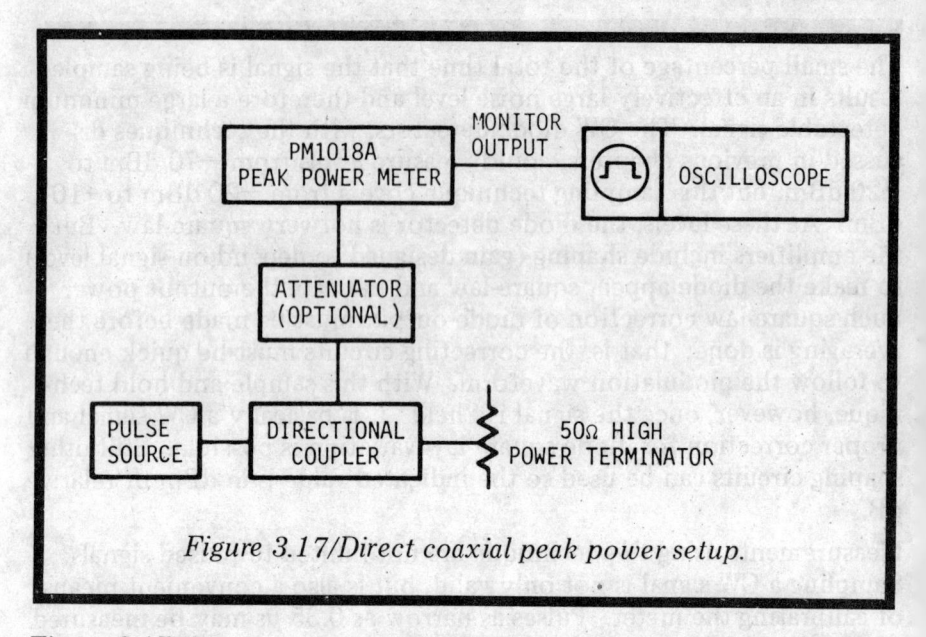

Figure 3.17 shows a power set-up in which the model 1018A RF Peak Power Meter could be used. One very important component is the directional coupler. The PM1018A by itself can read directly values of 10 microwatts to 10 milliwatts. However, a combination of external couplers or a coupler with an attenuator totalling 20, 40, or 60 dB permits readings of 10 milliwatts to 10 kilowatts. This region is the most useful, since most peak power measurements are in the hundred or kilowatt range.

Notice the great similarity between Figures 3.17 and, for example, Figure 3.5(a) — each has a source (pulse in Figure 3.17; CW in Figure 3.5); each has a directional coupler and an attenuator where power is coupled off the main line (effectively extending the range of the meter being used); and each has a 50 ohm termination. The only real variable is the power rating of each termination. The set-up in Figure 3.17 needs to have a higher power handling capability because of the increased power in the pulses. Off the coupled port in each figure there is a direct reading power meter. In Figure 3.17, we have the PM1018A peak power meter; in Figure 3.5, the HP432A (which we called a CW power meter, at the time). The real difference between the two set-ups is the addition of the oscilloscope in Figure 3.17. It is a very handy piece of equipment to have — it is much easier to know exactly where you are when you can see the pulse that you are measuring.

Power Measurements 133

Figure 3.18/PM1018 power meter.

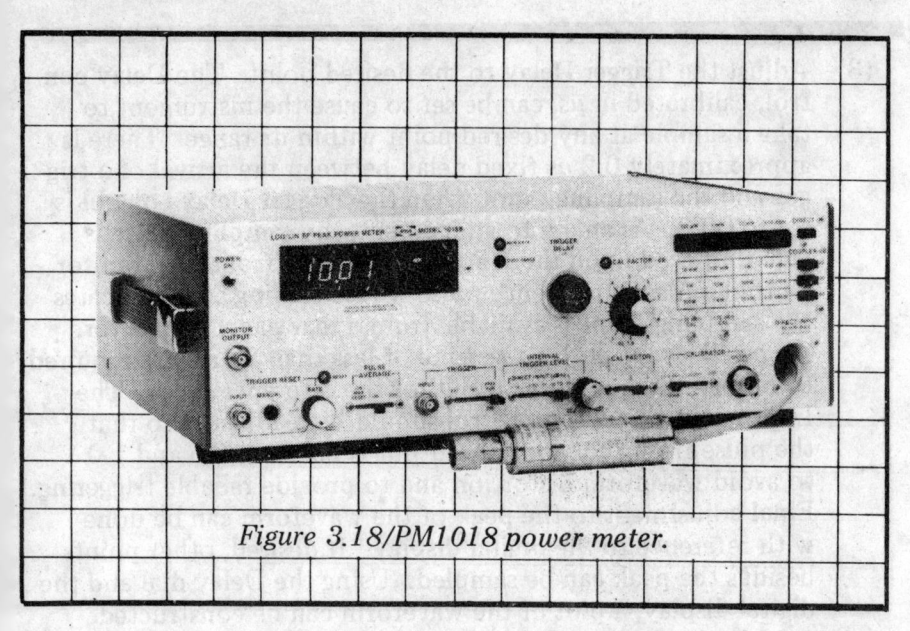

Figure 3.18 shows the front panel of a PM1018A; it helps in understanding what is being adjusted when the procedure is outlined. The following process is outlined because it is a general application — it can be used to measure CW or pulse:

(1) Connect the detector to the measurement point. Be sure that no more than 50 mW peak or CW is available at this point. If it is desired to measure power in excess of the +10 dBm capacity of the Model 1018A, connect suitable attenuators or directional couplers between the detector and the RF source. For example, to measure 1 kW, a 60 dB coupler would be appropriate. The test set-up is that of Figure 3.17.

(2) Set the Trigger selector switch as required. If pulse power is measured, either Internal or External triggering is necessary. For CW power, set it to Free Run. If external triggering is used, supply a pulse with 1 V or greater (positive) amplitude and a minimum duration of 200 nsec to the Trigger Input connector. If internal triggering is used, set the Direct Input (dBm) switch to the approximate level of the signal to be measured. Adjust the Level knob for reliable triggering. The Pulse Averaging switch should be set to "On Max Reset" to minimize the effects of noise.

(3) Adjust the Trigger Delay to the desired point. The Delay control, calibrated in µs, can be set to cause the instrument to take a sample at any desired point within its range. There is approximately 0.2 µs fixed delay between the arrival of a trigger and the sampling point when the Trigger Delay control is set to 00.0. As an aid to adjustment, the sampling pulse is superimposed upon the RF pulse envelope from the Monitor Output. Also, a reference pulse whose trailing edge indicates the sampling point is available from a **rear** panel connector. An oscilloscope with a rise time of less than 20 nsec is required to observe the monitor output, or the reference pulse. The Internal Trigger Level control should be readjusted so that the pulse from the Monitor Output is between 0.5 and 1.0 V to avoid waveform distortion and to provide reliable triggering. Final adjustment to the peak of the waveform can be done with reference to the digital display. If desired, other points besides the peak can be sampled. Using the Delay dial and the digital display, a plot of the waveform can be constructed (within the rise time limitations of the instrument).

(4) If the display properly indicates that you are reading the portion of the pulse that you need to read, you can take the peak power directly off the digital display.

Calibration Note — As mentioned above, to get the greatest accuracy from your test set-up, any components used to extend the range of the meter should be calibrated along with the meter. That is, if you are going to use a 20, 40, 60 dB coupler or any attenuation to measure, say, a 1 kW peak power, they should be calibrated with the meter.

The test set-up shown in Figure 3.17 has many advantages. The most important is that it is simple and easy to use. The main portion of the set-up (the meter and its associated coupler) is fully calibrated; you can see an accurate picture of the pulse you are measuring (provided you have the proper type of oscilloscope); and you can read the peak power directly without estimating between graticules on a meter. This type of meter and set-up finds wide application in those areas that deal with pulse systems, whether at a high or low power level.

Chapter Summary

The subject of power is one that must be considered very carefully. We have covered low level CW power, where attenuators are usually

Power Measurements

not used; medium power, where the power ratings of components become increasingly important; high power, where the danger of both heat and RF burns is present; and peak power, where up to 100 kW of power is measured at very short intervals. You can see that there are precautions and considerations in every area of microwave power, just as in other measurements. The one fact that stands out in this case is that if these precautions are not followed, you just might feel the effects of your mistake.

Noise Measure

4

ments

4. Definition

The noise characteristics of an amplifier, mixer, or complete system determine how well that component or system will perform. The ultimate sensitivity of a detector system is set by the noise presented to the system with the signal. In addition, any detection system inserts extra noise in its detector and amplification process. In RF microwave communications, radar, and electronic warfare systems, the weakest signal that can be detected is usually determined by the amount of noise added by the receiving system.

The list could go on and on, but the conclusion would be the same — noise is one of the most important parameters to characterize in a component or system. The question that now arises is: "Why is noise so important?" To answer that question, we will provide a definition of the term *noise*.

You can begin to understand the concept of noise if you relate it to interference. In this regard, consider an example. Suppose you are on the beach looking out over the ocean. You see a ship some distance offshore and are unable to tell what country it is from, since you cannot clearly see the flag it is flying. There are, on the beach, two telescopes you can use to try and make out the flag. You look through the first one and you can just barely see the ship because the lens has not been taken care of. It is dirty, scratched, and misaligned so badly that the telescope will not focus properly. As you look through the first telescope and focus it as well as you can, you will probably see a flag outline and maybe make out some colors on it, but you will not get a clear picture to allow you to identify it.

When you become totally discouraged in your efforts to identify the flag, you then try the second telescope. This one will focus very easily and has only a very few light lens scratches which are of little significance. You now can see the ship, the flag, and even the fact that the captain has a red beard. Your task can now be performed with a minimum amount of effort, since you have a system with a minimum amount of interference in it.

The example above illustrates the phenomena of interference as it applies to the eye of a human being. When there is a lot of interference, you cannot see very well, or maybe not at all. When there is little interference, the eye can see very well, very distinctly, and a very great distance. The same is true of a microwave component or system. If there is a lot of interference with its ideal operation (noise within the component or system), the device cannot "see" the microwave signals very well. That is, the noise masks the signal and the component or system does not know that it is there, and consequently cannot perform an operation on it (amplify, mix, etc.). This is why it is desirable for an amplifier or receiver to have very little internal noise. You can see that if there is an excess of internal noise (interference), your system will not be sensitive to low level microwave signals. This is how sensitivity and noise in a system are related. The higher the noise level, the worse the sensitivity.

We should be at a point where we can put together our information and present a definition of noise:

> *Noise is an internally generated interference which causes the circuit operation to be degraded from theoretical predictions.*

To further understand this definition and how it applies to individual components or systems, we will investigate two terms used to express noise: *noise factor* and *noise figure*.

The ideal limit of a system's sensitivity is set by the noise present at its input. In practice, however, the sensitivity is often limited by noise generated within the system itself, as we have previously discussed. The number used to indicate how closely the ideal is approached is called the *noise factor*, F. It is defined as the ratio of input signal-to-noise ratio to output signal to noise ratio. Expressed mathematically, it is:

Noise Measurements

$$F = \frac{S_i/N_i}{S_0/N_0}$$

where:

F	=	Noise Factor
S_i/N_i	=	Input Signal-To-Noise Ratio
S_0/N_0	=	Output Signal-To-Noise Ratio

The noise factor indicates the change in signal-to-noise ratio which occurs as a signal passes through a device. Thus, F is a *figure of merit* (ideally equal to one) which can be used to compare different amplifiers and receivers. Being a dimensionless quantity independent of bandwidth, noise factor is a much better means for comparison of receivers than sensitivity. Furthermore, with knowledge of a system noise factor and bandwidth, we can predict its sensitivity and how it might be improved by the addition of preamplifiers.

Noise figure is the logarithmic equivalent of *noise factor* and is expressed as:

$$NF = \log_{10} F$$
$$= 10 \log \frac{S_i/N_i}{S_0/N_0}$$

The term *noise figure* is a much more widely used term when characterizing noise. Noise figure, expressed in dB, is also referred to as a *figure of merit* of a device or system. This is easily understood, since it is merely a logarithmic representation of noise factor (which we have also called a figure of merit). The idea of a figure of merit is a very accurate appraisal, since the noise figure of a device is a measure of how the device handles its internal noise in order to produce an output that is as noise-free as possible. If the noise is dealt with effectively, it is a measure of merit of a device.

You will see the term *noise figure* used mostly to describe the input to a low level amplifier. Such areas as satellite communications have very strict requirements for pre-amplifier noise figures. This can be understood more easily when you consider that a satellite is in the neighborhood of 22,000 miles above the earth and the signal reaching the receiving antenna is not very large in amplitude. To complicate matters even further, the beamwidth of most of these systems is very narrow. Therefore, an amplifier in the front end of a receiver must have an excellent noise figure to be able to function properly in such

140   HANDBOOK OF MICROWAVE TESTING

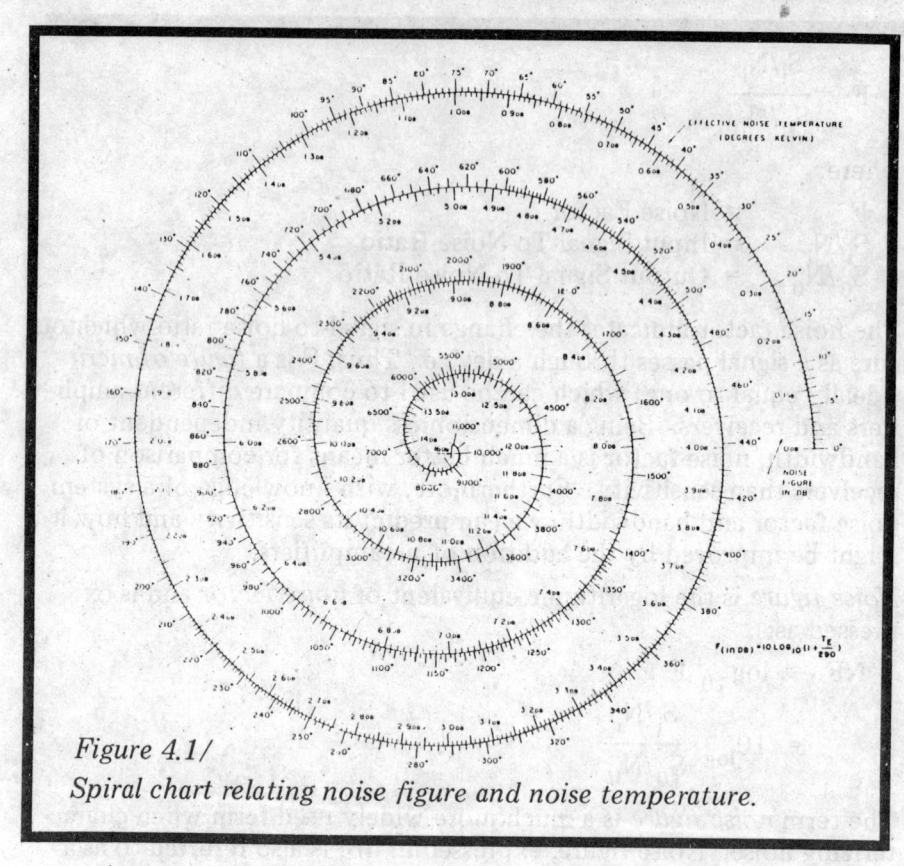

Figure 4.1/
Spiral chart relating noise figure and noise temperature.

an environment. Though noise is usually expressed in dB, it can be termed a *noise temperature*, in °K. A formula describes the relationship between *noise figure* and *noise temperature*.

where:

$$NF = 10 \log_{10}\left(1 + \frac{T_N}{290}\right)$$

NF = Noise Figure in dB
T_N = Noise Temperature in °K

A 40-inch spiral chart which lists an assortment of noise figure/noise temperature determinations found from the formula above is shown in Figure 4.1. The chart is rolled into a spiral to combine accuracy and convenience; it is most accurate at the low noise end of the scale. The next time someone refers to a noise temperature of 1100°, 300°, or 500°, check the spiral chart to convert that specification to the more familiar noise figure.

Noise Measurements 141

We now have three methods of expressing noise: noise factor, noise figure, and noise temperature. To illustrate their relationship to one another, consider the following example.

Suppose we are given an amplifier with a noise temperature (T_N) of $865°K$. What we need is to know the noise figure and the noise factor so that we can perform further calculations on a given system.

(1) From the formula above we know that

$$NF = 10 \log \left(1 + \frac{T_N}{290}\right)$$

$$NF = 10 \log \left(1 + \frac{865}{290}\right)$$

$$NF = 10 \log (1 + 2.982)$$

$$NF = 10 \log (3.982)$$

$$NF = 10 (0.6)$$

$$NF = 6.0 \text{ dB}$$

(2) Similarly, from discussions above we know that

$$NF = 10 \log F \text{ (Where F = Noise Factor)}$$

Transposing the equation, it becomes:

$$F = \log^{-1} \frac{NF}{10}$$

$$F = \log^{-1} \frac{6}{10}$$

$$F = \log^{-1} 0.6$$

$$F = 4$$

To summarize our calculations, we can say that the noise characteristics of the particular amplifier would be expressed as:

Noise Temperature = $865°K$
Noise Figure = 6.0 dB
Noise Factor = 4

Note the units that are associated with each of these expressions. Noise factor is merely a ratio of two numbers, therefore, it has no units; noise figure is a definite quantity and is expressed in dB; and noise temperature is a comparison with a standard $290°K$ and is, therefore, expressed in $°K$. Each of these expressions has a definite place in noise measurements.

There are many methods used to measure noise. Some are manual methods, some are automatic. It is possible to take swept noise readings as well as CW readings. Although not all methods of noise measurement can possibly be covered here, we will attempt to cover those most widely used and those that will produce the most meaningful results. Before getting into the actual measurement procedures we will become familiar with the pieces that make up a noise measurement setup — namely, the noise sources and the noise figure meters.

4.1 Noise Sources

Throughout the preceeding chapters we have presented many test setups. The one area that all of them had in common was that they had some sort of signal generator. It may have been a CW oscillator, a sweep generator, or a synthesized microwave generator, but all of them had a generator of one type or another. The noise measurement setups to follow are no exception to the trend. They also have a signal generator — a *noise source*.

Just as the signal generator discussed earlier varies in construction, so too do the noise sources that are used. There are basically four types of noise sources to be covered here:

- Thermal Noise Source
- Diode Noise Source
- Gas-Discharge Noise Source
- Solid-State Noise Source

The *thermal noise source* (also referred to as a hot/cold standard) is the most accurate of all the noise sources. It is used many times for the calibration of other types of noise sources. With the proper type of instrumentation, accuracies of better than ± 0.1 dB are readily attainable. The noise source standard employs two resistive terminations. One is immersed in liquid nitrogen (77.5°K, −195.7°C, −320°F), the other is in a proportionally controlled oven set to the temperature of boiling water (373°K, 100°C, 212°F). Figure 4.2 is a schematic representation of a typical thermal noise source. The terminating impedances shown in the figure are carefully controlled so that they track one another, in both magnitude and phase, over the full range of the instrument. For a typical hot/cold standard operating DC to 9 GHz which is commercially available, the impedance tracking is stated as follows:

The distance between the complex impedance points of the hot

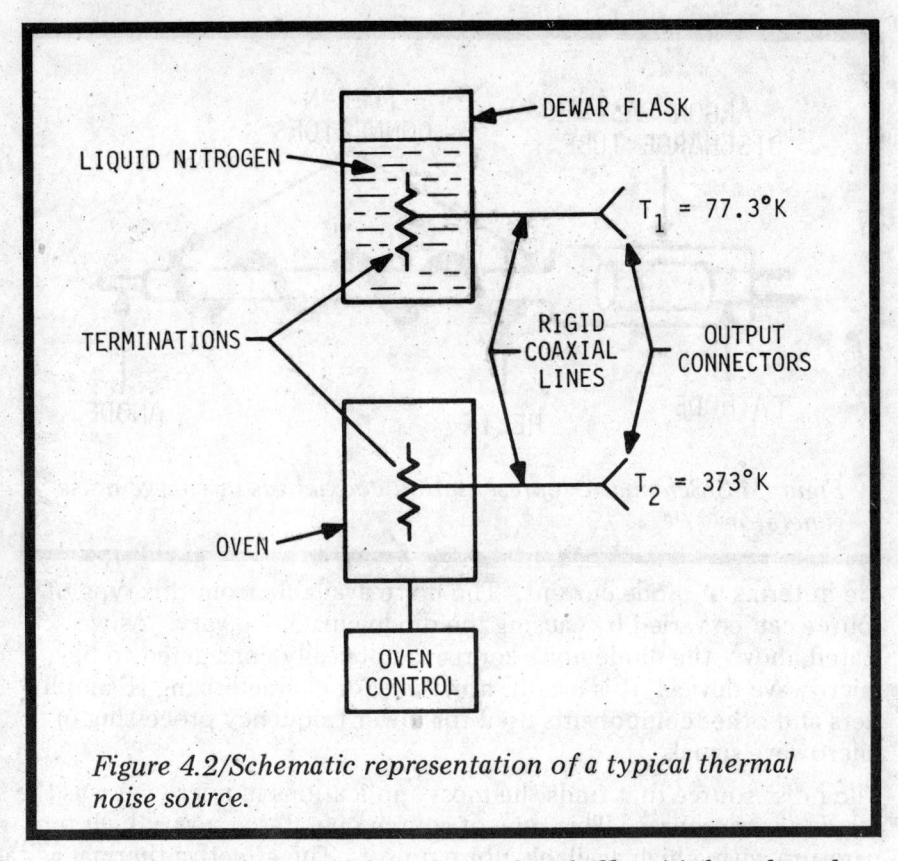

Noise Measurements

Figure 4.2/Schematic representation of a typical thermal noise source.

and cold terminations (plotted on a Smith Chart) is less than the diameter of a circle corresponding to an SWR of 1.05 from DC to 7 GHz and 1.10 from 7.1 to 9 GHz.

You can see from such a close track of impedances that mismatch uncertainties are virtually eliminated. This is of great importance in performing accurate noise figure measurements, especially at low noise figures.

The next type of noise source is the *temperature-limited diode source*. It is questionable if this particular type of source can be put into a microwave classification. This is because these particular sources are good at the very low end of the spectrum (below 1.0 GHz). The sources consist of a diode mounted in parallel with a resistive load equal to the source resistance for which the receiver, or device, to be tested was designed (for example, 50 ohm). The diode is operated temperature-limited (effectively current-limited), and a meter is placed in the plate circuit of the diode and is calibrated directly in noise fig-

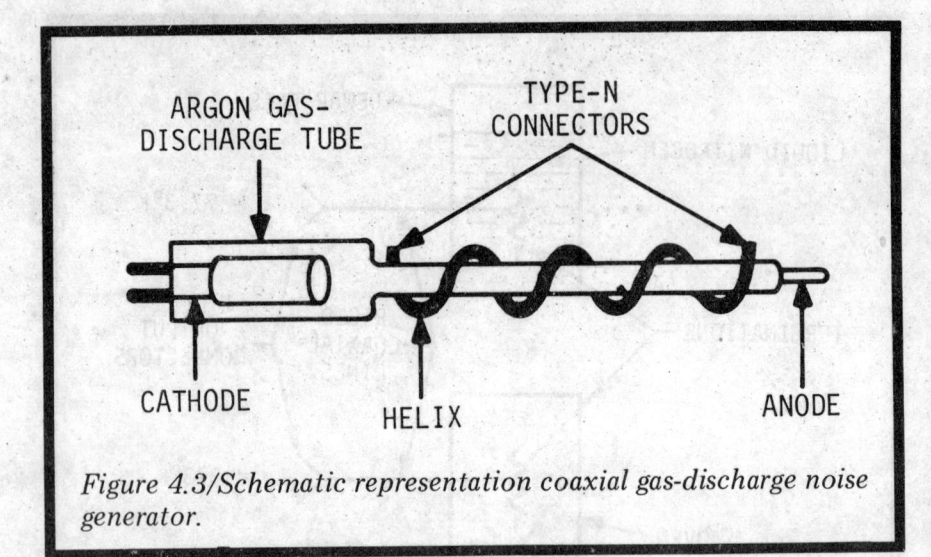

Figure 4.3/Schematic representation coaxial gas-discharge noise generator.

ure in terms of diode current. The noise available from this type of source can be varied by causing the diode current to vary. As we stated above, the diode noise source is not really considered to be a microwave device. It is useful, however, for characterizing IF amplifiers and other components used for lower frequency processing of microwave signals.

The noise source that finds the most applications in microwaves is the *gas-discharge* source. This type of source operates at a very high temperature with a high available noise power. The effective thermal agitation of an argon tube noise source (the most common), for example, represents an equivalent temperature of approximately $10,000°K$ (a noise figure of 15.8 dB), compared with a normal room temperature of $290°K$. The gas-discharge source can be either a coaxial or waveguide form. When used in a waveguide system, the gas tube (which is usually argon but could be neon) is matched to the structure by mounting it so that it extends through the broad sides of the waveguide at an angle of approximately 10 degrees.

Figure 4.3 is a schematic of a coaxial gas-discharge noise source. You will note that its construction is similar to that of a Backward-Wave Oscillator (BWO) tube. The major difference between the two is that the BWO has the helix inside the tube and the gas-discharge noise source has the helix wound on the outside of the tube with connection at each end of the coil. One end is terminated while the other end is the output.

Coaxial gas-discharge noise sources generally have an upper limit of

Noise Measurements 145

operations of between 4 and 5 GHz. You can justify this by referring again to Figure 4.3 and noting the cathode/anode construction of the device. It is relying on a specific transit time for its operation and, as such, it limits its high frequency operation.

The waveguide gas-discharge noise source has an upper limit of approximately 40 GHz. This restriction is placed on it simply because the waveguide dimensions become too small if we go any higher in frequency. Thus, the limitation is the result of physical size rather than electrical performance.

Solid-state noise sources are available up to 18 GHz. These devices are small in isze and high in reliability. The noise-generating diode that is used is usually isolated from the output by 12 to 20 dB; consequently, there are almost no mismatch errors as the source is switched on and off and thus, no destructive transients are present at the output. The Excess Noise Ratio (ENR) is normally 15 to 15.5 dB. However, ENRs of up to 42 dB are possible with commercially available units.

With each of the four types of noise sources presented, the list below shows some characteristics of each of them. Some are advantages; some are disadvantages. Each is a characteristic which must be considered when choosing a noise source.

1 - Thermal Noise Source
- The most accurate of all sources.
- Broad frequency coverage.
- Very low cold temperature (around 77°K), which allows them to be used for low noise receiver measurements.
- Close matching of impedance characteristics in both phase and magnitude.
- Low hot to cold temperature; therefore, they are not useful for high noise measurements.
- Require handling of liquid nitrogen — for use in labs only.

2 - Diode Noise Source
- Limited to low frequencies.
- A simple direct relationship between the Excess Noise Ratio (ENR) and dc plate current.
- A variable ENR because of the direct relationship mentioned above.
- No mismatch change from cold to hot.

3 - Coaxial Gas-Discharge Noise Source
- Noise temperature is high — in the range of 10,000°K to 20,000°K (15.3 to 18.4 dB noise figure).
- Relatively broadband (typical units go from 2 to 5 GHz).
- Convenient to use.
- Suitable for systems use.
- Rugged and economical.
- Significant mismatch change from cold to hot.
- Significant coupling corrections required at the band edges.

4 - Waveguide Gas-Discharge Noise Source
- Frequency range restricted by the waveguide size.
- Coupling corrections are small.
- Mismatch changes less than coaxial but still are larger than a diode or hot/cold noise source.
- No pulse leakage.

5 - Solid-State Noise Source
- High Excess Noise Ratio (ENR).
- Small size.
- Very high in reliability.
- Relatively high frequency range.
- Very low mismatch error.

4.2 Noise Meters

We have presented a means of generating the noise necessary for characterizing components or systems — the noise source. We now need a means of processing this generated noise and displaying it, which is the noise meter. In order to understand how a noise meter works, it is first necessary to understand what it is that the meter must measure. This will be our first order of business. (Since all of the meters available indicate noise figure in dB, we will refer to the noise meters as *noise figure meters* from this point on.)

If a known level of broadband noise can be introduced at the input of a device under test, a differential power measurement at the output would indicate a gain-bandwidth product of a device. It can be seen

Noise Measurements

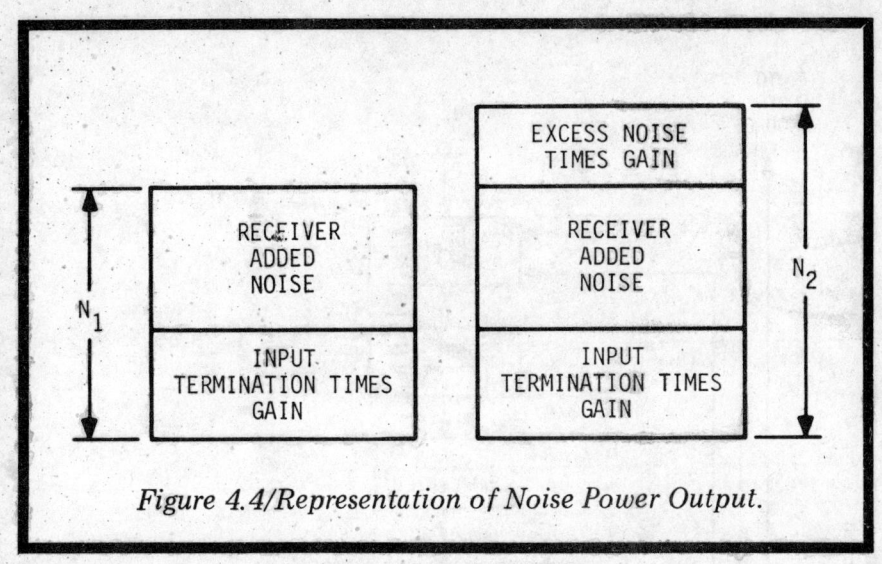

Figure 4.4/Representation of Noise Power Output.

that the gain-bandwidth product is a very important part of a noise figure calculation when we examine the formula:

$$NF = \frac{N_1}{kT_0BG}$$

when:

- NF = noise figure
- N_1 = measured noise power output of the system
- k = Boltzmann's constant, 1.374×10^{-23} joule/°K
- T_0 = absolute temperature (290°K)
- B = bandwidth of the system
- G = gain of the system

A measurement of the device output both with and without an additional noise power input will give an indication of gain-bandwidth product; it is possible to compute noise figure with no further measurements, since all terms in the above formula are known.

Using a system consisting of an input termination, an excess noise source, a receiver to be tested, and an output power detector, it is possible to measure N_1 (noise power output) with the noise source "cold" (off) and to measure a quantity N_2 with the excess noise source on (fired). Figure 4.4 illustrates the make up of N_1 and N_2. N_1 is shown to consist of the amplified input termination noise plus the

Figure 4.5/Noise figure meter block diagram.

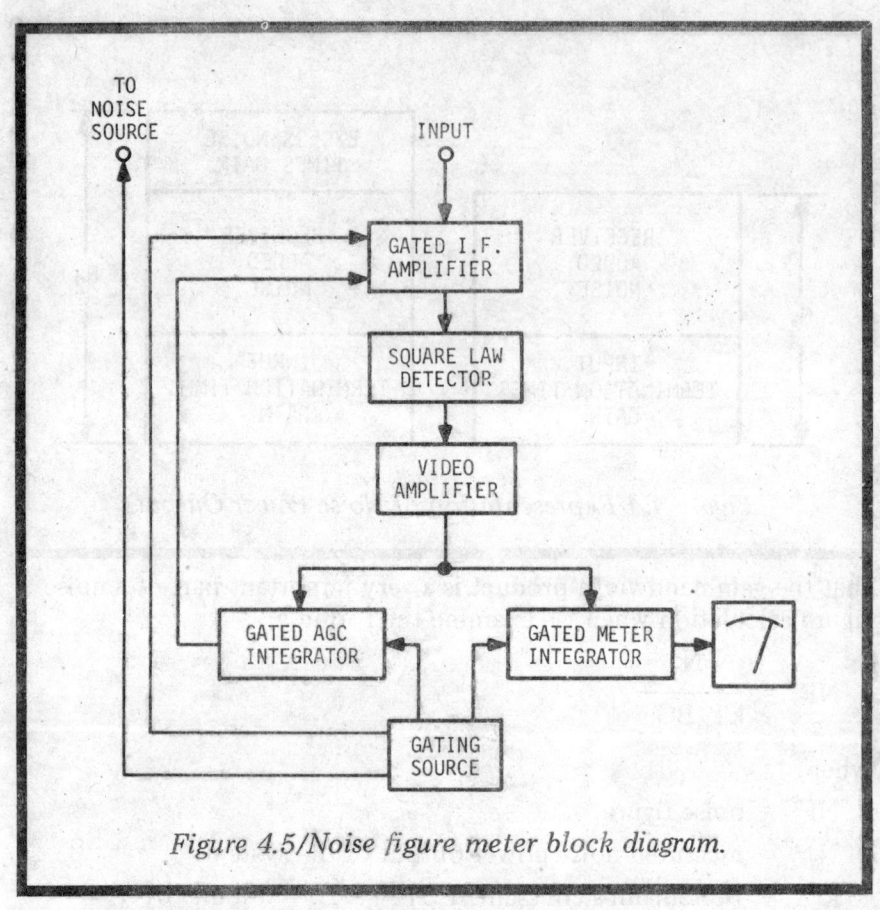

noise generated within the receiver. N_2 is shown to consist of all the components of N_1 plus the amplified excess noise power which appear at the output of the receiver. This, very basically, describes the task of the noise figure meter. It must read N_1, use it as a reference, and then compare it with N_2 to result in a noise reading of the system you are measuring.

Figure 4.5 is an example of a noise figure meter that accomplishes the task listed above. This is a commercially available meter that uses an excess noise figure reading of 15.2 dB as a reference. A gating process is used to measure N_1 and N_2 as follows:

In operation, the gating source pulses the noise source at a rate of 500 Hz; N_1 and N_2 pulses arrive at the IF amplifier. Noise sources have a finite noise build-up time, so the IF amplifier is gated to pass only the final amplitudes of N_1 and N_2 to the square law detector. The detect

Noise Measurements

N_2 pulse is switched to an AGC integrator, where a voltage for gain control of the IF amplifier is derived. The time constant of this circuit is made long enough to control the IF amplifier gain even when the N_1 pulse is passing through it. Since the AGC action keeps the detected N_2 pulse at a constant level, a measurement of the detected N_1 pulse is, in effect, a measurement of the pulse ratio. The N_1 pulse is measured by switching it to the meter integrator and meter.

Convenient internal calibration of the meter is accomplished by artificially creating readings of "$+\infty$" and "$-\infty$". By pulsing the noise source during both the N_2 and N_1 time periods, we obtain a condition of $N_2 = N_1$. In the formula $F_{db} = 15.2 - 10 \log(N_2/N_1 - 1)$, this condition results in a noise figure of $+\infty$. The artificial condition of $F = -\infty$ would correspond to an "N_1" value of "0". This can be created by gating "off" the IF amplifier during the "N_1" time period. If the metering circuit is designed to be a linear indicator of the power of "N_1" (square law detector), and the meter minimum position is calibrates as $-\infty$ and the full scale deflection as $+\infty$, all other points on the meter face can be calculated by the formula $F_{db} = 15.2 - 10 \log(N_2/N_1 - 1)$. For example, an "$N_1/N_2$" ratio of 1/2 would bring about a mid-scale reading. From the formula, this mid-scale reading is calculated to be 15.2. In a similar fashion, the balance of the scale is calibrated. (The 15.2 dB figure is arrived at from the fact that this is the excess noise of the average source as mentioned earlier.)

Figure 4.6 is a block diagram of another commercially available noise figure meter. This figure shows the basic elements of a precision automatic noise figure meter. As in the case of the meter above, the first requirement for this type of system is a method of periodically turning the noise generator on and off. This is accomplished by a modulator driven by a gate generator, usually a free running squarewave generator (multivibrator) operating at a low audio frequency. The output of the receiver consists of squarewave modulated noise. The two levels of this squarewave represent the device-under-test (DUT) noise (N_1) and the DUT plus generator noise (N_2). This signal is amplified in the IF amplifier and detected in a square-law detector. A sample of this modulated noise signal is also detected in a second detector which is gated to provide an output only during the time the noise generator is turned off. When the noise generator is turned on, the detector is gated off. Its output is then converted to a dc signal proportional to this noise-off condition. This dc signal is then amplified and applied to the IF amplifier in such a manner as to keep the noise-off level at the output of the IF amplifier constant. The square-law detector provides an out-

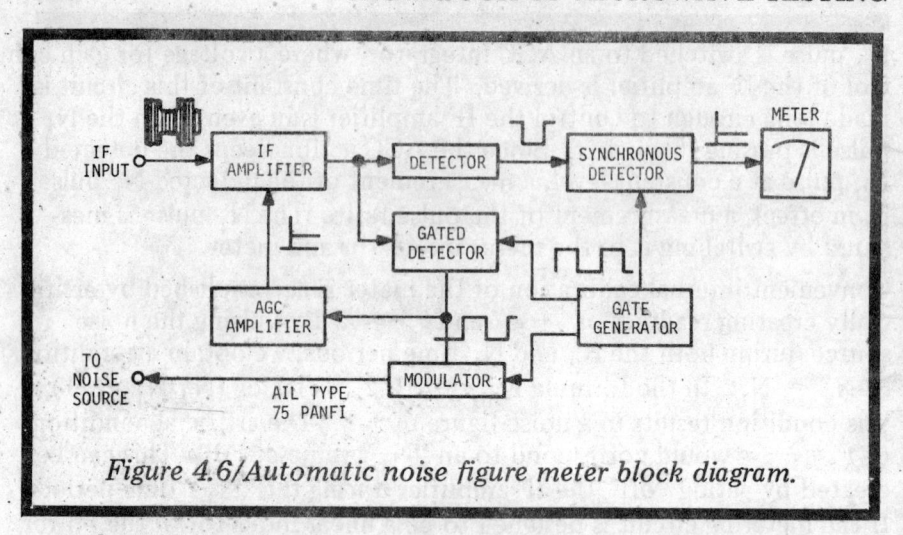

Figure 4.6/Automatic noise figure meter block diagram.

put voltage proportional to the input power. This output consists of two voltage levels proportional to the two levels of noise power at the input to the IF amplifier. One level (V_1) is for the noise-on voltage, and the other (V_2) is for the noise-off voltage. The peak-to-peak amplitude of the squarewave is measured by the synchronous detector and the resultant voltage is indicated by the meter. This voltage is inversely proportional to the noise figure of the device under test since, by mathematical expression, the excess noise of the system is known and N_1 remains constant throughout your measurements. This is highly desirable since lower noise figures read close to full scale and greater resolution is achieved. Also, because a square-law detector is used, the meter scale, when calibrated in dB, becomes logarithmic, thus producing a natural expansion on the upper half of the scale (when low noise figures are indicated).

It should be fairly obvious at this point that these meters are geared to low noise readings. That is not to say that they will not read higher noise figures. The first unit will read noise figures (using the proper noise source) from 0.2 dB to in excess of 25 dB with reasonable accuracy. The second unit will read noise figures, again from 0.2 dB to around 33 dB (once again dependent on the noise source). The emphasis was placed on the low noise area since it is the area where accuracy can decline so easily. This is why the major effort in designing such a meter is placed on the low noise figure end.

We have, to this point, covered noise in regard to noise factor, noise figure, and noise temperature. We have also discussed the various types of noise sources and two similar noise figure meters. We should

Noise Measurements

now have all the necessary tools to perform an efficient and accurate noise measurement on a receiver system or component. The only remaining decision to be made is whether a manual or automatic measurement setup is to be used. That is the job of the next two sections: to describe each of these methods, point out advantages and disadvantages, and recommend the proper use of each.

4.3 Manual Noise Measurements

A convenient and widely used formula for calculating noise figure presented earlier is repeated below:

$$NF_{dB} = 10 \log \frac{(T_2 - T_0)}{T_0} - 10 \log \left(\frac{N_2}{N_1} - 1\right)$$

where:

$\frac{(T_2 - T_0)}{T_0}$ is a measure of the relative excess noise power available from a noise source and is specified by the manufacturer. (In the case of argon, this number is 15.2 dB.)

N_2 = the noise power with the noise source fired (ON)

N_1 = the noise power with the noise source "cold" (OFF)

The use of this formula and manual techniques will yield the greatest accuracies in noise measurements. The method used for manual measurement can vary according to specific applications. Two procedures that are used are the "twice-power" method and the "Y-factor" method. These two will be covered as a means of comparison.

4.3.1 Twice-Power Noise Figure Measurement

In actually measuring the N_1 and N_2 of the above equation, if N_2 was set to be twice N_1, the equation now becomes:

$$NF_{dB} = 10 \log \frac{(T_2 - T_0)}{T_0}$$

With the proper equipment, the condition $N_2 = 2N_1$ can be established by varying the relative excess noise power of the noise source. A setup to perform this operation is shown in Figure 4.7. A basic procedure for use with this setup would be as follows:

HANDBOOK OF MICROWAVE TESTING

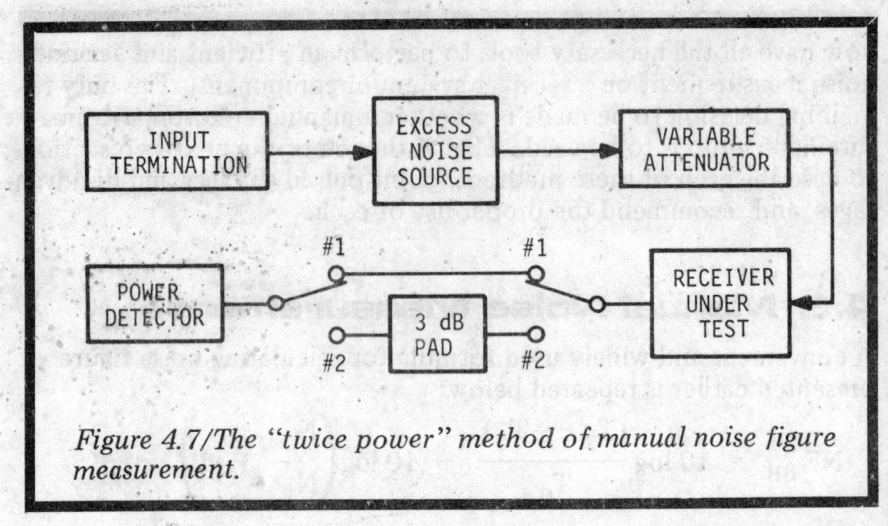

Figure 4.7/The "twice power" method of manual noise figure measurement.

(1) Set a convenient reference level on the power detector with the excess noise source "cold" (OFF) and the setup in position #1 (3 dB pad out of the circuit). This is N_1. Record this value

(2) Insert the 3 dB pad (position #2) and fire the excess noise source (turn it on).

(3) Adjust the variable attenuator until the original power detector reference point is reached. This creates a condition where the fired power level is twice that of the cold level ($N_2 = 2N_1$).

To understand how the 3 dB pad accomplishes a 2:1 ratio between N_1 and N_2, consider the basic formula for power in dB:

$$P_{dB} = 10 \log \frac{N_2}{N_1}$$

What we want is a relationship where $N_2 = 2N_1$, so that we can see how big the pad must be. If we substitute this into the formula:

$$P_{dB} = 10 \log \frac{2N_1}{N_1}$$

$$P_{dB} = 10 \log 2$$

$$P_{dB} = 10 (0.3)$$

$$P_{dB} = 3 \text{ dB}$$

So you can see that the 3 dB results in a 2:1 power ratio between N_2 and N_1. If you need further verification, you can work the formula

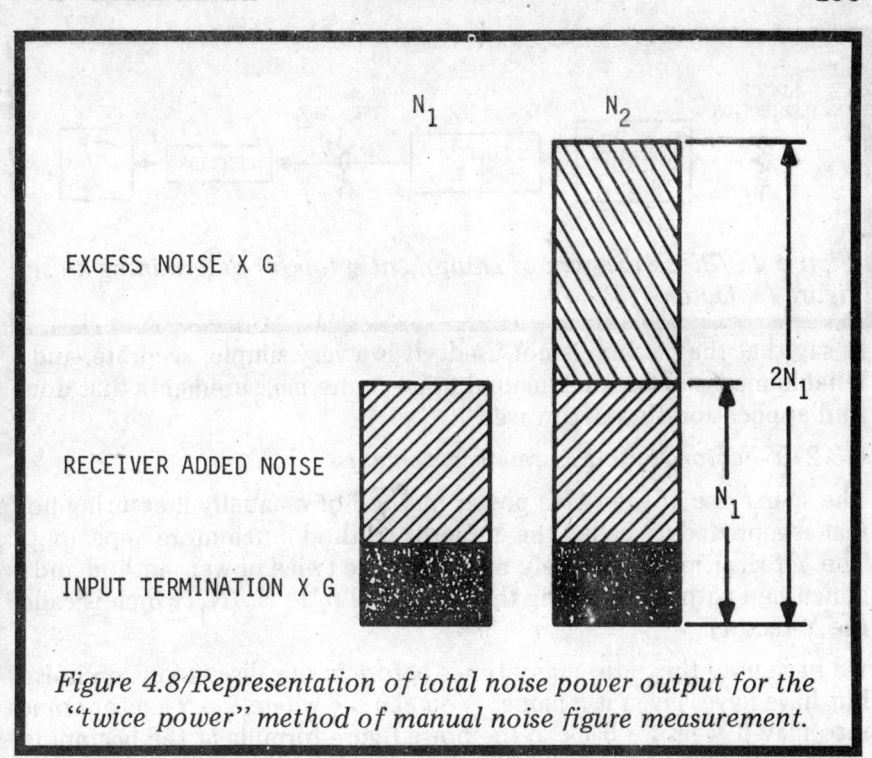

Noise Measurements 153

Figure 4.8/Representation of total noise power output for the "twice power" method of manual noise figure measurement.

backwards, beginning with P_{dB} = 3 dB and finding the N_2/N_1 ratio. You should be able to see quite readily that the ratio is 2.

The condition of $N_2 = 2N_1$ is illustrated in Figure 4.8. You can see how the output noise power contributed by the excess noise source exactly equals the sum of the amplified input termination noise plus the receiver noise contribution (N_1). Since this excess noise ratio was adjusted with the attenuator to be equal to input termination noise plus receiver noise (resulting in the relationship $N_2 = 2N_1$), from our noise figure equation it can be seen that the attenuated excess noise ratio is equal to the noise figure of the receiver. In the case of an argon source, it can be read as 15.2 dB minus the attenuator setting.

While the attenuator reduces the amount of excess noise injected into the system, it has no effect on input termination noise power if the termination and attenuator are at the same temperature, since, regardless of the amount of attenuation, when the excess noise source is cold the receiver input is still looking at a matched input at temperature T.

The twice power method is not used much, since most people do not want to insert or remove components from a test setup. This is not

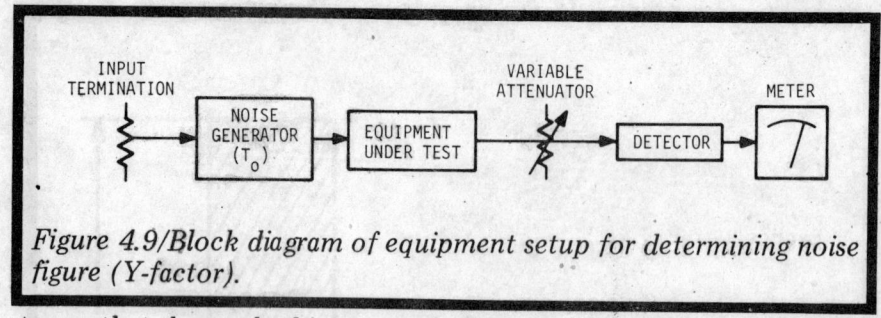

Figure 4.9/Block diagram of equipment setup for determining noise figure (Y-factor).

to say that the method is not used. It is a very simple, accurate, and reliable means of making manual noise figure measurements that does find applications in microwaves.

4.3.2 Y-Factor Noise Figure Measurements

The sparse use of the twice power method of manually measuring noise makes a procedure called the Y-factor method much more appealing. The Y-factor method closely resembles the twice power method and is concerned with determining the numerical ratio N_2/N_1 (which is called the Y-factor).

We have used this ratio many times before in our discussions on noise, but have never given it a name. You can see where the Y-factor comes into play if we refer back to the noise figure formula at the beginning of this section:

$$NF_{dB} = 10 \log \frac{(T_2 - T_0)}{T_0} - 10 \log \left(\frac{N_2}{N_1} - 1 \right)$$

The final term could actually be written as $10 \log (Y - 1)$ since $N_2/N_1 = Y$. The first term of the equation has been seen before. This term is called the *excess noise ratio* and is usually expressed as either ENR or T_{ex}. It represents the relative increase in noise power at the network input when the noise figure can be computed. It is important to note here that is is not necessary to measure absolute levels when measuring for Y-factor, only a ratio of the two quantities is necessary. So the major task before us is to measure the power ratio, and thus, the Y-factor. There are several methods of measuring Y-factor and, indirectly, noise figure. We will discuss a more common one here.

Figure 4.9 is a block diagram of a common Y-factor test setup. The noise generator used can be any one of a number of devices with a known excess noise ratio (ENR). We have covered noise sources in a previous section and any one of those covered can be used. The source used depends on a particular application or requirement, but

Noise Measurements

the most common for a setup of this type are the gas-discharge and the solid state noise source. If a high degree of accuracy is needed, there is no reason why a hot/cold source could not be used with a high degree of success.

The variable attenuator is a precision IF attenuator. This component aids greatly in the accuracy of the noise measurement. It is the attenuator that is used to establish the N_2/N_1 ratio, or Y-factor. Its accuracy should be reasonable even though only relative readings are being taken. The detector is used to provide a dc level to the meter which is proportional to the noise levels with the source "cold" and "fired". This can be a separate detector and indicating meter, or may be a commercially available receiver designed to be used in noise figure measurement setups. It may also be a noise figure meter in a manual mode.

When the noise generator is off (cold) in the setup, the input of the receiver (or device) under test sees a passive termination and the detector level meter indicates some value which is called N_1. When the noise generator is turned on (fired), the output level will increase because of an increase in noise at the input. This new level is called N_2. If the variable attenuator is increased until the detector level is the same as that when the noise source was off (N_1), then this increased attenuation is the Y-factor (N_2/N_1) expressed in dB. By converting the number to a power ratio and substituting in the above mentioned formula, this noise figure of the device under test can be completed.

Graphs are also available for specific values of ENR with the coordinates calibrated in Y-factor and noise figure. Such a graph with 8 values of ENR is shown in Figure 4.10. To illustrate how much a graph works, consider the following examples:

(1) You have a noise source with an excess noise ratio (ENR) of 15.9 dB.

(2) You have run through a measurement procedure and found a Y-factor of 5.4. What is the noise figure?

By referring to Figure 4.10 we can see that curve #5 represents an ENR of 15.9 dB. If we find a Y-factor of 5.4 dB on the bottom of the graph and move up until we interact the #5 curve, we will find point A. By moving to the left we can see that the noise figure in this case is 12 dB. To summarize we can say:

ENR   = 15.9 dB
Y-Factor   = 5.4 dB
Noise Figure = 12.0 dB

For those of you who have a tendency not to trust graphs and charts, we can prove that these results are valid by substituting into our basic formula presented earlier. We will proceed as follows:

Given:

$$Y\text{-Factor } (N_2/N_1) = 5.4 \text{ dB}$$

$$dB = 10 \log (N_2/N_1)$$

$$5.4 = 10 \log (N_2/N_1)$$

$$0.54 = \log N_2/N_1$$

$$\log^{-1} 0.54 = N_2/N_1$$

$$N_2/N_1 = 3.47 = Y = 5.4 \text{ dB}$$

also:

$$ENR = 15.9 \text{ dB}$$

Substituting into the noise figure formula:

$$NF_{dB} = 10 \log \frac{(T_2 - T_0)}{T_0} - 10 \log (N_2/N_1 - 1)$$

$$NF_{dB} = ENR - 10 \log (Y - 1)$$

$$NF_{dB} = 15.9 - 10 \log (3.47 - 1)$$

$$NF_{dB} = 15.9 - 10 \log (2.47)$$

$$NF_{dB} = 15.9 - 10(.39)$$

$$NF_{dB} = 15.9 - 3.9$$

$$NF_{dB} = 12.0 \text{ dB}$$

This noise figure corresponds to that read on the graph in Figure 4.10. We stated at the beginning of this section that the Y-factor method of manual noise figure measurement is more appealing than the twice-power method. Some of the reasons that this method is preferred are:

- The signal level entering the detector (usually a test receiver or a noise figure meter in the manual mode) is kept at a constant level. Nonlinearities, etc., are all held constant. In fact, the receiver may be operated in a nonlinear, or partially saturated condition. The only effect this has upon the measurement is a reduction in the resolution of the meter.

Noise Measurements

CURVE	ENR (dB)
1	14.7
2	15.0
3	15.3
4	15.6
5	15.9
6	16.2
7	17.6
8	17.9

$$F = \frac{T_{ex}}{Y-1}$$

Figure 4.10/Y-factors vs noise figure.

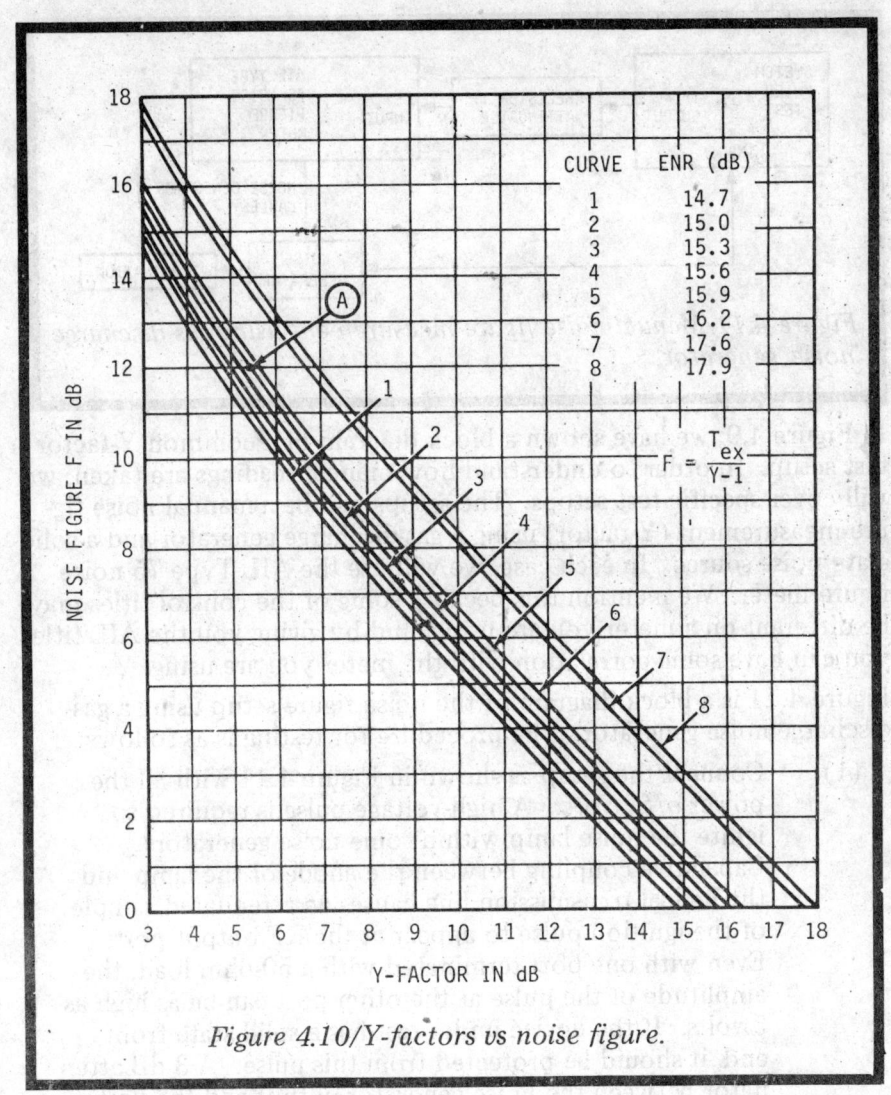

- The noise figure of the detector (or receiver) and the insertion loss of the attenuator make no contribution to the noise figure of the device under test. Again, the only effect a high insertion loss and high receiver noise figure have upon the measurement is a loss of resolution.

- If the meter change was read (in dB or any convenient ratio) instead of keeping its level constant, then the system must be checked very accurately for nonlinearities. Also, the meter must be calibrated over the range it is expected to operate.

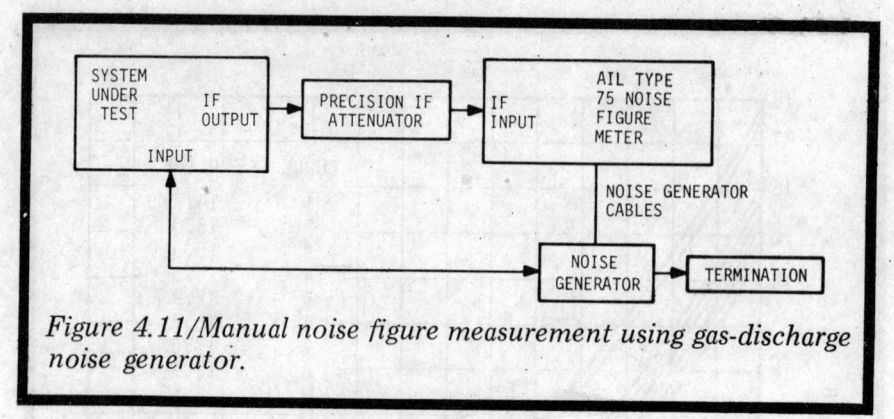

Figure 4.11/Manual noise figure measurement using gas-discharge noise generator.

In Figure 4.9, we have shown a block diagram of a common Y-factor test setup. In order to understand how manual readings are taken, we will cover specific test setups. The setups will be a manual noise figure measurement (Y-factor) using a gas-discharge generator and a solid-state noise source. In each case, we will use the AIL Type 75 noise figure meter. We mention this because some of the control titles may be different on a meter you are using, and by giving you the AIL titles you can have some correlation with the meter you are using.

Figure 4.11 is a block diagram of the noise figure setup using a gas-discharge noise generator. The procedure for testing is as follows:

(1) Connect the setup as shown in Figure 4.11 with all the power *off*. *Note:* A high-voltage pulse is required to ignite the noise lamp within some noise generators. Capacitive coupling between the anode of the lamp and the helical transmission line causes an attenuated sample of the ignition pulse to appear at the RF output ports. Even with one port terminated with a 50-ohm load, the amplitude of the pulse at the other port can be as high as 5 volts. If the device under test has a solid state front end, it should be protected from this pulse. A 3 dB attenuator between the noise generator output and the unit under test will usually suffice, although higher values of attenuation may be used if desired. The value of the attenuator (in dB) must then be subtracted from the measured value of noise figure (in dB) in order to obtain the true noise figure of the device under test.

(2) Set the *gas-diode* switch to the *gas* position.

Noise Measurements

(3) Set the *generator current* control fully counterclockwise.

(4) Depress the *manual off* button.

(5) Set the scale selector to *generator current* position.

(6) Turn on the AC power to the noise figure meter and the system under test. (If a separate amplifier is being tested, apply dc to the circuitry.)

(7) Depress the *manual on* button and adjust the *generator current* control for the meter current setting recommended by table supplied with the noise sources. An example of these settings is shown below:

Noise Generator	ENR (Excess Noise Ratio)	Generator Current (ma)
AIL 7012	15.75 dB	175
AIL 7048	15.35 dB	200
AIL 7053	16.15 dB	150

(8) Depress the *manual off* button. The scale selector switch may be set to any position without affecting the accuracy of the manual measurement.

(9) Adjust the *gain* control and the precision attenuator for a convenient meter reading.
Note: All attenuator settings should be obtained using counterclockwise rotation to eliminate backlash.

(10) Note the meter reading and the precision attenuator setting.

(11) Depress the *manual on* button and increase the precision attenuator setting until the meter returns to the setting noted in step 10.
Note: Do not disturb the *gain* control setting.

(12) Subtract the attenuator reading in step 10 from that in step 11. This is the Y-factor (in dB).

(13) To check for possible reading errors, depress the *manual off* button and readjust the precision attenuator to obtain the meter reading noted in step 10. Note that the attenuator reads the same as in step 10.

The noise figure of the device under test can be computed from:

$$F = \frac{\frac{T_2}{T_0} - 1}{Y - 1}$$

where:

T_2 = effective temperature of the noise generator in the *on* condition

T_0 = 290°K (standard reference temperature)

Y = ratio of the noise output of the device under test with the noise generator *on* to that with the noise generator *off*

The noise figure (in dB) is given by:

$$F_{dB} = 10 \log\left(\frac{T_2}{T_0} - 1\right) - 10 \log(Y - 1)$$

where:

$\frac{T_2}{T_0} - 1$ is the excess noise ratio (ENR) of the noise generator (in dB). This formula was used many times in our previous discussions.

A second manual noise figure measurement setup is shown in Figure 4.12. This setup uses a solid-state noise source instead of the gas-discharge device used above. Once again, the AIL Type 75 noise figure meter will be used for the setup. The test procedure is as follows:

(1) Connect the test setup as shown in Figure 4.12. Do not apply any ac to the noise figure meter or the system to be tested. (If an amplifier or other single component active device is being tested, do not apply dc to is as of yet.) *Caution:* The solid state noise generators must be driven by their proper modulators for a correct noise figure reading If uncertain of the diode gate (modulator) output, consult the appropriate manual.

(2) Set the *gas-diode* switch to the *diode* position.

(3) Depress the *manual off* button.

(4) Set the scale selector to the 0-dB position.

(5) Depress the AC power button on the noise figure meter and apply the appropriate ac or dc to the system under test.

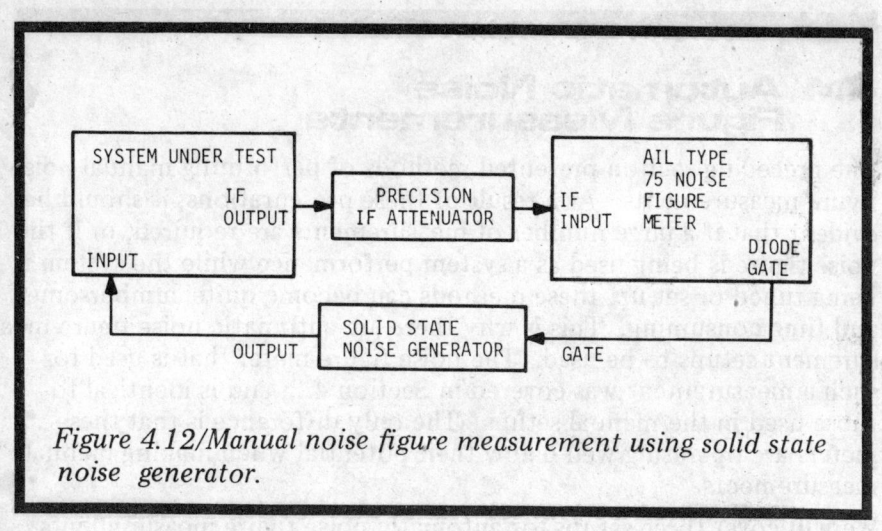

Noise Measurements

Figure 4.12/Manual noise figure measurement using solid state noise generator.

(6) Adjust the *gain* control and the precision attenuator for a convenient reference reading on the noise figure meter.

(7) Note the attenuator reading.

(8) Depress the *manual on* button.

(9) Increase the precision attenuator setting until the noise figure meter returns to the reference noted in step 6. Note this attenuator reading.

(10) Subtract the attenuator reading in step 7 from that in step 9; this is the Y-factor (in dB).

To check for possible reading errors, depress the *manual off* button and readjust the attenuator to obtain the same reading noted in step 6 on the noise figure meter. The noise figure of the device under test can be computed by the methods described in the preceeding test setup. The ENR of the Solid State Noise Generators is indicated on the noise generator case.

These are the manual methods of measuring the noise figure of a system or device. They are designed for use in a situation where you have either a single frequency to consider or relatively few frequencies (3 or 4 would be all you would want to handle with setups such as the ones described above). You can see how large the task would be if you were required to measure an octave band with one of these manual setups. That is why there are automatic methods for broadband readings and manual methods for single (or few) frequency readings. The two types each have their applications and must co-exist in the world of noise measurements.

4.4 Automatic Noise Figure Measurements

The preceding section presented methods of performing manual noise figure measurements. As a result of these presentations, it should be evident that if a large number of measurements are required, or if the noise figure is being used as a system performance while the system is being tuned or set up, these methods can become quite cumbersome and time consuming. This is why there are automatic noise figure measurement setups to be used. The noise figure meter that is used for such a measurement was covered in Section 4.2, and is identical to those used in the manual setups. The only difference is that these meters are operating well below their potential when making manual measurements.

We will cover three setups for automatic noise figure measurements. The first two will once again use the AIL type 75 automatic noise figure meter and the third will use the HP340B noise figure meter. Each has its own characteristics of operation, but it is also interesting to note the similarities in the two meters and their setups.

The first setup is for an automatic noise figure measurement using a gas-discharge noise generator. A block diagram of the test setup is shown in Figure 4.13. The test procedure is as follows:

(1) Connect the test setup as shown in Figure 4.13, with no ac power on the system or amplifier to be tested.
Note: In the manual measurement setup, we presented a precautionary note concerning the gas-discharge noise generator. It is of significant importance to be repeated here. A high voltage pulse is required to ignite the noise lamp within some noise generators. Capacitive coupling between the anode of the lamp and the helical transmission line causes an attenuated sample of the ignition pulse to appear at the RF output ports. Even with one port terminated with a 50 ohm load, the amplitude of the pulse at the other port can be as high as 5 volts. If the device under test has a solid state front end, it should be protected from this pulse. A 3 dB attenuator between the noise generator output and the unit under test will usually suffice, although higher values of attenuation may be used if desired. The value of the attenuator (in dB) must then be subtracted from the measured value of noise figure (in dB) in order to obtain the true noise figure of the device under test.

Noise Measurements

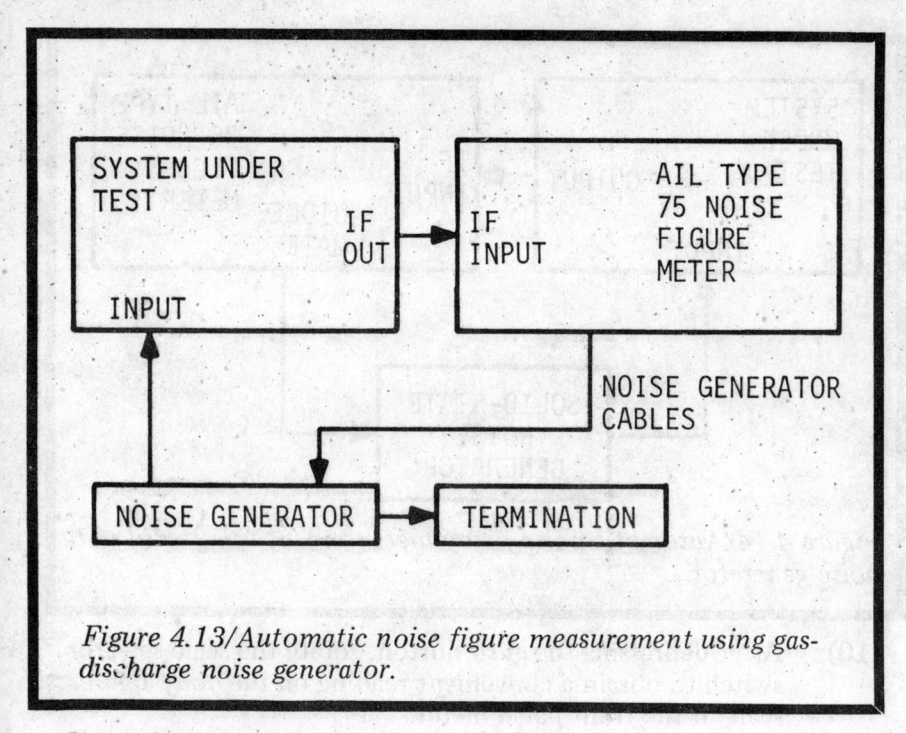

Figure 4.13/Automatic noise figure measurement using gas-discharge noise generator.

(2) Set *gas-diode* switch to *gas* position.

(3) Set *generator current* control fully counterclockwise.

(4) Depress *manual off* button.

(5) Set scale selector to *generator current*.

(6) Depress the *AC power* button on the noise figure meter and the system supply. Apply any dc power, also, that is required for circuit operation.

(7) Depress *manual on* button and adjust *generator current* control for the meter current setting recommended by tables supplied with the noise source.

(8) Depress CAL button and rotate the scale selector switch to any position except *generator current;* note that the AGC front panel indicator is illuminated. This indicates that the input signal is above the minimum required level. (If it is not, the gain of the device under test may be too low.)

(9) While observing the *calibrate* scale (green band) of the front panel meter, rotate the CAL ADJ control to obtain the appropriate excess noise ratio for the noise generator being used. (This is listed in the noise generator manual.)

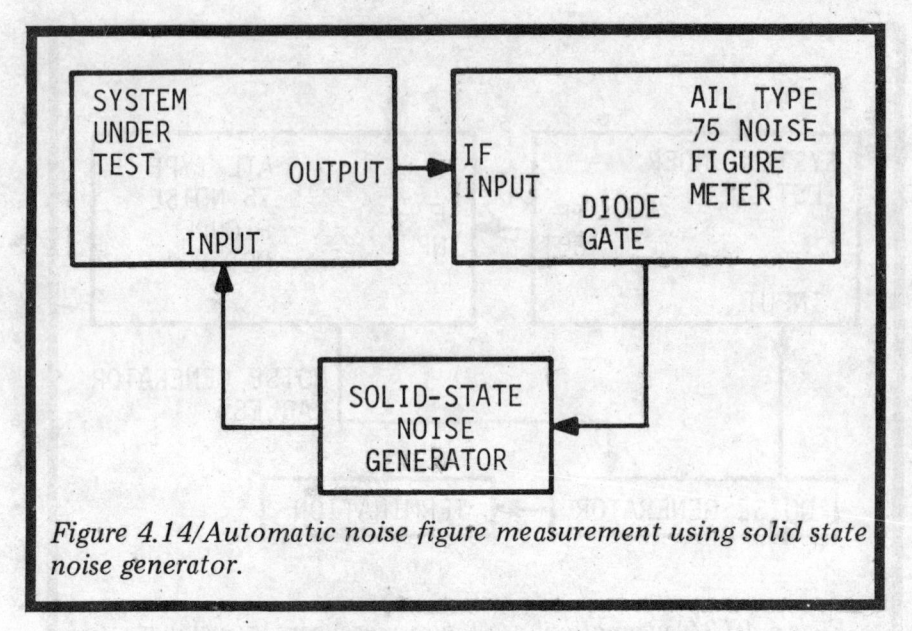

Figure 4.14/Automatic noise figure measurement using solid state noise generator.

(10) After depressing the *auto* button, rotate the scale selector switch to obtain a convenient reading on the *noise figure* scale of the front panel meter.
Caution: The AGC indicator must be on for all automatic noise figure readings; if it is not on, the gain of the device may be too low.

(11) Read the noise figure (in dB) of the system under test, on the meter.

It is obvious from the lack of formulas that this method of measurement lends itself to making multiple measurements. Since the noise figure is read directly, and no calculations are necessary, it is simply a matter of reading numbers from the front panel meter.

The second automatic setup is one that uses a solid-state noise generator as a source. Figure 4.14 shows the very simple block diagram of a measurement using the solid state noise source. It consists of a noise figure meter, a solid state noise generator, and the system to be tested. This is a very basic block with the basic components inserted for illustration. It may be necessary, however, to provide matching components if the system is sensitive to either input or output mismatches. The matching components could be circulators or attenuators, depending on the task to be performed. There also may have to be some sort of mixing circuitry if the output of the device being tested is not at a frequency that is compatible with the input of the noise figure meter.

Noise Measurements 165

(This area will be covered in the next section, where the HP340 noise figure meter is examined.)

The test procedure for Figure 4.14 is as follows:

(1) Connect the test setup as shown in Figure 4.14. *Do not apply ac power* to the system or device to be tested.
    Caution: The Solid State Noise Generators must be driven by their proper modulators for a correct noise figure reading. If uncertain of the diode gate (modulator) output, consult the appropriate manual.

(2) Set the *gas-diode* switch to the *diode* position.

(3) Depress *manual off* button.

(4) Set scale selector to any convenient noise figure position.

(5) Depress *ac power* button and the *manual on* button. Apply any external dc power that a device under test may require.

(6) Depress the CAL button. The AGC light must be on. If it is not, the input signal to the noise figure meter is below the minimum level required.

(7) Use the CAL ADJ control to set the front panel meter to the excess noise ratio (on the *calibrate* scale, in dB) of the solid state noise source. The scale selector may be in any position except *noise generator*. The system is now calibrated.

(8) Depress the *auto* mode button and set the scale selector for an on-scale meter reading.
    Caution: The AGC indicator light must be on for all automatic noise figure readings.

(9) Read the noise figure (in dB) of the device under test, on the meter.

Once again, you can see how easy it is to make and read noise figures with an automatic setup. You should also be able to see how multiple readings can be taken with speed and accuracy so that your particular band can be covered with a minimum amount of effort.

A third type of automatic noise figure measurement is that of a swept frequency setup. Figure 4.15 shows this type of setup. This setup, although primarily a swept setup, can also be used for single frequency measurements simply by eliminating the X-Y recorder and putting the sweeper in the CW mode.

The noise source shown in the test setup is usually a gas-discharge type. The diode noise source or the solid state noise source could also be

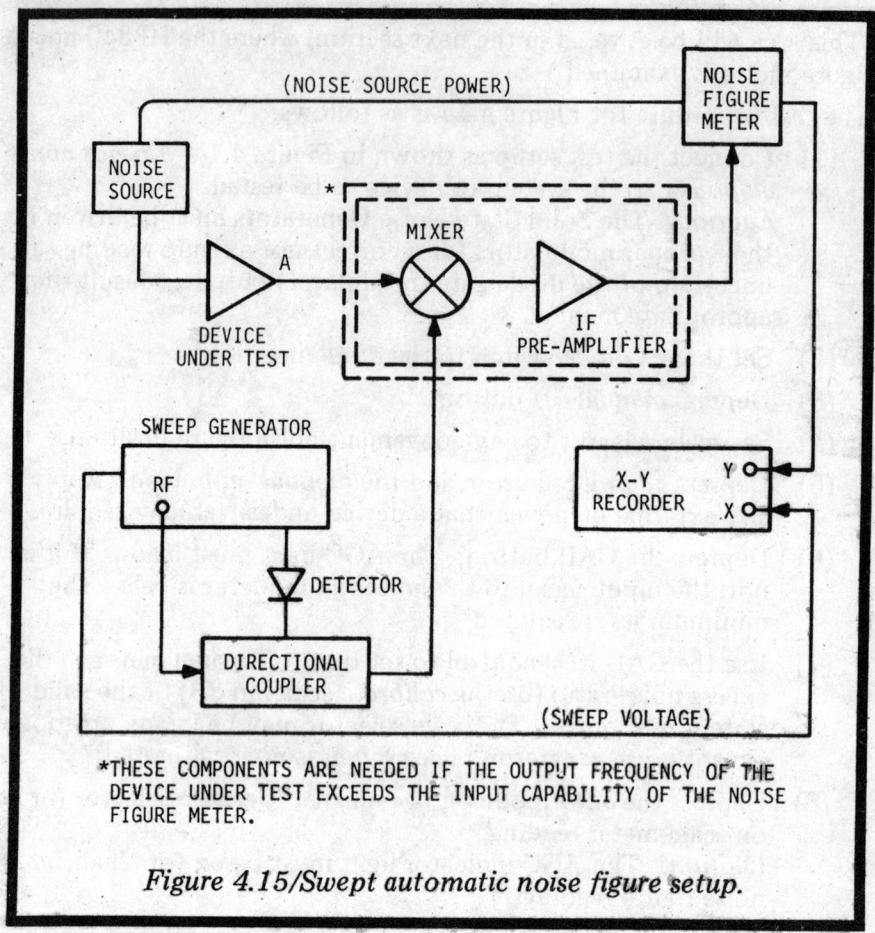

*THESE COMPONENTS ARE NEEDED IF THE OUTPUT FREQUENCY OF THE DEVICE UNDER TEST EXCEEDS THE INPUT CAPABILITY OF THE NOISE FIGURE METER.

Figure 4.15/Swept automatic noise figure setup.

used, depending on your particular application. The mixer and IF pre-amplifier, as stated in Figure 4.15, are used when the output frequency of the device under test exceeds the input capability of the noise figure meter. The automatic noise figure meter has a tuned input; that is, it cannot operate at just any frequency. The input could be set steps (30, 60, 120, 200 MHz) or a band of acceptable frequencies (10 - 1000 MHz). Regardless of how the input is tuned, however, the absolute maximum input frequency on currently available noise figure meters is 1 GHz. Therefore, the majority of the microwave spectrum has to undergo some sort of frequency conversion — the mixer, and its associated amplifier, provide that change. The amplifier also helps to increase the overall setup gain to the proper level, which may be indicated by a light or a meter reading of infinity (∞).

Noise Measurements

The sweep generator and its associated leveling circuitry provide the second signal to the mixer. If you are taking a single frequency measurement, it is simply a source for a second signal to be used for mixing; if you are taking swept frequency noise figure measurements, it also supplies the calibration lines on the X-Y recorder. To allow the system to have calibration lines, a step attenuator can be placed at point A to allow the insertion of various levels.

The noise figure meter has the ability to automatically display measured values while operating with either gas discharge, temperature-limited diode, or solid-state sources. Two important settings are to be made when using an automatic noise figure meter. The first is a zero setting, which sets a reference for all of your noise readings. The second is a check by which you can determine if there is a high enough level into the meter (a high enough gain in the system) to operate the meter properly, which may be a visual indication in the form of a light on the front panel or an "∞" setting on a meter. No meaningful data can be obtained unless the input level to the meter is sufficient.

The X-Y recorder, of course, is used to plot the noise figure as a function of frequency. Levels of "0" and "∞" can be used for initial calibration.

We will briefly outline a procedure for making a noise figure measurement; it is very general, since each noise figure meter made by a different manufacturer is unique in the way it is set up. It would be wise to consult the manual for your specific meter and learn its operation.

(1) Connect the setup as shown in Figure 4.15. Turn on all ac power.

(2) Adjust the sweeper for its proper frequency range and the noise source for its proper current (which can be found in the manual). If the mixer and preamplifier are used, be sure that an output frequency that is compatible with the input of the meter is produced.

(3) Set the meter "zero" and "∞" controls.

(4) Turn the noise figure meter to *noise figure* and read the value. If the mixer and preamplifier are used, some calculation is necessary because you are not only looking at the device under test but also at extra components (mixer and preamplifier) with specific noise figures and gains.

The formula used to calculate noise figure is

$$NF_s = NF_1 + \frac{(NF_2 - 1)}{G_1} + \frac{(NF_3 - 1)}{G_2} + \ldots \text{etc.}$$

where:

NF_s = noise figure as read on the meter
NF_1 = noise figure of the device under test
NF_2 = noise figure of the mixer
NF_3 = noise figure of the preamplifier
G_1 = gain of device under test
G_2 = gain of mixer
G_3 = gain of preamplifier

It can be seen that the overall noise figure of any cascaded amplifying system depends primarily on that of the first stage. The effects of subsequent stages are reduced by the gain up to that point. On the other hand, the use of a passive stage with gain less than 1 increases the importance of the subsequent stage's noise figure.

To illustrate how important the first stage noise figure is, and how the subsequent stage's noise figure contributes, we will consider an example.

Consider the following quantities:

Parameter	Noise Figure (dB)	Noise Factor
NF_1	13	20
NF_2	8	6.32
NF_3	6	4

G_1 = 20 dB (power ratio = 100)
G_2 = 10 dB (power ratio = 10)
G_3 = 20 dB (power ratio = 100)

We will now substitute the noise factor and power ratio numbers into the equation and solve for NF_s that we should read on the meter.

$$NF_s = \frac{(NF_2 - 1)}{G_1} + \frac{(NF_3 - 1)}{G_2}$$

Noise Measurements 169

$$NF_s = 20 + \frac{(6.32 - 1)}{100} + \frac{(4 - 1)}{(-10)}$$

(Note: The G_2 figure is negative since it is a loss in the mixer.)

$$NF_s = 20 + \frac{5.32}{100} + \frac{3}{(-10)}$$

$NF_s = 20 + 0.0532 - 0.3$

$NF_s = 19.7532$ (Noise Factor)

$NF_s = 10 \log \text{Noise Factor} = 10 \log 19.7532$

$NF_s = 10 (1.2956)$

$NF_s = 10 (1.2956)$

$NF_s = 12.95$ dB

You can see very clearly from the calculations above that the preceding stages have very little effect on the overall system noise figure. A device with a 13 dB noise figure registers 12.95 dB when used in conjunction with a mixer with an 8 dB noise figure and a 4 dB amplifier with 20 dB of gain. This is why there is such great care taken in choosing a preamplifier to be used in a receiver system. The noise figure of that device basically determines the noise figure of the system and thus, the sensitivity of the receiver.

4.5 Errors and Accuracy

As in any system, there are factors which decrease the overall accuracy of your measurements. Actual measurement techniques must consider the possibility of system errors caused by mismatch, temperature, and image and spurious responses. The instrument accuracy of the noise source and noise figure meter should also be taken into account.

Mismatch

Noise power obeys all power transfer laws but, since it is random in phase, mismatches cause ambiguous errors rather than known amounts of power loss.

The critical matching situation, then, involves the excess noise source. Noise sources are rated in available excess noise power, thus mismatches cause an ambiguous amount of excess noise power to be coupled to the system.

The difference in SWR of the noise generator between the hot (or on) condition and cold (or off) condition may have a significant effect upon the gain and noise figure of some high-gain, negative-input-resistance devices (parametric amplifiers, masers, etc.). The change in gain due to source impedance variations for high-gain amplifiers (greater than 15 dB) can be significant. One solution to the problem would be to use a 20 or 30 dB coupler. To provide good results, the test signal power should be at least 20 dB greater than the noise power at the amplifier input, but not large enough to cause limiting in the system.

A second solution would be to provide isolation, by means of a circulator or isolator, between the amplifier input and the noise source. This would reduce the impedance variation seen by the amplifier. The insertion loss of the isolator would add directly (in dB) to the noise figure of the device under test. This would be a low value, however — in the order of 0.3 to 0.5 dB for most microwave isolators.

Temperature

To obtain a value of excess noise, the standard temperature (T_0) is assumed to be 290°K (62.6°F). Manufacturers know the value of the noise source and rate it in terms of this standard value. A variation of 20° from the assumed 290°, for example, causes an error of about 0.3 dB in the measured noise figure. If you are trying to read a 3 dB noise figure, you have already introduced a 10% error.

Image and Spurious Response

When using a broadband excess noise source, an automatic noise figure meter measures the true noise figure of the total pass band of the device under test. If in its operation the device does not utilize the full pass band for signal information (as would be the case with a radar receiver with an image response), its operating noise figure is higher than that measured. This apparent value can be calculated by the equation:

Operating NF = NF (reading) + 10 log (B_t/B_u), where B_t is the total bandpass of the device, and B_u is the operational bandpass.

This equation is a convenient simplification that assumes constant gain in the device under test. Therefore, great care should be taken in noise figure measurements to be sure that you are measuring the desired frequency and not an image or spur.

Noise Measurements

Excess Noise Source Accuracy

Noise figure measurements assume an accurate knowledge of ENR, the excess noise ratio of the noise source being used. In general, the noise power output is not constant over the entire operating frequency range, or the same for all noise generators, gas-discharge or solid state.

In the case of gas-discharge noise generators, the excess noise ratio is a function of various physical characteristics of the tube, and of the coupling of the microwave structure to the discharge. The production tolerance of the tubes is ± 0.25 dB with an uncertainty on the center value of ± 0.2 dB. The coupling of the microwave structure to the discharge is related to the fired and unfired insertion loss of the gas-discharge noise generator.

In the case of solid state noise generators, the excess noise ratio is a function of the semiconductor fabrication, the coaxial mount, and the bias (power supply). Most solid state noise generators are designed for constant noise power outputs across their frequency bands. They exhibit less than ± 0.5 dB (maximum) deviation from a constant ENR of 15.5 dB across their specified frequency spectrums.

Any uncertainty in the ENR of a noise generator will cause a direct uncertainty in the noise figure measured, and should be taken into account when computing the total accuracy of a noise figure measurement, whether manual or automatic. When greater accuracy than is normally obtained is required, manufacturers can calibrate gas-discharge and solid state noise generators to uncertainties as low as ± 0.11 dB at specific frequencies with traceability to the National Bureau of Standards (NBS). This degree of accuracy is sometimes needed but, generally, the normal device accuracy is more than acceptable.

Noise Figure Meter Accuracy

Automatic noise figure meters are specified as accurate within ± 1/2 dB over most of their range, ± 1 dB over the remainder. This accuracy figure includes the effects of meter tracking, variation from square law, and aging effects. With the current specifications of noise sources, and the possible errors caused by mismatch and temperature, such meter accuracy would seem to be consistent with that of the overall system.

For measuring low-noise devices, modified noise figure meters which provide increased resolution and accuracy are available. The modification expands a 4 dB (up to 6 dB) portion of the scale over the full range of the meter scale. Noise figure meter accuracy in the expanded mode of operation is ± 0.2 dB; either expanded or normal operation can be selected.

Two additional factors must be taken into consideration when discussing noise measurement errors and accuracy. They are *cable loss* and *system bandwidth*.

Cable Loss

Any loss introduced between the noise generator and the device under test will cause an apparent change in the excess noise ratio of the generator. These losses should be measured accurately and subtracted (in dB) from the ENR of the generator. This corrected ENR should be used when setting up during the calibration mode of operation, or in the calculations for a manual measurement.

System Bandwidth

The system bandwidth of a noise figure measuring setup includes the instrument making the measurement, as well as the device under test. Therefore, if an amplifier with a 5 MHz bandwidth were measured with an instrument having a 1 MHz bandwidth, the result will, in general, be an optimistic (or low) noise figure reading. With this condition, it is necessary to make several measurements so as to have measured the spot noise figure of the device under test completely across its band of operation. These numbers are then averaged to give the true noise figure. This error can be eliminated by using an instrument with a bandwidth equal to, or wider than, the device under test.

Chapter Summary

The subject of noise and noise figure is one of great importance in microwaves and microwave receivers. Whether this noise is measured by means of a manual setup, an automatic setup covering many frequencies or by means of a spectrum analyzer (covered in detail in the following chapter), the result must be reliable and accurate readings so that your system can be properly characterized. Careful choice of components for your system will keep the noise figure low and sensitivity high.

Referral back to our basic definition will result in a clear understanding of what we are attempting to measure when we assemble a noise figure setup. That definition stated:

> *Noise is an internally generated interference which causes the circuit operation to be degraded from theoretical predictions.*

Thus, what we have at the end of a noise measurement is a reading of electrical interference with our desired microwave signal.

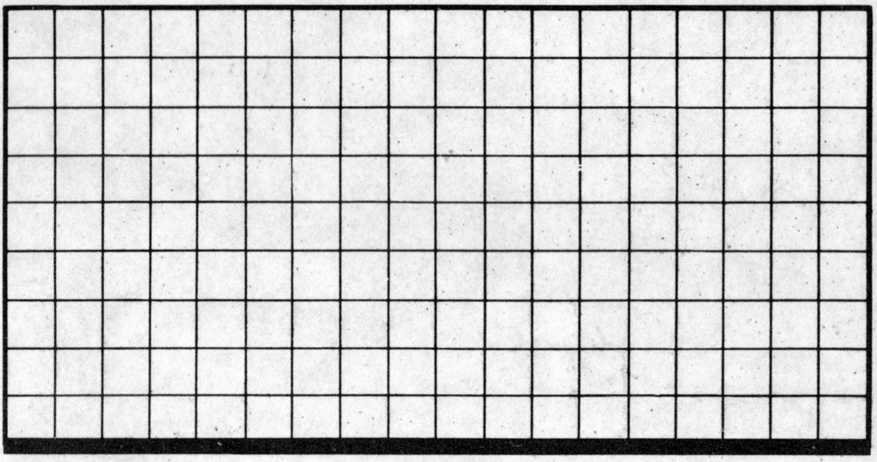

Spectrum Analyzer Measurements 5

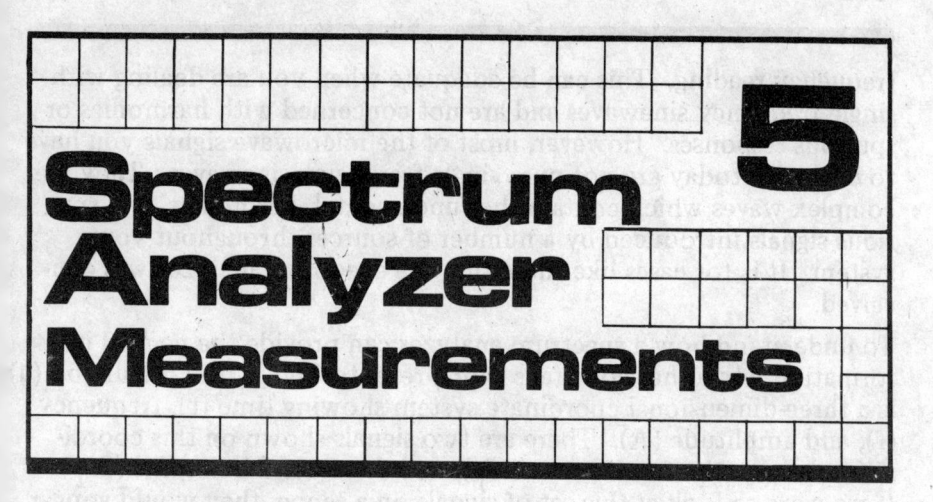

5. Definition

In Chapter 2, we referred to the spectrum analyzer as one of the most useful pieces of test equipment in microwaves today. In that chapter, the analyzer was covered very briefly, since our intention was simply to make you aware that such a piece of equipment existed. In this chapter, we will explain the analyzer and help you to realize that this opening description is a very large understatement.

The obvious questions that arise at this time are: "what is a spectrum analyzer?" and "why do we use one?" These are very fair questions to ask, and the answers will help to form a base for all of the discussions in this chapter.

First of all, there are probably many ways of defining a spectrum analyzer. One of the easiest ways is to compare it to an oscilloscope. You know that an oscilloscope displays the amplitude of a signal as a function of time. That is, the presentation represents a certain period of time. The start of the sweep (in the horizontal direction) is time zero and the remainder is calibrated in specific increments (milliseconds, microseconds, etc.), so that the duration of the signal can be read at any time. The vertical deflection on the face of the oscilloscope is, of course, proportional to the amplitude of the signal. This is also calibrated in a usable unit, volts/cm, for example.

The oscilloscope is very useful for measurements that are related to time. It does not, however, give much information with regards to frequency. It is true, of course, that you can measure the period of a sinewave display, take the reciprocal of the value, and obtain a frequency reading. This can be adequate when you are dealing with single frequency sinewaves and are not concerned with harmonics or spurious responses. However, most of the microwave signals you have to deal with today are not pure single frequency sinewaves. They are complex waves which contain the fundamental, harmonics, and spurious signals introduced by a number of sources throughout your system. It is for cases like these that the spectrum analyzer was conceived.

To understand how a spectrum analyzer can provide the needed information while the scope falls short, refer to Figure 5.1. Figure 5.1(a) is a three-dimensional coordinate system showing time (t), frequency (f), and amplitude (A). There are two signals shown on this coordinate system: the fundamental (f_1), and the second harmonic ($2f_1$). If we were to look at this set of signals on a scope, they would appear as shown in Figure 5.1(b). This figure shows three signals. They are: the fundamental (f_1), the second harmonic ($2f_1$), and the combination of the two ($f_1 + 2f_1$). In reality, the only signal you will see on the scope is $f_1 + 2f_1$. From this display you can plainly see that you are not looking at a single frequency signal. This becomes obvious, since the lower portion of the display is not a smooth sinewave. As we said, you realize that you are observing something other than a single frequency, but what other frequencies are present and how large are they? From the oscilloscope display, there is really no way of separating the harmonics from the fundamental frequency. This is also true when using other measuring instruments that either do not measure in the frequency domain or present only a narrow view of the frequency domain. One example of such a measuring device that can give a false reading is the power meter. There may be times when you are measuring a device that has little or no filtering at its output, thus producing many harmonics. The power sensor cannot distinguish the harmonics from the fundamental frequency, and thus reacts to the energy produced by the total of all of them. You will read a power level which is much higher than is actually present, simply because you are reading all of the harmonics power along with the fundamental power. Typical components for which you should not use a power meter as an output indicator are frequency multipliers and comb generators.

Another example of a device presenting only a narrow view of the frequency domain is the frequency counter. How many times have you obtained very erratic readings on a counter when you felt that you had a good signal going to the counter? Once again, you have a

Spectrum Analyzer Measurements

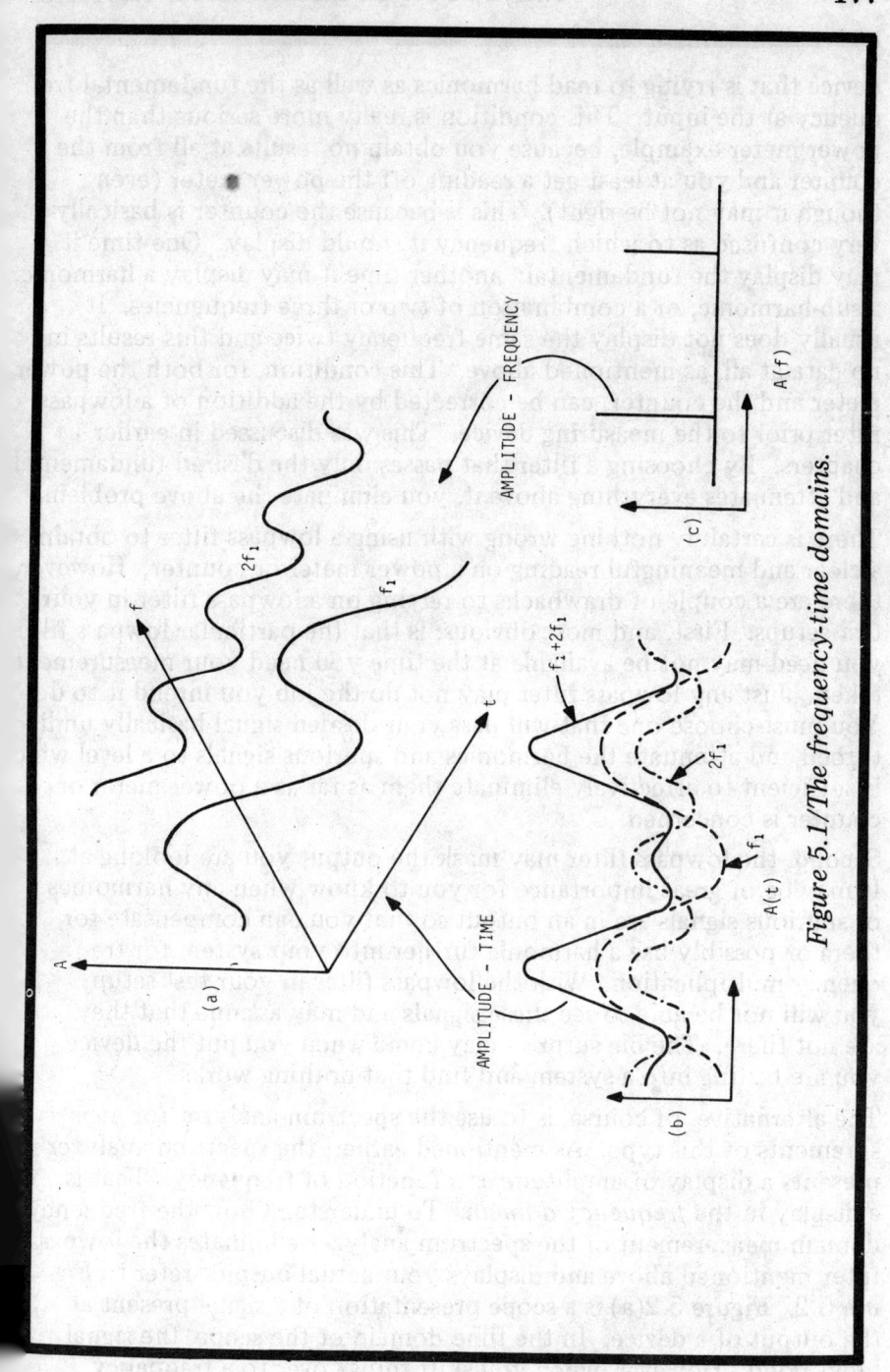

Figure 5.1/The frequency-time domains.

device that is trying to read harmonics as well as the fundamental frequency at the input. This condition is really more serious than the power meter example, because you obtain no results at all from the counter and you at least get a reading off the power meter (even though it may not be right). This is because the counter is basically very confused as to which frequency it should display. One time it may display the fundamental; another time it may display a harmonic, a sub-harmonic, or a combination of two or three frequencies. It usually does not display the same frequency twice and this results in no data at all, as mentioned above. This condition, for both the power meter and the counter, can be corrected by the addition of a lowpass filter prior to the measuring device. This was discussed in earlier chapters. By choosing a filter that passes only the desired fundamental and attenuates everything above it, you eliminate the above problem.

There is certainly nothing wrong with using a lowpass filter to obtain a clear and meaningful reading on a power meter or counter. However, there are a couple of drawbacks to relying on a lowpass filter in your test setups. First, and most obvious, is that the particular lowpass filter you need may not be available at the time you need your measurement taken. Just any lowpass filter may not do the job you intend it to do. You must choose one that will pass your desried signal basically undisturbed and attenuate the harmonics and spurious signals to a level which is sufficient to effectively eliminate them as far as a power meter or counter is concerned.

Second, the lowpass filter may mask the output you are looking at. It may be of great importance for you to know when any harmonics or spurious signals are in an output so that you can compensate for them or possibly use a harmonic further into your system for frequency multiplication. With the lowpass filter in your test setup, you will not be able to see these signals and may assume that they are not there. The big surprise may come when you put the device you are testing into a system and find that nothing works.

The alternative, of course, is to use the spectrum analyzer for measurements of this type. As mentioned earlier, the spectrum analyzer presents a display of amplitude as a function of frequency. That is, a display in the *frequency domain*. To understand how the frequency domain measurement of the spectrum analyzer eliminates the lowpass filter mentioned above and displays your actual output, refer to Figure 5.2. Figure 5.2(a) is a scope presentation of a signal present at the output of a device. In the time domain of the scope, the signal looks good. However, when you shift things over to a frequency

Spectrum Analyzer Measurements

domain by using a spectrum analyzer, there is a much different picture presented. With the analyzer you can see harmonic distortion associated with the signal that did not show up on the scope. This is shown in Figure 5.2(b).

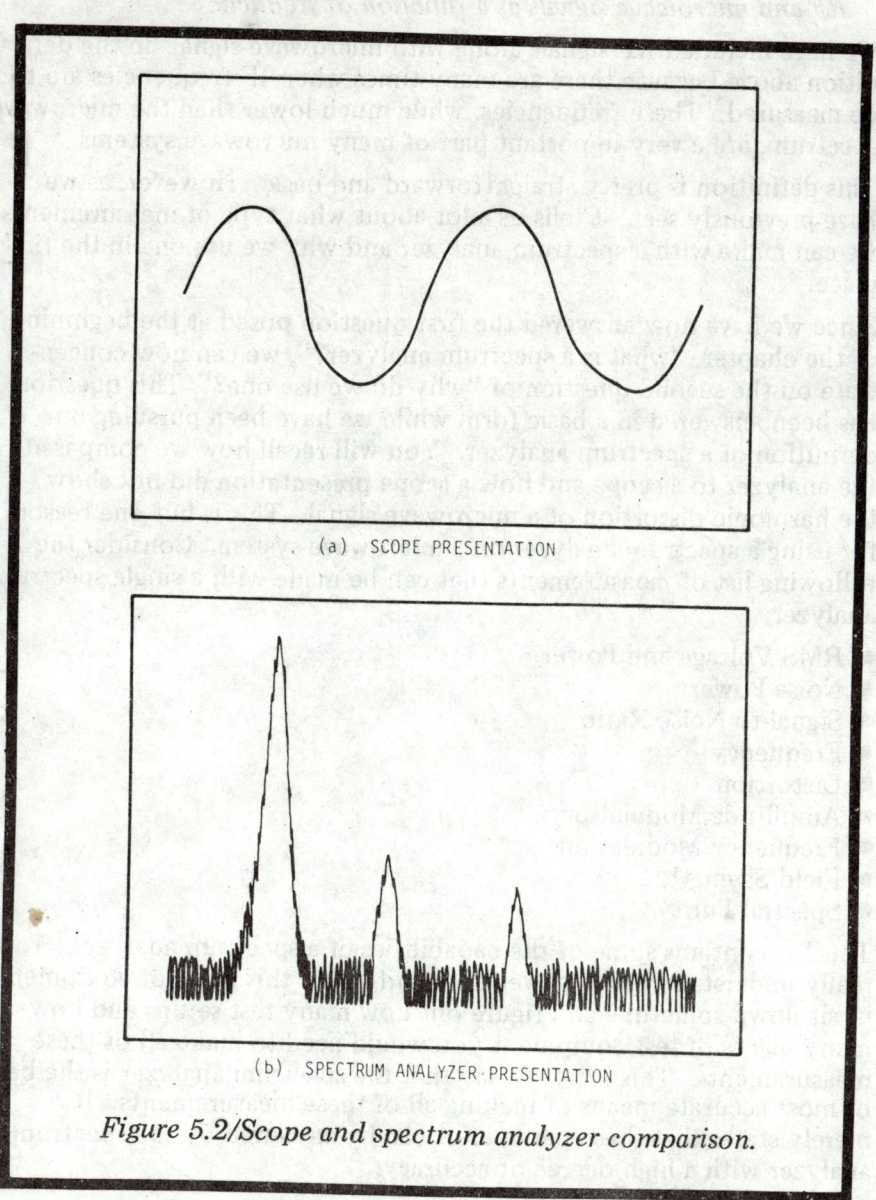

(a). SCOPE PRESENTATION

(b). SPECTRUM ANALYZER PRESENTATION

Figure 5.2/Scope and spectrum analyzer comparison.

We are at a point where a definition of the spectrum analyzer should be possible. It has actually been defined, indirectly, many times in our discussions above. To formalize it, we will state the following:

A spectrum analyzer is an instrument which is used to display RF and microwave signals as a function of frequency.

We have included RF signals along with microwave signals in the definition above because there are many times when IF frequencies are to be measured. These frequencies, while much lower than the microwave spectrum, are a very important part of many microwave systems.

This definition is pretty straightforward and basic. However, as we have previously seen, it tells us a lot about what type of measurements we can make with a spectrum analyzer and why we use one in the first place.

Since we have now answered the first question posed at the beginning of the chapter, "what is a spectrum analyzer?", we can now concentrate on the second question of "why do we use one?" This question has been answered in a basic form while we have been pursuing our definition of a spectrum analyzer. You will recall how we compared the analyzer to a scope and how a scope presentation did not show the harmonic distortion of a microwave signal. This is but one reason for using a spectrum analyzer on a microwave system. Consider the following list of measurements that can be made with a single spectrum analyzer:

- RMS Voltage and Power
- Noise Power
- Signal-to-Noise Ratio
- Frequency
- Distortion
- Amplitude Modulation
- Frequency Modulation
- Field Strength
- Spectral Purity

This list contains some of the capabilities of a spectrum analyzer. To really understand just how versatile and useful this piece of equipment is, sit down sometime and figure out how many test setups and how many pieces of test equipment you would need to make all of these measurements. This is not to say that the spectrum analyzer is the best or most accurate means of making all of these measurements. It merely states that these parameters can be measured by the spectrum analyzer with a high degree of accuracy.

Spectrum Analyzer Measurements

Obviously, a piece of test equipment designed to make only one specific measurement is much more accurate than one that is more of a general purpose device. If the spectrum analyzer could make all of the measurements listed above, with the same accuracy and ease as the equipment specifically designed for those measurements, the only instrument you would need in your lab would be a spectrum analyzer. This, of course, is not the case. The point that is being made here is that the analyzer is a very versatile piece of equipment that can be used to make a variety of measurements where specific instruments are not available due to previous commitments elsewhere or simply because a lab or company cannot afford to have all of them. And so, these are some of the reasons that we use a spectrum analyzer.

We have discussed what a spectrum analyzer is and why we use one. Now we should explain what parameters to look for in a good spectrum analyzer. By a *good* spectrum analyzer, we mean one that will do the job that we need done. Therefore, the order of importance of the parameters may vary according to your particular application. Some important parameters to consider when choosing your spectrum analyzer are:

- Frequency Range
- Resolution
- Stability
- Frequency Accuracy
- Amplitude Accuracy
- Dynamic Range
- Residual FM
- Noise

To aid in choosing the *right* analyzer for your application, we will investigate all of the parameters listed and show typical specifications available on units today.

Frequency Range

This must be your first consideration when choosing a spectrum analyzer. The analyzer you choose should cover the range in which you plan to take measurements. You should also consider an analyzer that will cover the harmonics which your circuit or device may generate as well as any IF that may be present on the low end of the spectrum (if receivers are being checked). This parameter can be likened to the process of buying a shirt. You first have to find the right size; considerations of color, style, etc., follow after that. The same is true of the spectrum analyzer; you first find the proper frequency range (size), and then consider the other factors.

The frequency ranges of commercially available spectrum analyzers vary depending on the manufacturer that you consult. Some typical analyzers are listed below:

Manufacturer	Frequency Range
A	1 MHz to 20 GHz (Extended to 40 GHz)
B	1 MHz to 1 GHz
C	10 MHz to 22 GHz (extended to 40 GHz)
D	100 MHz to 1.5 GHz
E	1 kHz to 1.8 GHz
F	15 GHz to 18 GHz (Extended to 60 GHz)

You will notice that some of the frequency ranges listed above have an additional note printed next to them. This note says "extended to 40 or 60 GHz". This is added because the spectrum analyzers built today are not capable of reading directly frequencies in the millimeter wave bands. Therefore, there must be a means of extending the range of the analyzer if you plan to use it in these ranges. The process of *harmonic mixing* is used to extend the range in most cases.

To understand harmonic mixing, we should first refer to a basic block diagram of a spectrum analyzer input section. This is shown in Figure 5.3 with the following relationships:

F_{LO} = Frequency of local oscillator

F_S = Frequency of input signal

F_{IF} = First IF

A response will appear on the analyzer whenever $F_{LO} - F_S = F_{IF}$ This is the basic tuning equation for such a spectrum analyzer. In this case, the first IF and the local oscillator frequencies are higher than any input signal to be observed. For example, a 0-1500 MHz spectrum analyzer might be constructed by using a 2 GHz first IF and a 2-3.5 GHz local oscillator.

To extend the frequency coverage of the analyzer, it would be necessary to extend the local oscillator frequency range. However, for coverage into millimeter wave regions, an extremely high local oscillator frequency would be required. Clearly, it would be impractical to build, say, a 40-80 GHz oscillator to allow 40 GHz coverage on a spectrum analyzer.

Spectrum Analyzer Measurements 183

Stability, accuracy, and technology are all limiting factors in high frequency oscillator design.

Well, then, how can frequency be extended? One way would be to remove the low pass filter at the input. In this way, the tuning equation could be modified to allow signals higher in frequency than the IF. That is, either equation below would describe the tuning of the analyzer.

$$F_S - F_{LO} = F_{IF}$$
$$F_{LO} - F_S = F_{IF}$$

Or, simplifying:

$$|F_S - F_{LO}| = F_{IF}$$

Then:

$$F_S - F_{LO} = \pm F_{IF}$$

and

$$F_S = F_{LO} \pm F_{IF}$$

Figure 5.3/Spectrum analyzer basic block design.

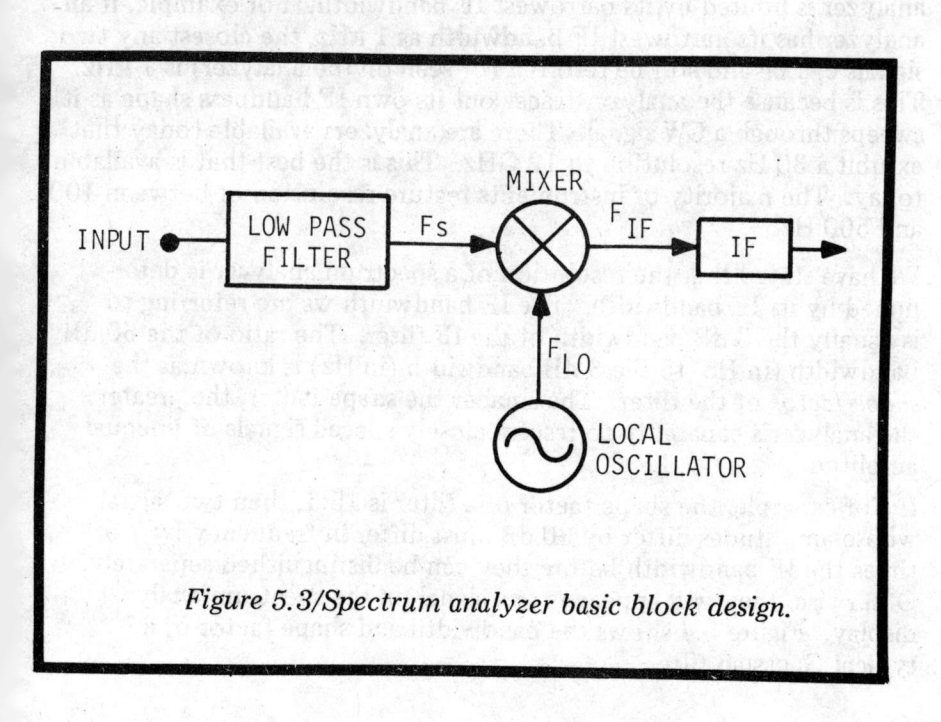

Although this more general equation allows extended coverage, there are still limitations due to the available range of local oscillator frequencies.

Consider the effect of creating harmonics of the local oscillator in the input mixer. This would allow high frequency local oscillator signals while using a lower frequency LO. For example, although a 2-4 GHz LO is used, the 10th harmonic goes from 20 to 40 GHz, thus satisfying the need for a high frequency LO. The tuning equation then becomes:

$$F_S = nF_{LO} \pm F_{IF}, \quad n = \text{the harmonic number}$$

This is the general tuning equation for a harmonic mixing spectrum analyzer. The input signal can mix the fundamental or any of the harmonics of the local oscillator to produce the proper IF.

Resolution

Before the frequency of a signal can be measured on a spectrum analyzer, it must first be resolved. Resolving a signal means, basically, distinguishing it from its nearest neighbors. The resolution of a spectrum analyzer is limited by its narrowest IF bandwidth. For example, if an analyzer has its narrowest IF bandwidth as 1 kHz, the closest any two signals can be and still be resolved (or seen on the analyzer) is 1 kHz. This is because the analyzer traces out its own IF bandpass shape as it sweeps through a CW signal. There are analyzers available today that exhibit a 30 Hz resolution to 12 GHz. This is the best that is available today. The majority of instruments feature resolution of between 100 and 500 Hz.

We have stated that the resolution of a spectrum analyzer is determined by its IF bandwidth. The IF bandwidth we are referring to is usually the 3 dB bandwidth of the IF filter. The ratio of the 60 dB bandwidth (in Hz) to the 3 dB bandwidth (in Hz) is known as the *shape factor* of the filter. The smaller the shape factor, the greater the analyzer's capability to resolve closely spaced signals of unequal amplitude.

If, for example, the shape factor of a filter is 15:1, then two signals whose amplitudes differ by 60 dB must differ in frequency by 7.5 times the IF bandwidth before they can be distinguished separately. Otherwise, they will appear as one signal on the spectrum analyzer display. Figure 5.4 shows the bandwidth and shape factor of a typical Gaussian filter.

Stability

Stability is usually characterized as either short term or long term and is important since the stability of the analyzer should be better than that of the signal you are measuring. Residual FM is a measure of the short term stability and is usually expressed in Hz peak-to-peak. The residual FM of an analyzer is actually a measure of the internal jitter of the instrument. Short term stability is also characterized by noise sidebands which are a measure of the spectral purity of the analyzer. Noise sidebands are specified in terms of dB down a specified number of Hz away from the carrier in a specific bandwidth. An example of noise sidebands and residual FM (jitter) is shown in Figure 5.5. Figure 5.5(a) shows a signal with noise sidebands indicating poor spectral purity. Figure 5.5(b) shows how the residual FM of the signal makes it move back and forth on the analyzer display. Typical values for these parameters would read as follows:

Residual FM — 150 Hz peak-to-peak
Noise Sidebands — > 70 dB down 50 kHz away in a 1 kHz bandwidth.

As previously stated, the Residual FM specification is presented as a certain number of Hz peak-to-peak (150 Hz peak-to-peak); and the Noise Sideband specification is presented as a dB figure down (70 dB) a specified number of Hz away from the carrier (50 kHz) in a specified bandwidth (1 kHz). These are very representative values for spectrum analyzers available today.

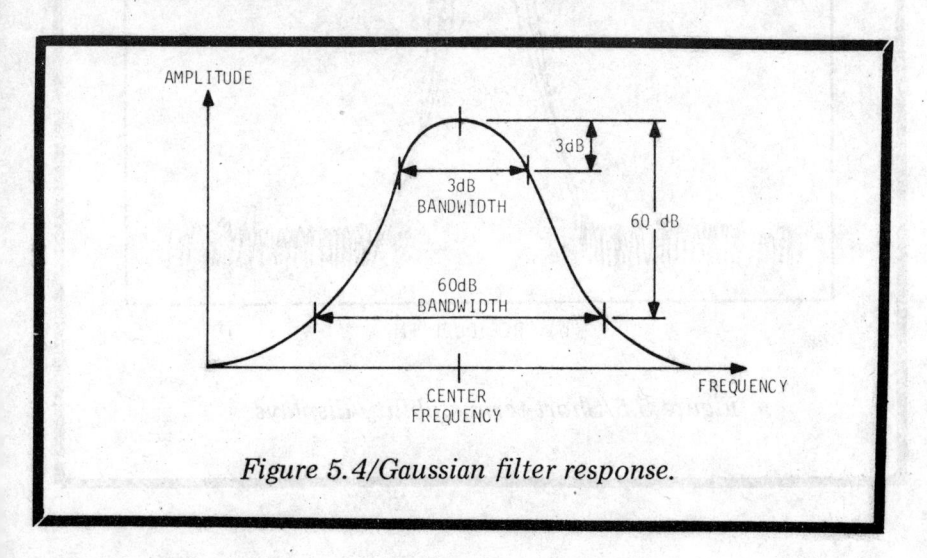

Figure 5.4/Gaussian filter response.

(a) NOISE SIDEBANDS

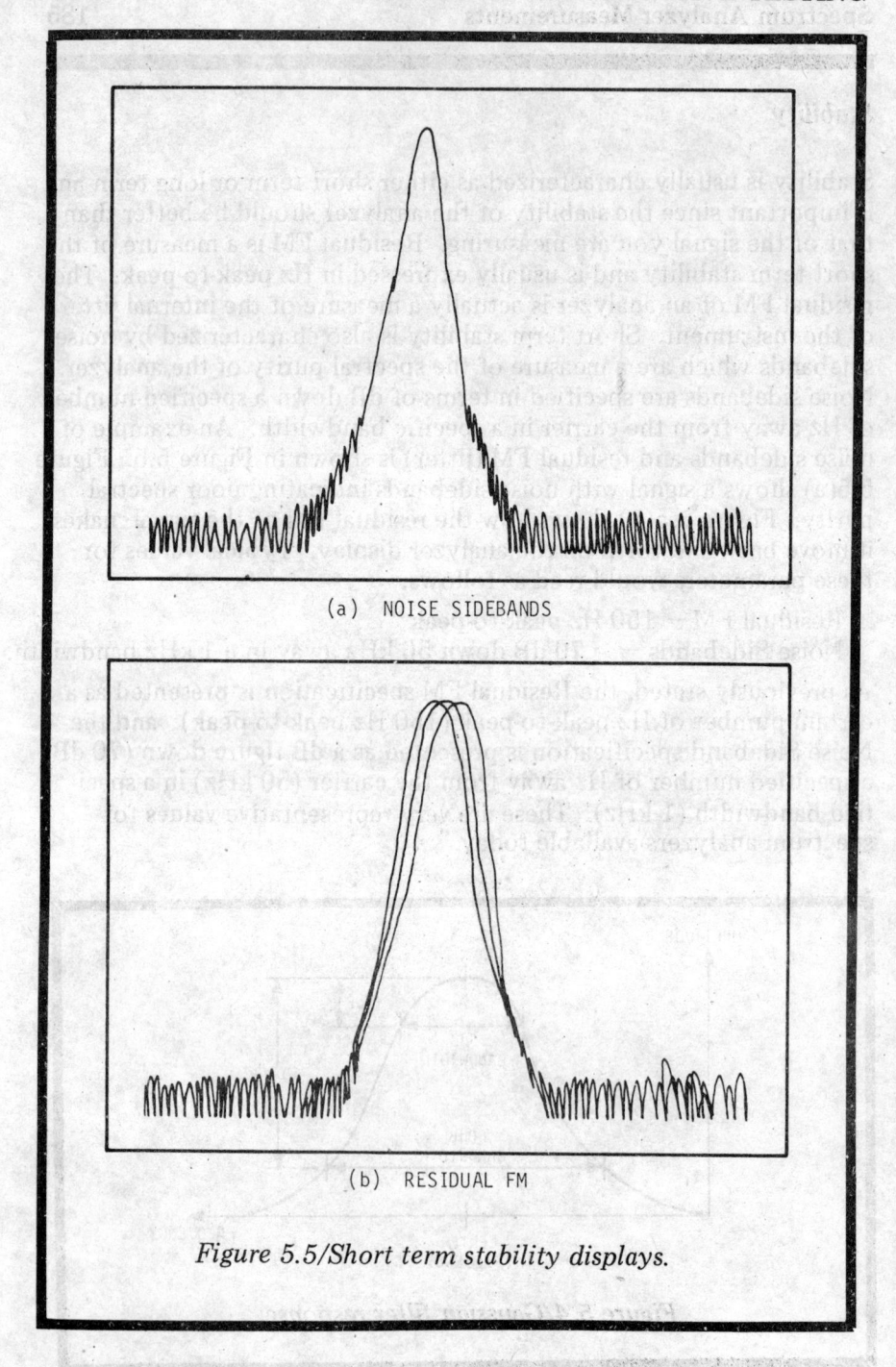

(b) RESIDUAL FM

Figure 5.5/Short term stability displays.

Spectrum Analyzer Measurements

Frequency Accuracy

In order to conduct meaningful frequency measurements, the accuracy of the instrument must be specified. The accuracy that the instrument exhibits must be better than the frequency (or frequencies) you are trying to measure. This stands to reason, since you certainly cannot possibly read an accuracy better than that of the instrument you are using for the measurement. The accuracy of a spectrum analyzer is usually expressed as a ± MHz figure, a percentage, or a combination of the two. The method used depends on the manufacturer and how he feels most comfortable in specifying his equipment. One manufacturer specifies frequency accuracy as:

± 6 MHz from 1 MHz to 1 GHz
± 8 MHz from 1 GHz to 2 GHz
0.2% from 2 GHz to 20 GHz

Another specifies it as: ± (5 MHz + 20% of Span), where the Span is the total range of frequencies being swept by the analyzer.

Another specifies it as: N × (± 15 MHz), where N is the mixing mode being used.

As we have said, the way that frequency accuracy is specified depends on individual manufacturers. The type that you choose for your particular job should be looked at very closely to ensure that the accuracy being specified is adequate for your requirements.

Amplitude Accuracy

Amplitude accuracy refers to the error in the amplitude calibration of the analyzer. The error is not a single quantity but, rather, a combination of errors from several components within the analyzer. Some of these errors are: system flatness, input attenuator characteristics, bandwidth characteristics, IF gain variations of log amplifiers, and errors in any built-in calibrator the analyzer uses.

If an IF substitution scheme is used in the instrument, the errors can be reduced to the sum of the frequency response variations and the IF gain accuracies mentioned above.

The importance of amplitude accuracy is realized most when absolute power readings are needed. It is important to know exactly what level you are at and what tolerance you can expect your readings to maintain.

When a spectrum analyzer can make both absolute and relative power level measurements, it is said to be absolute amplitude calibrated. To be absolute amplitude calibrated, the analyzer must satisfy these requirements:

(1) The input attenuator must be flat to maintain the overall frequency response of the system.

(2) The input mixer must be flat or gain compensated over the frequency ranges of the input and LO.

(3) The IF attenuator must be accurate for proper amplitude display.

(4) The log/linear amplifier must be extremely accurate.

(5) The sweep times must be slow enough to allow the IF and video filters to respond fully, or an uncalibrated situation will result.

(6) Uncalibrated situations should be indicated by an uncal warning light, or avoided by automatic adjustment of the sweep time for any setting of the video or IF bandwidth or frequency sweep width.

(7) It must indicate the absolute signal level represented by some point on the CRT for any control settings, and an amplitude internal reference must be supplied.

Typical values of amplitude accuracies can range from ± 0.5 dB to ± 1.5 dB for typical analyzers available today.

Dynamic Range

Dynamic range, as it applies to spectrum analyzers, is defined as the ratio of the largest signal to the smallest signal that can be displayed simultaneously with no analyzer distortion products. There is an optimum signal level associated with dynamic range; reducing the signal level at the input mixer by 10 dB increases dynamic range by at least 10 dB, assuming there is sufficient sensitivity. This definition of dynamic range means distortion-free display range. All signals displayed on the CRT are real signals and not analyzer distortion or spurious mixing products. This definition is not the same as the ratio of the largest signal that can be applied without serious distortion (which usually means 1 dB amplitude compression) to the smallest signal that can be detected. Likewise, it is not the ratio of the largest to the smallest signal that can be measured by the analyzer.

To illustrate how dynamic range is characterized, refer to Figure 5.6.

Spectrum Analyzer Measurements

The maximum input level to the spectrum analyzer is the damage level or burn-out level of the input circuit. This is typically +13 dBm for the input mixer and +20 and +30 dBm for the input attenuator.

Before reaching the damage level of the analyzer, the analyzer will begin to gain compress the input signal. This gain suppression is not considered serious until it reaches 1 dB. The maximum input signal level which will always result in less than 1 dB gain compression is called the linear input level. Above 1 dB gain compression the analyzer is considered to be operating nonlinearly because the signal amplitude displayed on the CRT is not an accurate measure of the input signal level.

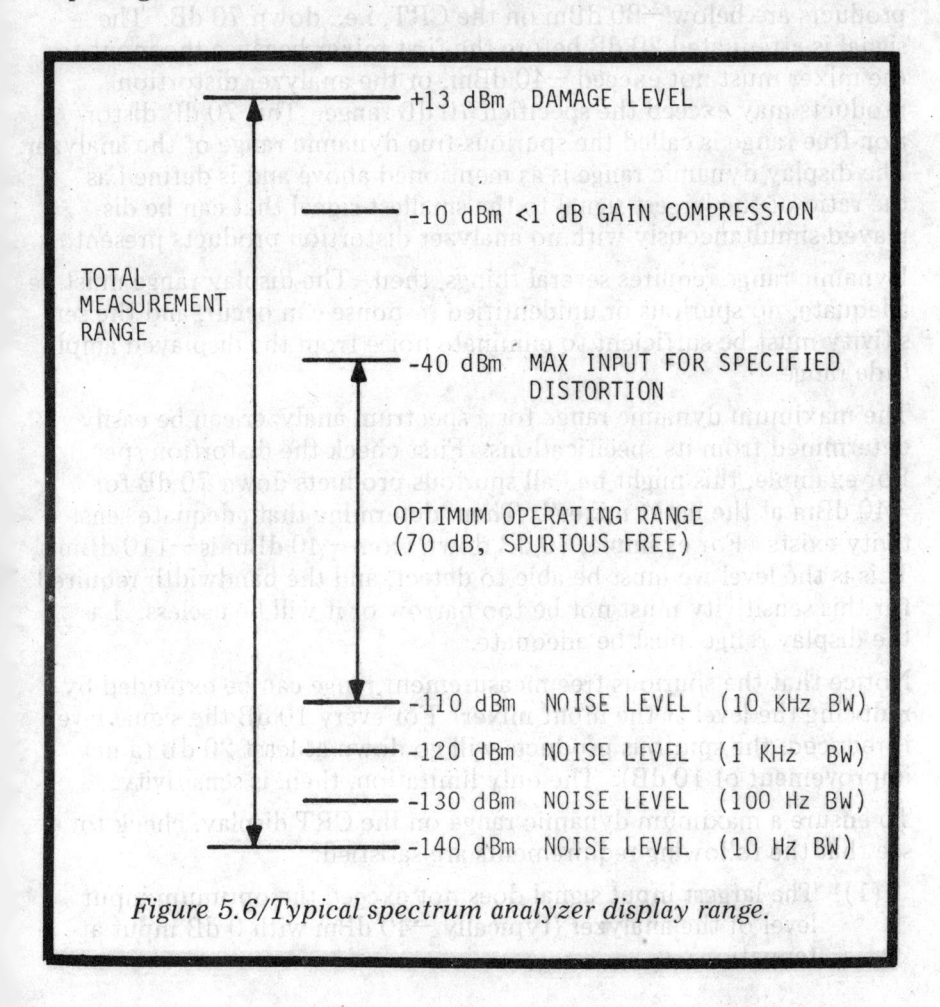

```
TOTAL
MEASUREMENT
RANGE
```

+13 dBm   DAMAGE LEVEL

−10 dBm   <1 dB GAIN COMPRESSION

−40 dBm   MAX INPUT FOR SPECIFIED DISTORTION

OPTIMUM OPERATING RANGE
(70 dB, SPURIOUS FREE)

−110 dBm   NOISE LEVEL   (10 KHz BW)
−120 dBm   NOISE LEVEL   (1 KHz BW)
−130 dBm   NOISE LEVEL   (100 Hz BW)
−140 dBm   NOISE LEVEL   (10 Hz BW)

Figure 5.6/Typical spectrum analyzer display range.

Whenever a signal is applied to the input of the analyzer, distortion products are produced within the analyzer itself. These distortion products are usually produced by the nonlinear behavior of the input mixer. They are typically 70 dB below the input signal level for signal levels not exceeding −10 dBm at the input of the first mixer. To accommodate larger input signal levels, an attenuator is placed in the input circuit before the first mixer. The largest input signal that can be applied at each setting of the input attenuator, while maintaining the internally generated distortion products below a certain level, is called the optimum input level of the analyzer. For example, a −20 dBm optimum level setting means that all analyzer distortion products are below −90 dBm on the CRT, i.e., down 70 dB. The signal is attenuated 20 dB before the first mixer because the input to the mixer must not exceed −40 dBm, or the analyzer distortion products may exceed the specified 70 dB range. This 70 dB distortion-free range is called the spurious-free dynamic range of the analyzer. The display dynamic range is as mentioned above and is defined as the ratio of the largest signal to the smallest signal that can be displayed simultaneously with no analyzer distortion products present.

Dynamic range requires several things, then. The display range must be adequate, no spurious or unidentified response can occur, and the sensitivity must be sufficient to eliminate noise from the displayed amplitude range.

The maximum dynamic range for a spectrum analyzer can be easily determined from its specifications. First check the distortion spec. For example, this might be "all spurious products down 70 dB for −40 dBm at the input mixer". Then, determine that adequate sensitivity exists. For example, 70 dB down from −40 dBm is −110 dBm. This is the level we must be able to detect, and the bandwidth required for this sensitivity must not be too narrow or it will be useless. Last, the display range must be adequate.

Notice that the spurious-free measurement range can be extended by reducing the level at the input mixer. For every 10 dB the signal level is reduced, the spurious products will go down at least 20 dB (a net improvement of 10 dB). The only limitation, then, is sensitivity.

To ensure a maximum dynamic range on the CRT display, check to see that the following requirements are satisfied:

(1) The largest input signal does not exceed the optimum input level of the analyzer (typically −40 dBm with 0 dB input attenuation).

Spectrum Analyzer Measurements   191

(2) The peak of the largest input signal rests at the top of the CRT display (reference level).

Typical values for dynamic range are from 60 to 100 dB, depending on the frequency range of the analyzer and the manufacturer. You are usually able to obtain 70 dB or greater from just about any analyzer you can buy.

Residual FM and Noise

These two terms are grouped together, since they both are a measure of short term stability and have been covered previously. They will be only briefly mentioned again so that you may be aware of the importance they have in choosing a spectrum analyzer.

You will recall that the residual FM is also called internal jitter of the analyzer and is specified in Hz peak-to-peak. Typical values are specified at 150 Hz peak-to-peak.

Noise sidebands are a measure of spectral purity of the analyzer. This term is specified as dB down a certain number of Hz away in a specified bandwidth. Typical values are 70 dB down 50 kHz away in a 1 kHz bandwidth.

We have now presented a basic definition of a spectrum analyzer, explained why we use one, and have shown important parameters to consider when choosing an analyzer. We can now proceed with the measurements that this valuable piece of equipment can perform.

5.1 Power Measurements

Most spectrum analyzers available today are capable of measuring both relative and absolute power. A relative power measurement is basically a comparison reading. A reference level is set up (usually one that is convenient for the indicating device being used), the component inserted into the test set-up, and the new level is compared to the reference. This type of measurement is expressed in dB which is a relative measure of power.

An absolute measurement is one which has a tangible quantity related to it. In the case of the power measurement this is the dBm. For absolute readings, it is only necessary to ensure that the analyzer is calibrated. You then can apply the signal to be measured and read its absolute power level directly off the CRT (in dBm).

To understand the difference in relative and absolute power, we should provide a definition of the units used to measure these powers.

The dB (decibel) is a unit used to express the ratio between two amounts of power, P_1 and P_2, existing at the two points. This is a measure of relative power between the two points and is dimensionless because the units of both powers are watts. Expressed methematically:

$$\text{dB (power)} = 10 \log_{10} \frac{P_1}{P_2}$$

There are also relationships for voltage and current expression in dB. They are:

$$\text{dB (voltage)} = 20 \log_{10} \frac{E_1}{E_2}$$

$$\text{dB (current)} = 20 \log_{10} \frac{I_1}{I_2}$$

The use of dB has two advantages. First, the range of numbers commonly used is more compact; for example, +63 dB to −153 dB is more concise than 2×10^6 to 0.5×10^{-15}. The second advantage is apparent when it is necessary to find the gain of several cascaded devices. Multiplication is then replaced by the addition of the power gain in dB for each device.

The dB discussed above was referred to as a relative parameter, since it is the ratio of two powers, each of which could vary. The dBm does away with this relative status and allows absolute power expression. The formula for dBm looks very similar to that used for dB:

$$\text{dBm} = 10 \log_{10} \frac{P_1}{1 \text{ mW}}$$

The one difference, which makes absolute readings possible, is that P_1 is the only quantity that can vary. The denominator (formally a P_2 variable) is a constant reference of 1.0 milliwatt. An example of how the formula gives absolute readings is an amplifier that has an output of 13 dBm. By solving for P in the above equation, the power output is found to be 20 mW (an absolute reading of power). Notice the difference from saying that an amplifier has 13 dB of gain, for example. All this is saying is that there is an increase in signal from input to output of 13 dB (a factor of 20). It says nothing about how much power is coming out of the amplifier, it only tells you that there is 20 times more (13 dB) coming out than you put in. If, for example, you specified that this same amplifier had an input applied at 0 dBm (this is 1.0

Spectrum Analyzer Measurements

mW if you run through the formula), you would then know that the absolute output level of the 13 dB amplifier was +13 dBm (or 20 mW). So you can see that dBm actually means "dB above 1 milliwatt," but a negative number of dBm is to be interpreted as "dB below 1 milliwatt."

An example of a negative dBm number would be when you are testing an attenuator. Suppose we have a 20 dB attenuator and apply a +7 dBm (5.0 mW) input to it. At the output we would have a signal that is 20 dB lower, or −13 dBm (0.05 mW). This means that the output is 13 dB below 1 milliwatt or 1/20 of a milliwatt; which it is.

You can see from the examples above the advantages of dBm and how they are identical to those listed for dB. The only difference to remember is that dB results in relative readings and dBm in absolute readings.

With all of the necessary definitions and guidelines set down, let us proceed with the main objective of this chapter — the measurement of power with a spectrum analyzer.

There are times when you will only need a relative indication of how a particular device or system is performing. It is at times like these that you will set a reference level, insert the device to be tested, and compare the new reading to the reference level. We have very basically outlined a relative power measurement test procedure. Let us now explain in more detail how a relative measurement would be taken with a spectrum analyzer.

To illustrate a procedure for making relative power measurements, we will present an example and follow it through. The problem we will address in this example is to characterize a 10 dB attenuator in the 2 to 4 GHz frequency range. Equipment that is available for the characterization includes a sweep generator, a spectrum analyzer, and an assortment of lab attenuators. We should emphasize here that these are only typical, everyday lab attenuators, and not a collection of high priced bureau of standards calibrated attenuators. This is not to say that calibrated attenuators could not be used if you have some available. We are just saying that calibrated attenuators are not needed since the setup will provide overall system calibration.

Figure 5.7 shows the evolution of a test setup to accomplish our goal of characterization of the attenuator. Figure 5.7(a) is a very basic test setup that probably would be the first choice of many people because of the availability of the equipment and the setup simplicity. Although this setup would work for attenuator measurements, there are disadvantages in using it. The main one is the mismatches that can result. To understand how these mismatches can come about, consider what each of the pieces of equipment involved looks like at its respective input or output. A typical microwave sweeper will have a VSWR of < 1.5:1 while the spectrum analyzer usually runs from 1.35 (best) to 1.5:1. A more important VSWR than the sweeper, however, is that of the coupler. This can range from a best of 1.1 to 1.4.

With the numbers above in mind, consider what would happen if the spectrum analyzer was connected directly to the sweeper as shown in Figure 5.7(a). The chart below shows the resultant VSWR when the spectrum analyzer is connected directly to the sweeper with a leveling loop.

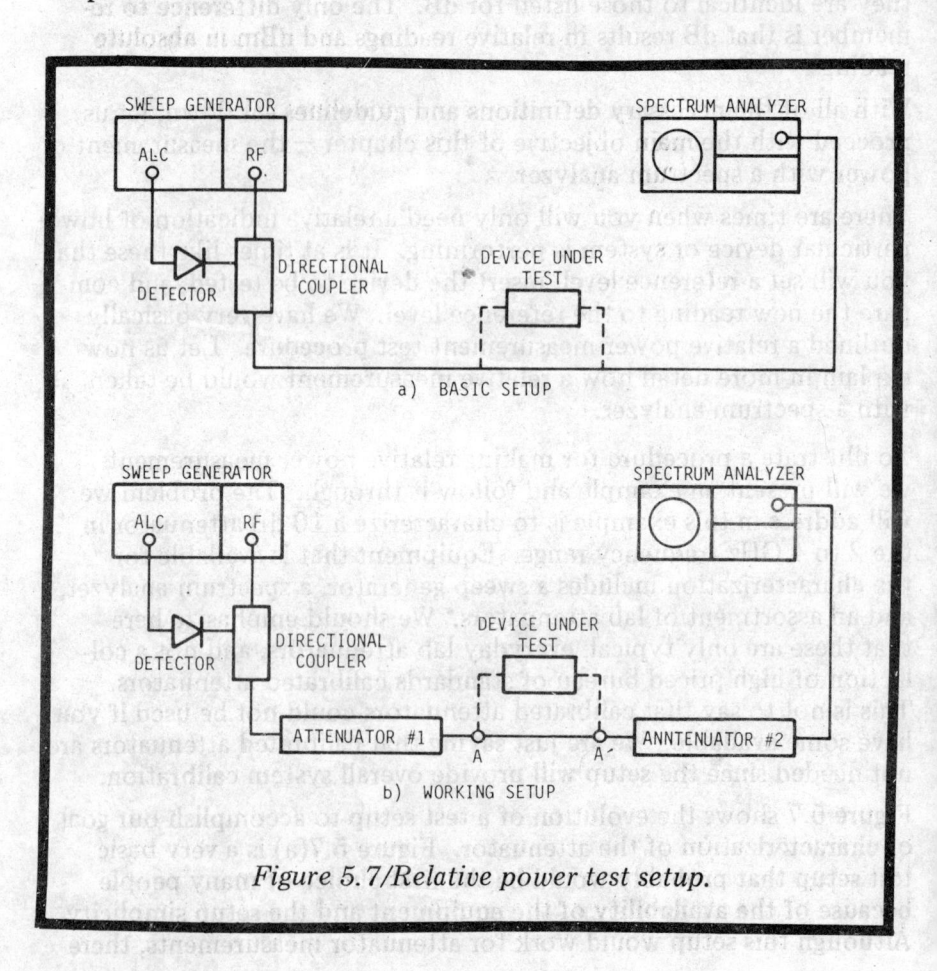

Figure 5.7/Relative power test setup.

Spectrum Analyzer Measurements

Coupler VSWR	Analyzer VSWR	Resultant VSWR
1.1	1.35	1.48
1.1	1.40	1.52
1.1	1.50	1.66
1.25	1.35	1.69
1.25	1.40	1.75
1.25	1.50	1.87
1.40	1.35	1.88
1.40	1.40	1.96
1.40	1.50	2.1

From the chart above you can see that if you had the best analyzer and the best coupler, the best resultant VSWR that the source would see would still be nearly 1.5:1. This will work satisfactorily, but if the source can see a 1.1 or 1.2:1, it will operate much better and result in more accurate readings.

In order to work toward this 1.1 or 1.2:1 VSWR, we will use Figure 5.7(b), which makes use of attenuators for better impedance matching. The question is — how large an attenuator will it take to do the job? A VSWR of 1.1:1 is present when we have a 26 dB return loss at the output of the ALC loop. A VSWR of 1.2:1 is present with 21 dB of return loss at the output. We, therefore, must have a total of between 21 and 26 dB of return loss to accomplish our goal. This is the total return loss of the attenuator and the specturm analyzer.

You recall that the spectrum analyzer specifications we have been using for VSWR are from 1.35 to 1.50:1. The return losses that correspond to these numbers are 16.5 dB and 14 dB, respectively. To obtain the minimum return loss figure of 21 dB (1.2:1 VSWR) we would need 4.5 dB or 7 dB additional loss, depending on the spectrum analyzer used. To eliminate any further confusion as to which analyzer is used, we will assume a worse case condition and say that we have a spectrum analyzer with a VSWR of 1.5:1. With this fact in mind, we will need 7 dB additional attenuation in the return loss path between the coupler and the attenuator. This would mean a 3.5 dB attenuator since we are talking about energy coming from the source and being reflected back. Obviously, we are not going to find a 3.5 dB attenuator around the lab so we had better carry our investigation a step farther.

Standard attenuators are 3, 6, 10, and 20 dB in values. We have already ruled out the 3 dB attenuator since our minimum attenuation is 3.5

dB. We, therefore, have a choice of 6, 10, or 20 dB. We can probably rule out the 20 dB attenuator since this would mean 40 dB additional return loss, and would result in a 54 dB reading which translates into a VSWR approaching the theoretical limit of 1.0:1 (1.004:1). The 10 dB attenuator would result in 34 dB of return loss and the 6 dB attenuator amazingly results in a 26 dB return loss, or a 1.1:1 VSWR. We would, therefore, probably use a 6 dB attenuator following the ALC loop coupler (attenuator #1 in Figure 5.7(b)). This is not to say that the 10 dB attenuator could not be used. It surely could be used if a better match is desired or would be helpful to your measurements. Simple calculations reveal that the use of the 10 dB attenuator would result in 34 dB return loss and a 1.04:1 VSWR. The choice of attenuators comes down to what you have available and what level generator output is needed to obtain a meaningful reading on the spectrum analyzer. (The meaningful reading referred to is one that is not down in the noise of the instrument, but up at a level to distinguish it from other signals.) It should be pointed out that the attenuators also have a VSWR. This is usually very low and is negligible in our application.

With attenuator #1 defined, it is usually a standard practice to duplicate its value in attenuator #2. You would, therefore, use a 6 dB attenuator if 6 dB was chosen for attenuator #1, and 10 dB if 10 dB was chosen. This does not rule out the possibility of having different attenuator values in your setup. Once again, the availability of attenuators and the proper level to the analyzer are to be considered. The latter of the two reasons is probably the more important of the two for attenuator #2. You need enough attenuation in the setup to provide matching and protect the analyzer input, but you should not have so much in the setup that the signal cannot be recognized on the screen. This is one reason why using the same value as attenuator #1 is a good idea and provides safety for the analyzer while providing an adequate signal level.

We should have the setup characterized to a point where we are now able to calibrate it and run the tests on that 10 dB attenuator from 2 to 4 GHz. We will proceed with the testing in two steps: calibration and measurement.

Calibration

(1) Connect the test setup as shown in Figure 5.7(b) with a short cable connecting points A-A'.

(2) Turn on the ac power to the sweep generator and spectrum

Spectrum Analyzer Measurements 197

analyzer. Be sure the sweeper is in a *standby* or *RF off* position and the analyzer *RF attenuator* has some attenuation set into it.

(3) Set the sweeper to the low (2 GHz) and high (4 GHz) frequency settings and place the controls to CW.

(4) Turn on the *RF power* to the sweeper and note the level on the analyzer display.

(5) Adjust the analyzer display for a convenient level by using the analyzer *RF input attenuator, IF attenuator,* and the *RF level* adjust on the sweep generator. An expanded scale on the analyzer such as a 2 dB/division setting may be set in at this point to give resolution.

(6) Manually sweep the sweep generator from the low frequency to the high frequency and note any variations in level as you go through the frequencies. (A spectrum analyzer with storage capability is excellent for a test such as this. The variations will remain stored on the screen and can be recorded or photographed to preserve a record of the spectrum calibration.)

(7) Place the sweep generator in a *standby* or *RF off* position and *do not change any other switches or adjustments.*

Measurement

(1) Place the device to be tested in place of the short cable at points A-A'. Set the sweep generator on CW at the low end of the frequency band to be tested (2 GHz in our case).

(2) Turn on the *RF power* to the sweep generator and note the new level on the spectrum analyzer.

(3) Manually sweep the sweep generator through the desired frequency range and note the level of the signal throughout the sweep. Data can now be compared to the calibration run and recorded point-by-point or a photograph can be taken as the device to be tested is automatically (or manually) swept through the range to be tested.

(4) When your data is obtained, place the sweep generator once again in a *standby* or *RF off* position and remove the device being tested.

For the example we have been using, we should have data that covers frequencies from 2 to 4 GHz with some amplitude fluctuations around the 10 dB line on the spectrum analyzer. If a 2 dB/division scale is used the amplitude should be right around the fifth division down from the top of the display. Variations should be no more than ± 0.5 dB for an average attenuator. A value of ± 0.2 dB would be very good.

This has been a description of a relative power measurement. It is achieved by setting up a convenient reference, inserting the component to be measured, and noting the difference between the reference level and the new level. We can also make a power measurement that will give an absolute reading of power at the output of a device. We will now set up such a measurement for absolute power. The interesting point is to note the similarities in the measurement compared to the relative power method.

One of the areas where it is necessary to know an absolute power level is at the output of a frequency multiplier. The spectrum analyzer is an absolute must for characterizing such a device. You have a situation where the input to the device is at a certain frequency and power level, while the output is at a different frequency and usually a lower level if no amplification is built in. From our previous discussions on frequency domain and time domain, you can see how the analyzer would be ideal for viewing both the input and output of the multiplier so that you could see both the frequency and absolute amplitude of the signals.

In the example we will use to set up an absolute power measurement, we will assume that we have a frequency multiplier which requires a 2 GHz input at 100 milliwatts (+20 dBm), and an output at 6 GHz (a X3 multiplier) with a minimum output power of 25 mW (+14 dBm).

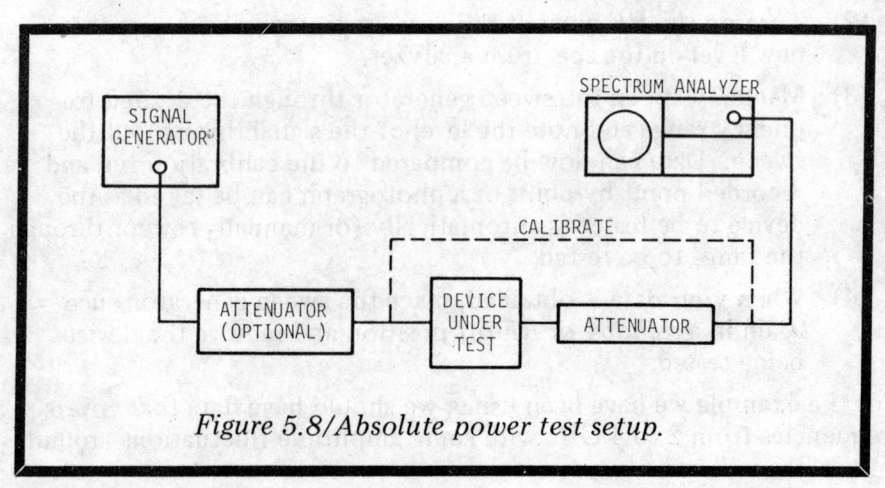

Figure 5.8/Absolute power test setup.

Spectrum Analyzer Measurements

The test setup to be used is shown in Figure 5.8 and is similar to that used in the relative power test setup. The main differences are that this setup uses a CW generator and that the input attenuator is listed as "optional". The CW generator could, of course, be replaced by a sweep generator if you require a swept response of a device or system. For our application we only need a single frequency, so a simple CW generator is used. There is also another advantage in using a single frequency generator, and that is that you can obtain a high power from a single frequency source than one that sweeps a band of frequencies.

The input attenuator is listed as "optional" since the generator you use may not have enough power to drive the multiplier circuit we will be testing if an attenuator is used. The attenuator is normally inserted to aid in impedance matching between the generator and the device under test. The input of a multiplier usually has a fairly good VSWR and the attenuator will not really improve the match between the generator and the multiplier. There are, however, some devices where this attenuator is needed for matching. In these cases, you should be sure that the attenuator will not reduce the power level too much. If an attenuator is not practical you may be able to use a circulator or isolator to aid in matching.

The attenuator at the output is another story in the case of our multipliers. The output of a frequency multiplier consists of many different frequencies. There is usually a bandpass filter supplied to cause only the desired frequency to be passed. This, however, does not mean that the output VSWR is necessarily a low value so that it will interface with an instrument such as the spectrum analyzer. You will recall that even when the spectrum analyzer and the component being measured each had a 1.4:1 VSWR, the resultant VSWR was nearly 2.0:1 (1.96:1). So, it should be considered a standard part of your test setup to place the attenuator between the device (multiplier) output and the spectrum analyzer. When this component is inserted in the test setup, you must remember that the output of the device being tested (the multiplier in our case) is actually higher than the analyzer display shown by the amount equal to the attenuator value. If, for example, you were to use a 6 dB attenuator in your setup and the analyzer shows a signal at a +13 dBm level, the output of the device under test would actually be +19 dBm (or close to the value, depending on the attenuator calibration value at the frequency we are using).

To illustrate some of the points we have been making, let us go through a test procedure for an absolute power reading.

Calibration

(1) Connect the test setup as shown in Figure 5.8 with a cable in the *calibrate* position. (For this setup we will use the input attenuator.)

(2) Apply ac power to the signal generator and spectrum analyzer. Set the signal generator to the proper frequency.

(3) When RF power is available, set the proper input level needed for proper operation of the device being tested (+20 dBm in our case).

Note: The calibration setup does not include the output attenuator, since its inclusion would result in an inaccurate reading of device input power.

Measurement

(1) Place the device to be tested (multiplier) and the output attenuator in the test setup as shown in Figure 5.8.

(2) Apply any dc supply voltage or bias necessary for device operation.

(3) Tune the analyzer to the device output frequency (6 GHz in our example) and read the power level of that signal. (Do not forget that this is not the actual output power. The actual power is the analyzer reading *plus* the output attenuator value.

A tabulation of parameters and results using such a setup as described above for our multiplier example might read as follows:

Input Frequency = 2 GHz
Input Power = +20 dBm
Output Frequency = 6 GHz
Power Read on Analyzer = +8 dBm
Output Attenuator Value = 6 dB*
Actual Output Power = +14 dBm

We have made basic power measurements for both relative and absolute power using the spectrum analyzer. The use of storage displays aids greatly in making these measurements. However, there are units available (normalizers) that provide digital storage and normalization of swept responses that will adapt to the spectrum analyzer to aid

---

*This attenuator is very important in regards to its own individual calibration. You must know exactly what attenuation it exhibits at the frequencies you will be using it. Otherwise, you will not have an accurate absolute power reading.

Spectrum Analyzer Measurements 201

even further in the measurement you may need to make. These units vary according to manufacturers and the one suited for your particular use should be chosen.

A device such as the normalizer is necessary because swept measurements of network characteristics generally include the response of the test system itself and all of its components. These effects must be taken into consideration when determining what portion of your results is the actual device and which is the test system. This compensation is often accomplished by drawing the systems response on the display with a grease pencil or plotting grid lines with an X-Y recorder during the calibration phase. As mentioned above, the normalizer eliminates this by storing the system's response in a memory and subtracting it from the measured data for a direct display of the actual characteristics of the device alone. This greatly reduces the possibility of results being erroneously interpreted. These normalizers are compatible with many spectrum analyzers. When they are not directly compatible with an analyzer, a normal low-frequency oscilloscope can be used to obtain digitally stored CRT displays. By utilizing the proper interconnections, you can obtain a digitally stored and/or normalized spectrum analyzer display. With this setup, you can have the advantage of both a conventional and digitally-stored display.

You can see from the section presented that the spectrum analyzer can be a very useful piece of equipment for measuring power — whether relative or absolute. Although the setups presented are very similar, they each have their own individual way of obtaining the desired results. The main points emphasized concern the attenuators and when and where they are used. An additional point should be brought up before leaving this section on power measurements. That point concerns medium and high power measurements. Usually power levels greater than one watt are not even considered when speaking of spectrum analyzers, and that is a good rule to follow. However, when at the one watt level, you should be concerned about attenuator power handling capability and what the maximum input level to the analyzer is for a particular RF attenuator setting. These precautions cannot be over-emphasized, because they will, if observed, aid in more accurate and reliable power measurements for you.

5.2 Frequency Measurements

Until recently, frequency measurements using a spectrum analyzer were only used as a last resort — that is, if no other means was available. The new generation of analyzers, with more sophisticated markers, higher resolution, and digital frequency readouts (either for designated markers, center frequency, or at the sweep start), make accurate and reliable frequency measurements an every-day occurrence. There are two basic means of measuring frequency on a spectrum analyzer: relative measurements and absolute measurements.

Relative frequency measurements require a linear frequency span. This means that the space between every division on the analyzer must represent the same frequency difference. By measuring the relative separation of two signals on the display, the differences can be determined.

Figure 5.9 presents a way of making a relative frequency measurement. A signal of known frequency is displayed on the analyzer (2 GHz in this case) and positioned at a convenient spot. The unknown signal is then displayed for measurement. The number of divisions between the known signal frequency and the unknown signal peak is multiplied by the frequency scan per division (scanwidth) and then added to or subtracted from the known signal frequency. The example in Figure 5.9 would result in the following:

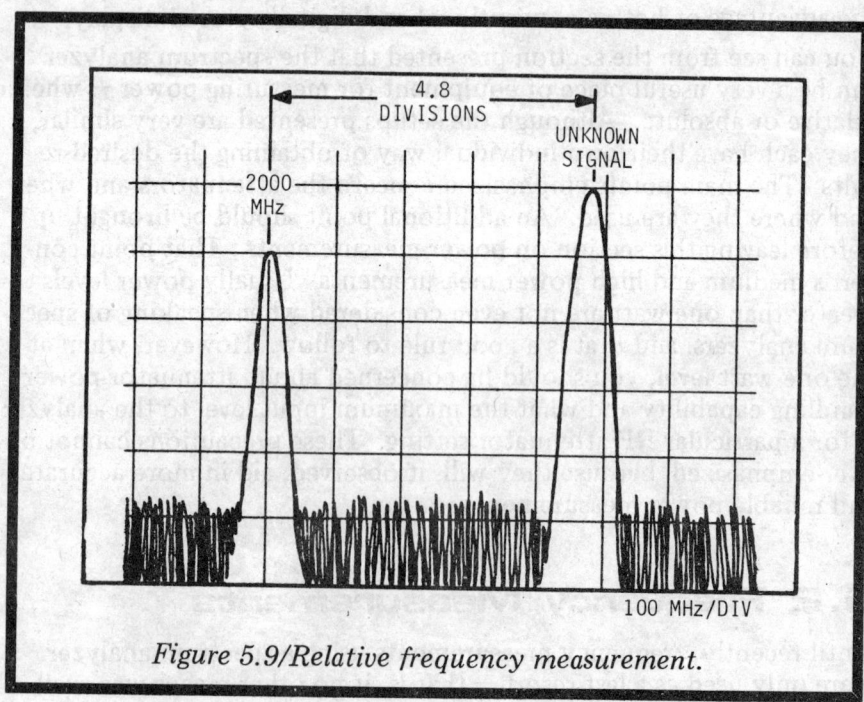

Figure 5.9/Relative frequency measurement.

Spectrum Analyzer Measurements   203

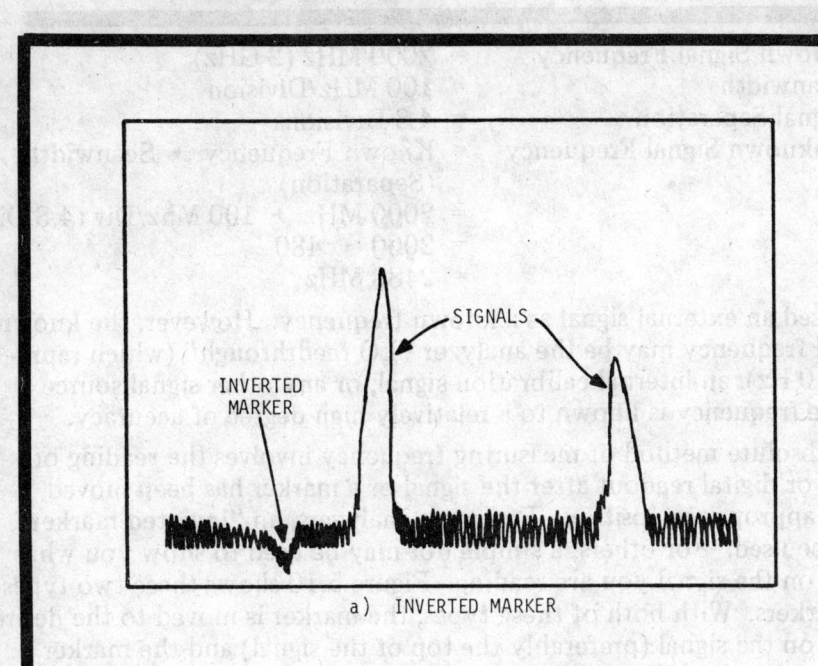

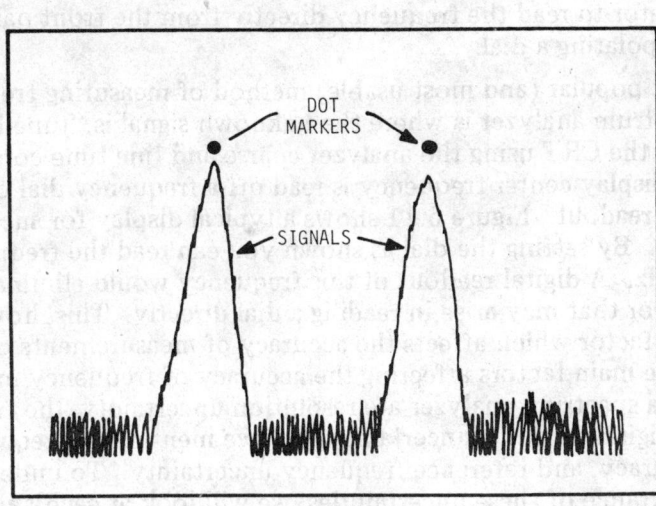

a) INVERTED MARKER

b) DOT MARKER

Figure 5.10/Frequency markers.

Known Signal Frequency	= 2000 MHz (2 GHz)
Scanwidth	= 100 MHz/Division
Signal Separation	= 4.8 Divisions
Unknown Signal Frequency	= Known Frequency + Scanwidth (Separation)
	= 2000 MHz + 100 Mhz/Div (4.8 Div)
	= 2000 + 480
	= 2480 MHz

We used an external signal as a known frequency. However, the known signal frequency may be the analyzer "LO feedthrough" (which represents 0 Hz), an internal calibration signal, or any other signal source whose frequency is known to a relatively high degree of accuracy.

The absolute method of measuring frequency involves the reading of a dial or digital readout after the signal or a marker has been moved to an appropriate position. For some analyzers, an "inverted marker" may be used. For others, a simple dot may be used to show you what point on the signal you are reading. Figure 5.10 shows these two types of markers. With both of these types, the marker is moved to the desired point on the signal (preferably the top of the signal) and the marker position is used. The frequency (or marker position) can be read on a frequency dial, or a series of lever switches on some analyzers enable the operator to read the frequency directly from the front panel without interpolating a dial.

The most popular (and most usable) method of measuring frequency on a spectrum analyzer is where the unknown signal is "tuned" to the center of the CRT using the analyzer coarse and fine tune controls, and the display center frequency is read off a frequency dial or built-in digital readout. Figure 5.11 shows a typical display for such a measurement. By setting the dial as shown you can read the frequency as 2.240 GHz. A digital readout of this frequency would eliminate much of the error that may arise in reading a dial directly. This, however, is only one factor which affects the accuracy of measurements of this type. The main factors affecting the accuracy of frequency measurement on a spectrum analyzer are resolution uncertainty, the frequency dial (or digital readout) uncertainty we have mentioned, frequency span accuracy, and reference frequency uncertainty. To understand the importance of these uncertainties, we will look at each and find reasons for them.

Resolution Uncertainty — The uncertainty surrounding the designation of one particular point as the "peak" of a displayed signal is defined as "resolution uncertainty". Resolution (IF) bandwidth is the window through which an analyzer looks at frequencies present at its input. It is the shape of this IF filter which is traced out on the display that represents the "signal". Since points whose frequencies are close to the center frequency also appear on the curvature at the peak of the trace, it is difficult to determine the "center" frequency (see Figure 5.11). The uncertainty in defining a signal peak is important when measuring frequency because it limits the ability of a user to tune an unknown signal to the center of the CRT or to determine the distance between a reference point and the displayed signal. Resolution uncertainty is typically less than 30% of the bandwidth setting with 10 dB/division vertical scaling (20% for 2 dB/division and 10% for linear scaling). When measuring the distance between two signals, resolution uncertainty is two times that of a lone signal.

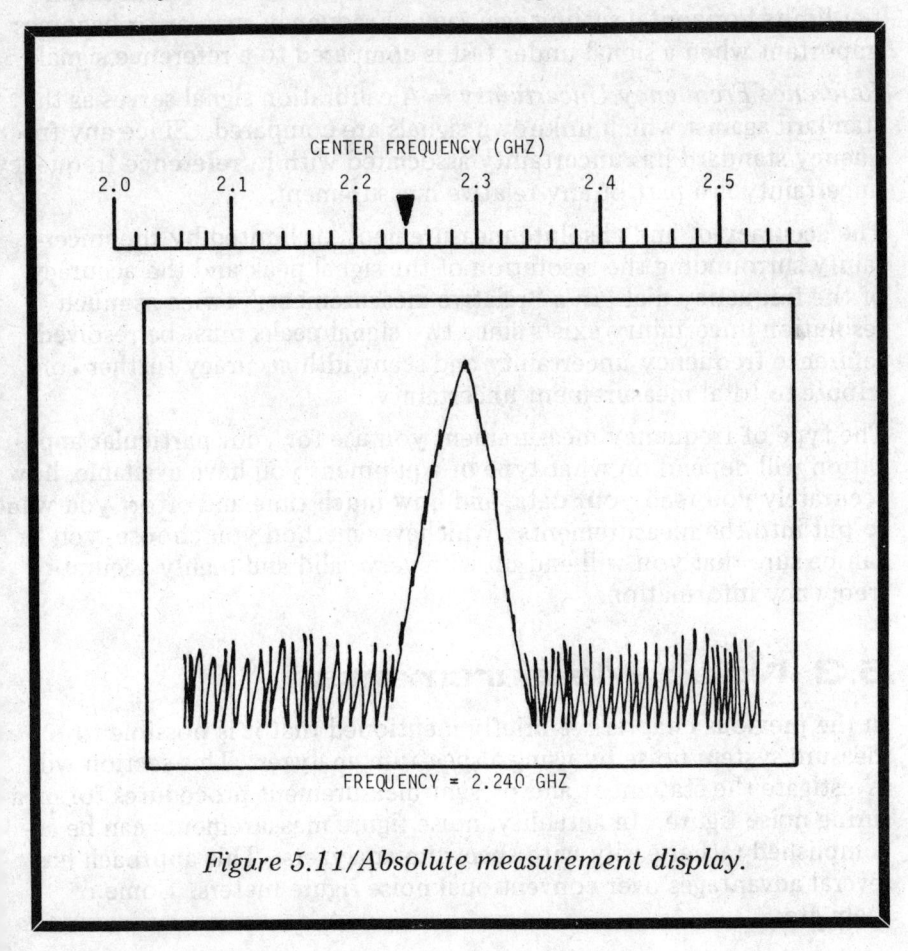

Figure 5.11/Absolute measurement display.

Frequency Dial Uncertainty — Frequency dial uncertainty refers to how accurately the dial (or digital readout) reflects the display center frequency. The input signal frequency is deduced from the swept LO (mechanically in the case of a dial, and electronically for a digital readout). Uncertainty is caused by limited dial/display resolution and error in the deduction process. It limits a user's ability to determine the center frequency from the analyzer dial with a high degree of accuracy. This source of uncertainty is particularly important when the "absolute measurement technique" is being employed.

Frequency Span Accuracy — Frequency span per division (scanwidth) determines the horizontal scaling of the analyzer. It is controlled by attenuating the sawtooth that drives the first LO. Uncertainty in the attenuation and non-linearity of the oscillator and horizontal amplifier limits horizontal scaling accuracy. Frequency span error becomes important when a signal under test is compared to a reference signal.

Reference Frequency Uncertainty — A calibration signal serves as the standard against which unknown signals are compared. Since any frequency standard has uncertainty associated with it, reference frequency uncertainty is a part of any relative measurement.

The accuracy of an "absolute measurement" is limited by the uncertainty surrounding the resolution of the signal peak and the accuracy of the frequency dial. In a "relative measurement," twice as much resolution uncertainty exists since two signal peaks must be resolved; reference frequency uncertainty and scanwidth accuracy further contribute to total measurement uncertainty.

The type of frequency measurement you use for your particular application will depend on what type of equipment you have available, how accurately you read your data, and how much time and effort you want to put into the measurements. Whichever method you choose, you can be sure that you will end up with very valid and highly accurate frequency information.

5.3 Noise Measurements

In the previous chapter we briefly mentioned that it is possible to measure system noise by using a spectrum analyzer. This section will investigate the statement and present measurement procedures for measuring noise figure. In actuality, noise figure measurements can be accomplished rather easily with spectrum analyzers. This approach has several advantages over conventional noise figure meters. Some of them are:

Spectrum Analyzer Measurements

- Noise figure can be measured at any frequency within a spectrum analyzer's multi-decade frequency ranges. This enables measurement at the device's operating frequency without changes in the test setup.
- They provide frequency selective noise figure measurements independent of device's bandwidth or spurious responses.
- Standard spectrum analyzers can make a variety of frequency domain measurements (power, frequency, distortion, dB, etc), as we have previously seen, as well as noise figure.

These advantages do not mean that you should throw away your noise figure meter. The sensitivity and accuracy of a spectrum analyzer become limiting factors in many measurements of noise figure, just as with any other measurement. Nevertheless, the advantages listed above may make the analyzer the best choice for your particular noise figure measurement. If nothing else, the spectrum analyzer should be given a great deal of consideration when planning such a measurement.

It may not be readily apparent how a spectrum analyzer can give you a reading of noise figure on a device. To aid in your understanding, we should return to the basic equations for noise factor (F) of a device or system. You will recall that noise factor was defined as the input to output signal-to-noise ratio:

$$F = \frac{S_i/N_i}{S_o/N_o}$$

where

S_i/N_i = input signal-to-noise ratio

S_o/N_o = output signal-to-noise ratio

The number, F, indicates the change in signal-to-noise ratio which occurs as a signal passes through a device. Thus, F is a figure of merit (ideally equal to one) which can be used to compare different amplifiers and receivers. So, being a dimensionless quantity independent of bandwidth, noise factor is a better basis for comparison of receivers than sensitivity. Furthermore, with knowledge of a system's noise factor and bandwidth, we can predict its sensitivity and how it might be improved by the addition of pre-amplifiers.

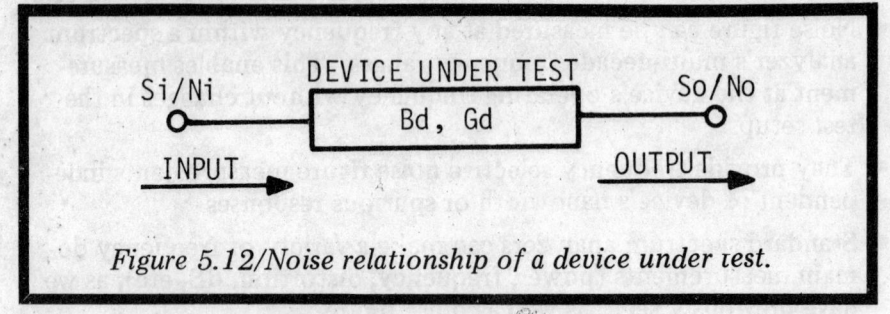

```
┌─────────────────────────────────────────────────┐
│              DEVICE UNDER TEST                  │
│   Si/Ni     ┌─────────────┐    So/No            │
│        o────┤   Bd, Gd    ├────o                │
│             └─────────────┘                     │
│     INPUT ──────►            OUTPUT ──────►     │
│                                                 │
│   Figure 5.12/Noise relationship of a device under test. │
└─────────────────────────────────────────────────┘
```

Figure 5.12 shows a device that can be tested and its noise relationships. The terms presented are:

S_i/N_i = input signal-to-noise ratio

S_o/N_o = output signal-to-noise ratio

B_d = noise bandwidth of the device

G_d = gain of the device;

With Figure 5.12 and the terms defined, we can rearrange the equation as follows:

$$F = \frac{S_i/N_i}{S_o/N_o} = \frac{S_i/kTB_d}{S_iG_d/N_o} = \frac{N_o}{G_d kTB_d}$$

where

N_o = noise power output (delivered to a matched load with the input terminated in its characteristic impedance)

k = Boltzman's constant, 1.374×10^{-23} joule/°K

T = 290°K (room temp.) $kT = 3.98 \times 10^{-23}$ watts/Hz, equivalent to -174 dBm in a 1 Hz bandwidth

B_d = device noise power bandwidth in Hertz

G_d = device gain

You will recall from discussions in Chapter 4 that noise figure is the logarithmic equivalent of noise factor and is expressed as:

NF - 10 log F

We will once again rearrange the noise figure equations to reflect the logical order we would think of characterizing noise. That would be that the noise figure is the noise output minus the gain minus the noise

Spectrum Analyzer Measurements

input. If all of these factors are considered, we would be left with the noise of the device you are measuring. The expression now becomes:

NF = Noise Output − Gain − Equivalent Noise Input in Bandwidth B

NF = $10 \log N_o - 10 \log G_d - (-174 \text{ dB} + 10 \log B)$

Note that the number used for B in the equation relates to noise power bandwidth. This is because there is a difference in the total noise power which passes through a real vs ideal filter of bandwidth B as shown in Figure 5.13. The noise power bandwidth for the measurement can be set by selecting the resolution bandwidth, B, of the spectrum analyzer sufficiently narrow that the analyzer determines the system bandwidth. The noise power bandwidth for some analyzers' filters are typically in the order fo 1.2 times the resolution bandwidth indicated on the front panel bandwidth control. We will designate this correction factor, A, as shown in Figure 5.13. If the correction factor is known to be in the order of 1.2, we can use the resolution bandwidth setting for B in the above equation provided we use a 0.8 dB (10 log 1.2) term to compensate for it.

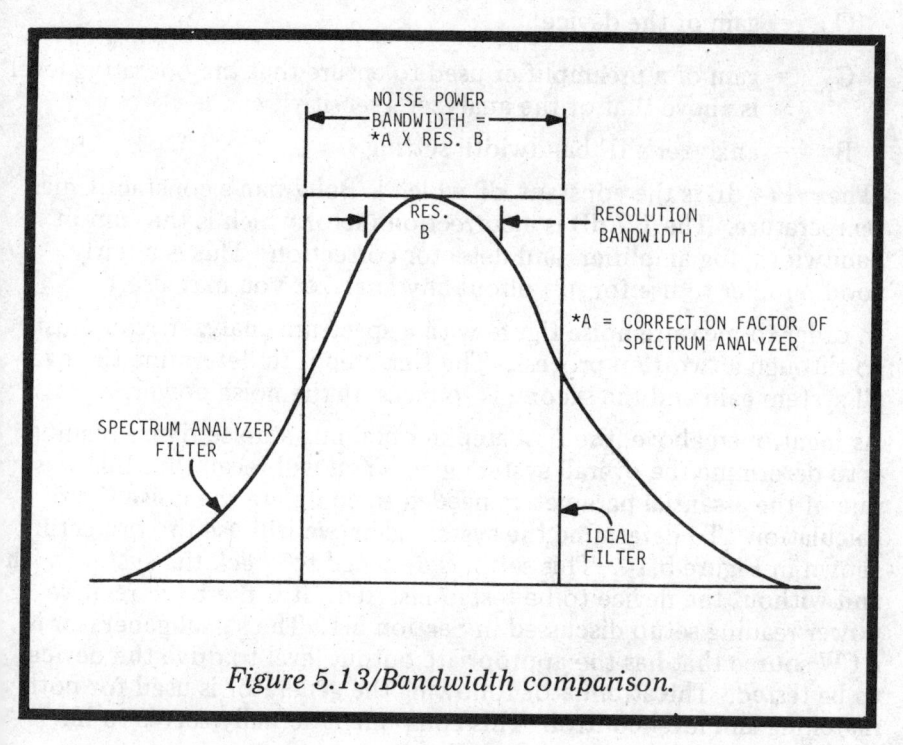

Figure 5.13/Bandwidth comparison.

Next, the device's gain is measured by noting the effect it has on the power of a signal as displayed by the spectrum analyzer set to the desired frequency.

Finally, noise power measurements on a spectrum analyzer must account for the random fluctuations of its power with time. For that reason, the detected noise should be averaged by suitable video filtering (video bandwidth $\leq$ 0.01 IF bandwidth) so the average noise power can be read from the display as a smooth line.

To this point we have shown that noise figure can be measured once the device gain, bandwidth, and noise power output are known. It is now time to see how the spectrum analyzer can measured these values. Before beginning an explanation of the measurement procedure, we will present the noise figure equation in the form we can most easily use:

$$NF = N - (G_d + G_p) - 10 \log B + 174 \text{ dB} + 1.7 \text{ dB}$$

where:

N = the noise level measured on the analyzer

G_d = gain of the device

G_p = gain of a preamplifier used to ensure that the operating level is above that of the analyzer's sensitivity

B = analyzer's IF bandwidth setting

(The -174 dB is the constant kT which is Boltzman's constant times temperature. The 1.7 dB is a correction factor which is the sum of bandwidth, log amplifier, and detector correction. This is a fairly good number to use for just about any analyzer you may use.)

In order to measure noise figure with a spectrum analyzer, you must go through a two-step process. The first step is to determine the overall system gain and the second is to measure the noise power.

As mentioned above, the first step in obtaining a noise figure reading is to determine the overall system gain. You will recall that this was one of the essential parameters needed to complete the noise figure calculation. To determine the system gain, we will use the test setup shown in Figure 5.14. This setup is designed to check the system with and without the device to be tested inserted. It is the basic relative power reading setup discussed in Section 5.1. The signal generator is a CW source that has the appropriate output level to drive the device to be tested. The attenuator following the generator is used for both matching and level control. This component actually serves to im-

Spectrum Analyzer Measurements

prove the system match regardless of which position the switch is in. When the switch is in the #1 (Reference) position, it helps to improve the match between the generator and the spectrum analyzer. You will recall the lengthy discussion of generator-to-analyzer matching which took place in Section 5.1, and how important the matching was to set-up operations. With the switch in position #2, the attenuator provides matching between the generator and the device to be tested. This ensures that there will be no erroneous readings obtained and that the signal at the output is a true representation of what the device is actually doing.

The preamplifier could probably be considered to be an optional component. Its use depends on the noise output of the device being tested, N_o. If N_o is below the sensitivity of the spectrum analyzer, the level must be increased by a low-noise, high-gain (G_p) preamplifier. With the amplifier in the setup, the noise level measured by the spectrum analyzer is greater than the noise power of the device by the gain of the preamplifier (G_p). Once again, this ensures that a true and accurate reading is being displayed on the analyzer. As mentioned above, this device is not always necessary. It should only be used when the noise power output of the device being tested is below the sensitivity level of the analyzer. If the noise power level is sufficiently high to give good analyzer readings, it should not be used. If the amplifier is used when the output level is high, there is a danger

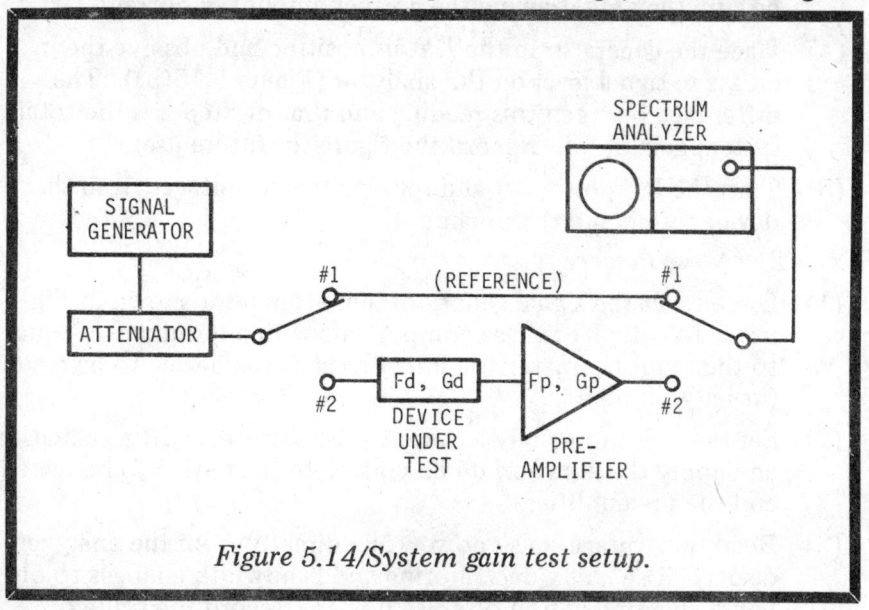

Figure 5.14/System gain test setup.

of overdriving the analyzer and introducing spurious responses that interfere with your readings. Therefore, take a few minutes to investigate the levels that will be present and make the decision whether or not to use a preamplifier according to these levels.

With the major components of the test setup defined and explained, let us proceed to perform the task of making a measurement.

Step #1 — System Gain

(1) Connect the test setup as shown in Figure 5.14 with the spectrum analyzer and signal generator AC turned on. The signal generator should be in a *standby* or *RF off* position and the setup should be in the *reference* position. This could be accomplished by either using RF switches or by placing a cable directly from the attenuator to the spectrym analyzer.

(2) Turn *on* the RF on the signal generator. By means of the generator level adjust, the external attenuator, and the analyzer RF and IF attenuators, adjust the analyzer display for a signal with as low a noise level as possible. Figure 5.15(a) shows a typical signal.

(3) Place the generator in the *standby* or *RF off* position and insert the test device and the preamplifier, if needed. Apply any dc voltage needed for the preamplifier and device to be tested and be sure they are drawing the correct amount of current.

(4) Place the generator in the *RF on* position and observe the increase in signal level on the analyzer (Figure 5.15(b)). The difference between this reading and that of Step 2 is the total system gain in dB. Record the figure for future use.

(5) Turn the RF power *off* and remove the dc voltages from the device tested and the preamplifier.

Step #2 — Noise Power

(1) Disconnect the signal generator and attenuator shown in Figure 5.14. Replace these components with a termination equal to the input characteristic impedance of the device to be tested (usually 50 ohm).

(2) Set the spectrum analyzer input attenuator to 0 dB attenuation and apply the appropriate dc voltage to the device to be tested and the preamplifier.

(3) Read the average noise power, N_o, directly from the analyzer display. Use any video filtering and bandwidth changes to obtain a clear indication of noise power. Record the value of

Spectrum Analyzer Measurements

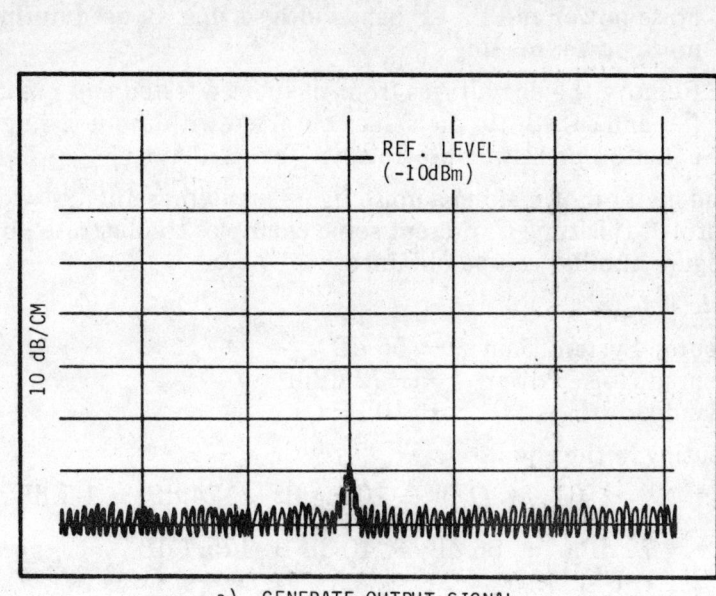

a) GENERATE OUTPUT SIGNAL

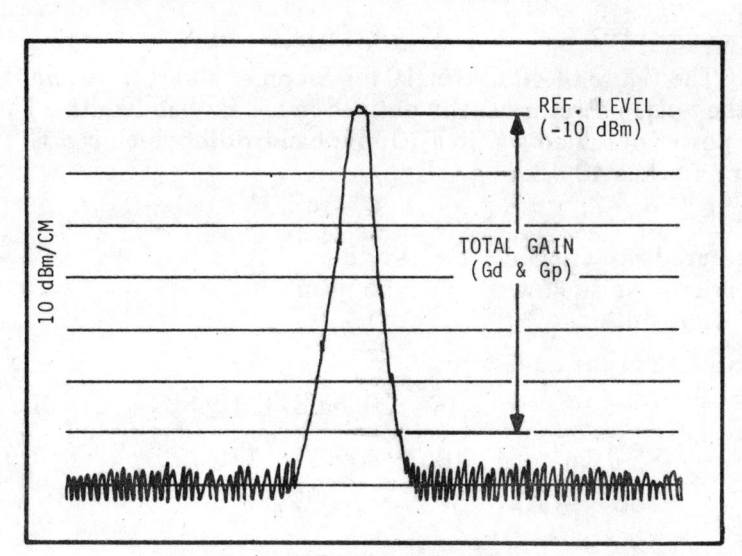

b) OVERALL SYSTEM GAIN

Figure 5.15/System gain displays.

noise power and the IF bandwidth setting, B, used during the noise power reading.

(4) Remove the dc voltages from the device tested and preamplifier and substitute the values obtained with the noise figure equation presented previously.

With the method of making a noise figure setup presented, it would be helpful at this time to present some examples to illustrate how a noise figure number can be obtained.

Example # 1

Measured System Gain	— 56 dB
Measured Noise Power	— −72 dBm
IF Bandwidth	— 10 kHz

Substituting in the equation:

NF = $N_o - (G_d + G_p) - 10 \log B + 174$ dB + 1.7 dB

NF = −72 dBm − 56 dB − 40 dB + 175.7 dB

NF = −168 + 175.7

NF = 7.7 dB

Note: The figure of 40 dB for 10 log B comes about from the fact that the noise power is usually defined in a 1 Hz bandwidth. The noise power obtained was in a 10 kHz bandwidth which is a 10^4 factor and thus 40 dB.

Example # 2

Measured System Gain	— 48 dB
Measured Noise Power	— −85 dBm
IF Bandwidth	— 1 kHz

Substituting in the equation:

NF = $N_o - (G_d + G_p) - 10 \log B + 174$ dB + 1.7 dB

NF = −85 dBm − 48 dB − 30 dB + 175.7 dB

NF = −163 + 175.7 dB

NF = 12.7 dB

You can readily see from Example # 2 above how a lower system gain and narrower bandwidth on the analyzer has resulted in an increased noise figure. This is true even though the measured noise power went from −72 dBm in Example # 1 to −85 dBm in Example # 2. It is for

Spectrum Analyzer Measurements

cases just such as this that we have been emphasizing accurate system gain measurements, and that you know in exactly what bandwidth you are obtaining your measurement.

This section has concentrated on a noise figure reading using the spectrum analyzer. This is but one type of noise and noise measurement which can be made using a spectrum analyzer. Other types of noise measurements have been covered previously and will be covered in following sections, such as oscillator spectral noise (purity) measurements in a following chapter.

5.4 Receiver Measurements

The spectrum analyzer probably finds more than 90% of its application in receiver measurements. Many of the measurements in receiver application have been covered in previous sections. There are three, however, that have not been covered and are very important in the characterization of receivers. These three are modulation, distortion, and pulsed RF. It is the objective of this section to cover these three measurements so that you may become aware of the procedures necessary to obtain good data.

5.4.1 Modulation

Modulation, very basically defined, is a variation. When a CW carrier signal is *modulated*, its characteristics are *varied* from what they were for the CW signals. The variation may be in amplitude, frequency, or phase, depending on the particular system you have. The quantity that is to be measured for such a parameter is the percent of modulation (or how much variation there is).

In this discussion, modulation will be treated as two separate types: AM and FM. Such topics as pulse-code modulation (PCM) or time or frequency multiplexing will not be covered here. Actually, in order to cover angular modulation in its entirety one should cover both FM and phase modulation. We have chosen to cover only FM since it is felt that a good presentation of FM will provide more than adequate coverage of the topic of angular modulation.

To understand what is taking place in our modulation measurements to be presented, the following definitions are presented:

Carrier — An RF wave which is unvarying in amplitude, frequency, and phase.

Modulating Wave — A lower frequency wave which causes some characteristics of the carrier to vary.

Modulated Wave — The combination of a carrier and modulating wave which causes variation (frequency, amplitude, or phase) which results in an RF signal with intelligence superimposed on it.

With these definitions set down and a basic understanding of what modulation is, let us proceed with our first type of modulation measurements — that of amplitude modulation.

The most recognized form of amplitude modulation is that shown in Figure 5.16(a). This, of course, is a display in the *time domain* (oscilloscope). The display that we are interested in is shown in Figure 5.16(b) This is the same signal as shown in Figure 5.16(a), except it is now show in the frequency domain. You can see from this display that the carrier is shown in the center (F_c) with an upper sideband ($F_c + F_m$) and a lower sideband ($F_c - F_m$). It is the amplitude ratio of F_c to F_m that is measured to determine the percent of modulation for a specific system. The percent of modulation can be read from either an oscilloscope or a spectrum analyzer. To illustrate this, we will perform a measurement on the same signal using an oscilloscope and a spectrum analyzer. The carrier signal used is, of course, not a microwave signal, but is an IF output of a microwave receiver. This particular IF is at 10 MHz. This range of frequencies is why it can be displayed directly on an oscilloscope.

Before making our measurements, we should establish some basic relationships which will be very helpful in our measurements. First, in the time domain (oscilloscope) the percent of AM modulation is computed using a factor, K. This derived as:

$$K = \frac{E_{max}}{E_{min}}$$

where E_{max} and E_{min} are the maximum and minimum voltages, respectively, of the modulated signal. Figure 5.17(a) shows how these voltages are measured.

When the K factor is determined, the percent of modulation, m, can be calculated as follows:

$$m = \frac{K - 1}{K + 1}$$

Spectrum Analyzer Measurements

Referring once again to Figure 5.17(a), we can now calculate the percent of modulation of our AM signal:

$$K = \frac{E_{max}}{E_{min}} = \frac{3}{1} = 3$$

$$m = \frac{K-1}{K+1} = \frac{3-1}{3+1} = \frac{2}{4}$$

$$m = 50\%$$

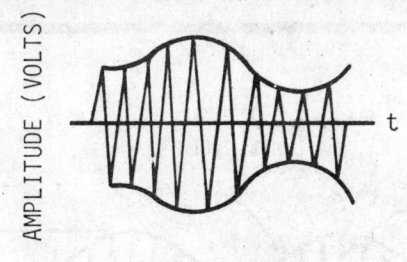

a) TIME DOMAIN (OSCILLOSCOPE) REPRESENTATION

F_c = CARRIER FREQUENCY.

F_m = MODULATING FREQUENCY

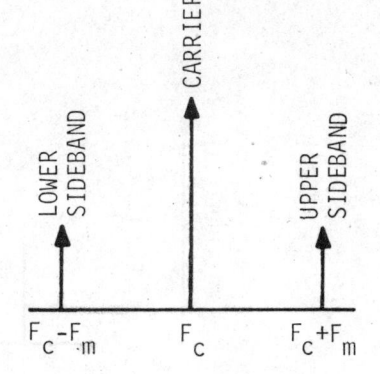

(b) FREQUENCY DOMAIN (SPECTRUM ANALYZER) REPRESENTATION

Figure 5.16/Amplitude modulation.

When the spectrum analyzer is used, there is also a relationship set up to calculate the percent of modulation. This relationship says:

$$m = \frac{2 E_s}{E_c} \times 100$$

where E_s and E_c are voltages of the sideband and carrier, respectively, as shown in Figure 5.17(b). (The ratio that is set up in E_s/E_c and is multiplied by 2 since there are double sidebands.)

From Figure 5.17(b) we can see that there is a difference between the carrier and the sidebands of approximately 12 dB. In order to use

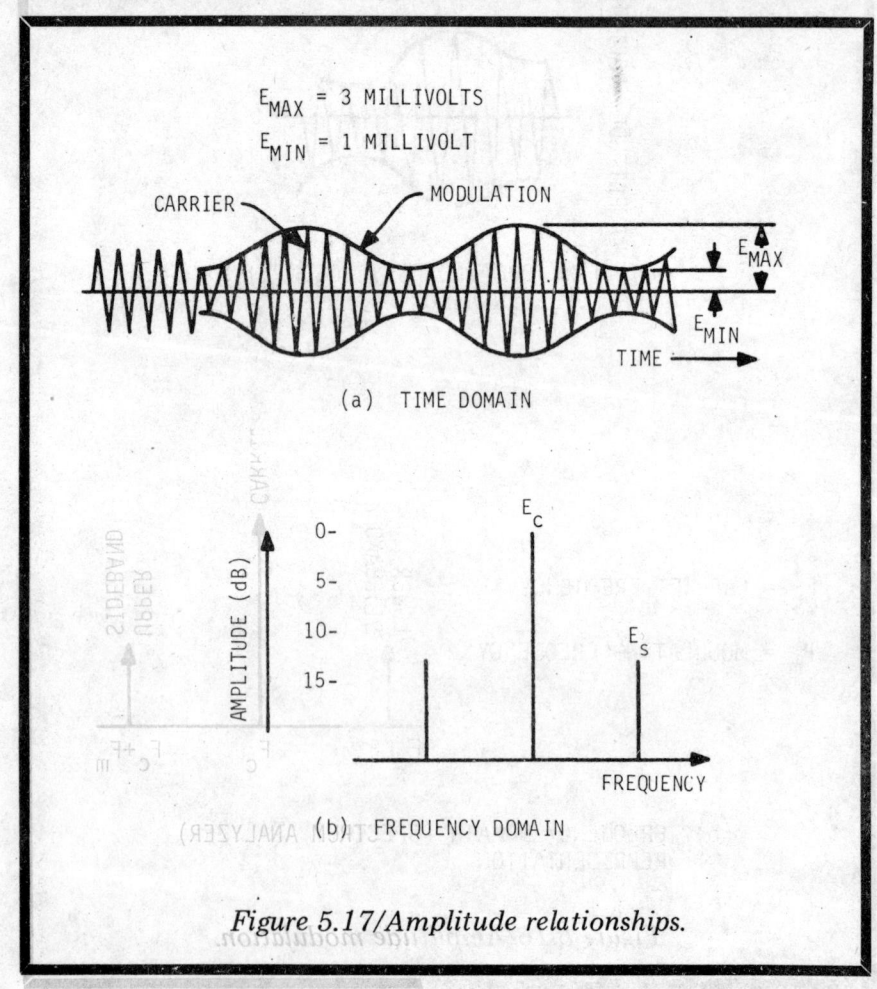

Figure 5.17/Amplitude relationships.

Spectrum Analyzer Measurements

this number for this example, the 12 dB must be converted to a voltage ratio. Substituting into the dB formula we have:

dB = 20 log (Voltage Ratio)
12 = 20 log (Voltage Ratio)
0.6 = log (Voltage Ratio)
Voltage Ratio = $\log^{-1} 0.6$
Voltage Ratio = 4

From this result, we have that the carrier to sideband ratio is 4:1. Therefore, we can say that $E_s = 1$ and $E_c = 4$. The equation presented above now becomes:

$$m = \frac{2 E_s}{E_c} \times 100$$

$$m = \frac{2(1)}{4} \times 100$$

$$m = 0.5 \times 100$$

$$m = 50\%$$

As can be seen above, it is relatively easy to calculate percent of modulation (m) from a linear presentation in frequency or time domain. There are many applications, however, where the logarithmic display on the spectrum analyzer offers advantages, especially at low modulation percentages ($< 10\%$). The high dynamic range of the typical spectrum analyzer (> 70 dB) allows accurate measurements of modulation percentages as low as 0.06% if necessary. To understand what an advantage you have with a spectrum analyzer for measuring low modulation percentages, refer to Figure 5.18. This figure is for an AM signal with a modulation percentage of 2%. First, consider displaying this signal on an oscilloscope as shown in Figure 5.18(a). You will recall that the percent of modulation is first calculated by measuring E_{max} and E_{min} and then using the K factor to find a percentage.

You will have to agree that it could be rather difficult to distinguish very accurately the difference between E_{max} and E_{min} from the oscilloscope display. If you work the numbers backwards (solve for K when given m) you will find that the required voltage ratio is 1.04. This

a) TIME DOMAIN (OSCILLOSCOPE)

b) FREQUENCY DOMAIN (SPECTRUM ANALYZER)

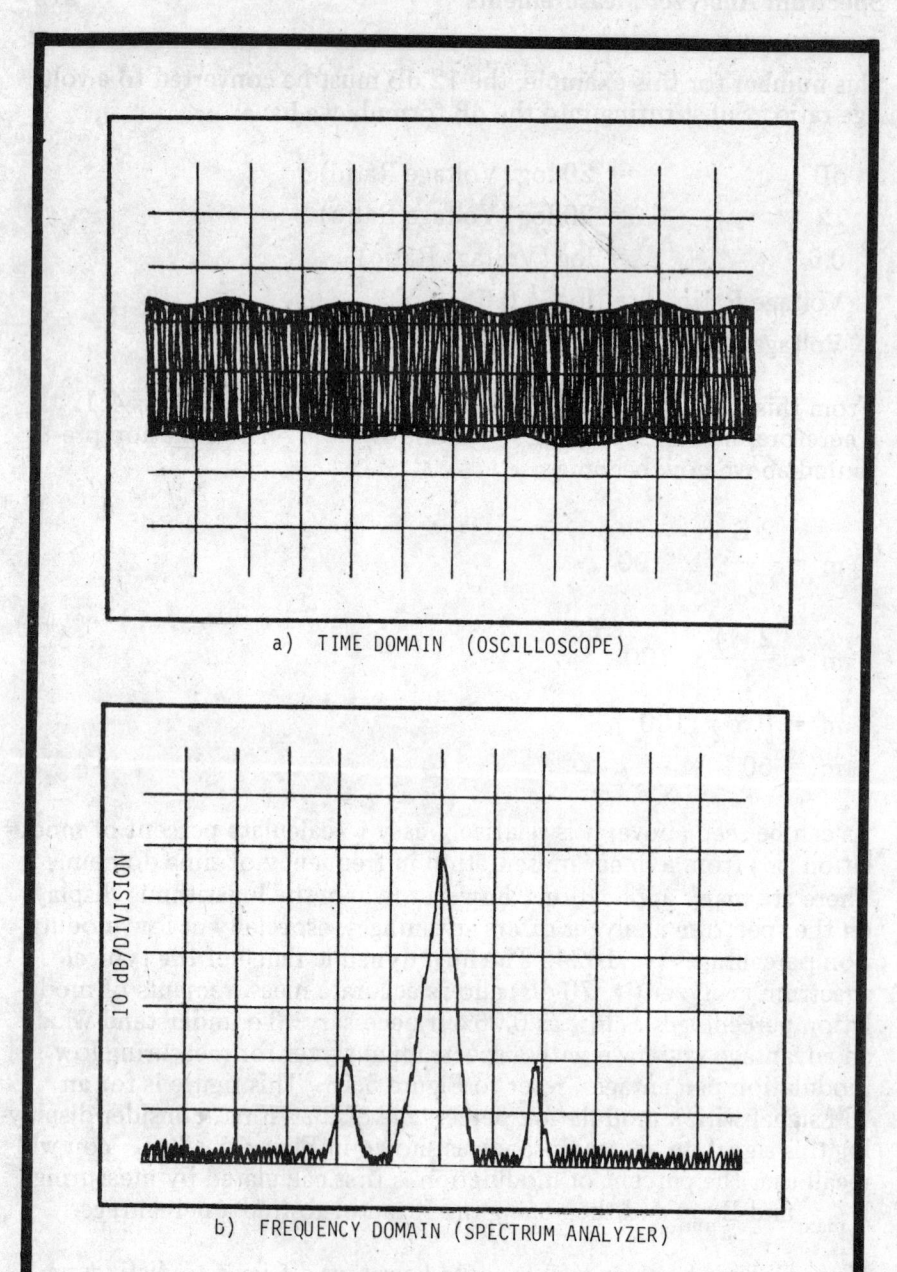

Figure 5.18/Low modulation percentage.

Spectrum Analyzer Measurements

could be rather difficult to measure on a scope. Typical ratios are shown below for various percents of modulation.

m (percent)	K (ratio)
50	3.00
40	2.33
30	1.85
20	1.50
10	1.22
2	1.04
1	1.008

From the chart above, you can readily see why the spectrum analyzer is much more accurate when the percent of modulation is 10% or below. Figure 5.18(b) serves to strengthen this statement even more. This is a frequency domain representation of the same 2% AM signal displayed in Figure 5.18(a). You will notice that there is very clear distinction between the carrier and sidebands. The differences can be very easily read, even at this very low modulation percent. If we go through the numbers once again we find that the amplitude difference between the carrier and the sidebands is 40 dB. When this is substituted into the previous formula for use in the frequency domain, it results in a 2% figure for m. To eliminate the numerous calculations that you may have to make, the chart in Figure 5.19 is presented. This is a graph of modulation level versus sideband level. To check is authenticity, you can use the example presented above. We had the sidebands down 40 dB from the carrier and said that this corresponded to a percent modulation of 2%. If we find 40 dB on the bottom of the chart; follow this point up until it intersects the line; move to the left to the Y axis; we can read the figure "2" which corresponds to 2%. To further verify the chart, you can check the following points:

$E_{SB} - E_c$ (dB)	m (%)
10	60
12	50
16	30
26	10
46	1
60	0.2

There are other types of amplitude modulation in addition to the most common type we have discussed. We know that a change in the degree of modulation of a particular carrier does not change the amplitude of the center component itself. It is the amplitude of the sidebands that is changed, thus changing the amplitude of composite wave. Since the amplitude of the carrier component does not change, all the transmitted information must be contained in the sidebands. Therefore, the rather considerable power transmitted in the carrier is essentially wasted. For improved power effi-

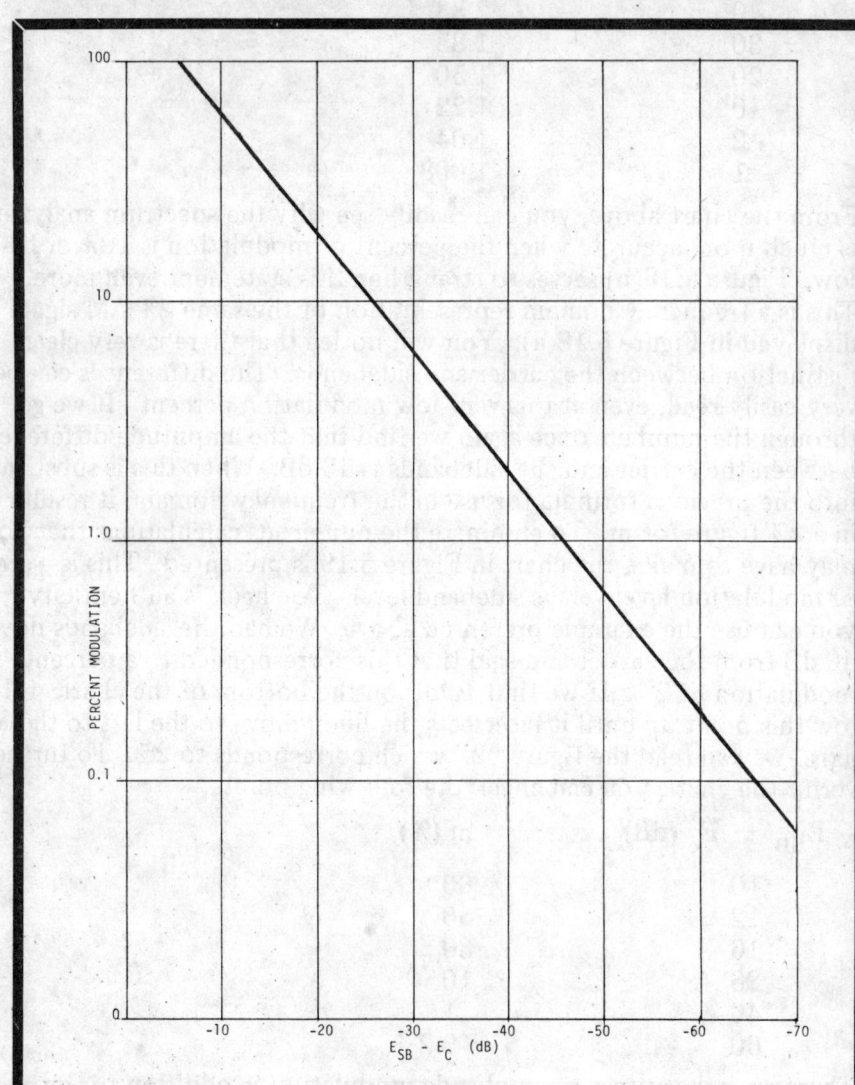

Figure 5.19/Modulation percent vs sideband level.

a) TIME DOMAIN (OSCILLOSCOPE)

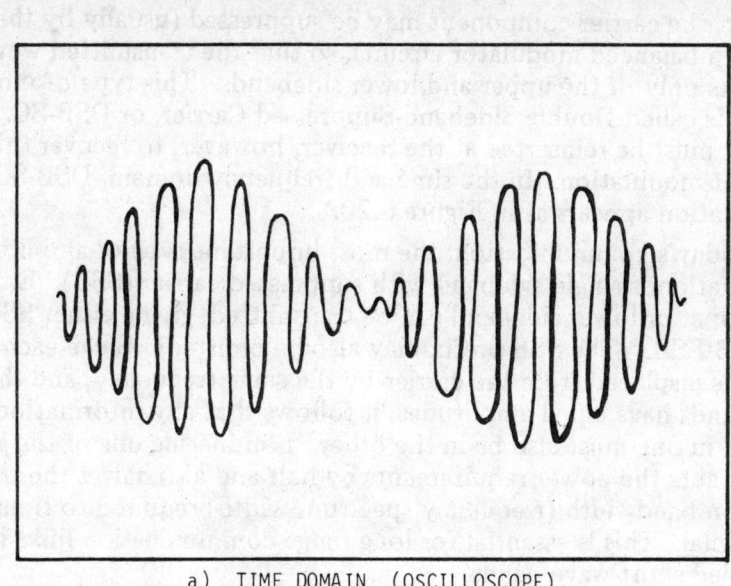

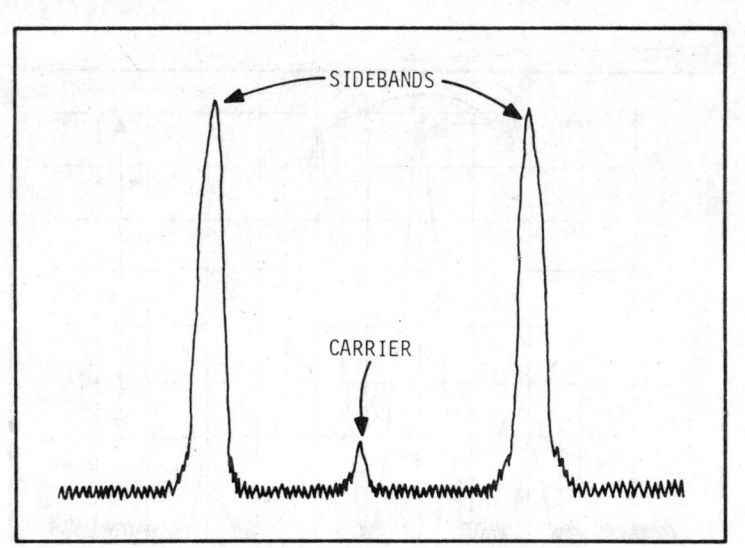

b) FREQUENCY DOMAIN (SPECTRUM ANALYZER)

Figure 5.20/Double sideband — suppressed carrier modulation.

ciency, the carrier component may be suppressed (usually by the use of a balanced modulator circuit), so that the transmitted wave consists only of the upper and lower sidebands. This type of modulation is called Double Sideband-Suppressed Carrier, or DSB-SC. The carrier must be reinserted at the receiver, however, to recover this type of modulation. In the time and frequency domain, DSB-SC modulation appears as in Figure 5.20.

For today's communication, the most important type of amplitude modulation is single sideband with suppressed carrier (SSB). Either the upper or lower sideband can be transmitted, giving either SSB-USB or SSB-LSB. (The SSB prefix may also be omitted). Since each sideband is displaced from the carrier by the same frequency, and the two sidebands have equal amplitudes, it follows that any information contained in one must also be in the other. Eliminating one of the sidebands cuts the power requirements by half and also halves the transmission bandwidth (frequency spectrum width) required to transmit the signal. This is essential for long range communication links in the crowded short-wave bands.

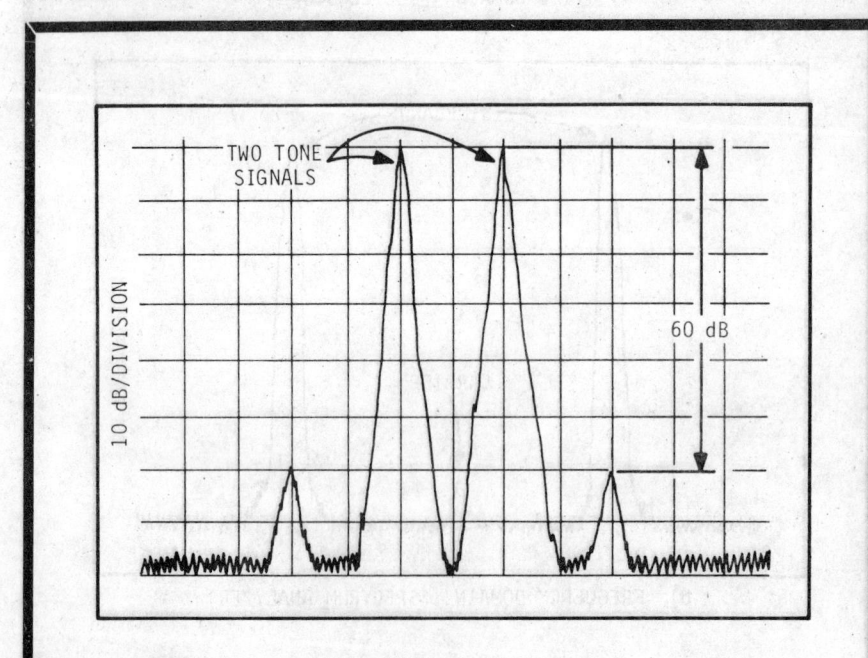

Figure 5.21/"Two-tone" intermodulation display.

Spectrum Analyzer Measurements

A measurement which is of particular interest in single sideband is that of intermodulation distortion (IM). The measurement is performed by modulating the single sideband transmitter (or generator) with two audio tones and checking the output with a spectrum analyzer for additional sidebands. Here a spectrum analyzer having good IM immunity is essential. The spectrum analyzer displays all intermodulation products of the receiver simultaneously, thus decreasing measurement and alignment time substantially. The analyzer display for a typical "Two Tone" test is shown in Figure 5.21. The number that is of importance for measurement purposes is how far down from the two main signals the intermods are. In our case, they are 60 dB down. Since most spectrum analyzers have greater than 70 dB of IM rejection, this number of 60 dB represents the actual intermodulation level of the system we are testing. Details of Intermodulation Measurements are covered in Chapter 6.

Frequency modulation is subject to some basic relationships and rules, much the same as amplitude modulation was when we discussed it above. Some of these relationships are:

(1) Frequency modulation is a constant power process. The power of the modulated wave does not change as the degree of modulation changes.

(2) The frequency-domain representation of an FM wave consists of a carrier and sidebands spaced in frequency around the carrier. The spacing between frequency components is equal to the modulating frequency (f).

(3) Theoretically, the FM wave contains an infinite number of sidebands. The sideband energy, however, falls off very rapidly outside the peak frequency deviation. Deviation is measured with respect to the carrier frequency.

(4) The amplitudes of the various frequency components, including the carrier component, change as the deviation changes. This is a consequence of the requirement that the total power remain constant regardless of the deviation.

(5) The information of interest in FM is: the carrier frequency (F), the modulating frequency (f), and the deviation (ΔF). The carrier frequency, F, is obtained by reading the spectrum-analyzer center-frequency dial; and the modulating frequency, f, is obtained by calculating the frequency spacing between two adjacent components by use of the calibrated dispersion. The deviation (ΔF) can, however, not be determined directly. First,

one obtains the modulation index, from which the deviation is then calculated.

(6) The modulation index is defined as: $m = \dfrac{\Delta F}{f}$

(7) For modulation indices less than 0.2, the modulation index can be measured using the linear approximation:

$$m = 2V_s/V_c$$

when V_s is the voltage of the sidebands and V_c is the voltage of carrier. In the log mode:

$$V_s(dB) - V_c(dB) = 20 \log M/2$$

(8) The relative amplitude of the frequency components are in the same relationship as the relative amplitudes of Bessel functions of the first kind. Bessel functions of the first kind are designated by the letter "J". The complete characterization of the frequency component amplitudes is

$$J_p\left(\dfrac{\Delta F}{f}\right)$$

where p is called the order and represents the frequency component number (p = 0 for the carrier, p = 1 for the first sideband, etc.), and $\Delta F/f$ is called the argument and represents the modulation index. The modulation index (m) has previously been defined as ΔF (peak frequency deviation) divided by f (modulating frequency).

(9) Bessel functions are the solution to a certain differential equation, just as the standard trigonometric functions (sine and cosine) are the solution to a specific differential equation. Graphs and tables of Bessel functions of the first kind are very readily available. Figure 5.22 shows a graph of Bessel functions. This graph shows the relation between the carrier and sideband amplitudes of the modulated wave as a function of the modulation index. Note that the carrier component (J_o) and the various sidebands (J_1, J_2, J_3, etc.), go to zero amplitude at specified values of modulation index. (The carrier zeros are emphasized in Figure 5.22).

Spectrum Analyzer Measurements

From the curves in Figure 5.22 we can get the amplitude of the carrier and sideband components in relation to an unmodulated carrier. Consider an example where m = 3. Check Figure 5.22 and you will find the following values (the sign is of no significance since the spectrum analyzer reads only absolute values):

Carrier (J_o)	= 0.27
First Order Sideband (J_1)	= 0.33
Second Order Sideband (J_2)	= 0.48
Third Order Sideband (J_3)	= 0.33
Fourth Order Sideband (J_4)	= 0.13

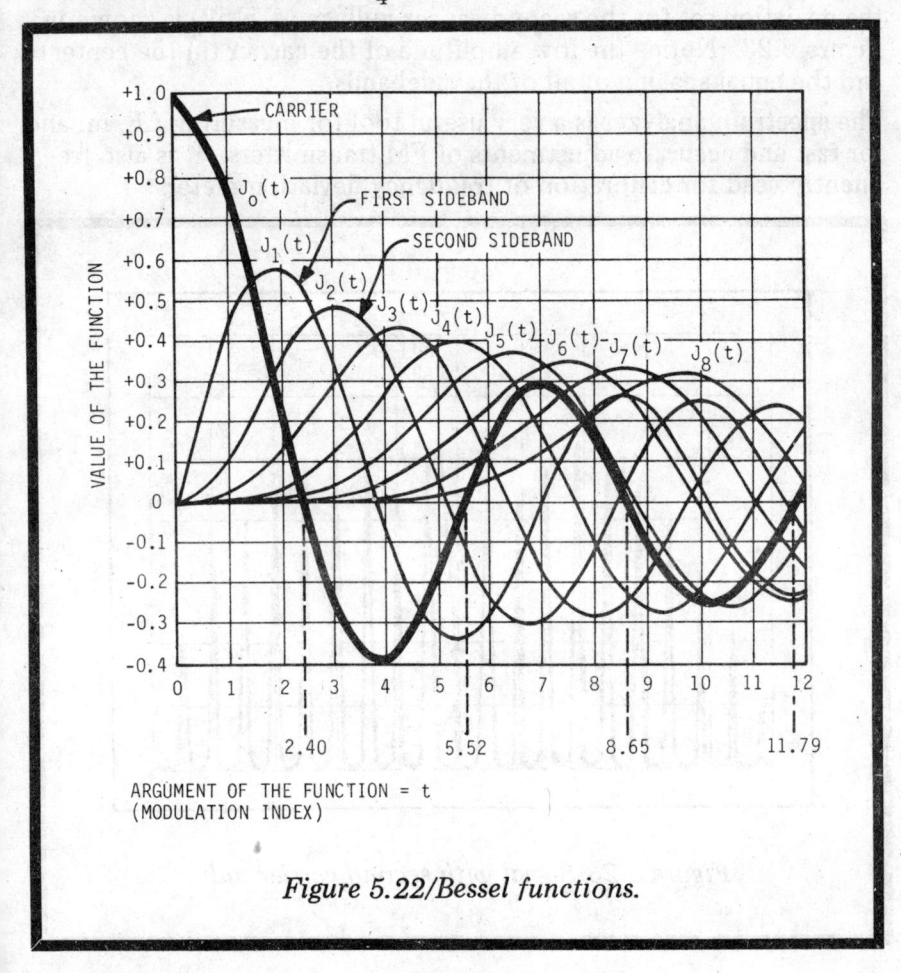

Figure 5.22/Bessel functions.

The exact values for the modulation index corresponding to each of the carrier zeros can be seen in Figure 5.22. They are listed below:

Order of Carrier Zero	Modulation Index
1	2.40
2	5.52
3	8.65
4	11.79
5	14.93
6	18.07
n(n > 6)	18.07 + π(n − 6)

An example of the presentation on the analyzer of an FM signal with the deviation set for the second carrier null (m = 5.52) is shown in Figure 5.23. Notice the low amplitude of the carrier (in the center) and the equal spacing of all of the sidebands.

The spectrum analyzer is a very useful tool for measuring ΔF, m, and for fast and accurate adjustments of FM transmitters. It is also frequently used for calibration of frequency deviation meters.

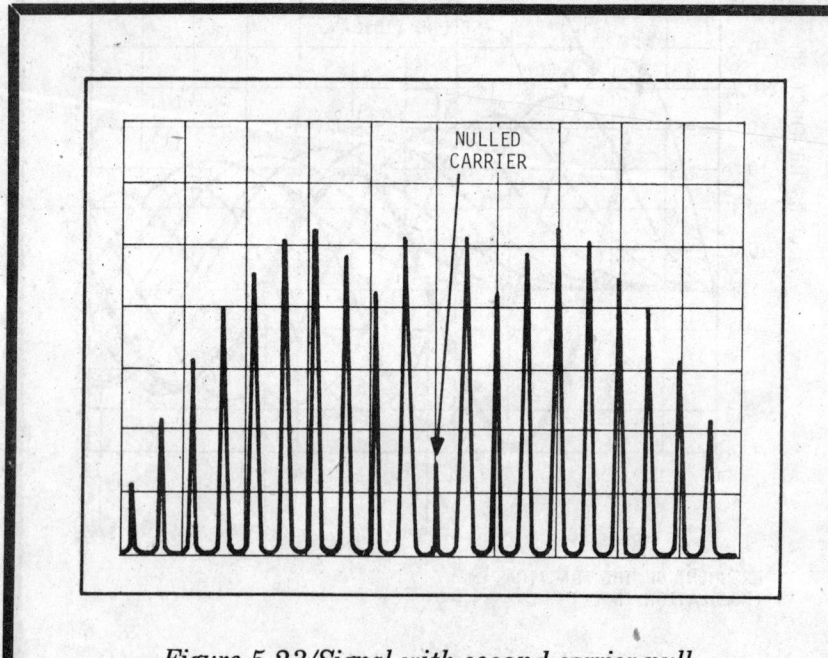

Figure 5.23/Signal with second carrier null.

Spectrum Analyzer Measurements 229

A signal generator or transmitter is adjusted to a precise frequency deviation with the aid of the spectrum analyzer using one of the carrier zeros and selecting the appropriate modulating frequency. As an example, a modulation frequency of 10 kHz and a modulation index of 2.4 (first carrier null) necessitate a carrier peak frequency deviation of exactly 24 kHz. Since the modulation frequency can easily be set accurately with the aid of a frequency counter, and the modulation index is also known accurately, the frequency deviation thus generated is equally accurate.

Table 5.1 is a useful chart that provides the modulation frequency to be set on the counter for commonly used values of deviation for the various orders of carrier zeros.

The procedure for setting up a known deviation is as follows:

(1) Select the column with the appropriate deviation required, for example, 250 kHz.

(2) Select an order of carrier zero number which gives a frequency in the table that is commensurate with the normal modulation circuit is provided in the 250 kHz example above, it will be necessary to go to the 5th carrier zero to get a modulating frequency within the audio passband of the generator (16.74 kHz).

(3) Set the modulating frequency to 16.74 kHz. Monitor the generator output spectrum on the analyzer and adjust the amplitude of the audio modulating signal until the carrier amplitude has gone through four zeros and stop when the carrier is at its fifth minimum. With the modulating frequency of 16.74 kHz and the spectrum at its fifth zero, then a unique 250 kHz deviation is being provided by the setup. The modulation meter may then be calibrated. You can make a quick check by moving to the adjacent carrier zero and resetting the modulating frequency and amplitude (i.e., 13.84 kHz at the sixth carrier zero in the above example).

Other intermediate deviations and modulation indexes are settable using various orders or sideband zeros, but these are influenced by incidental amplitude modulation. Since it is known that amplitude modulation does not cause the carrier to change but instead puts all the modulation power into the sidebands, incidental AM will not affect the carrier zero method above.

If it is not possible or desirable to alter the modulation frequency to get a carrier or sideband null, there are other ways to obtain usable information about frequency deviation and modulation index. One

TABLE 5.1
Peak Deviation and Modulation Index for Various Carrier Zeros

Commonly Used Values of FM Peak Deviation

Order of Carrier Zero	Modulation Index	7.5 kHz	10 kHz	15 kHz	25 kHz	30 kHz	50 kHz	75 kHz	100 kHz	150 kHz	250 kHz	300 kHz
1	2.40	3.12	4.16	6.25	10.42	12.50	20.83	31.25	41.67	62.50	104.17	125.00
2	5.52	1.36	1.81	2.72	4.53	5.43	9.06	13.59	18.12	27.17	45.29	54.35
3	8.65	.87	1.16	1.73	2.89	3.47	5.78	8.67	11.56	17.34	28.90	34.68
4	11.79	.66	.85	1.27	2.12	2.54	4.24	6.36	8.48	12.72	21.20	25.45
5	14.93	.50	.67	1.00	1.67	2.01	3.35	5.02	6.70	10.05	16.74	20.09
6	18.07	.42	.55	.83	1.88	1.66	2.77	4.15	5.53	8.30	13.84	16.60

method is to calculate m by using the amplitude information of five adjacent frequency components in the FM signal. These five measurements are used in a recursion formula for Bessel functions to form three calculated values of a modulation index. Averaging yields m with practical measurements errors taken into consideration. Because of the number of calculations necessary, this method is only applicable when using a computer.

A somewhat easier method consists of two measurements. First, the sideband spacing of the modulated carrier is measured by using a sufficiently small IF bandwidth (BW), thus getting the modulation frequency f. Second, the peak frequency deviation, ΔF, is measured by selection of a convenient scan width and an IF bandwidth wide enough to cover all significant sidebands. Modulation index m can then be calcualted easily by using the convenient formula:

$$m = \frac{\Delta F}{f}$$

Since both ΔF and f have been measured, the only remaining task is a simple calculation. Figure 5.24 shows displays for measuring ΔF and f. In Figure 5.24(a), the measurement of ΔF (peak deviation) is shown. This, as mentioned above, is accomplished by displaying the entire spectrum and then increasing the IF bandwidth of the spectrum analyzer sufficiently to provide the display shown. By adjusting the bandwidth and scan width, the signal may be centered on the display to enable a reading of peak deviation.

Figure 5.24(b) is a conventional FM display with no carrier suppression. With a display such as this you can read directly the modulating frequency (f) by observing what the scan width of the analyzer is set to and the number of divisions the sidebands are from the carrier.

There are other forms of modulation which are not clear cut AM or FM. One of these is a combination of AM and FM. Combined AM and FM is usually an accidental, or incidental, occurrence. The desired modulation is usually AM, with the FM modulation an incidental byproduct of an imperfect AM modulator. Combined AM and FM is characterized by two sidebands of unequal amplitude. This is because the AM sidebands are of the same phase while the FM sidebands are of opposite phase. A measurement technique for combined AM and FM is shown in Figure 5.25. From the unequal sideband amplitudes we conclude that the signal contains both AM and FM. Next, since the signal is supposed to be purely AM, we assume that the AM sidebands are larger than the FM sidebands. Except in very unusual circumstances

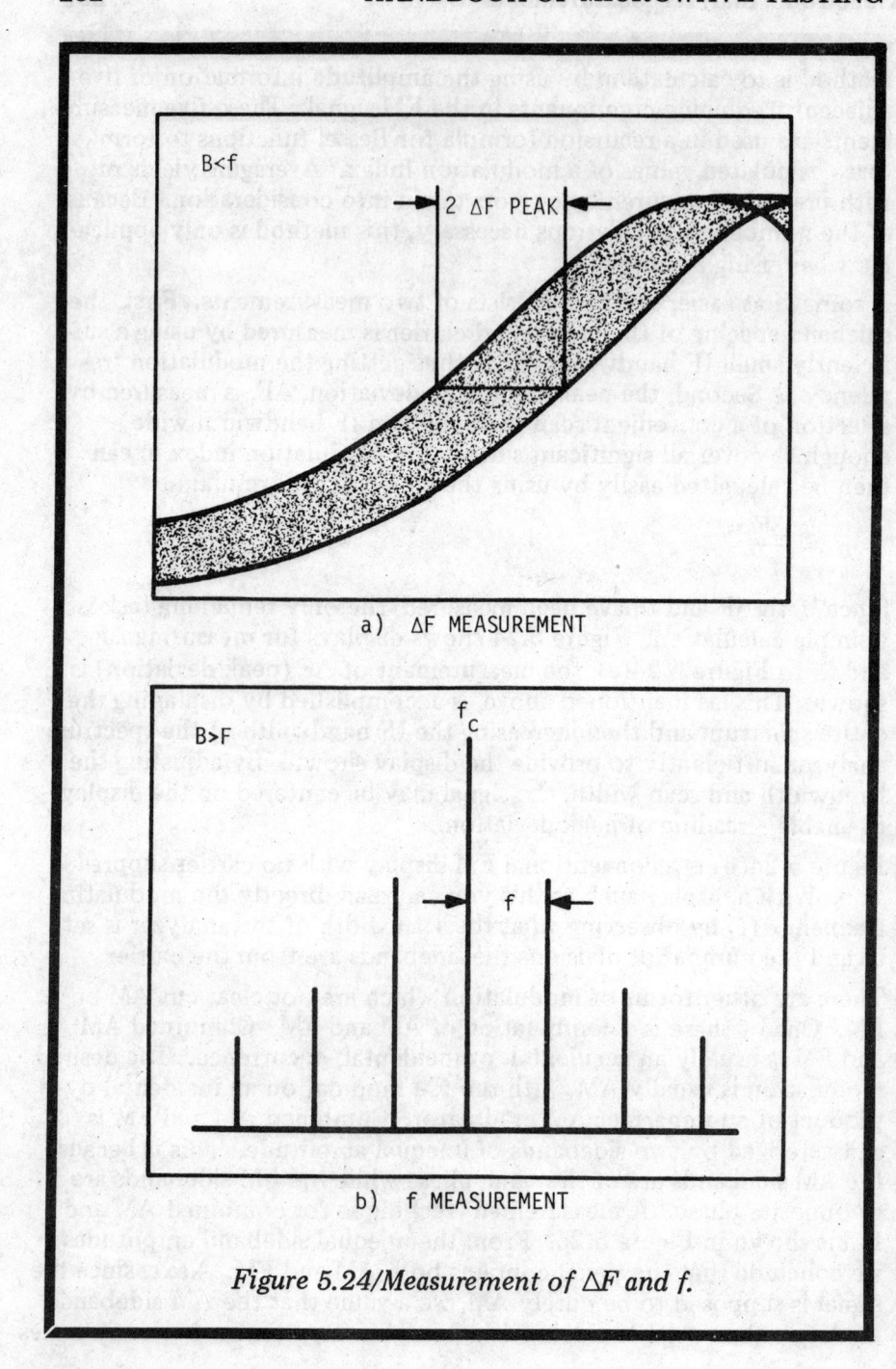

Figure 5.24/Measurement of ΔF and f.

Spectrum Analyzer Measurements

this will always be the case. A further verification of the small size of the FM sidebands is the fact that the combined signal has only one significant sideband. We now compute the amplitudes of the individual AM and FM sidebands, using the fact that, in the combined spectrum, one sideband consists of the sum of an AM and FM sideband while the other sideband consists of the difference between an AM and FM sideband. From Figure 5.25, one sideband is about 2.1 cm high while the other is about 1.7 cm high. This leads to the conclusion that the AM sidebands are 1.9 cm high and the FM sidebands are 0.2 cm high. From the above information, we can compute the percent of AM modulation and FM modulation index.

$$\% \text{ AM} = \frac{2 \text{ (Sideband)}}{\text{Carrier}} \times 100$$

$$\% \text{ AM} = \frac{2 (1.9)}{6.5} \times 100$$

$$\% \text{ AM} = 58.5\%$$

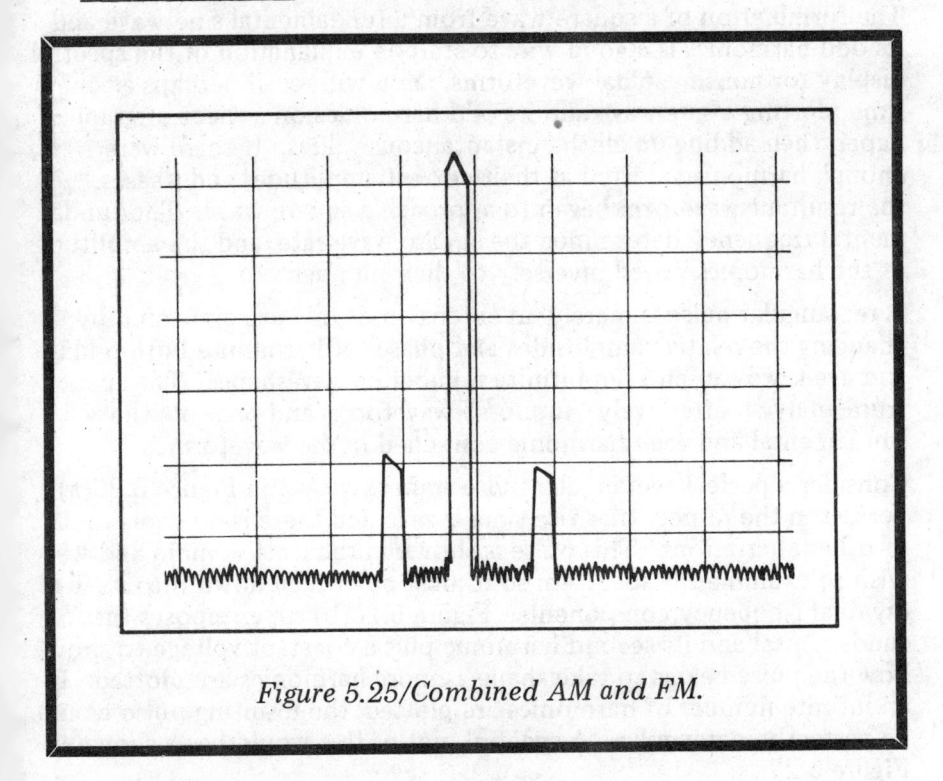

Figure 5.25/Combined AM and FM.

$$\text{FM Modulation Index (m)} = \frac{2\,(\text{Sideband})}{\text{Carrier}}$$

$$m = \frac{2\,(0.2)}{6.5}$$

$$m = 0.0615$$

5.4.2 Pulsed RF

The spectrum analyzer was originally designed to look at the output of radar transmitters. A pulse radar signal is a train of RF pulses with a constant repetition rate, constant pulse width and shape, and constant amplitude. By looking at the characteristic spectra, all important properties of the pulsed signal, such as pulse width, occupied bandwidth, duty cycle, peak and average power, etc., can be measured easily and with high accuracy. This is a change from other measurements, which were all based on sinusoidal waveforms.

The formulation of a square wave from a fundamental sine wave and its odd harmonics is a good way to start an explanation of the spectral display for nonsinusoidal waveforms. You will recall perhaps at one time plotting a sine wave and its odd harmonics on a sheet of graph paper, then adding up all the instantaneous values. If there were enough harmonics plotted at their correct amplitudes and phases, the resultant waveform began to approach a square wave. The fundamental frequency determined the square wave rate, and the amplitudes of the harmonics varied inversely to their number.

A rectangular pulse is merely an extension of this principle, and by changing the relative amplitudes and phases of harmonics both odd and even, we can plot an infinite number of waveshapes. The spectrum analyzer effectively "unplots" waveforms and presents the fundamental and each harmonic contained in the waveform.

Consider a perfect rectangular pulse train as shown in Figure 5.26(a), perfect in the respect that rise time is zero and there is no overshoot or other aberrations. This pulse is shown in the time domain and we wish to examine the spectrum so it must be broken down into its individual frequency components. Figure 5.26(b) superimposes the fundamental and its second harmonic plus a constant voltage to show how the pulse begins to take shape as more harmonics are plotted. If an infinite number of harmonics are plotted, the resulting pulse would be perfectly rectangular. A spectral plot of this would be as shown in Figure 5.27.

Spectrum Analyzer Measurements

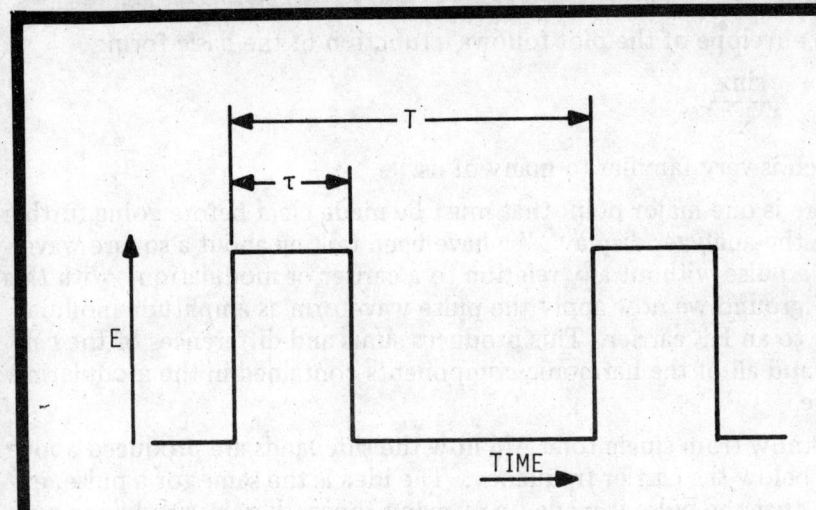

(a) PERIODIC RECTANGULAR PULSE

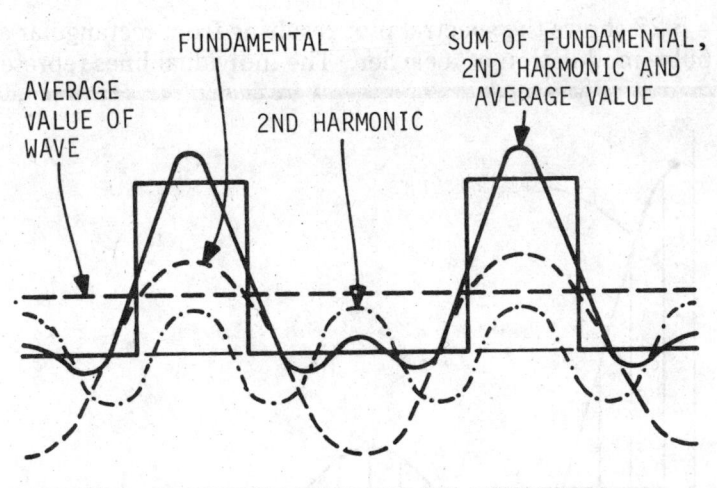

(b) FUNDAMENTAL SIGNAL AND HARMONICS

Figure 5.26/Rectangular pulses.

The envelope of the plot follows a function of the basic form:

$$y = \frac{\sin x}{x},$$

which is very familiar to many of us.

There is one major point that must be made clear before going further into the analyzer display. We have been talking about a square wave and a pulse without any relation to a carrier or modulation. With this background we now apply the pulse waveform as amplitude modulation to an RF carrier. This produces sums and differences of the carrier and all of the harmonic components contained in the modulating pulse.

We know from single tone AM how the sidebands are produced above and below the carrier frequency. The idea is the same for a pulse, except that the pulse is made up of many tones, thereby producing multiple sidebands which are commonly referred to as spectral lines on the analyzer display. In fact, there will be twice as many sidebands or spectral lines as there are harmonics contained in the modulating pulse.

Figure 5.28 shows the spectral plot resulting from rectangular amplitude pulse modulation of a carrier. The individual lines represent the

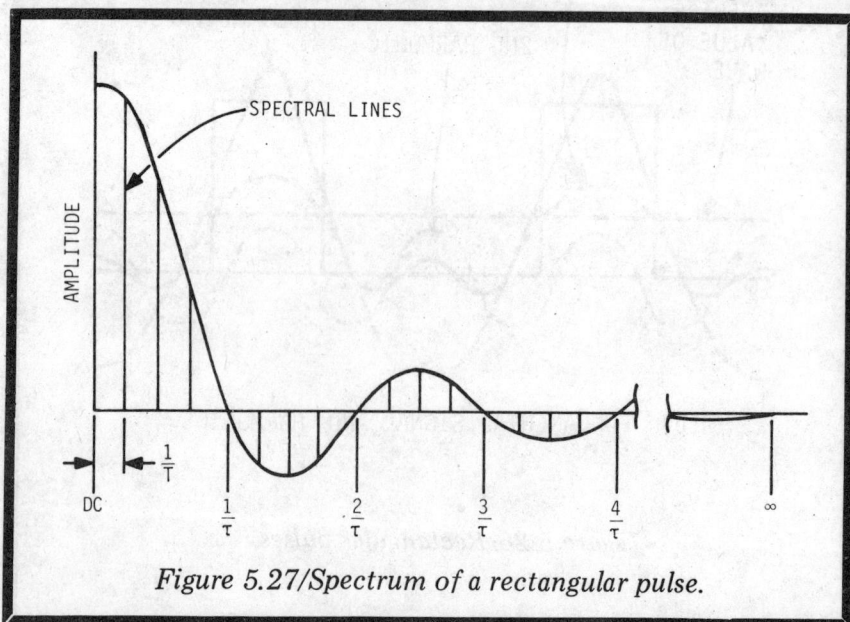

Figure 5.27/Spectrum of a rectangular pulse.

Spectrum Analyzer Measurements

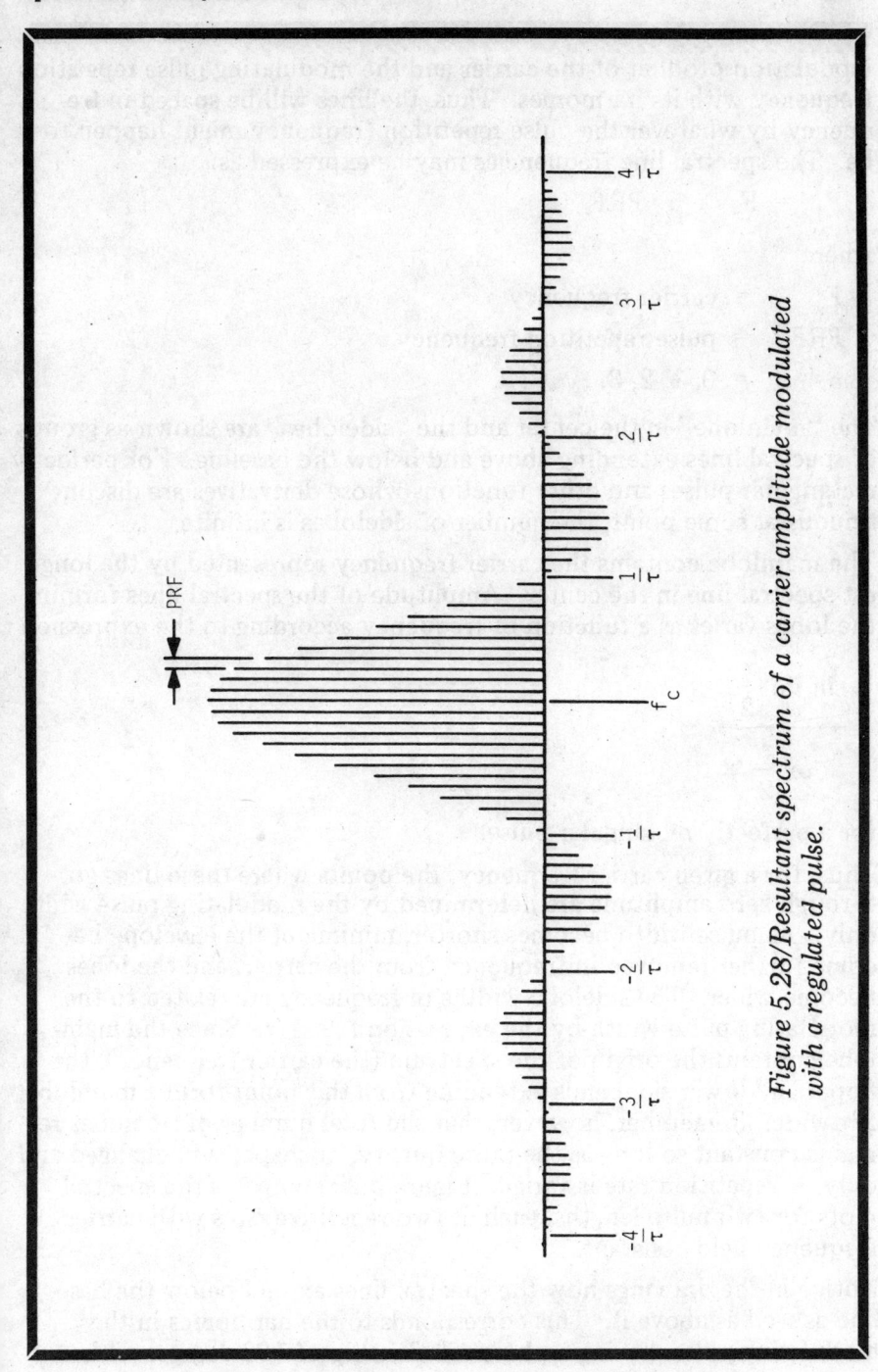

Figure 5.28/Resultant spectrum of a carrier amplitude modulated with a regulated pulse.

modulation product of the carrier and the modulating pulse repetition frequency with its harmonics. Thus, the lines will be spaced in frequency by whatever the pulse repetition frequency might happen to be. The spectral line frequencies may be expressed as:

$$F_L = F_r \pm n \cdot PRF$$

where

F_r = carrier frequency
PRF = pulse repetition frequency
n = 0, 1, 2, 3........

The "mainlobe" in the center and the "sidelobes" are shown as groups of spectral lines extending above and below the baseline. For perfectly rectangular pulses and other functions whose derivatives are discontinuous at some point, the number of sidelobes is infinite.

The mainlobe contains the carrier frequency represented by the longest spectral line in the center. Amplitude of the spectral lines forming the lobes varies as a function of frequency according to the expression

$$\frac{\sin \omega \frac{\tau}{2}}{\omega \frac{\tau}{2}}$$

for a perfectly rectangular pulse.

Thus, for a given carrier frequency, the points where these lines go through zero amplitude are determined by the modulating pulse width only. As pulse width becomes shorter, minima of the envelope become further removed in frequency from the carrier, and the lobes become wider. The sidelobe widths in frequency are related to the modulating pulse width by the expression $f = 1/\tau$. Since the mainlobe contains the origin of the spectrum (the carrier frequency), the upper and lower sidebands extending from this point form a mainlobe $2/\tau$ wide. Remember, however, that the total number of sidelobes remains constant so long as the pulse quality, or shape, is unchanged and only its repetition rate is varied. Figure 5.29 compares the spectral plots for two pulse lengths, each at two repetitive rstes with carrier frequency held constant.

Notice in the drawings how the spectral lines extend below the baseline as well as above it. This corresponds to the harmonics in the modulating pulse, having a phase relationship of 180 degrees with

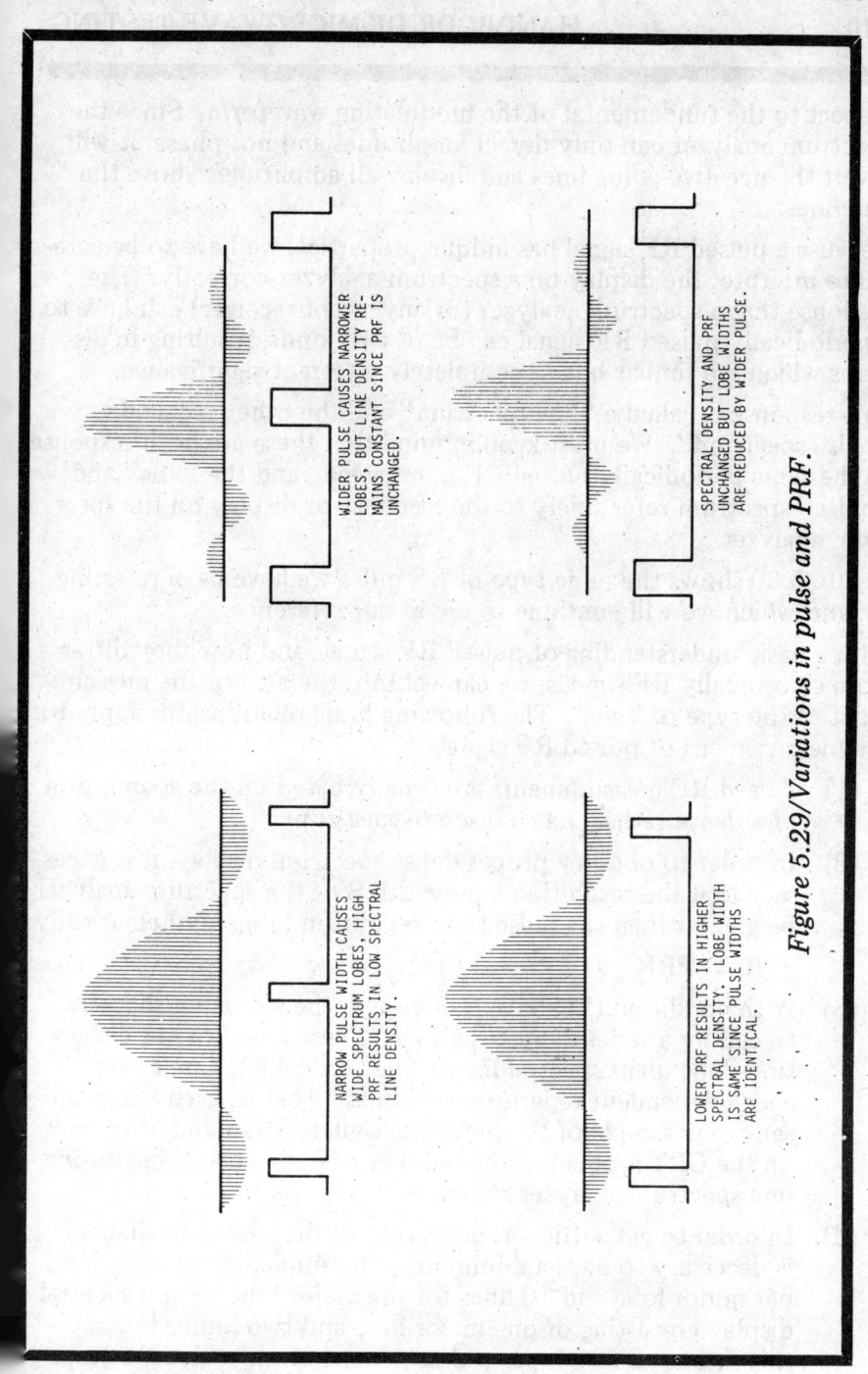

Spectrum Analyzer Measurements

WIDER PULSE CAUSES NARROWER LOBES, BUT LINE DENSITY REMAINS CONSTANT SINCE PRF IS UNCHANGED.

SPECTRAL DENSITY AND PRF UNCHANGED BUT LOBE WIDTHS ARE REDUCED BY WIDER PULSE.

NARROW PULSE WIDTH CAUSES WIDE SPECTRUM LOBES, HIGH PRF RESULTS IN LOW SPECTRAL LINE DENSITY.

LOWER PRF RESULTS IN HIGHER SPECTRAL DENSITY. LOBE WIDTH IS SAME SINCE PULSE WIDTHS ARE IDENTICAL.

Figure 5.29/Variations in pulse and PRF.

respect to the fundamental of the modulating waveform. Since the spectrum analyzer can only detect amplitudes and not phase, it will invert the negative-going lines and display all amplitudes above the baseline.

Because a pulsed RF signal has unique properties, we have to be careful to interpret the display on a spectrum analyzer correctly. The response that a spectrum analyzer (or any swept receiver) can have to a periodically pulsed RF signal can be of two kinds, resulting in displays which are similar but of completely different significance.

One response is called a "line spectrum" and the other is called a "pulse spectrum". We must keep in mind that these are both responses to the same periodically pulsed RF input signal, and the "line" and "pulse" spectrum refer solely to the response or display on the spectrum analyzer.

Figure 5.30 shows the basic type of RF pulse we have been referring to, and which we will continue to use as our reference.

With a basic understanding of pulsed RF signals and how they differ from sinusoidally RF signals, we can get into the area of the measurement of the type of signal. The following basic relationships apply to the measurement of pulsed RF signals:

(1) Pulsed RF measurements are usually based on the assumption of a *dense* rather than a *discrete* spectrum.

(2) In order to obtain a proper dense spectrum display, it is necessary that the resolution bandwidth B of the spectrum analyzer be greater than the pulse-train repetition rate. Mathematically:

$$B \geqslant PRR$$

(3) With condition (2) above established, the spectrum shape is traced by a series of vertical lines. These lines are not dispersion-dependent spectral lines. The vertical lines are sweep-speed-dependent repetition-rate lines. That is, each line represents one sample of the incoming signal. The number of lines on the CRT is equal to the number of pulses occurring during one spectrum-analyzer sweep.

(4) In order to get sufficient definition of the spectrum shape, it is necessary to have a minimum of 5 sample, or rep-rate, lines per minor lobe and 10 lines for the major lobe. For a spectral display consisting of one major lobe and two minor lobes, this means twenty input pulses per spectrum-analyzer sweep.

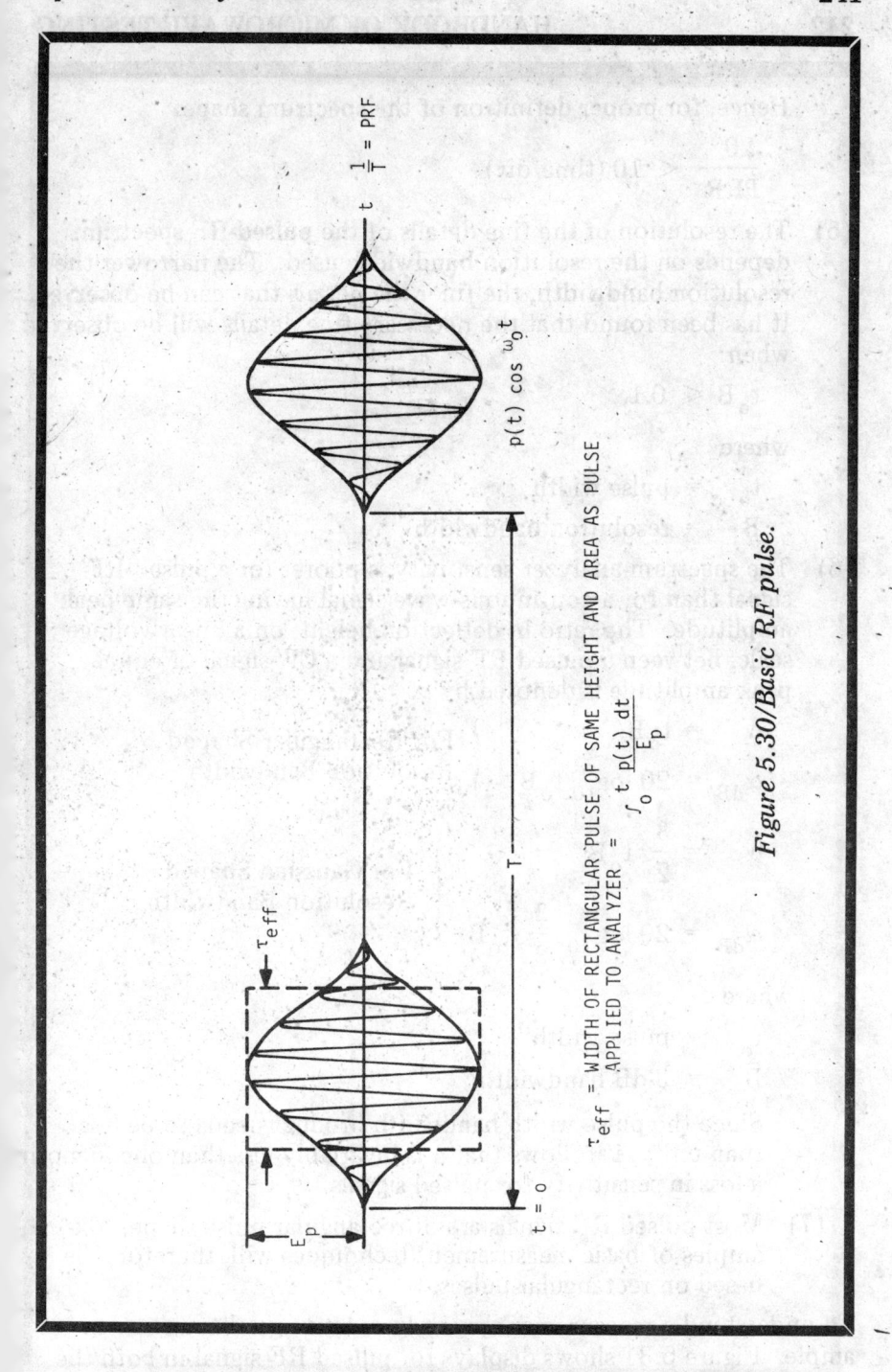

Spectrum Analyzer Measurements

$p(t) \cos \omega_0 t$

τ_{eff} = WIDTH OF RECTANGULAR PULSE OF SAME HEIGHT AND AREA AS PULSE APPLIED TO ANALYZER = $\dfrac{\int_0^t p(t)\, dt}{E_p}$

$t = \dfrac{1}{T} = PRF$

Figure 5.30/Basic RF pulse.

Hence, for proper definition of the spectrum shape:

$$\frac{20}{\text{PRR}} < 10 \text{ (time/div)}$$

(5) The resolution of the fine details of the pulsed-RF spectrum depends on the resolution bandwidth used. The narrower the resolution bandwidth, the finer the details that can be observed. It has been found that the necessary fine details will be observed when:

$$t_o B \leq 0.1,$$

where

t_o = pulse width

B = resolution bandwidth

(6) The spectrum-analyzer sensitivity is poorer for a pulsed-RF signal than for a continuous-wave signal having the same peak amplitude. The ratio in deflection height, on a linear voltage scale, between a pulsed-RF signal and a CW-signal of equal peak amplitude is denoted by α.

$$\left. \begin{array}{l} \alpha = t_o B \\ \alpha_{dB} = 20 \log_{10} t_o B \end{array} \right\} \text{For Rectangular Shaped Resolution Bandwidth}$$

$$\left. \begin{array}{l} \alpha = \frac{3}{2} t_o B \\ \alpha_{dB} = 20 \log_{10} \frac{3}{2} t_o B \end{array} \right\} \text{For Gaussian Shaped Resolution Bandwidth}$$

where

t_o = pulse width

B = 3 dB bandwidth

Since the pulse width bandwidth product needs to be less than unity, it follows that α is invariably less than one, denoting a loss in sensitivity for pulsed signals.

(7) Most pulsed RF signals are of rectangular pulse shape. Examples of basic measurement techniques will, therefore, be based on rectangular pulses.

To understand a measurement procedure, let us go through an example. Figure 5.31 shows displays for pulsed RF signal in both the

Spectrum Analyzer Measurements

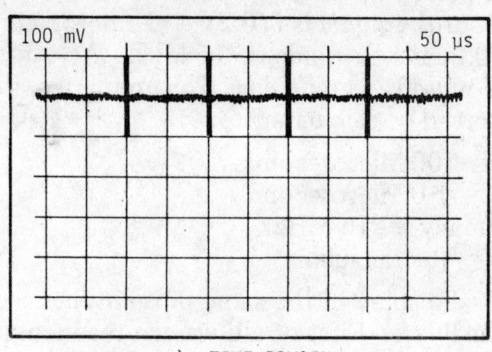

a) TIME DOMAIN

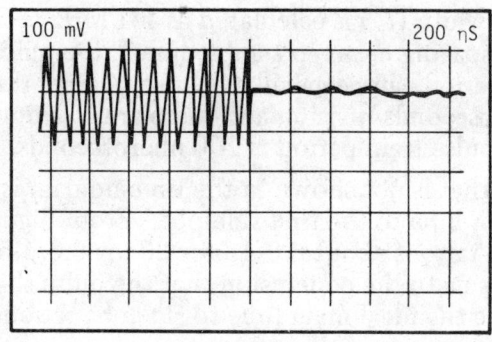

b) TIME DOMAIN

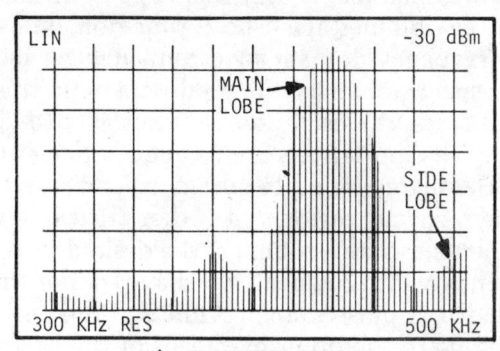

c) FREQUENCY DOMAIN

Figure 5.31/Measurement example.

time-domain and the frequency-domain. From Figure 5.31(a), we can see that the period of the pulse train is 2(50) = 100 microseconds. The pulse width of the signal is 5(0.2) = 1 microsecond and the carrier frequency is 3/0.2 microseconds = 15 MHz. We can also tell from the display that the pulse is rectangular. To summarize, the following has been either observed or calculated:

- Pulse Train = 100 Microseconds
- Pulse Width = 1.0 Microsecond
- Carrier Frequency = 15 MHz
- Pulse Shape = Rectangular

Figure 5.31(c) is a display of the same pulse in the frequency-domain on a spectrum analyzer. First of all, we can read from the spectrum analyzer center frequency dial that the carrier frequency is 15 MHz. From the width of the sidelobes (2 divisions) (500 kHz/division) = 1 MHz, the pulse width (t_o) is calculated as 1/1 MHz = 1 microsecond. From the time spacing between sample line 1, the pulse train repetition rate (PRR) or period can be calculated. We observe that the display is set at 500 microseconds/division and there are 5 samples per division. Therefore, the pulse train period is 100 microseconds.

One parameter that is not shown in the time-domain is the amplitude ratio of the main lobe to the first sidelobe. From Figure 5.31(c), we can see that the ratio is about 6.6/1.5 or 20 log 6.6/1.5 = 13 dB. This figure leads us to the conclusion that the pulse shape is essentially rectangular since the ideal main lobe to sidelobe ratio is 13.2 dB for a rectangular pulse.

Based on the above example, it may seem apparent that the same basic information can be obtained from both time-domain oscilloscope measurements and frequency-domain spectrum-analyzer measurements. The reason why spectrum analyzers predominate in this area is that, except in special cases, the oscilloscope is not able to display the signals in question. The limitations are frequency range (pulsed-RF signals in the GHz region are quite common) and sensitivity (many signals are in the picowatt region). Another difficulty with oscilloscope measurements is that frequently the desired information is the occupied frequency width or spectrum shape, rather than the time-domain pulse width or pulse shape. While it is theoretically possible to convert from one to the other by means of the Fourier mathematics, the task can be quite difficult.

Most pulsed RF measurements occur in radar systems. These involve both the determination of what a radar set is transmitting and the adjustment or calibration of the radar set so that it gives the required

Spectrum Analyzer Measurements

output. Spectrum analyzers are also frequently used in testing components such as pulsed magnetrons. The data of interest usually involves the following:

(1) Carrier Frequency (F)
(2) Pulse Width (t_o)
(3) Pulse Repetition Rate (PRR), Interpulse Interval (T)
(4) Pulse Shape
(5) Occupied Signal Bandwidth
(6) Percentage Missing Pulses
(7) Carrier On/Off Ratio
(8) Presence of FM

Since radar sets usually put out much more power than the spectrum analyzer can accommodate, the signal connection is typically made through a directional coupler. Even then it may be necessary to add attenuation to the signal path to maintain spectrum-analyzer operation within the linear region. A typical test setup is shown in Figure 5.32(a). An alternate procedure, especially helpful in radar set tests as opposed to alignment, is to receive the transmitted signal with a second antenna. Such an arrangement calls for a transportable spectrum analyzer. Care should be taken in antenna placement to prevent reflections from nearby objects. Figure 5.32(b) shows this type of test setup.

The use of pulsed RF signals in microwave technology finds many applications in radar, electronic countermeasures (ECM), and electronic warfare (EW) equipment. With applications such as these, there must be a means of characterizing such pulses — this means is the spectrum analyzer.

5.4.3 Distortion

Distortion can be defined as an undesired change in the waveforms of the original, resulting in an unfaithful reproduction of the required signal. Minimizing this distortion has a very high priority in design and production. The main order of business when distortion is present is to have it accurately characterized.

Distortion falls into two general areas: spurious and predictable. Spurious signals are unrelated to the carrier signals present. For example, parasitic oscillations may occur in an amplifier at a frequency unrelated to the input. Harmonic and intermodulation distortion are common, "predictable" problems which fall at frequencies directly related to the input signal frequencies. We will concentrate only on harmonic distortion in this section. The problem of intermodulation will be covered in the following chapter on active devices.

Figure 5.32/Radar pulsed RF setup.

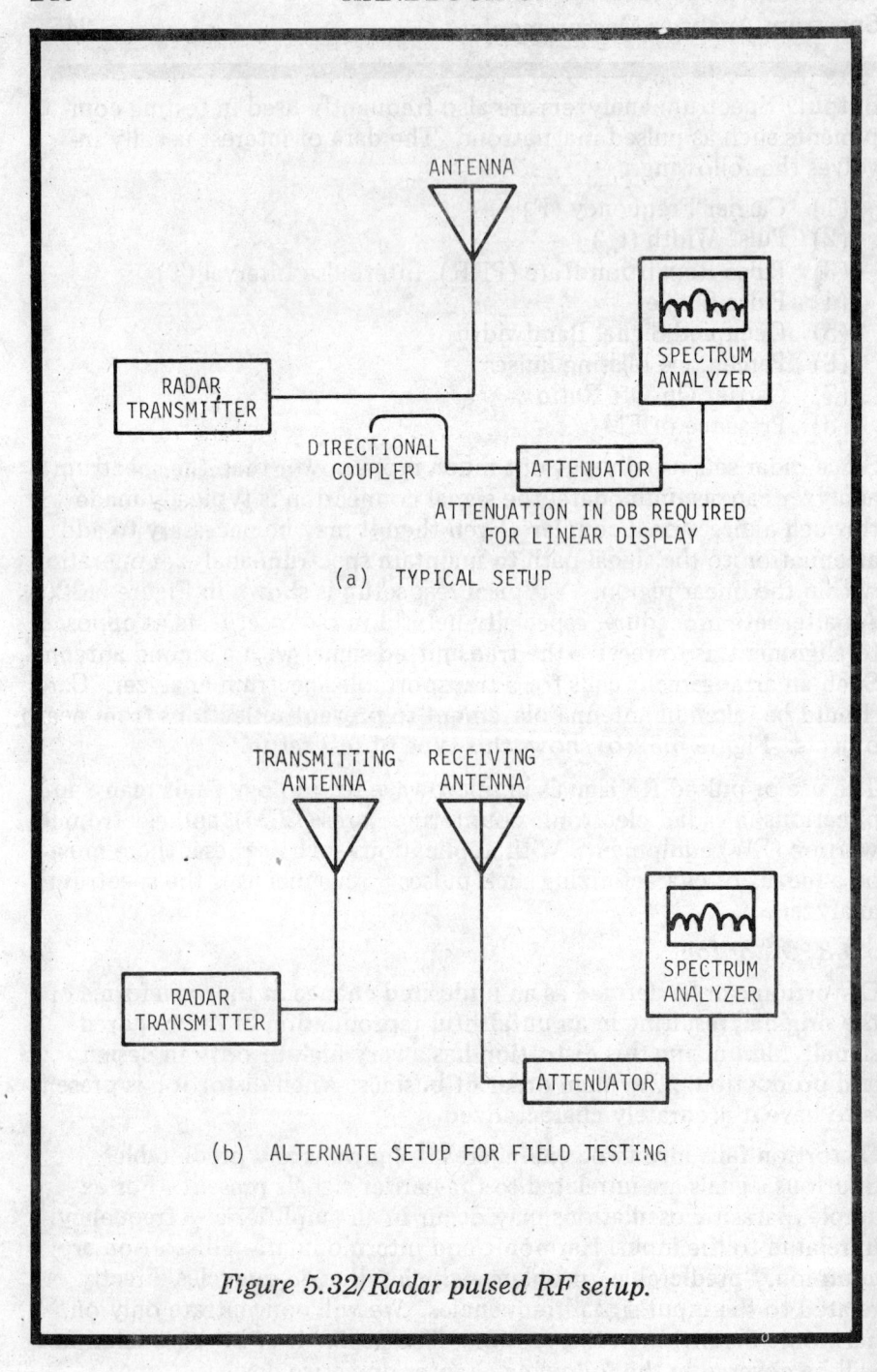

Spectrum Analyzer Measurements

Harmonic distortion is directly related to a fundamental frequency signal and its integer multiples, called harmonics. It is a measure of the relation amplitudes of the harmonics and fundamental signals. Often, interest is not in each harmonic's effect, but the *total harmonic distortion* or THD. Figure 5.33 shows two cases of harmonic distortion. Each is shown in both the time and frequency domains. Figure 5.33(a) shows a circuit that is severely overdriven and produces noticeable harmonic distortion, even on the time-domain presentation (scope). In Figure 5.33(b), the drive has been reduced and the time domain signal appears to be pure undistorted sine wave. However, the frequency domain display (spectrum analyzer) shows that there are still harmonics present which cause some degree of distortion.

Harmonic distortion must be calculated when you use a spectrum analyzer for a display. The calculations are as follows:

$$\text{THD} (\%) = 100 \times \frac{\sqrt{(A_2)^2 + (A_3)^2 + \ldots + (A_n)^2}}{A_1}$$

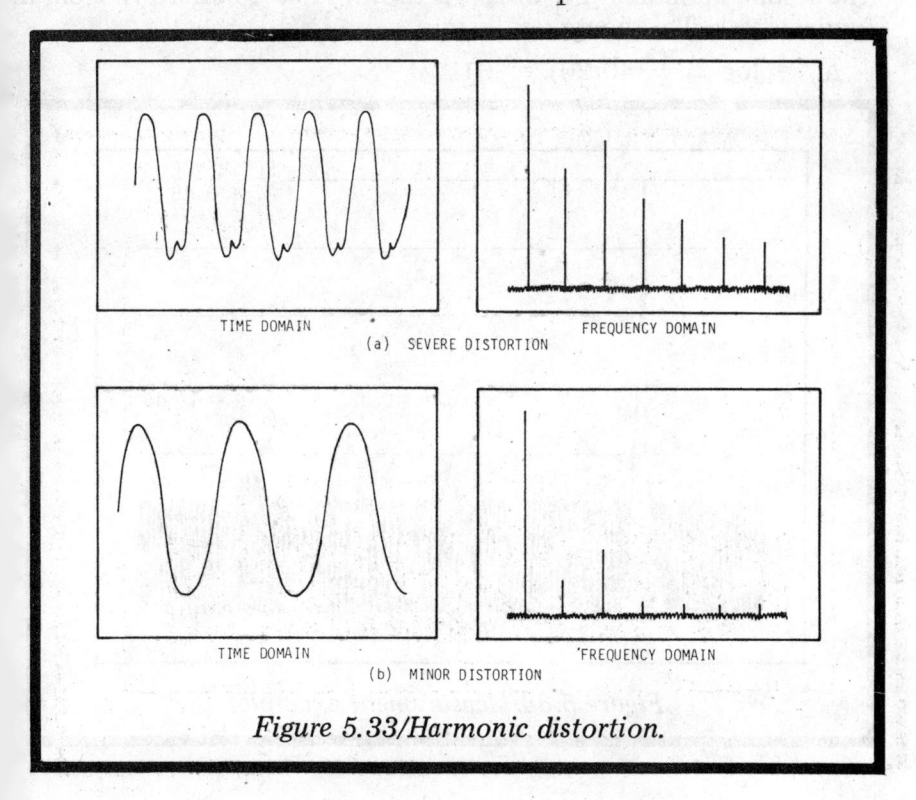

Figure 5.33/Harmonic distortion.

where

A_1 = fundamental amplitude (volts)
A_2 = second harmonic amplitude (volts)
A_3 = third harmonic amplitude (volts), etc.

The spectrum analyzer, as used for this measurement, displays signals in logarithmic form (dB). All of the "A_n" terms are in linear voltage, and the log display of the analyzer must be converted for use. To do this easily and avoid absolute voltage calculations, all terms are measured relative to the fundamental. This sets A_1 to unity reference ($\log^{-1}(0/20) = 1$). For example, if the second harmonic component is 40 dB below the fundamental, then A_2 (linear) = $\log^{-1}(-40/20)$, or $A_2 = 0.01$. If this is the only distortion present, our THD formula yields 1.0% total harmonic distortion.

To illustrate this procedure, we will go through a measurement example. Figure 5.34 shows a signal with two harmonics associated with it. The reference has been set as the fundamental equal to 0 dB.

The second harmonic ($2f_1$ or A_2) is shown to be 40 dB down from the fundamental. This figure results in:

$A_2 = \log^{-1}(-40/20) = .01$

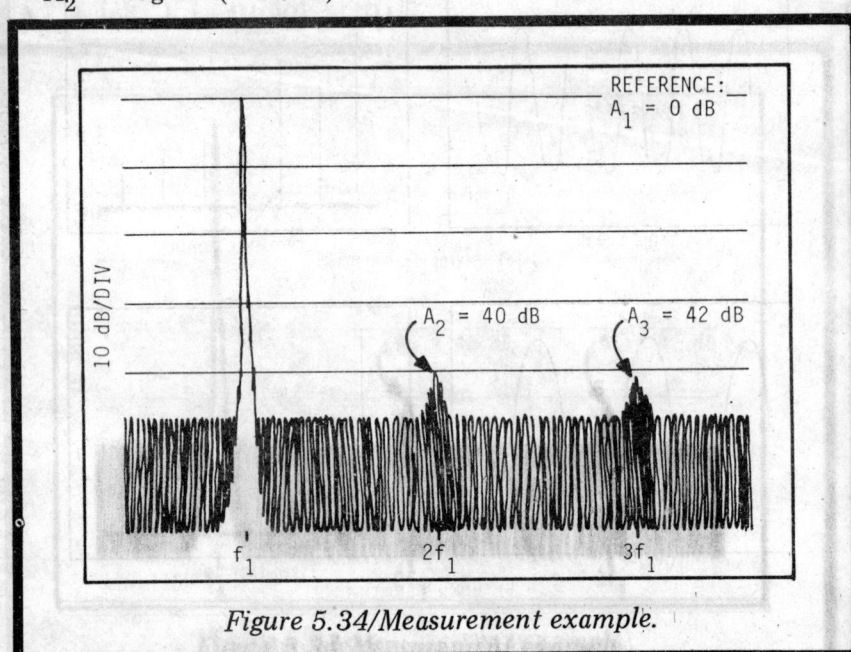

Figure 5.34/Measurement example.

Spectrum Analyzer Measurements 249

The third harmonic ($3f_1$ or A_3) is shown to be 42 dB down from the fundamental. This figure results in:

$A_3 = \log^{-1} (-42/20) = .0079$

With the individual components identified we can now substitute in the formula above and obtain the total harmonic distortion.

$$THD = 100 \frac{\sqrt{(A_2)^2 + (A_3)^2}}{A_1}$$

$$THD = 100 \frac{\sqrt{(.01)^2 + (.0079)^2}}{1}$$

$THD = 1.28\%$

Another method which can be used to perform distortion measurements is by using the RMS-sum method. This is done by taking the square root of the sum of the squares of the individual distortions.

We can illustrate this method by using the same example. This is very similar to the process used above except that no logs are needed.

You will recall that the harmonics were 40 dB and 42 dB down from the fundamental. When we convert these to voltage ratios we get 1/100 and 1/126, respectively, as the amplitude of the harmonics with respect to the fundamental. These numbers now become 0.01 and 0.0079. By multiplying by 100, we obtain the percent of distortion for each harmonic (1% and 0.79%). We can now compute the total distortion as follows:

$THD = \sqrt{(HD_2)^2 + (HD_3)^2}$

$THD = \sqrt{(1)^2 + (.79)^2}$

$THD = \sqrt{1.624}$

$THD = 1.28\%$

You can see how either method yields the same percentage of distortion.

The harmonic distortion examined in this section can be kept to a minimum by careful circuit design and close attention to drive levels for active circuits. It is usually an active device being overdriven which causes circuit nonlinearities which result in harmonics distortion at the output. The more linear a device can be kept, the less harmonic distortion is present.

5.5 Auxiliary Equipment

The spectrum analyzer is a very useful microwave tool by itself. Its performance can be improved, however, by the use of what we will call "auxiliary" equipment. This equipment consists of storage normalizers, tracking preselectors, and tracking generators.

The storage normalizer is the first device to be mentioned. Swept measurements of network characteristics generally include the response of the test system itself and all of its components, making it necessary for the operator to take the systems effects into account. This is often done by drawing the system's responses on the display with a grease pencil or plotted grid line with an X-Y recorder. Now the storage normalizer can store the system's response in memory and then subtract it from the measured data for direct display of the test device alone. This greatly reduces the possibility of results being erroneously interpreted. The normalizer is compatible with many spectrum analyzers. The normalizer digitally displays system data, updates the analyzer display, and normalizes out the system so that only the test device is shown on the analyzer.

The storage normalizers are usually specified to operate with particular types of equipment. It is advisable that you check to see what inputs and outputs are available on your particular analyzer and then match the specification to a storage normalizer right for you.

The tracking preselector (which could also be termed a tracking filter) is a YIG tuned filter which is used with spectrum analyzers to ensure that they see only pure and uncluttered signals at their input. The preselector eliminates harmonic mixing image and multiple responses. Preselectors are available up to 18 GHz. Their lower limit is usually in the 1.5 to 1.8 GHz range simply because a YIG filter is usually not practical at frequencies below these. Most units use a fixed low-pass filter rather than a tracking filter to cover from dc to 1.5 or 1.8 GHz.

With an arrangement such as a YIG tuned filter ahead of your analyzer you can expect a very wide measurement range and will no longer need to use signal identification. If a signal is on the display it is real because everything else has been filtered by the preselector. Rejection on a preselector such as these is in the order of 70 dB.

The insertion loss of this type of preselector must be considered. Since a YIG device is used, insertion losses will run from 7 to 10 dB. This is

Spectrum Analyzer Measurements

a figure which must be taken into consideration if an absolute power reading is being made.

Another area for consideration when using a tracking preselector is the maximum input power level. You will recall from our previous discussions that some spectrum analyzers will handle up to 2 watts (+33 dBm). The tracking preselector cannot handle anything close to these numbers. The usual numbers for maximum input are around 100 mW (+20 dBm). Beyond this point the input circuitry will be damaged. So, when a clean, broadband presentation is required, the tracking preselector should be considered.

The spectrum analyzer, as we have stated earlier, is a very versatile instrument for displaying the amplitude and frequency component of a signal, but it does have limitations. If you need a frequency readout to better than 1% accuracy, the analyzer's CRT will not help you. Also, if you want to measure a crystal filter for a microwave receiver application with a response over a 100 dB range, the analyzer's mixer will only guarantee distortion components in the area of 70 dB down.

The solution to the limitations is to use a tracking generator. Because the tracking generator ties into the internal processing circuitry of the analyzer, it bypasses the primary source of distortion — the spectrum analyzer mixer. Since a tracking generator can in essence amplify a weak signal located by the spectrum analyzer's marker, it is also possible to obtain counter accuracy on otherwise impossible frequency measurements.

The analyzer-tracking generator has three attractive features:
   (1) large spurious-free measurement range
   (2) excellent frequency accuracy
   (3) wide to narrow scan widths

To understand the advantages of a tracking generator, consider a conventional signal generator operating at some fixed frequency. As the analyzer sweeps across the frequency range, it detects the fundamental and then the harmonics which may or may not be formed in the analyzer itself. These spurious responses may limit the usable measurement range more than the noise level. So our dynamic range is limited if not by sensitivity then by spurious responses.

What happens if instead we move the generator's frequency so that it always corresponds to the frequency the analyzer is tuned to? In this way, the generator would keep pace with, or track, the analyzer. As the analyzer tunes to each new frequency it sees the fundamental output of the generator — not its harmonics. But what about the spurious analyzer responses? Where are they?

Since the analyzer is tuned to only one frequency at a time, in this case the fundamental, it never sees anything else. This happens during the whole sweep so that as the tracking generator and analyzer scan together, the analyzer tunes only to the fundamental and never to any harmonics or spurious responses.

The process by which the tracking generator works in conjunction with the spectrum analyzer can be seen in Figure 5.35. You will note how the tracking generator uses the same local oscillator as the spectrum analyzer. Thus, it always knows where the analyzer is tuned.

Let's look at some typical applications where a tracking generator can prove its worth. The wide measurement range and low residual FM permits detailed characterization of filters. This application points out another feature: the frequency marker which can be positioned anywhere on the response curve to measure the cutoff or center frequencies. The marker is the result of a momentary pause of the analyzer scan which allows the internal counter to trigger and count before continuing the scan. Since the counter is measuring the applied signal frequency, the points on the response curve will be measured correctly even when there is tracking error.

Connect a coupler at the generator's output and you have a system for measuring return loss or SWR. These measurements, where the output of the generator passes through the device under test and on to the analyzer, are called *closed loop swept measurements*. Care must be taken to sweep slow enough so the device being tested has sufficient time to respond. Remember that the "UNCAL" light only monitors the internal functions of the analyzer. When performing a swept measurement, slowing the sweep until there is no change in the response insures that the device under test will be fully responding.

In another type of test, *open loop measurements*, we transfer the sensitivity and selectivity of the analyzer to a counter for accurate frequency measurement. Conventional frequency counter measurements are limited to about −30 dBm sensitivity and will only count the strongest signal in a group of others. Since we know that the tracking generator is coupled to the local oscillators of the analyzer, its output always follows the input of the analyzer. Once again, the marker is placed on the desired signal and the frequency corresponding to that point on the trace is indicated on the digital display.

For tracking generators without a counter or analyzers with tracking generators built in, the generator output can drive a counter, completing the last step in transferring the analyzer's abilities to a counter.

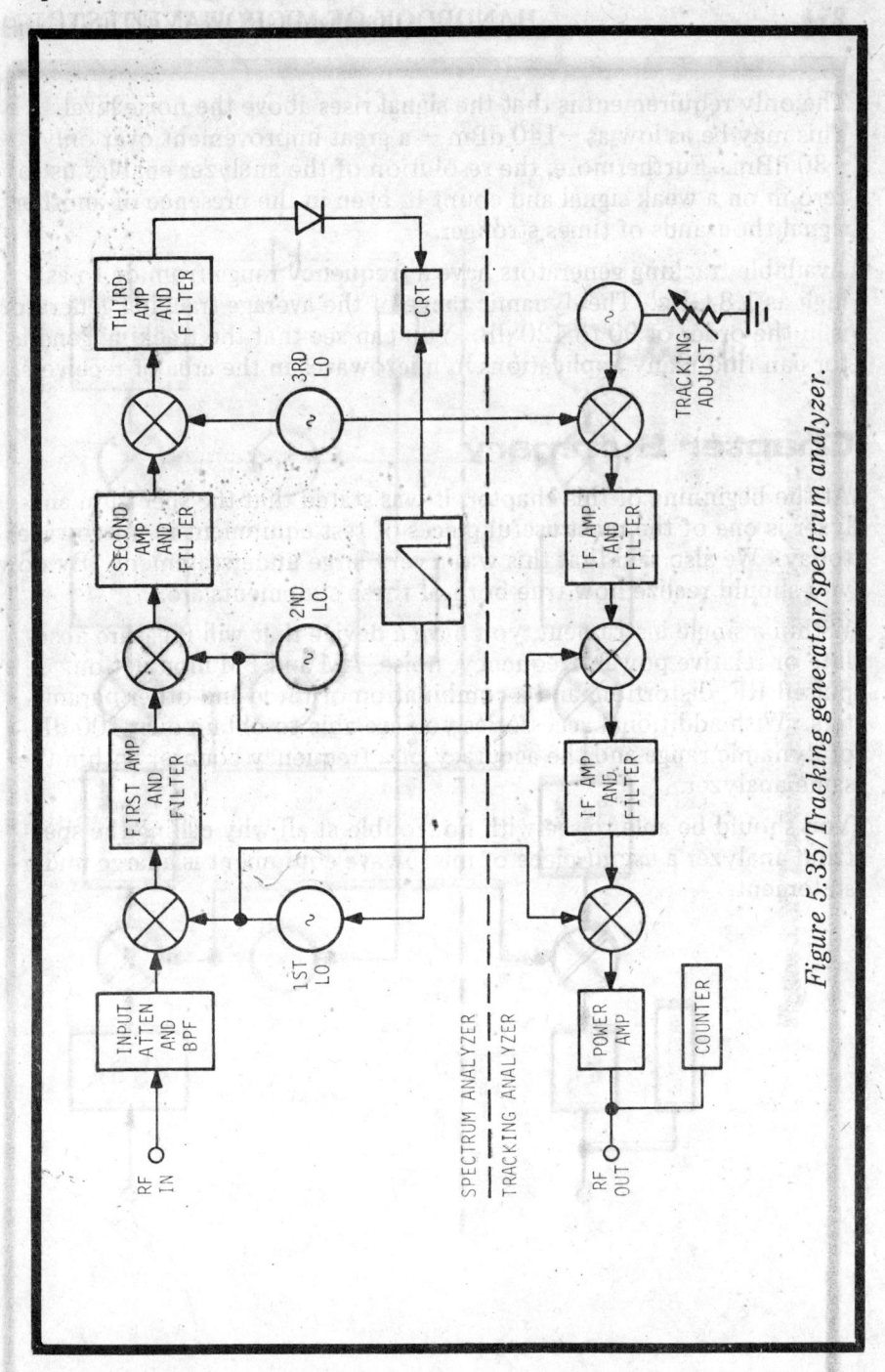

Spectrum Analyzer Measurements

Figure 5.35/Tracking generator/spectrum analyzer.

The only requirement is that the signal rises above the noise level. This may be as low as −140 dBm — a great improvement over only −30 dBm. Furthermore, the resolution of the analyzer enables us to zero in on a weak signal and count it, even in the presence of another signal thousands of times stronger.

Available tracking generators have a frequency range from dc to as high as 1.8 GHz. The dynamic range of the average tracking generator is in the order of 90 to 120 dB. You can see that the tracking generator can find many applications in microwaves in the area of receivers.

Chapter Summary

At the beginning of this chapter, it was stated that the spectrum analyzer is one of the most useful pieces of test equipment in microwaves today. We also said that this was a very large understatement. By now you should realize how true both of these statements are.

Within a single instrument, you have a device that will measure absolute or relative power, frequency, noise, AM and FM modulation, pulsed RF, distortion, and a combination of these and other parameters. With additional accessories you are able to obtain over 100 dB of dynamic range and the accuracy of a frequency counter within the same analyzer.

You should be able to see with no trouble at all why calling the spectrum analyzer a *useful* piece of microwave equipment is a large understatement.

Active

6. Testing

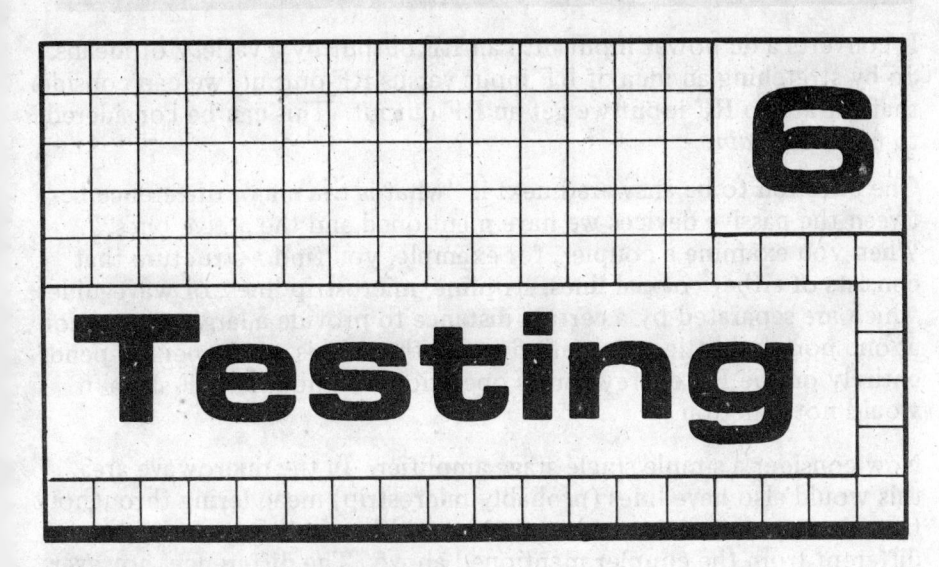

6. Definition

Measurements to this point have been directed, for the most part, toward *passive* components — that is, components that did not require ac or dc power. There have been times, however, when passive components could not be used. In these instances, specific instructions and precautions were presented regarding the application and removal of supply voltages.

This chapter is directed toward the components that are categorized as *active* devices. The question that arises at this point is "how do we distinguish between *active* and *passive* devices?" You will notice from previous measurement setups that whenever we had a passive component to be tested, it was a coupler, or an attenuator, or filter, or a similar component. There was one thing that each of these components had in common — they all have had less RF power coming out of them than they have had going in. That is, they all exhibited an RF *loss*.

Now consider the components that we should group in the category of *active* devices. For our purposes, we will consider two — the amplifier and the oscillator. In contrast to all of the passive components we have listed above, each of these has more RF power coming out than is going in. That is, they exhibit an RF *gain*. This is fairly easy to see when you consider the case of an amplifier. An RF signal at one level is applied at the input and a higher level appears at the output. This concept of gain can be grasped rather easily. It is not, however, as easy to see a gain with an RF oscillator. An oscillator actually has no RF input.

It converts a dc power input into an RF output by a variety of means. So by stretching an idea of RF input versus RF output, we can consider that for a zero RF input we get an RF output. This can be considered to be an RF *gain*.

One question to be answered next is "what is the major difference between the passive devices we have mentioned and the active ones?". When you examine a coupler, for example, you find a structure that consists of either coaxial lines, stripline, microstrip lines, or waveguides which are separated by a certain distance to provide a large attenuation at one port and little attenuation to another. This component depends entirely on the RF energy for its operation. Without an RF signal it would not function.

Now consider a simple single stage amplifier. In the microwave area, this would also have lines (probably microstrip) meandering throughout the circuit and at different lengths and widths. In this sense it is no different from the coupler mentioned above. The difference, however, is noticed in between the sets of lines that are printed on the board. This difference is a microwave transistor that acts like a switch between the input and the output. With dc voltage properly applied to the transistor and an RF input, the output level will be higher than the input (the amount depends on the particular transistor that is used and how well its input and output are matched by the microstrip or other form of circuitry). In other words, it will exhibit a *gain*. With an RF input applied and no dc voltage, the transistor is turned off and the output level is lower than the input — the circuit exhibits a *loss*. This circuit thus depends on both RF and dc for its operation, whereas the coupler relied on only RF power for its operation.

We should be at a point where we are able to state a definition for *active* devices. From our discussions above, we can say that *active* devices are:

> *Those components which rely on an external dc energy in addition to RF energy for their proper operation. This external energy causes an overall RF gain in the component.*

This definition will form the basis for the following measurements on active devices. Such areas as gain, 1 dB compression, intermodulation, and intercept point are all related to dc power, RF power, and the gain of the component — in other words, they are all associated with *active devices*.

Active Testing

6.1 Gain

This parameter is a very appropriate one to begin our discussion on active device testing because it was directly mentioned in our definition. One thing that must be mentioned here is that gain is a relative measurement. It is a comparison of the output level to the input level. The readings that are obtained are in dB (not dBm) and are the result of the following relationship presented in earlier chapters:

$$dB = 10 \log \frac{P_1}{P_2}$$

where:

P_1 = output power of the device

P_2 = input power of the device

Note: If we were measuring a loss through a device, the ratio of P_1/P_2 would be a number less than 1. For gain measurements, however, this number should be greater than 1.

The most basic test setup for measuring the gain of a device is shown in Figure 6.1. It consists of a CW generator, matching attenuators

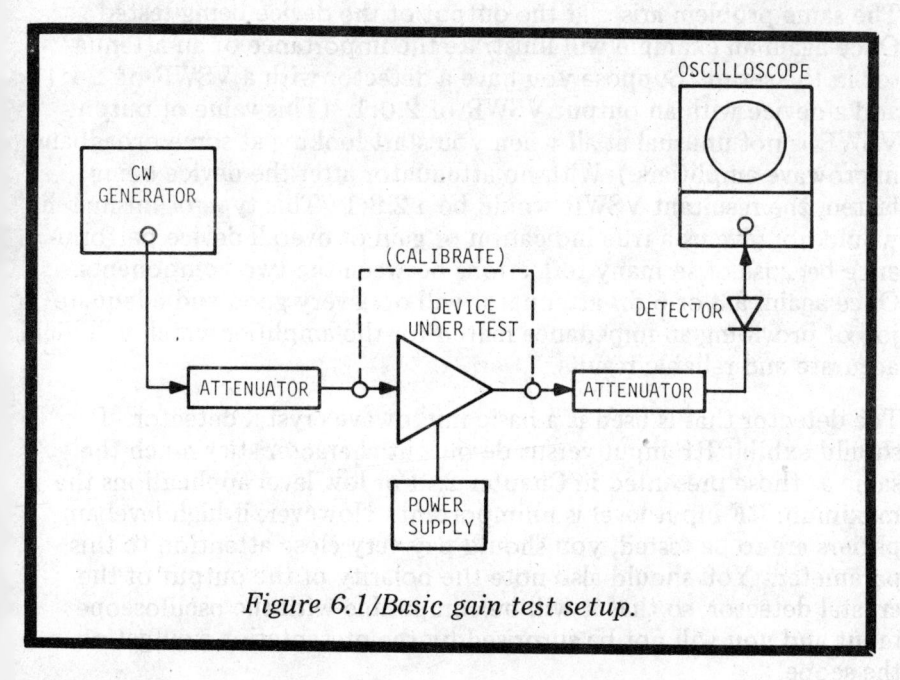

Figure 6.1/Basic gain test setup.

(optional usage), a detector, and an oscilloscope. The only requirements that are placed on the generator are that it will operate at the required frequency and provide the necessary power output.

The attenuators in the setup are designated as optional components. They are listed as such because they are not always necessary. Their function is to provide a good match between the generator and the device under test, and between the device under test and the diode detectors.

You will recall lengthy discussions and numerous examples of just such matching problems that were presented in previous chapters. To refresh your memory, consider the example where the device to be tested has a VSWR of 1.5:1 (a fairly good input match) and the generator has an output VSWR of 1.3:1 (also a fairly good match for a microwave generator). If the attenuator is not inserted, the generator will see an equivalent VSWR of 1.95:1, which is not a true picture of what is taking place within the circuit being tested. An attenuator as small as 3 or 6 dB will aid the situation greatly and give the generator a good impedance match to look into.

The same problem arises at the output of the device being tested. Once again an example will illustrate the importance of an attenuator in the setup. Suppose you have a detector with a VSWR of 1.4:1 and a device with an output VSWR of 2.0:1. (This value of output VSWR is not unusual at all when you start looking at some broadband microwave amplifiers.) With no attenuator after the device being tested, the resultant VSWR would be a 2.8:1. This type of mismatch would not give us a true indication of gain or overall device performance because of so many reflections between the two components. Once again, a 3 or 6 dB attenuator will do a very good and adequate job of providing an impedance match for the amplifier which will yield accurate and reliable results.

The detector that is used is a basic microwave crystal detector. It should exhibit RF input versus dc output characteristics much the same as those presented in Chapter 2. For low level applications the maximum RF input level is unimportant. However, if high level amplifiers are to be tested, you should pay very close attention to this parameter. You should also note the polarity of the output of the crystal detector, so that it will be compatible with the oscilloscope input and you will not be surprised by the presentation you get on the scope.

Active Testing

The oscilloscope used should be one that has an adequate dynamic range to display the output of the device under test. This could be accomplished by your choice of plug-in units for the vertical portion of your scope.

To illustrate how a basic gain measurement is made, we will go through a procedure using the setup just described.

(1) Connect the test setup as shown in Figure 6.1 in the *calibrate* position with ac power applied to the CW generator and oscilloscope. Be sure that the generator is in a *standby* or *RF off* position.

(2) Place the generator in an *RF on* position following any built-in delays in the generator and/or oscilloscope.

(3) Set up a convenient level on the oscilloscope and record this as your reference level. This may be done by recording the voltage from the scope or by tracing the response on the face of the scope with a grease pencil.

(4) Place the generator in the *standby* or *RF off* position and insert the device to be tested.

(5) Apply the necessary dc voltages and check to see that the current drain is at the proper level.

(6) Place the generator in an *RF on* position once again and note the "change" in level on the scope. This is the gain of the device under test. This change is a voltage change and may be converted to dB by using the formula: dB = 20 log (voltage change).

Note: To obtain a direct reading from such a setup you can place a variable attenuator at the output of the device in place of the fixed attenuator. With a variable attenuator in the circuit you can set a level on the scope during calibration; note the level by marking the face of the scope; insert the device to be tested; and adjust the variable attenuator until the display returns to the original calibration level. The value read off the attenuator is the value of the gain for the device under test.

Either method listed above results in about the same degree of accuracy. Each method relies on your ability to read the display on the oscilloscope.

The setup discussed above can also be adapted to be used as a swept gain setup. Figure 6.2 shows such a swept gain setup. There are two basic changes between this setup and that shown in Figure 6.1. First, the CW generator has been replaced with a sweep generator. This is a logical change since we need the capability of covering wide frequency bands within a short time. With a CW generator you, of course, would not be able to do this. Second, a leveling loop (ALC loop) has been added. The ALC loop, you will recall, consists of a directional coupler to sample the RF energy from the sweep generator and a detector to convert the RF power to dc for use in controlling the leveling circuitry within the sweep generator. The addition of a leveling loop ensures that any variations present on the scope will be due almost entirely to the device under test. The procedures for running a gain test would be as follows:

(1) Connect the setup as shown in Figure 6.2 with the device under test out of the circuit and a cable (or adapters) in place of the device.

(2) Apply ac power to the sweep generator and oscilloscope. Set the sweep generator on *standby* or *RF off* and the oscilloscope in the *ext. horizontal* mode.

(3) Following any needed time delays, place the sweep generator in the *RF on* position and be sure that the system is leveled. This indication is usually on the sweeper itself and when a light is on it means the system is not leveled. Reduce the RF level until the light goes out.

(4) Adjust the oscilloscope vertical and horizontal gains and position and the generator RF level and sweep time for a satisfactory display. (A satisfactory display is one that is within the display area of the scope and is fast enough to eliminate any flickering).

(5) Trace the level you have set as a calibration level either with a grease pencil or by storing it with a storage display. This is the measurement reference.

(6) Place the sweep generator in the *RF off* position and insert the device to be tested as shown in Figure 6.2. Apply the necessary dc voltages and check to see that the current drain is proper.

(7) Turn the sweeper to the *RF on* position and note the *change* in the display level. This is a voltage change.

(8) The value of gain in dB can be calculated by using the formula: dB = 20 log (voltage change)

Active Testing

Note: Once again, you can read specific points more accurately by replacing the output fixed attenuator with a variable attenuator. A level is set during calibration; the device to be tested is inserted; and the attenuator is adjusted to match the original calibration level. The value is then read directly from the attenuator.

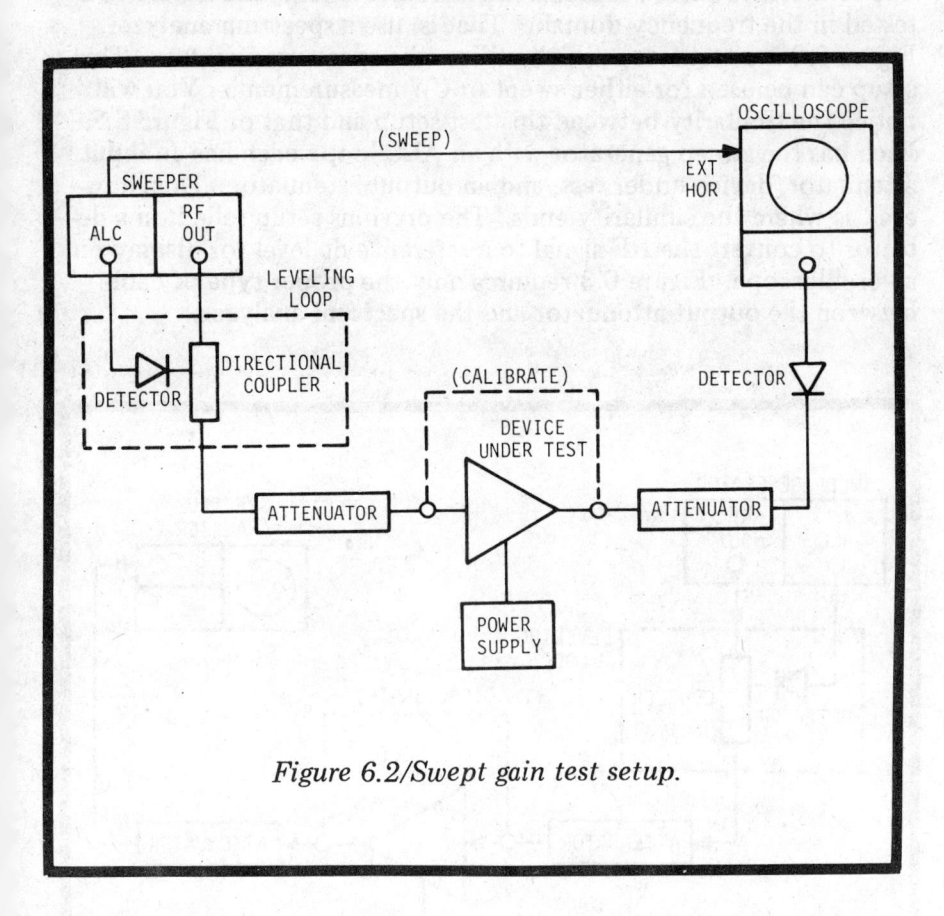

Figure 6.2/Swept gain test setup.

The two previous setups are both concerned with the use of an oscilloscope and an RF detector for display of an RF gain figure. These are useful setups but have certain limitations. One is the response of the oscilloscope both vertically and horizontally. First of all, the measurement range is limited and displays only a time-domain picture of the device. Secondly, we are only sweeping the frequency band of interest and cannot see either below or above this band.

A second limitation is the crystal detector. This device exhibits a logarithmic response and limits the level of devices to be tested as well as dynamic range.

The solution to these problems lies in characterizing the device to be tested in the frequency domain. That is, use a spectrum analyzer. Figure 6.3 is a test setup which utilizes the spectrum analyzer. This setup can be used for either swept or CW measurements. You will notice the similarity between this test setup and that of Figure 6.2. Each has the sweep generator with an ALC loop; each has an input attenuator, device under test, and an output attenuator. This, however, is where the similarity ends. The previous setup relied on a detector to convert the RF signal to a reference dc level for display on an oscilloscope. Figure 6.3 requires only the proper type of cable between the output attenuator and the spectrum analyzer.

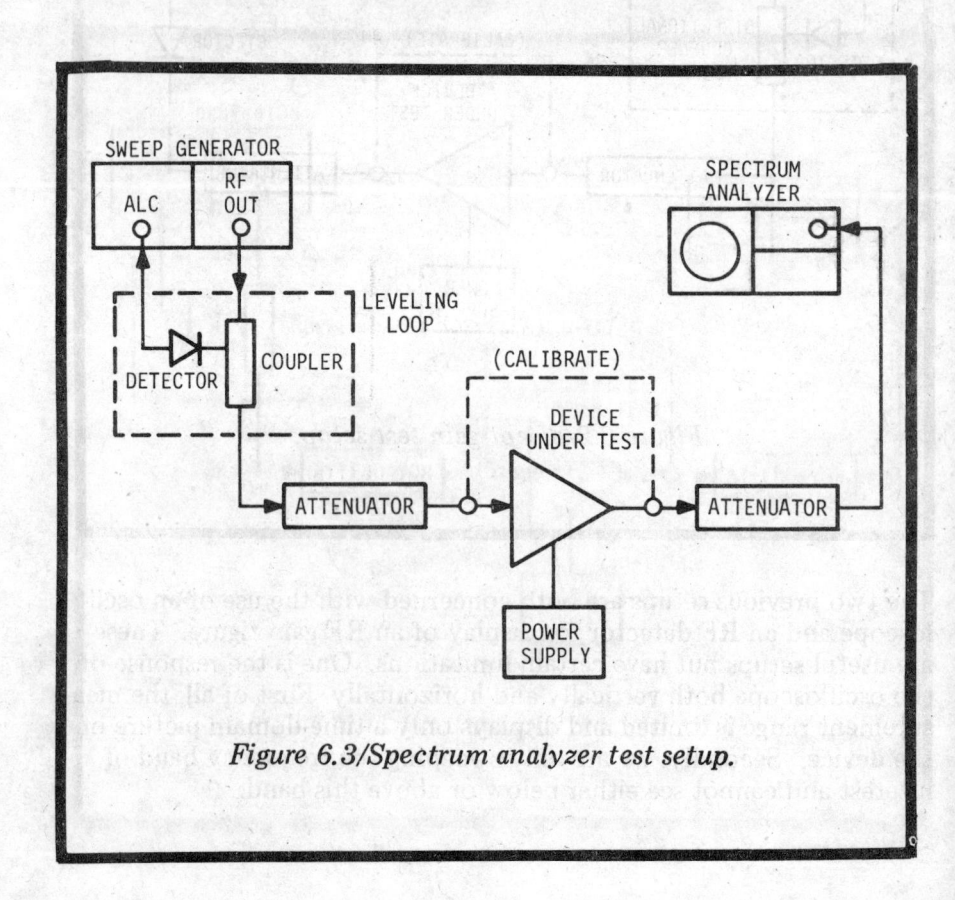

Figure 6.3/Spectrum analyzer test setup.

Active Testing

A measurement procedure would be as follows:

(1) Connect the setup as shown in Figure 6.3, with the setup in the *calibrate* mode.

(2) Apply ac power to the generator and spectrum analyzer. Leave the generator in the *standby* or *RF off* position.

(3) After any time delays in the generator and/or spectrum analyzer, place the generator in the *RF on* position.

(4) If the measurement to be made is a CW measurement, set the single frequency to a convenient level on the analyzer and note its level (it is suggested that you set it to a convenient *absolute* level such as 0 dBm, −10 dBm, etc.).

(4a) If the measurement to be made is a swept measurement, sweep the generator (either manually or at a slow automatic rate) and record the level that the generator is sweeping. (Once again, you should use an *absolute* power for a reference level.)

(5) Place the generator in the *standby* or *RF off* position and insert the device to be tested as shown in Figure 6.3.

(6) Apply the necessary dc voltages and be sure that the current the device is drawing is accurate.

(7) Place the generator in the *RF on* position.

(8) For CW measurements, note and record the new absolute reading. The difference between this reading and the reference reading is a direct reading of the gain of the device.

(8a) For swept measurements, note and record the level of the envelope of frequencies being swept. The difference between these readings and those of the reference level are direct readings of the gain of the device. If one particular frequency is needed, the CW measurement above can be used.

(9) To disconnect the setups you should first place the generator in the *RF off* or *standby* position, remove the dc voltage from the device, and then disconnect the components.

The signal generator may be replaced by a signal source which tracks the spectrum analyzer tuning frequency. A tracking generator used with a spectrum analyzer provides a quick, accurate display of frequency response for the device under test. Spurious responses are not seen due to the tracking nature of the signal source and spectrum analyzer.

In order to examine fine grain variations in gain, the 2 dB per division log display may be used. The 10 dB per division log display is ideally suited for measurements outside the normal frequency range of the amplifier. Wide range displays are especially important in feedback amplifiers for stability prediction.

IF Substitution

For highest accuracy in gain measurement, you can use the IF substitution technique. The gain of the spectrym analyzer's IF amplifier is substituted for the gain of the amplifer under test. A precision attenuator in the IF amplifier enables you to accurately measure the difference in IF gain. The signal from the signal generator is first coupled directly into the spectrum analyzer and the output display is adjusted to a convenient reference line in the log mode. The test amplifier is then inserted between the signal generator and the spectrum analyzer. The IF attenuator settings for the two conditions is the gain of the amplifier. The measurement is accurate to the accuracy of the IF attenuator.

The HP8755 Frequency Response Test Set is another means of measuring the gain of an active device. Figure 6.4 shows a setup that can be used to make gain measurements. The measurement procedure is as follows:

(1) Connect the setup as shown in the *calibrate* position. Apply ac power to the sweep generator and 8755. Leave the sweeper in the *standby* or *RF off* position. Set up the frequency band you wish to sweep.

(2) After any required equipment time delays place the sweeper in the *RF on* position. Adjust the Channel A *offset CAL* potentiometer for a 0 dB gain. This should coincide with the A channel position line.

(3) Place the sweeper in the *standby* or *RF off* position and insert the device to be tested.

(4) Apply the necessary dc voltage(s) to the device and check to see that the device is drawing the proper current.

(5) Place the sweep generator in the *RF on* position and select the A/R display. Adjust the *dB/DIV* and *offset dB* controls to provide a centered display with the desired resolution. The gain can be used directly from the face of the 8755.

Active Testing

Note: If you required an absolute power output reading, it can be obtained by simply selecting the A display and adjusting the display to the center. The absolute output power can then be used directly off the face of the scope on the 8755.

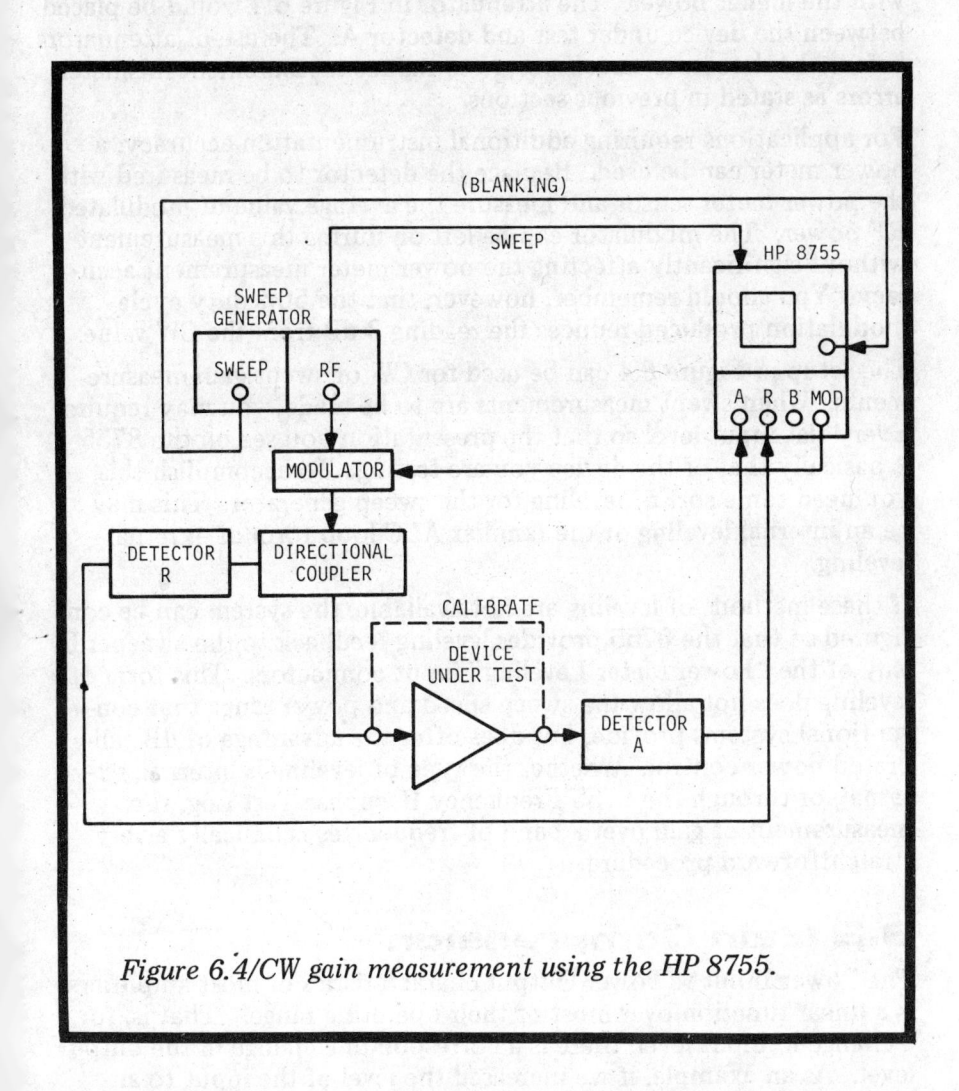

Figure 6.4/CW gain measurement using the HP 8755.

Also, the input to the device under test can be checked at any time All that needs to be done is to select the R display and you can read the input power directly in dBm.

The major source of inaccuracy in the system described is mismatch between components. The instrument accuracy for ratio measurements (such as the gain measurement above) can be optimized by insuring that the absolute powers at each detector are relatively equal. This can be accomplished by placing an attenuator before the detector with the higher power. The attenuator in Figure 6.4 would be placed between the device under test and detector A. The use of attenuators before the detectors can also serve to reduce measurement mismatch errors as stated in previous sections.

For applications requiring additional instrumentation accuracy, a power meter can be used. Replace the detector to be measured with the power meter sensor and measure the average value of modulated RF power. The modulator can be left on during this measurement without significantly affecting the power meter measurement accuracy. You should remember, however, that the 50% duty cycle modulation produced reduces the reading 3 dB from the CW value.

The setup in Figure 6.4 can be used for CW or swept gain measurements. When swept measurements are to be made, you may require a very flat input level so that the presentation you see on the 8755 is basically that of the device you are testing. To accomplish this, you need some sort of leveling for the sweep generator. This may be an internal leveling or the familiar ALC loop form of external leveling.

If these methods of leveling are not available, the system can be configured so that the 8755 provides leveling feedback to the sweeper by way of the "Power Meter Leveling" input connectors. This form of leveling does not allow the sweep speed and power range that conventional systems provide, but does offer the advantage of dB calibrated power control. Whether the type of leveling is internal, external, or through the 8755 Frequency Response Test Log, the measurement of gain over a band of frequencies is basically a very straightforward procedure.

6.2 Gain Compression

The Power Input vs Power Output characteristics of most amplifiers is a linear function over most of their operating range. That is, for a change in input level, there is a corresponding change in the output level. As an example, if we increased the level of the input to an amplifier by 3 dB, the output level would indicate an additional 3 dB if we were still in the linear region of the amplifier characteristics.

Active Testing

There is a limit, however, to how high an input an amplifier can take and still produce a corresponding change in the output. When the output change no longer follows the input, the amplifier is said to be in *compression*. That is, as you increase the input to the amplifier, the output level is being compressed and remains at a set level, since the amplifier circuit is only capable of putting out a certain level and that level has been reached. No matter how you keep increasing the input to the amplifier, the output will not increase. The only thing you will succeed in doing is decreasing the gain (since gain is P_{out}/P_{in} and P_{out} is remaining the same as P_{in} increases) and eventually you will stress the transistor to such a level that you will destroy it. So careful attention should be given to where an amplifier goes into gain compression.

Procedures for both CW and swept gain compression testing will be presented so that basically every application can be covered. The instrument that makes the variety of measurements possible is the HP8755 Frequency Response Test Set. This instrument will be used in both the CW and swept measurements.

Before some of the modern testing systems were designed, there was only one way to characterize gain compression, and that was by doing a series of point-to-point CW measurements. Only the most basic of setups was required. That setup was a signal generator with a calibrated output level and a power meter. This setup is shown in Figure 6.5. To obtain a gain compression reading, you apply a low level

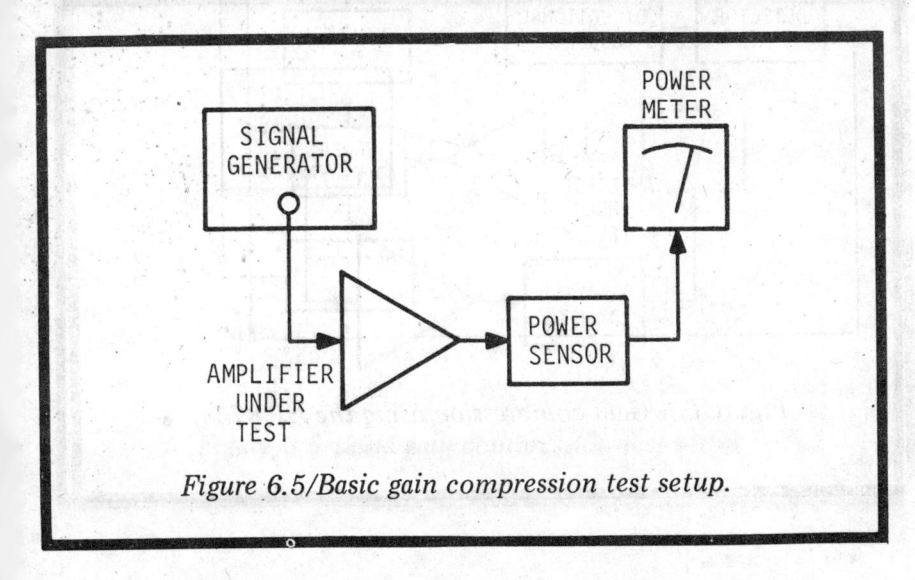

Figure 6.5/Basic gain compression test setup.

signal to the amplifier input and note the power output level on the power meter; increase the input by a certain increment (5 or 10 dB) and note the output change on the power meter; continue this incremental increase until the output reading is 1 dB less than that applied at the input. At this point, the power read on the meter is the power output at the 1 dB *compression point*. This is a very popular specification that is on many amplifier data sheets.

A more sophisticated CW measurement technique, shown in Figure 6.6, relies on a sweep generator and the 8755 Frequency Response Test Set. You will note that the setup is similar to the gain setup shown in the previous section. The setups differ only in their horizontal display positions.

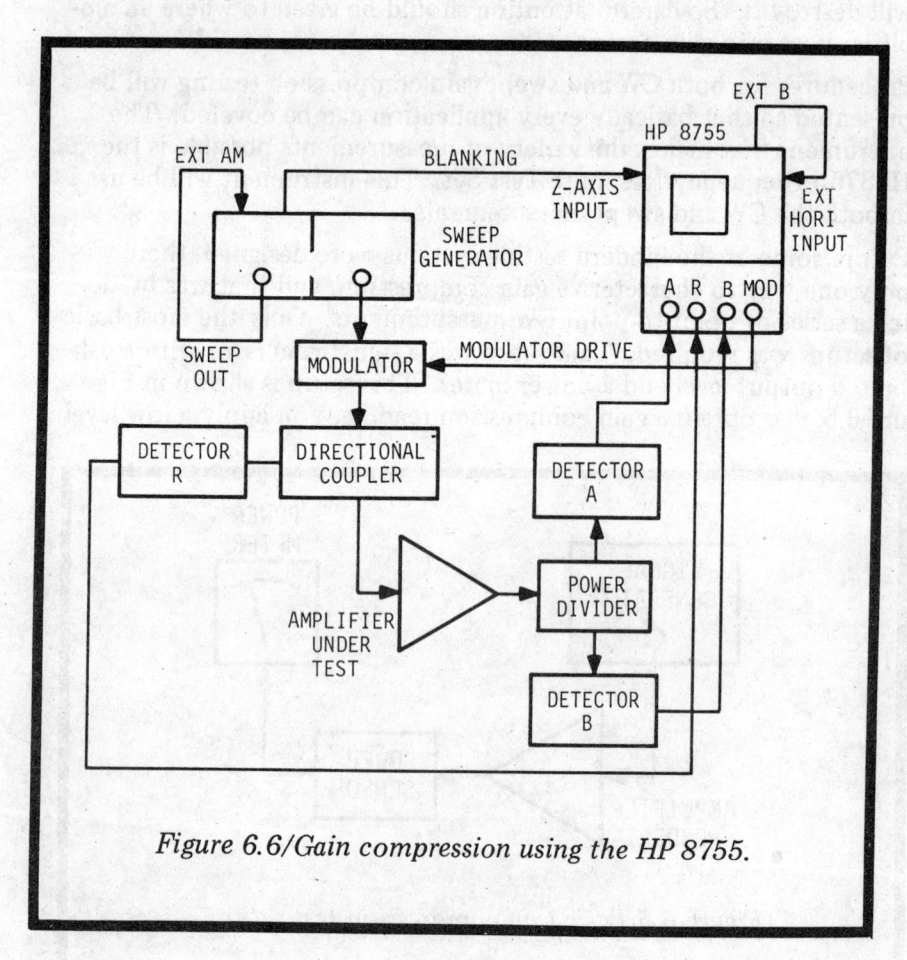

Figure 6.6/Gain compression using the HP 8755.

Active Testing 271

The measurement is made at a CW frequency with the output power varied by driving the sweeper internal modulator with a ramp voltage. One method to accomplish this is to place the sweeper in ΔF mode with the ΔF sweep width set to 0 and use the sweep output to drive the EXT. AM input. Another is to use the sweeper internal 1 kHz square wave to modulate the RF. The third method is to use a function generator to drive the EXT. AM input. To achieve widest possible dynamic range and most linear power variation, the function generator drive voltage should be offset to match the sweeper AM input characteristics. This can best be done with a function generator with dc offset capability.

The B channel of the 8755 is used only to display Power out vs. Power out for horizontal calibration. The horizontal gain control should be adjusted so this trace is a 45° display. The A channel then displays Gain vs. Power out with a horizontal display calibrated in power output dB. Figure 6.7 shows an amplifier measurement made at 4 GHz. *Note:* The diagonal trace of the B det (Power-out vs.

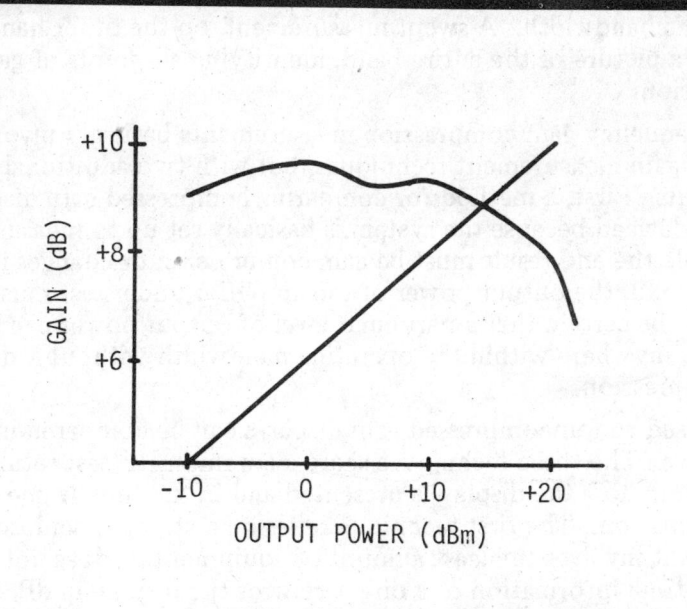

Figure 6.7/Gain vs power output.

Power-out) only serves to set up the horizontal display calibration for the Gain Power-out display. The Gain trace can be moved horizontally by adjusting the B channel *offset* dB. At low power output, i.e., power input noise becomes an increasing proportion of the R-detector signal causing widening of the Gain trace.

In Figure 6.7, you can see that the nominal gain is 9 dB. Therefore, when the gain drops to 8 dB that is the 1 dB compression point. You can also see from Figure 6.7 that this is a +20 dBm output. This test is saying that when you have a +20 dBm out of this particular amplifier, you are operating in a compressed mode. This is usually where you do not want to operate. Thus, if you need a +20 dBm output level, you had better look for another amplifier. Operation of this amplifier at +20 dBm will result in increased spurious outputs and higher noise.

Swept frequency gain compression measurements have a great advantage when the bandwidth of interest is very wide. Using a CW technique, many measurements must be made to insure that excessive gain compression does not occur at any frequency within the amplifier's bandwidth. A swept measurement, on the other hand, provides a picture of the entire band, identifying all points of gain compression.

Swept frequency gain compression measurements basically involve standard gain measurement techniques, but with two additional requirements. First, a method for comparing compressed gain displays must be deviced because the system is basically set up to measure gain, while the end result must be gain compression or changes in gain. Second, the output power of the amplifier under test must be leveled to be certain that a particular level of output power can be obtained anywhere within the operating bandwidth without 1 dB gain compression.

Compressed and uncompressed gain displays can be compared in three ways. The three techniques share a common RF test setup, but differ in how the display is presented and in the time frame of the information. The first technique relies on a storage oscilloscope. Although it involves the least amount of equipment, it does not porvide real-time information or a direct view of the display in dB. The second method modulates the RF with a square wave at one-half the sweep speed rate, so both the uncompressed and compressed displays are presented as real-time information. The final approach uses digital normalization to provide a direct view of gain compression in dB.

Active Testing

Figure 6.8 illustrates the basic configuration for the three measurement techniques. Channel 1 of the frequency response test set displays gain (the ratio of detectors A and B), while Channel 2 displays the output power of the amplifier (detector B). Note that power and gain levels will be displayed on the vertical axis and frequency is represented on the horizontal axis.

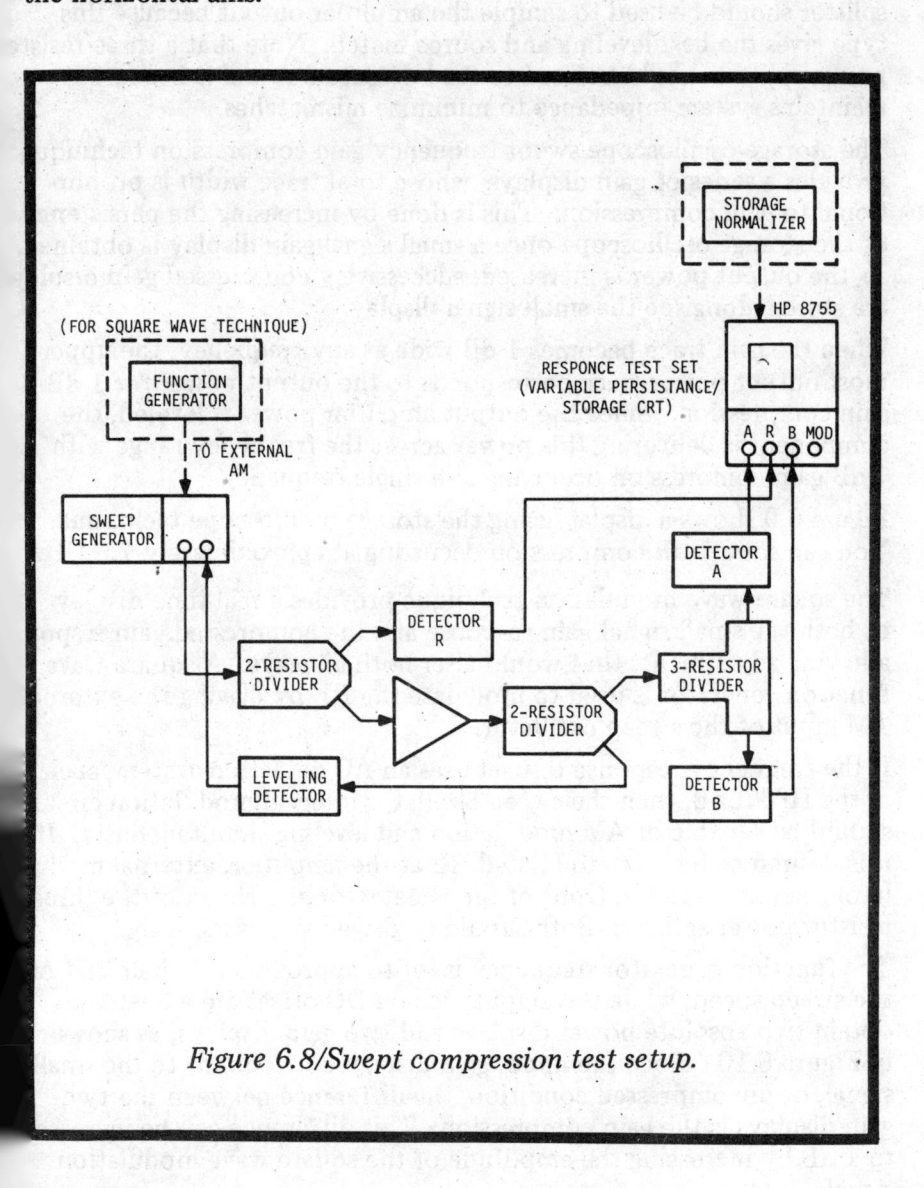

Figure 6.8/Swept compression test setup.

The external leveling capabilities of the sweeper can usually be used to meet the second measurement requirement, maintaining a constant amplifier output. The leveling loop gain should be adjusted to hold the output power constant with frequency while being varied from a small signal level to a fully compressed level. A two-resistor power splitter should be used to sample the amplifier output because this type gives the best leveling and source match. Note that a three-resistor power splitter should be used to feed detectors A and B because it maintains system impedance to minimize mismatches.

The storage oscilloscope swept frequency gain compression technique provides a series of gain displays, whose total trace width is proportional to gain compression. This is done by increasing the persistence of the storage oscilloscope once a small signal gain display is obtained. As the output power is increased, successively compressed gain displays are stored alongisde the small signal display.

When the gain trace becomes 1 dB wide at any frequency, the uppermost output power trace corresponds to the output power for 1 dB gain compression. Since the output amplifier power is leveled, the component is delivering this power across the frequency range with 1 dB gain compression occurring at a single frequency.

Figure 6.9 shows a display using the storage oscilloscope technique. You can see a 1 dB compression occurring at approximately 7.6 GHz.

The square wave modulation technique provides a real time display of both the small signal gain response and the compressed gain response allowing adjustments that would alter both displays. A square wave function generator is used to modulate the RF by driving the external AM input of the sweep oscillator.

If the frequency response test set uses an AC detection system, such as the HP8755B, then the sweep oscillator internal modulation circuit should be capable of AM modulation and leveling simultaneously. If it is desired to have unmodulated RF at the amplifier, external modulators can be placed in front of the R detector and ahead of the three-resistor power splitter. Both should be driven simultaneously.

The function generator frequency is set to approximately half that of the sweep speed, while the amplitude and DC offset are adjusted to obtain two absolute power displays and two gain displays, as shown in Figure 6.10. Since the upper gain display corresponds to the small signal, or uncompressed condition, the difference between the two gain displays is the gain compression. This difference can be increased to 1 dB by increasing the amplitude of the square wave modulation signal.

Active Testing

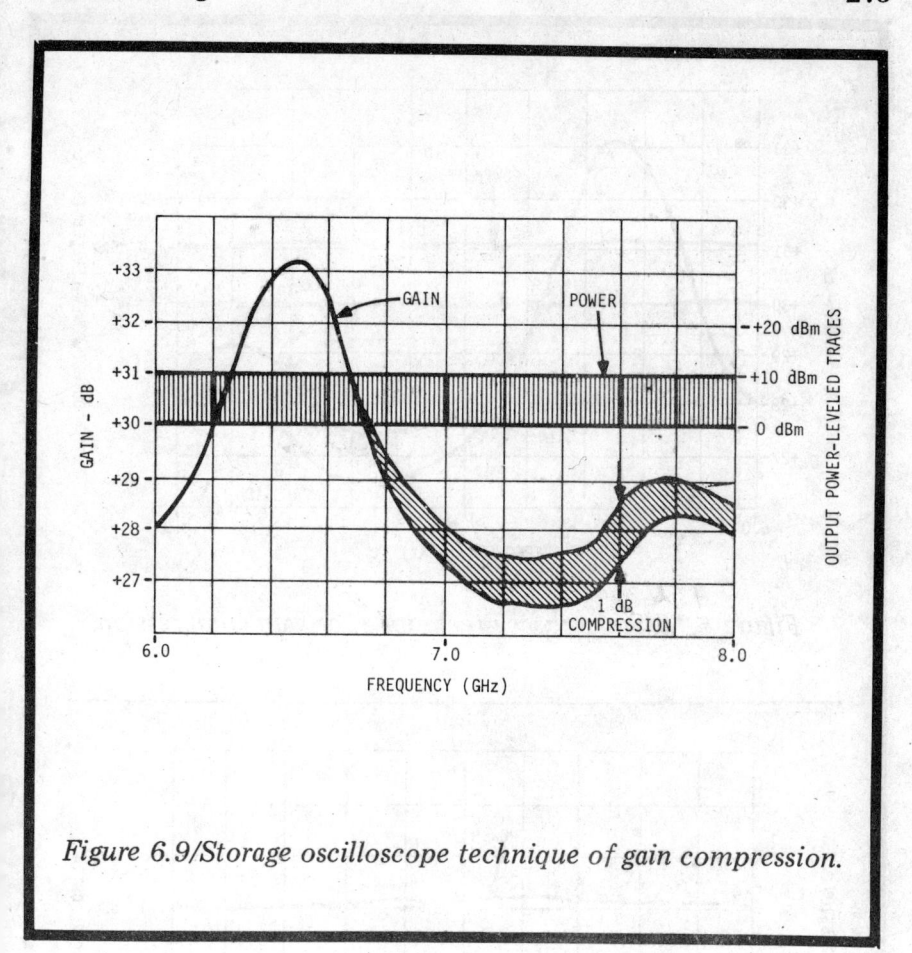

Figure 6.9/Storage oscilloscope technique of gain compression.

The final method is digital normalization. Here, a small signal, uncompressed gain display is digitized and stored. This characteristic is later subtracted from successively compressed gain displays. Changes in gain or gain compression are viewed directly in dB as the amplifier output power level is increased by changing the power level control on the sweep oscillator. Once the gain is compressed by 1 dB at any frequency on the display, the output power indicated by the B detector is the output power of the amplifier for 1 dB gain compression (see Figure 6.11). The initial storage process should be repeated if adjustments are made on the amplifier that would alter the small signal gain response. The normalizer referred to has been mentioned previously when discussing network and spectrum analyzers. The main point that was made at those times was that the need for grease pencils on the face of the scope was eliminated.

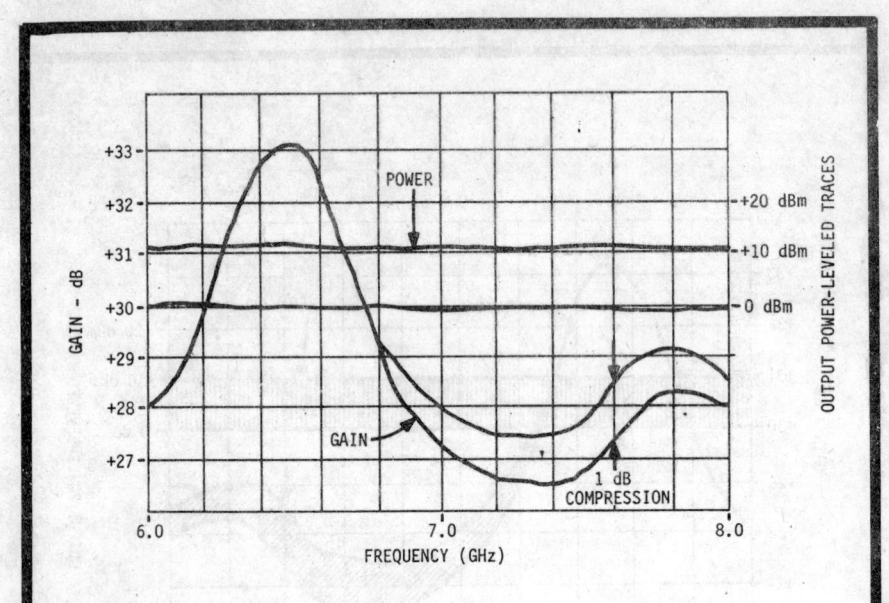

Figure 6.10/Square wave technique for gain compression.

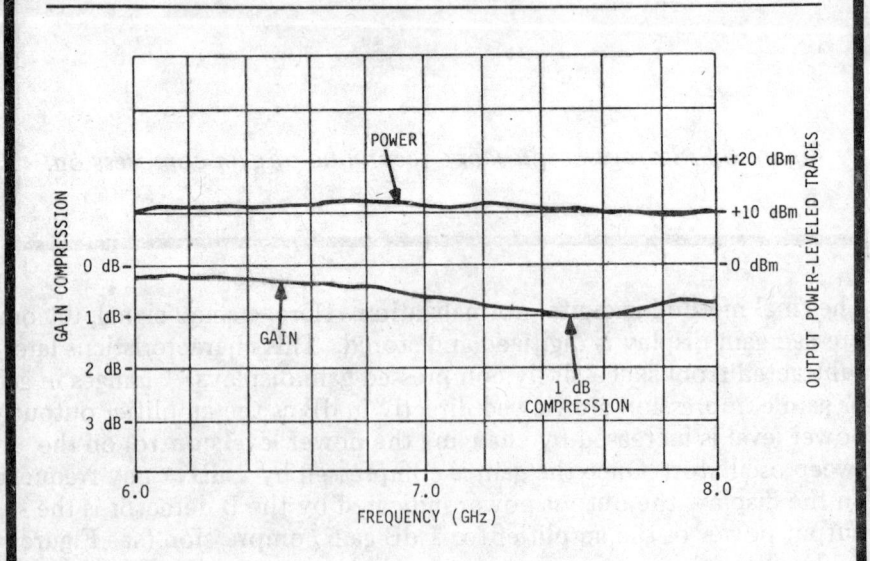

Figure 6.11/Digital normalization technique for gain compression.

Active Testing 277

Thus, we have four main methods that can be used to measure gain compression (specifically, 1 dB compression). There may be one that is best for your application and equipment availability. If so, you should use that one. If at all possible, you should investigate all of them to see which is best for you.

6.3 Intermodulation

Intermodulation is a distortion caused by two or more carrier signals and/or their harmonics creating additional (and unwanted) frequency components at the output. The behavior of an intermodulation product is characteristic of its "order".

The most common "order" found in microwave amplifiers is *third order intermodulation*. The main cause for the creation of third order intermodulation products is where two signals are present and strong second harmonic components are generated. The two signals (f_1 and f_2) mix with each other's second harmonic ($2f_1$ and $2f_2$) creating distortion products evenly spaced about the fundamentals ($2f_1 - f_2$ and $2f_2 - f_1$). This is actually called *two-tone third order IM*.

Figure 6.12 is a test setup used to check third order IM's. The two generators are adjusted a few MHz apart and adjusted so that they have the same level as they are applied to the amplifier. The response on the spectrum analyzer is then read to determine the third order IM responses. A more detailed procedure would be as follows:

(1) Connect the test setup as shown in Figure 6.12 with the device to be tested removed and a cable from the power divider to the spectrum analyzer.

(2) Apply ac power to the generators and the spectrum analyzer and be sure the generators are in the *standby* or *RF off* position.

(3) Following any necessary time delays, place the generators in the *RF on* position. Set the frequency of Generator #1 to a pre-determined frequency and the frequency of Generator #2 offset by an appropriate amount.

(4) Set the levels of the two signals just made available so that they appear of equal amplitude on the spectrum analyzer. (This sets the generator up so that the two signals go to the amplifier at the same level.)

(5) Place the generators in the *standby* or *RF off* position and insert the amplifier to be tested.

(6) Apply the necessary dc voltage(s) to the amplifier and check to see that it is drawing the right value of current.

(7) Place the generators in the *RF on* position and read the IM product levels on the analyzer as shown in Figure 6.13. This particular device has third order IM's down 56 dB with a specified input level and frequency spacing.

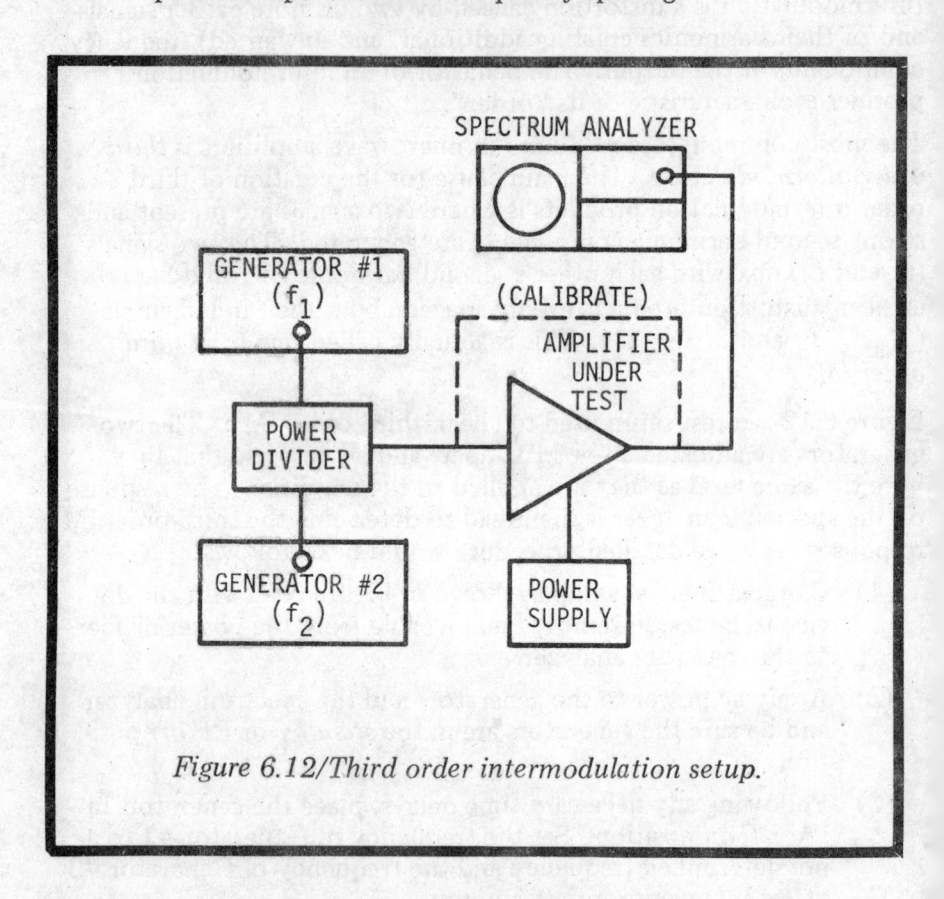

Figure 6.12/Third order intermodulation setup.

This is a very common measurement which does a great deal to characterize the operation of an amplifier. It does this by telling you what type of distortion is produced when two equal level, closely spaced signals are applied simultaneously to the input of your amplifier.

Active Testing   279

Figure 6.13/IM display.

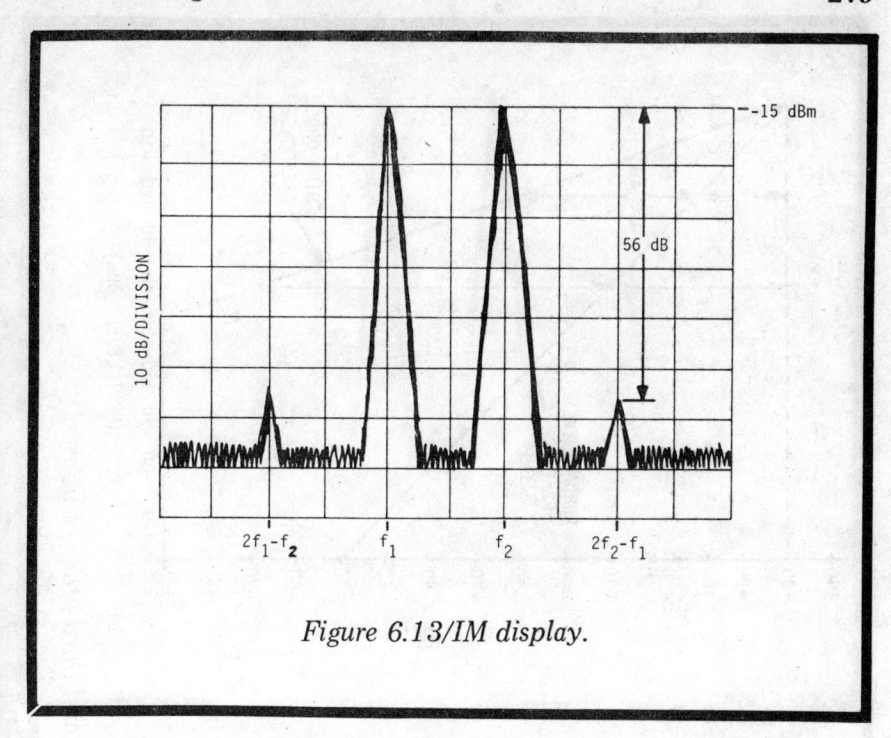

6.4 Third Order Intercept

Thus far, we have specified intermodulation distortion products by suppression, in dB, from the carriers. A problem here is that drive levels vary widely for different tests and devices, making the figures difficult to compare. An accepted method to normalize these differences is to define "intercept points." Intercept points are the theoretical points at which the fundamentals and intermodulation products have equal amplitude ("theoretical," because gain compression eventually limits the output power to less than the intercept point). Intercept calculation is only valid when extrapolated from the linear operation range of the device under test.

To determine intercept points, some information is necessary:
1. Order of distortion product
2. Device drive level in dBm
3. Distortion product suppression at that drive level

The order of the IM product is needed to determine its change in power for a change in the fundamental's power level. Intermodulation

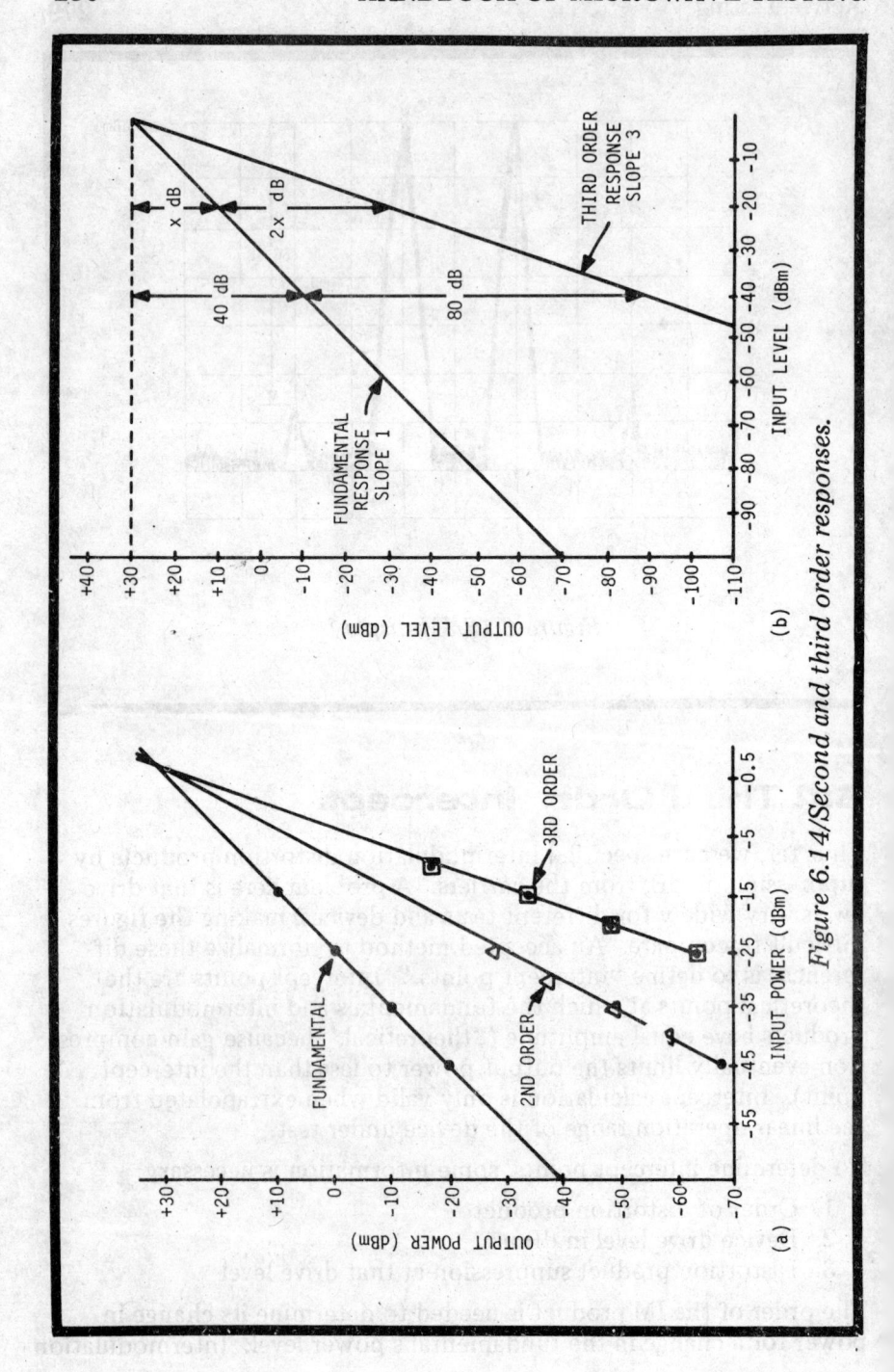

Figure 6.14/Second and third order responses.

Active Testing 281

products are found to have a slope equal to their order; that is, a third order IM product would have a 3:1 slope. Thus, a 1 dB reduction in the carrier levels results in a 3 dB drop in third order product power, a gain of 2 dB in relative suppression from the carrier. This relationship is shown in Figure 6.14. In Figure 6.14(a), you can see the relationship between the fundamental, second, and third order responses. Note the difference in slope of the curves. Figure 6.14(b) is for the third order only. A plot of relative suppression for an Nth order intermodulation product, then, would have a slope of $(N - 1):1$. The equation below allows Nth order intercept calculation from this information.

$$I_n \text{ (dBm)} = \frac{S}{N - 1} + P$$

where:

- I_n is the Nth order intercept point in dBm
- S is the relative suppression from carriers in dB
- N is the order of the intermodulation product
- P is the power level of the carrier tones, in dBm

As an example, consider Figure 6.13 with two -15 dBm tones and 56 dB suppression of third order IM products. The Third Order Intercept (TOI) would be $(56/2) - 15 = +13$ dBm.

If the intercept point is known, then the relative suppression of distortion products can be easily determined.

A monograph has been provided to allow simple correlation of intercept point, tone level and intermodulation product suppression. This is shown in Figure 6.15 and illustrates the example above.

You can see the value of the intercept point figure. The most common one that is specified on amplifier data sheets is the third order response. By adding the intercept point to our list of amplifier characteristics, gain and 1 dB compression, we now have a set of specifications that should tell you just how well a particular amplifier is operating and if it will do the job for your application.

HANDBOOK OF MICROWAVE TESTING

Figure 6.15/Intercept point monograph.

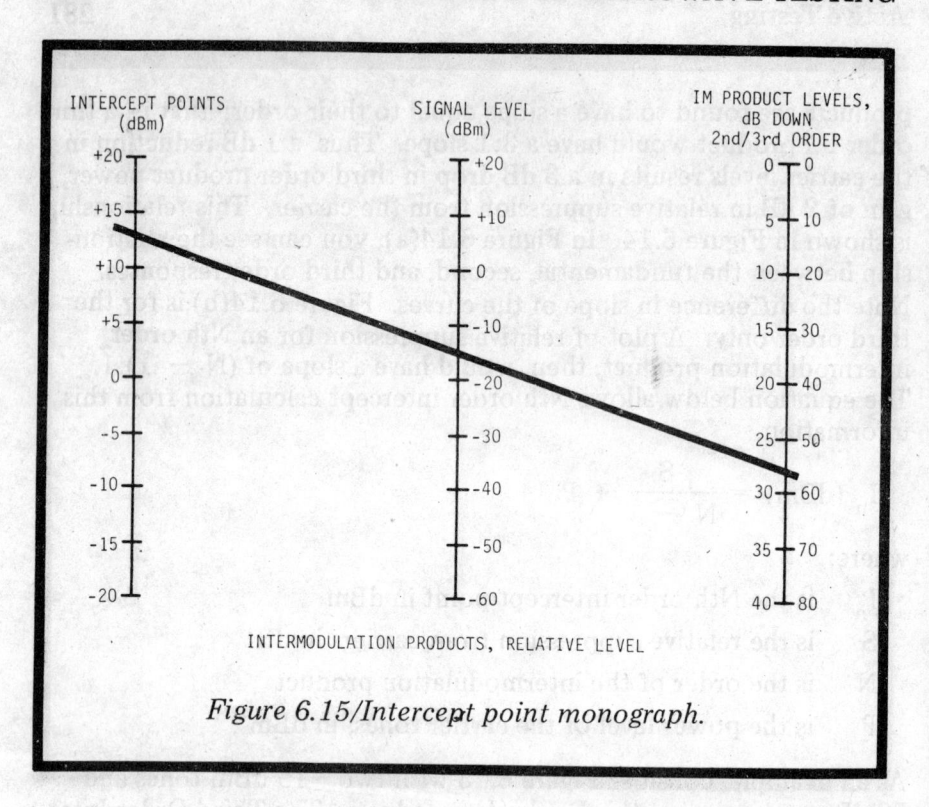

6.5 Spectral Purity

One aspect of oscillator evaluation is spectral purity; that is, how closely does the oscillator signal approach an ideal CW signal. The oscillator signal may carry frequency or amplitude noise. This is often referred to as residual FM and residual AM. In the frequency domain, this signal noise produces sidebands around the oscillator center frequency.

The sideband level indicates the magnitude of the noise. The oscillator may be evaluated by noting the level of the noise sidebands as related to the amplitude at the oscillator center frequency. The wide dynamic range of the spectrum analyzer makes accurate measurement simple. Automatic stabilization, featured on some spectrum analyzers, allows use of narrow bandwidths in order to resolve sidebands very close to the carrier. There is no complicated phase-locking procedure required to take advantage of the high resolution capability of the analyzer.

Active Testing 283

If the residual FM is at a rate lower than the analyzer bandwidth, the sidebands will not be resolved by the spectrum analyzer. However, the center frequency will appear to jitter back and forth on the CRT. With the variable persistence display section, the deviation limits may be maintained on the screen. From these limits, it is easy to determine the peak-to-peak signal deviation. The high stability of the analyzer ensures that any jitter observed is in fact due to instability of the oscillator. (Residual FM was covered in the previous chapter, thus residual AM will now be covered.)

The method of measurement for residual AM will consist of using an envelope detector to recover the AM sidebands while ignoring FM or phase modulation (PM) sidebands. The basic block diagram is as shown in Figure 6.16.

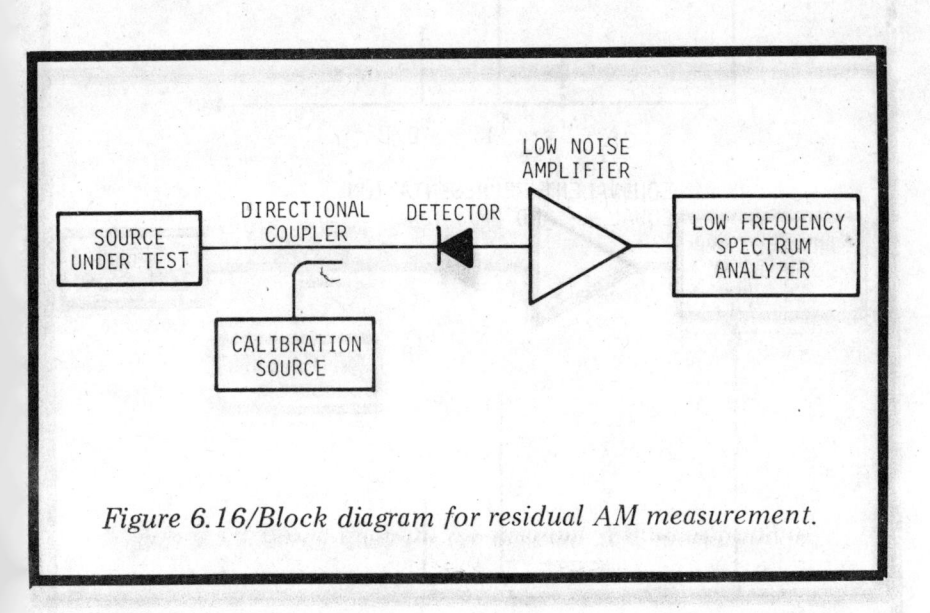

Figure 6.16/Block diagram for residual AM measurement.

The calibration source is used to create an accurately known modulation level to calibrate the system. This is accomplished by essentially using the detector as a mixer.

The source under test is for a level of around +15 to +17 dBm and the calibration source is set for a small frequency offset with a level 50 to 100 dB below the source under test. The output of the directional coupler looks like Figure 6.17 in the frequency domain.

When f_2 is more than 26 dB below f_1, the output represents an equal amount of AM and PM adding in phase. Thus, the lower sidebands cancel, and the upper sidebands add.

The actual level of the upper AM sideband is 6 dB below the calibration generator output. So if we set the calibration source 84 dB below

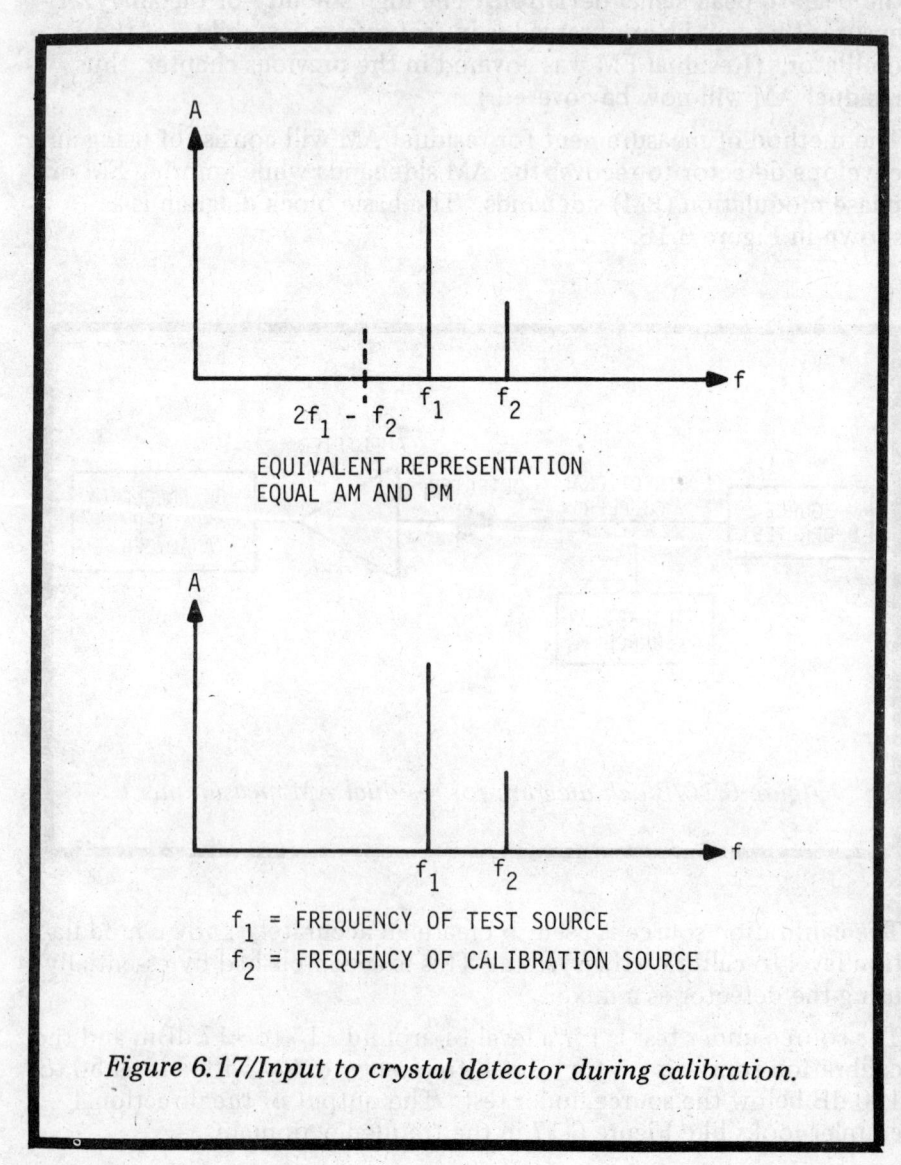

Figure 6.17/Input to crystal detector during calibration.

f_1 = FREQUENCY OF TEST SOURCE
f_2 = FREQUENCY OF CALIBRATION SOURCE

Active Testing

the test source, the equivalent output from the detector at $f_2 - f_1$ will represent AM 90 dB below the carrier. (Remember, the detector ignores any PM which is present.)

The high level of the source under test is sufficient to drive the detector into its linear region. We can check the operation of the detector circuit by changing the calibration source's frequency and amplitude. As the amplitude is changed in 10 dB steps, the output should change in 10 dB steps. Also, as the frequency is changed, the output frequency should change, but the amplitude should remain constant.

The amplifier should have a high impedance input to avoid loading the detector and the output impedance should be 50 ohms to interface to the spectrum analyzer without loss.

The measurement technique is quite simple. We first calibrate for a known carrier-to-sideband ratio. This is accomplished as outlined before. That is, the calibration source is set for a frequency slightly higher than the test source and a level 50 to 100 dB lower. The output from the amplifier is adjusted to a convenient reference on the CRT. For example, a calibration signal 74 dB below the test source is used, and the reference signal out represents AM 80 dB below the carrier.

Then, turn off the calibration source and read the noise sideband level. Remember to add the 2.5 dB correction factor and normalize to some bandwidth as for any random noise measurement. Discrete sidebands can be measured directly without corrections.

The overall sensitivity will depend on the detector noise figure, the amplifier noise figure and gain, the spectrum analyzer characteristics.

For discrete (narrowband) sidebands, the sensitivity will be related to the spectrum analyzer bandwidth. For noise sidebands, sensitivity close to the carrier will also be related to the bandwidth. Of course, as narrower bandwidths are used, narrower video filter settings are required. Further smoothing may be accomplished by using an X-Y recorder to display the output.

> *Note: Care must be taken to shield the low frequency portion of the system from radiated signals such as the AM broadcast band. Also, if you are working in an RF range where radiated signals may be present, these may radiate into the detector. Any such signal would appear as a spurious sideband. For maximum sensitivity measurements such as this, it may be desirable to operate in a shielded room.*

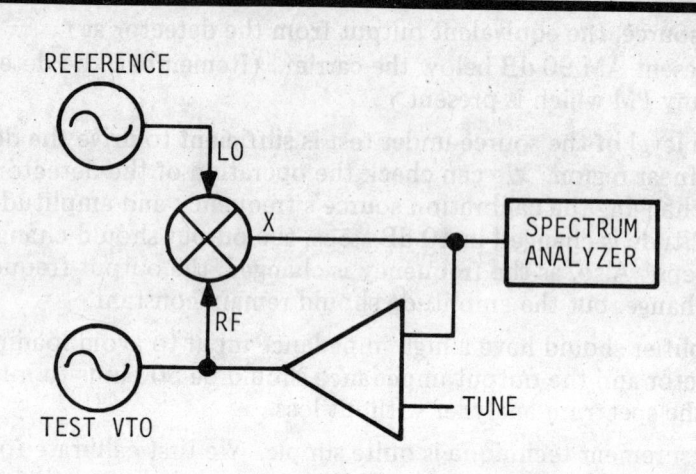

Figure 6.18/Test setup for phase modulation measurements on a VTO.

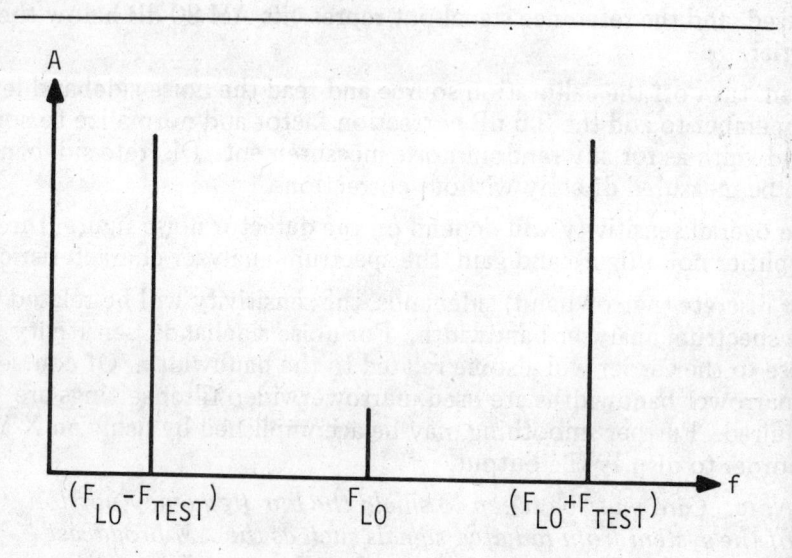

Figure 6.19/Output of the mixer when local oscillator and test oscillator are unlocked.

Active Testing

To measure a voltage-tuned oscillator, we will phase lock it to a stable reference. The block diagram is shown in Figure 6.18.

The gain crossover frequency of the lock loop must be a frequency lower than any to be measured. Calibration of the system is readily accomplished by tuning the VTO off frequency so that no lock occurs. The output of the mixer will then be as shown in Figure 6.19.

The signals at $F_{LO} + F_{Test}$ and $F_{LO} - F_{Test}$ represent sidebands 6 dB below the fundamental. The $F_{LO} - F_{Test}$ signal will appear in the range of measurement and should be set to the −6 dB graticule line on the spectrum analyzer. The log reference level will now read the equivalent carrier level.

The VTO is then tuned until lock occurs. At this point, we need to establish phase quadrature. The loop is adjusted so that the dc component of the error voltage is at a minimum to assure quadrature and true phase detection.

Now, applying the usual corrections, you can measure the phase noise sidebands.

The measurement technique for phase lock oscillators is quite similar to voltage-tuned oscillators except the lock loop is built into the oscillator unit. For this case, we will lock a spectrally pure synthesizer and the oscillator under test to the same reference to obtain a constant phase relationship. (Figure 6.20 shows the test setup.)

First, the test oscillator and the synthesizer are offset in frequency by a small amount, and the display is calibrated as for the case of the VTO; i.e., the output from the mixer represents a sideband 6 dB below the carrier.

Next, the mixer output is monitored on a dc coupled oscilloscope, and phase lock is broken by opening one loop. As the phase error crosses zero, the lock loop is closed. This may need to be done several times until the lock occurs with zero phase error. If some adjustment within the lock loop is available on the test oscillator, it may be possible to achieve phase quadrature in this manner without locking and unlocking the oscillators.

The resultant display will now be phase noise versus frequency, and the usual corrections can be applied.

For microwave oscillators, a delay line and phase shifter often provide the best technique for obtaining phase quadrature between the VTO and RF ports of the mixer. (Figure 6.21 shows a test setup.)

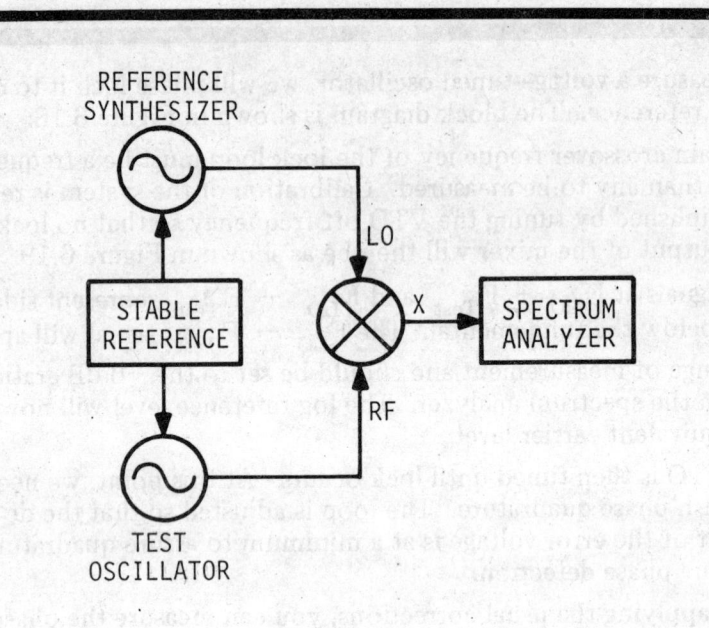

Figure 6.20/Block diagram for phase modulation tests on a phase-lock oscillator.

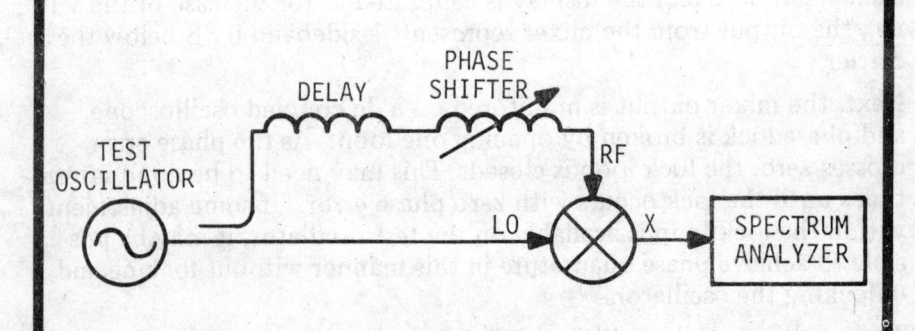

Figure 6.21/Block diagram for testing a microwave fixed oscillator.

Active Testing

The delay line assures that the random noise appearing at the two ports of the mixer does not cancel. (This would occur if identical length lines were used.) The phase shifter allows adjustment to phase quadrature by obtaining a zero dc component out of the mixer.

To calibrate, a second oscillator, slightly different in frequency and at the same level, is inserted into the mixer at the LO port in place of the oscillator under test. (The directional coupler should be left in the circuit and its through arm terminated in 50 ohms.) Here, too, the mixer output will represent a sideband 6 dB below the carrier.

To measure, reconnect as shown in the diagram, and adjust for zero dc output to obtain quadrature. The display will now be phase noise versus frequency from carrier. Again, apply the usual corrections.

To aid in the testing of oscillator spectral purity, the following suggestions are made:

- A low noise amplifier may be used to increase the spectrum analyzer sensitivity, if required. Be sure to account for its gain after calibration. (Do not attempt to calibrate with the amplifier in the circuit, since it will probably be overloaded.)
- The spectrum analyzer bandwidth determines the sensitivity for discrete sidebands, and it also affects sensitivity for close-in noise sidebands.
- In the case of the VTO and the fixed oscillator, the lock loop bandwidth must be lower in frequency than the lowest frequency of interest.
- For the case of the microwave fixed oscillator, the delay must be long enough to allow a random relationship between the two ports of the mixer. Normally, this should be about 100 cycles of the fundamental frequency.

Chapter Summary

The measurement of active devices requires procedures that are similar to those used for passive devices. The main area which is different is that dc voltages are needed for amplifier and oscillators. If you recall our definition of active devices, we said that active devices are those that relay on an external dc energy in addition to RF energy for their proper operations. Thus, careful considerations in both the RF and DC areas are the order of the day when dealing with *active* devices.

Antenna Measure

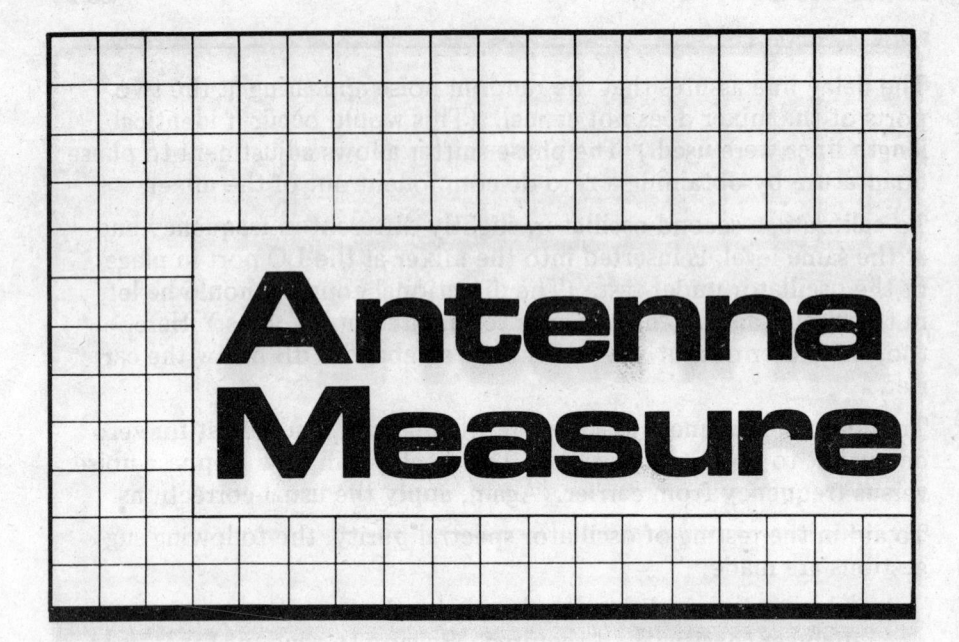

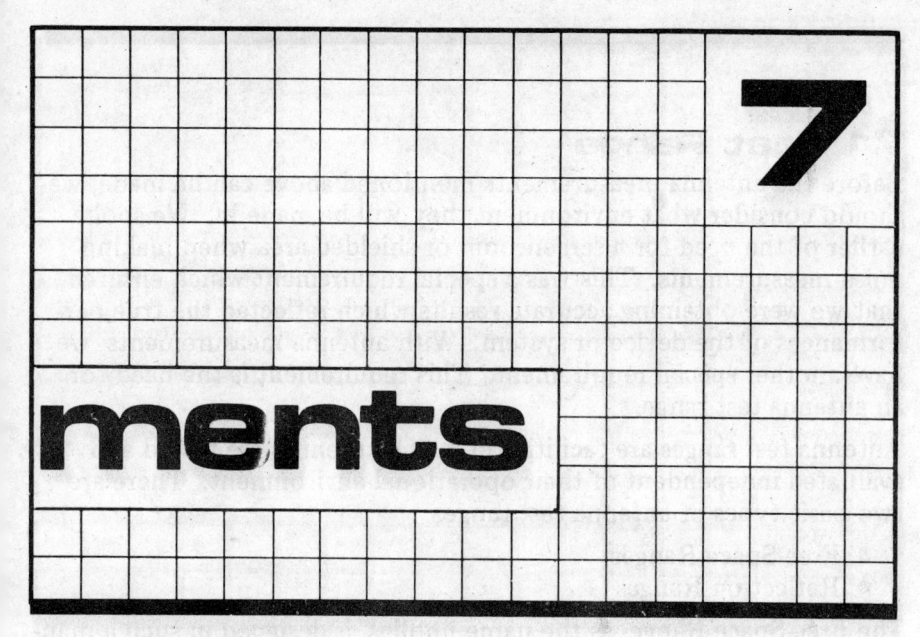

7. Definition

In any system that transmits or receives electromagnetic energy into or from a surrounding medium, an antenna is perhaps the single most important component. Its function is to couple energy from a transmission line into its environemnt (air, for instance), and vice versa. Basically, an antenna is a component which transfers energy either from transmitter output circuitry to a medium or from the medium to a receiver's circuitry. In a broad sense, an antenna is an impedance matching component — effectively, it matches the impedance of the transmitter or receiver circuitry to that of the medium used, so as to maximize the transfer of power.

There are certain measurements which need to be made to see how efficiently the antenna is doing the job we have outlined above. Measurements such as gain, directivity, and polarization will be covered to help explain the operation of microwave antennas. A Boresight measurement will also be covered to characterize the accuracy of directional antennas.

With a basic idea of antennas and some of the parameters which characterize them, we can now proceed with the actual measurement.

7.1 Test Range

Before the antenna measurements mentioned above can be made, we should consider what environment they will be made in. We spoke earlier of the need for a screenroom or shielded area when making noise measurements. This was a special requirement which ensured that we were obtaining accurate results which reflected the true performances of the device or system. With antenna measurements, we have another special requirement. This requirement is the need for an antenna test range.

Antenna test ranges are facilities in which antennas are tested and evaluated independent of their operational environment. There are two basic types of antenna test ranges:

- Free-Space Ranges
- Reflection Ranges

The Free-Space Range, as the name implies, is designed in such a manner as to suppress the effects of the surroundings. Typical Free-Space Ranges include the *elevated range, slant range, anechoic chambers*, and a recent development called the *compact range.*

Reflection Ranges are designed to make use of energy which is reradiated from the range surface(s) to create constructive interference with the direct-path signal. The two major types of reflection ranges in use are the *ground reflection range* and the *tapered anechoic chamber*

With the two basic types of ranges identified, we can now look into a brief explanation of each of the ranges lsited under the two categories.

Elevated ranges are generally designed over a flat area. The effects of the surroundings are suppressed by a careful choice of the source antenna with regard to directivity and sidelobe level, and by a redirection or absorption of the energy that would be reflected from the range surface and from any obstacles facing toward the test region. Special signal processing techniques, such as a predetermined modulation tagging of the desired return or a short-pulse technique, may also be used for more accurate antenna measurements.

Slant ranges are designed with the source antenna near the ground and the test antenna (along with its positioner) mounted on a non-conducting tower at a fixed height. An example of a slant range is shown in Figure 7.1. The source antenna is located and oriented so that its free space radiation pattern maximum points toward the center of the test

Antenna Measurements

Figure 7.1/Slant range diagram.

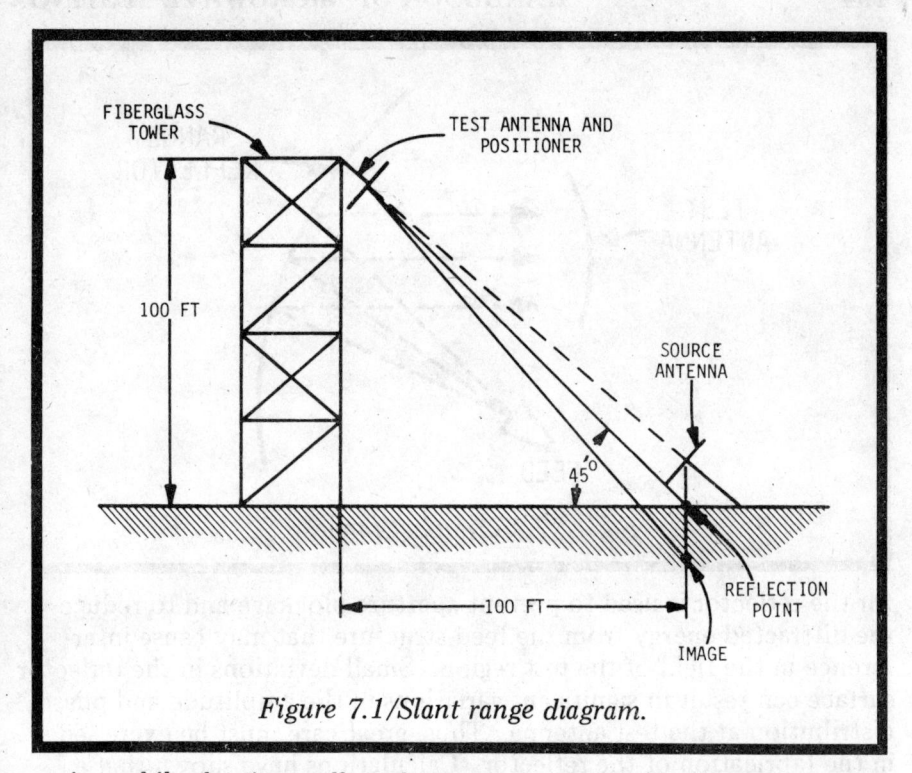

region while the first null of the pattern points toward the specular reflection point on the ground (shown in Figure 7.1). This careful pointing of the source antenna ensures that the reflected signals are suppressed. Slant ranges have been designed with adjustable towers so that the spacing between the source and test antennas may be varied. In general, the slant range requires less land for a given antenna spacing than does an elevated range. This stands to reason since the spacing on a slant range is varied by changing tower height — not by moving source and test towers physically apart.

The usefulness of the ranges discussed thus far is dependent on the fact that the spacing between source and test antennas is sufficient so that the spherical wave of the source antenna approximates a plane wave over the test region. In the compact range, the test antenna is illuminated by the collimated energy near the aperture of a larger point or line-focus antenna. In the compact range, a precision paraboloidal antenna can be used to collimate the energy as shown in Figure 7.2. The linear dimensions of the reflector must be at least three or four times that of the test antenna in order that the illumination at the test antenna sufficiently approximates a plane wave. An offset feed

Figure 7.2/Compact range.

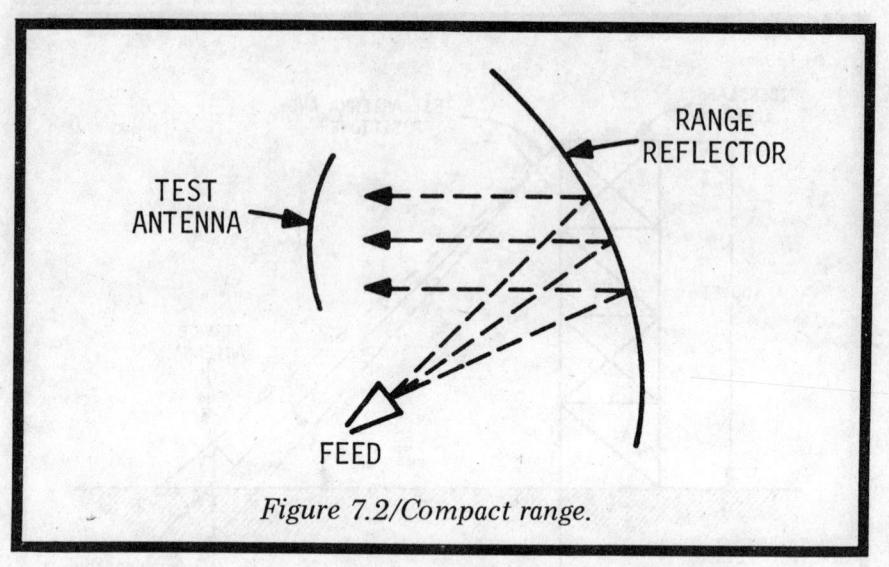

for the reflector is used to prevent aperture blockage and to reduce the diffracted energy from the feed structure that may cause interference in the field of the test region. Small deviations in the reflector surface can result in significant variations in the amplitude and phase distribution at the test antenna. Thus, great care must be exercised in the fabrication of the reflector. Calculations have shown that a half dB amplitude change will occur for a 0.007λ surface deviation.

The major sources that perturb the desired plane wave field include direct radiation from the feed of the source antenna into the test antenna, diffraction from the feed support, diffraction from the edges of the reflector, depolarization coupling between the two antenna systems, and room reflections. An offset reflector is used to prevent aperture blockage and to reduce the diffracted energy from the feed structure which may disturb the field in the test region. To further reduce the effects of the diffraction from the feed structure and also to suppress any direct radiation from the feed antenna in the direction of the test region, the reflector can be designed with a focal length long enough that the feed antenna can be mounted directly below the test antenna. This also allows the high-quality absorbing material to be placed between the test and feed antennas to absorb the unwanted radiation.

The use of a relatively long focal-length reflector also has the advantage that, for a given size reflector, the depolarization effect associated with curved reflectors is lessened. The effect of the diffraction from the edges of the reflector can be reduced by designing the reflector with

Antenna Measurements

serrations about the edges. These designs are empirical — that is, experimental; the lengths and positions of the serrations are adjusted so that the energy is diffracted in directions away from the test region for a broad band of frequencies. Usually the compact range is installed indoors and room reflections are reduced by lining the walls of the range with absorbing material.

Operation at frequencies up to 30 GHz has been reported. Improved surface tolerances and the reduction of edge diffraction on the reflector used as the source antenna are the two important parameters which must be controlled to obtain higher frequency operation.

Compact ranges have emerged from the development stage as shown by the results obtained over the last several years. Operations up to 18 GHz are now possible with commercially available ranges. Their excellent performance and compact size make them competitive with other antenna ranges for many applications.

The term *anechoic*, in simple terms, means echo-free. If you think back to the requirements set down for accurate antenna measurements, you will note that our main gaol is to reduce reflections (echoes) between the source antenna and the test antenna. Thus, an *anechoic chamber* is just what we are looking for.

The last two decades have seen a large increase in the number of anechoic chambers being installed. In part, the impetus for this development was sparked by the availability of commercially made RF absorbing material with improved electrical characteristics. This makes the advantages of anechoic chambers available to most potential users. These advantages include all-weather operation, security, and the protection of expensive systems under test such as complete-flight-qualified satellites, as well as absence from electromagnetic interference. Broadband operation with predictable and stable levels of performance as well as easy access are also advantages over outdoor ranges. The increased use of anechoic chambers, each tailored to customer requirements, has led to several methods for testing their performance. Unfortunately, there has been a significant lack of information in the open literature on the design of anechoic chambers.

The rapid advances in the development and production of radar systems at many RF frequencies led to the development of broadband absorbers in the late forties. The first significant step that paved the way for the development of present day chambers came from the production of "hairflex" in commercial quantities sufficient for use as wall liners in anechoic chambers. "Hairflex" was a material made of curled animal fibers sprayed or dipped with conducting carbon in neoprene. A reflection coefficient of −20 dB at 3 GHz to −30 dB at 25 GHz was obtained at normal incidence for pyramidal-shaped material 3.75 inches (9.5 cm) thick with pyramids 2.5 inches (6.4 cm) high and 3 inches (7.6 cm) wide at the base. It should be noted that the reflection coefficient increased for angles of incidence other than normal and it was also sensitive to the incident polarization (except for the normal incidence). This is true for all absorbers designed for lowest reflection at normal incidence.

Materials were improved steadily during the next two decades; data for present day absorbers indicate that a reflection coefficient of as low as −50 dB is achievable for a material four free-space wavelengths deep. At angles away from the normal, the number increases to about −25 dB for a 70° incident angle. Material has been developed to operate at 100 MHz with a reflection for normal incidence of −40 dB with a thickness of 12 feet (3.7 m).

Present day anechoic chambers consist of two basic types: the rectangular and tapered chamber. The rectangular anechoic chamber is usually designed to simulate free-space conditions. In order to decrease the reflected energy, high quality absorbing material is used on surfaces that reflect energy directly toward the test region. Even though the side walls, floor, and ceiling are covered with absorbing material, significant reflections can occur from these surfaces (Figure 7.3), especially for the case of large angles of incidence. The angles of incidence are usually limited to those for which the reflected energy is below a level consistent with the accuracy required for the measurements to be made in the chamber. Often, for high quality absorbers, this limit is taken to be a range of incidence angles of 0° to 70° (as measured from the normal to the wall). For the rectangular chamber, this leads to the criterion that the overall width and height of the chamber be restricted to values such that

$$W \geqslant \frac{R}{2.75}$$

in which R is the source and test antenna separation and W is the overall width or height of the chamber. The actual width and height chosen depend upon the magnitude of the errors that can be tolerated and upon the measured characteristics of the absorbing material used to line the walls. Additionally, the room height, width, and the size of the source antenna are chosen such that no part of the main lobe of the source antenna is incident upon the side walls, ceiling, and floor.

The design of rectangular anechoic chambers can be based on geometri-

Antenna Measurements

Figure 7.3/Rectangular anechoic chamber.

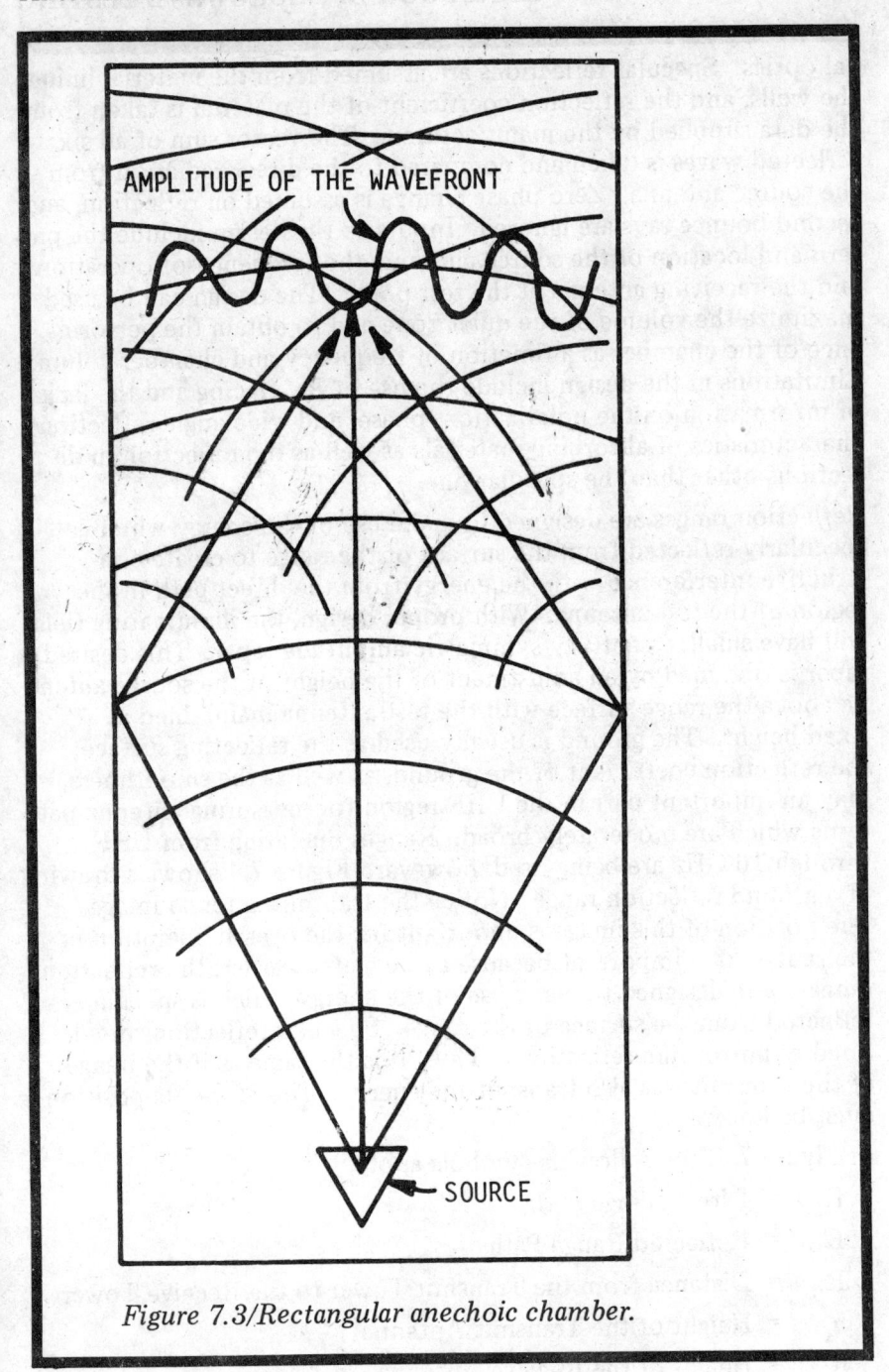

cal optics. Specular reflections are assumed from the material lining the walls, and the reflection coefficient of the material is taken from the data supplied by the manufacturers. The vector sum of all six reflected waves is taken and compared to the direct radiation from the source antenna. Zero phase change is assumed on reflection, and second bounce rays are ignored. Inputs to the design include the pattern and location of the source antenna, the frequency of operation, and the receiving antenna at the test point. The design can be used maximize the volume of the quiet zone and to obtain the performance of the chamber as a function of frequency and chamber volume. Limitations in the design include the use of ray tracing and the lack of information on the polarization, phase, and wide angle reflection characteristics of absorbing materials as well as the reflection in directions other than the specular one.

Reflection ranges are designed to make use of the energy which is specularly reflected from the surface of the range to create constructive interference with the energy from the direct path in the region of the test antenna. With proper design, the illuminating field will have small, essentially symmetric amplitude taper. This desired taper is obtained by an adjustment of the height of the source antenna above the range surface with the test antenna maintained at a fixed height. The ground is usually used as the reflecting surface; the reflection coefficient of the ground, as well as the smoothness, play an important part in the UHF region for measuring antenna patterns which are moderately broad. Ranges operating from UHF through 16 GHz are being used, however. Figure 7.4 shows a drawing of a ground reflection range. Notice the transmit antenna image. The position of this image is important for the overall operation of the range. It is important because as we stated earlier, the reflection ranges were designed to make use of the energy which is specularly reflected from the surface of the range. Specular reflections are defined as mirror-like reflections. Thus, it is the same as if the image of the transmit was also transmitting energy. Therefore, its position must be known.

In Figure 7.4, the following symbols apply:

R_D = Direct Range Path

R_R = Reflected Range Path

R_0 = Distance from the Transmit Tower to the Receive Tower

h_t = Height of the Transmit Antenna

h_r = Height of the Receive Antenna

Figure 7.4/Ground reflection range.

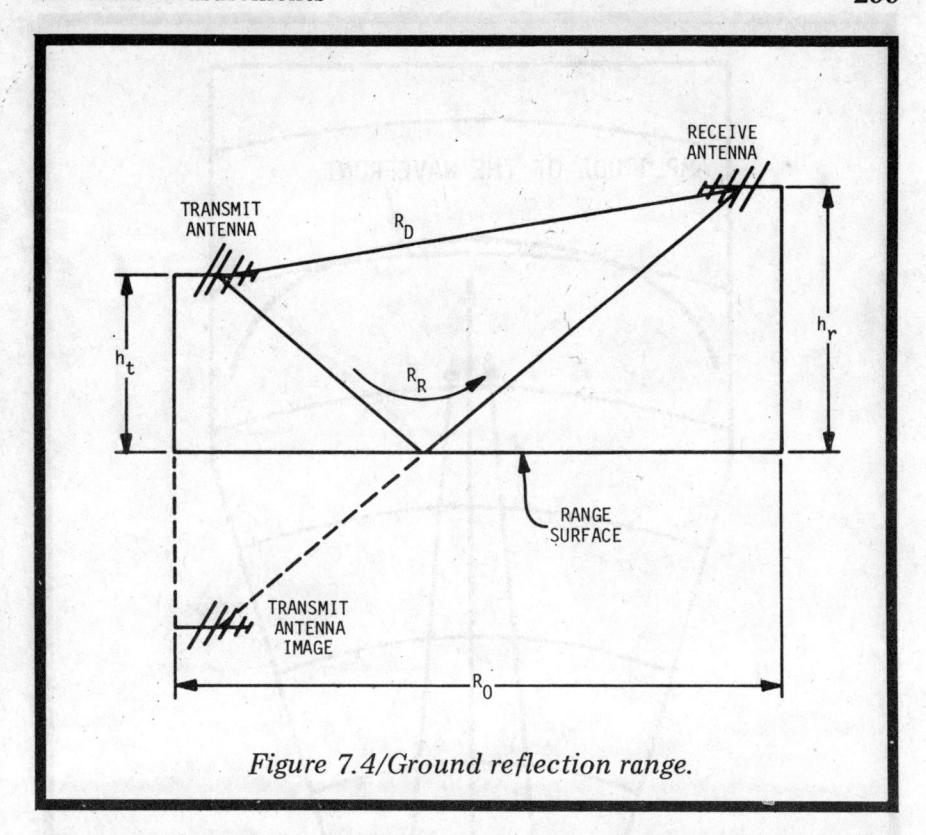

The tapered anechoic chamber is designed in the shape of a pyramidal horn that tapers from the small source end to a large rectangular test region as shown in Figure 7.5. This type of anechoic chamber has two modes of operation. At the lower end of the frequency band for which the chamber is designed, it is possible to place the source antenna close enough to the apex of the tapered section so that the reflections from the side walls, which contribute directly to the field at the test antenna, occur fairly close to the source antenna. Using ray tracing techniques one can show that, for a properly located source antenna, there is little difference in phase between the direct path and reflected rays at any point in the test region of the tapered chamber. The net effect is that these rays add vectorially in such a manner as to produce a slowly varying spatial interference pattern and hence a relatively smooth illumination amplitude in the test region of the chamber. It should be emphasized that the source antenna be positioned close enough to the apex for this condition to exist. The position is best determined experimentally, although a useful way of estimating its required position can be obtained by drawing an analogy between the tapered anechoic

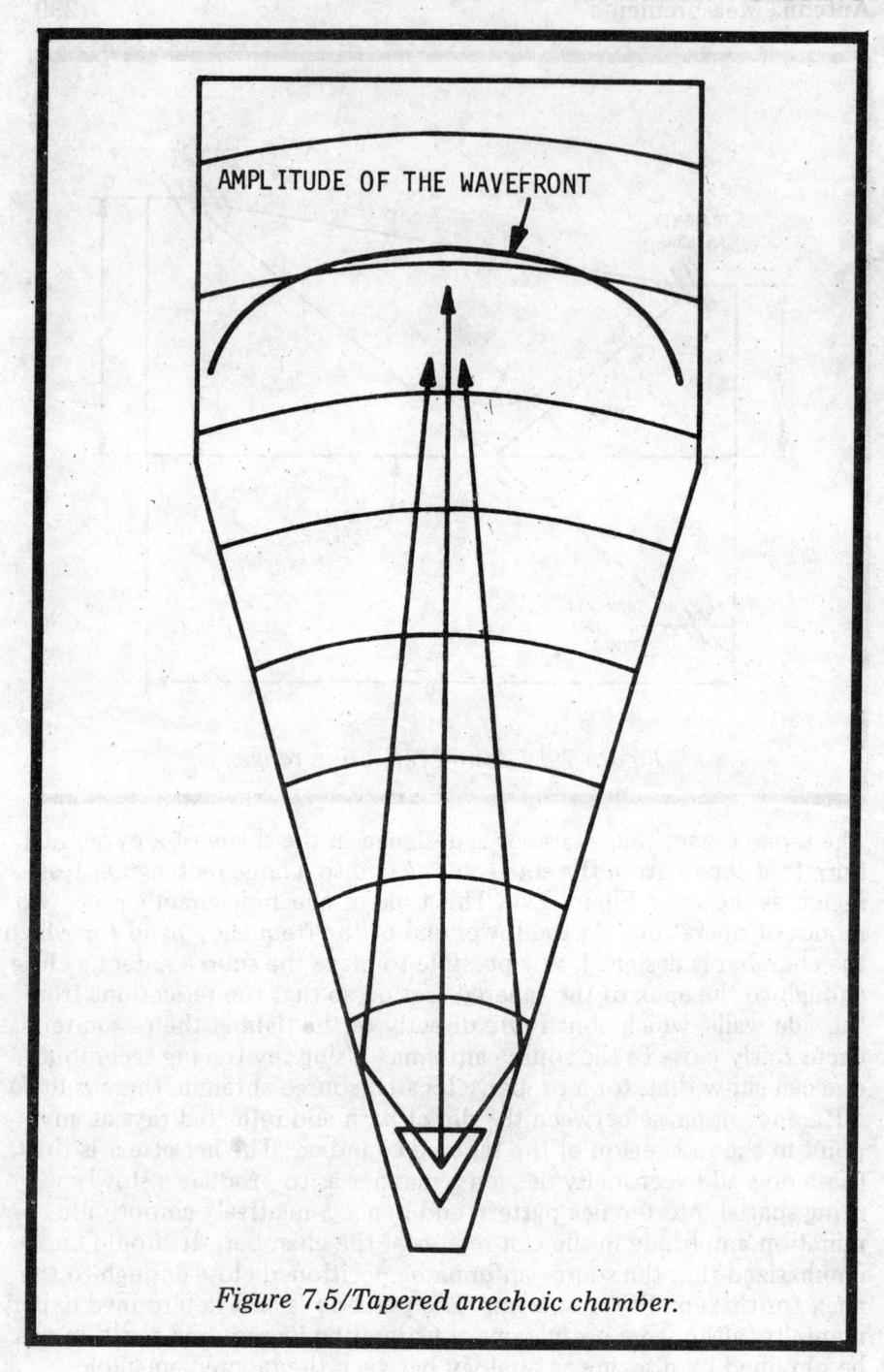

Figure 7.5/Tapered anechoic chamber.

Antenna Measurements

chamber and the ground reflection range. By use of this analogy, the perpendicular distance from the source antenna to the chamber wall, h_t, should satisfy the inequality

$$h_t < \frac{\lambda R}{4h_t}$$

in which λ is the wavelength, R is the separation between the source and test antennas and h_t is the perpendicular distance from the fixed test antenna to the side wall. If the source antenna is moved forward in the chamber, then the interference pattern becomes more pronounced with deep nulls appearing in the region of the test antenna. As the frequency of operation is increased, it becomes increasingly difficult to place the source antenna near enough to the apex. When this occurs, a higher gain source antenna is used in order to suppress reflections. It is moved away from the apex and the chamber is then used in the free-space mode similar to the rectangular chamber.

You can see from the previous discussions that there are many types of antenna ranges. We have covered six types which can be grouped under the major headings of *free-space* ranges and *reflection* ranges. The one that is right for your application depends on many factors. First of all, an antenna range is not like a signal generator. You cannot have a variety of them on hand for use in many different applications. You usually do the best you can with the range you have. Second, most facilities that do any amount of antenna work have an outdoor range and one or more anechoic chambers. This offers a variety of ways of testing antennas. Finally, if you do not have adequate facilities at your plant, you can usually arrange for those companies who do antenna work to either rent their range or have them make your measurements for you. So you can see that you should be able to make just about any antenna measurement needed whether you have antenna test range facilities within your company or not.

7.2 Gain Measurements

We talked about gain measurements in the preceding chapter on Active Device Testing. In that chapter, you will recall we made a calibration run of the system and then compared the performance of the test device to this calibration. The measurement of the gain of an antenna is accomplished in much the same way.

The term "Antenna Gain" can be defined in a number of different ways. No matter how you word the definition, however, it still is basically a comparison of one phenomenon to another. With this in mind, consider the following definition: *The gain of an antenna in a given direction is the ratio of the power radiated by the antenna in that direction to the power which would be radiated by a lossless isotropic radiator with the same power accepted through its input terminals.* The term "gain" is often taken to mean the maximum gain of the antenna, hence the gain at the peak of the main lobe.

Since an isotropic radiator, which is basically an *ideal* antenna, does not exist in practice, the obvious choice of measuring the gain of a test antenna by comparing it to a lossless isotropic radiator is immediately ruled out. The next most obvious choice is to compare the test antenna to an antenna whose gain is known. This is in fact usually done.

This gain comparison technique requires that an antenna be provided whose gain is known. This antenna is usually a dipole or standard horn antenna whose gain has been calibrated in a previous measurement or has been calibrated from the antenna geometry. Calculated gains suffice for some comparison measurements where high accuracy is not required. However, for precision measurements, a gain standard which has been previously calibrated must be utilized. The accurate calibration of this gain standard can be accomplished by the three antenna method or the two antenna method. In these methods, as in any extremely precise measurement, the utmost attention must be given to obtaining a proper test environment and to processing the errors from various sources to insure that the measured gain is indicative of the true gain of the antenna.

It should be fairly obvious from the discussion above that the calibration of the gain standards is a very important part of the gain measurement. We will, therefore, examine this area first and then proceed with the measurement of antenna gain using these gain standards as *calibrated references.*

The calibration of gain standards for use in comparison (gain) measurements is usually done by means of the two antenna method or the three antenna method. We will look at both of these methods.

Consider two antennas, A and B, separated by a distance R (Figure 7.6). The power transfer between the two, assuming them to be polarization matched, is:

$$P_r = P_o G_A G_B \left(\frac{\lambda}{4\pi R}\right)^2$$

Antenna Measurements

Figure 7.6/Two antenna test setup.

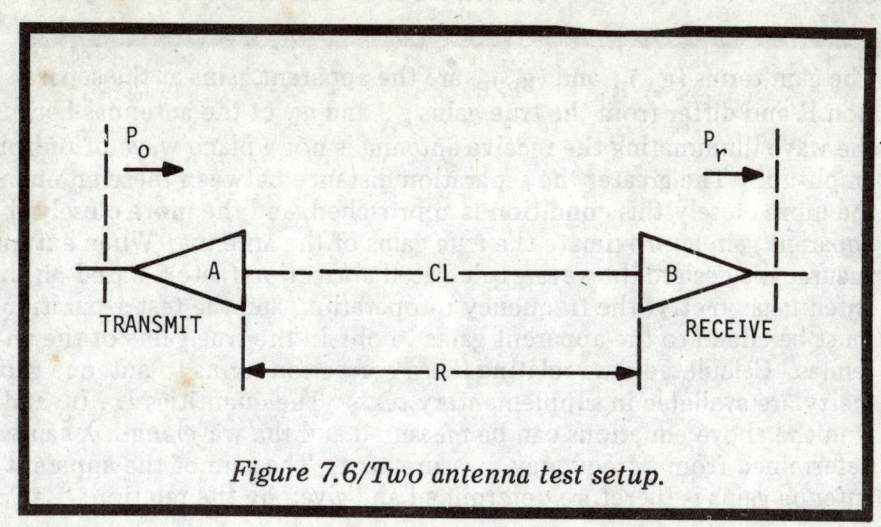

where

- P_r = Power Received
- P_o = Power Transmitted
- G_A = Gain of Antenna A
- G_B = Gain of Antenna B
- R = Separation Between Antennas
- λ = Wavelength of the Transmitted Signal

Written in a more convenient logarithmic form, the equation becomes:

$$L_r = L_o + (g_A)_R + (g_B)_R - 20 \log \frac{4\pi R}{\lambda}$$

where

- L_r = Power level at the output terminals of the receive antenna in decibels relative to a convenient reference such as 1 milliwatt, (dBm)
- L_o = Power level at the input terminals of the transmit antenna expressed in the same units as L_r
- $(g_A)_R$ = Apparent maximum decibel gain of antenna A at the test separation R
- $(g_B)_R$ = Apparent maximum decibel gain of antenna B at the test separation R
- λ = Wavelength of the transmitted wave

The gain terms $(g_A)_R$ and $(g_B)_R$ are the apparent gains at the separation R and differ from the true gains g_A and g_B of the antennas because the wave illuminating the receive antenna is not a plane wave of uniform amplitude. The greater the separation distance between these antennas, the more closely this condition is approached, and the more closely the apparent gain approximate the true gains of the antenna. When extreme accuracy is desired, however, calculated correction factors based on the antenna geometry, the frequency of operation, and the test separation must be added to the apparent gains to obtain the true gains of the antennas. Calculated data relating these correction terms to antenna geometry are available in supplementary texts. The quantities L_r, L_o and R in the above equations can be measured and the wavelength λ can be determined from a frequency measurement. The sum of the apparent antenna gains is therefore determined and given by the relation:

$$(g_A)_R + (g_B)_R = 20 \log \frac{4\pi R}{\lambda} - (L_o - L_r)$$

A condition of the two antenna method is that the antennas selected are identical. With this condition, the individual gains become:

$$(g_A)_R = (g_B)_R = 0.5 \left[20 \log \frac{4\pi R}{\lambda} - (L_o - L_r) \right]$$

Since the quantity $L_o - L_r$ is equivalent to the logarithmic relationship of P_o/P_r, the equations above may be written in a more recognizable form as follows:

$$(g_A)_R = (g_B)_R = 0.5 \left[20 \log \frac{4\pi R}{\lambda} - 10 \log \frac{P_o}{P_r} \right]$$

The three antenna method does not make use of the assumption that $(g_A)_R = (g_B)_R$. It is assumed that there can be enough difference between the relative gains of the two antennas to contribute an error greater than is compatible with the accuracy required of the measurement. The equations above, therefore, do not hold for these cases. Instead, a third antenna C replaces antenna B and the measurements are repeated. This now relates the gains of antennas A and C. A third set of measurements is then made with antennas B and C and a third equation is formed. The equations obtained are solved simultaneously, which results in the apparent gains of all three antennas. The apparent gains are then converted to the true gains by means of the correction factors mentioned earlier.

The block diagram shown in Figure 7.7 is a typical diagram of the in-

Antenna Measurements

strumentation required for antenna gain measurements using the two or three antenna methods. The instrumentation used should be highly stable with the source producing a single sinusoidal frequency output. The attenuators, couplers, and tuners used in the test setup must be able to operate in the desired frequency band and have sufficient power handling capability for the system in which they are being used. A typical test procedure for making an antenna gain measurement would be as follows:

(1) Connect the test setup as shown in Figure 7.7 and apply ac power to the source and frequency counter. Place the source in a *standby* or *RF off* position.

(2) After any needed time delays in the equipment, turn the *RF out* level to its minimum position and place the source in the *RF on* position.

(3) Place the frequency counter at the source output and slowly increase the RF level until the counter indicates the output frequency of the source. Adjust the source for the desired frequency. Disconnect the counter from the source.

(4) Place the source output at the input to the calibrated coupling network. By means of power meters at the *transmit test point* and *point # 1*, characterize (and calibrate) the coupling network. This can be done by setting the source RF level control so that the level at point # 1 is representative of the input level to antenna A under operating conditions. At this point you can now read a power that corresponds to that level at the *transmit test point*. (If a more convenient level is required than what is achieved, you can place a variable attenuator in place of a fixed attenuator and adjust the level to a spot you like). If only one level is to be applied to the antenna, you have completed the calibration of the coupling network. If additional levels are to be used you should adjust the source to each individual level at point # 1 and record the corresponding power readings at the *transmit test point*. You now have a calibration curve of *transmit test point power vs. antenna input level* for the setup being used.

(5) Place the coupling network and antenna A together again. Read the power levels at the *transmit test point* (P_o) and the *receive test point* (P_r). The relative power level P_o/P_r can now be determined.

(6) By substituting the relative power level figure, the wavelength of the signal applied, and the distance between the transmit and receive antennas into the formula:

$$\text{Gain} = 0.5 \left[20 \log \frac{4\pi R}{\lambda} - 10 \log \frac{P_o}{P_r} \right]$$

you can now obtain the antenna gain you desire. (This, of course, is the apparent gain. Corrections must be used to obtain the true gain.)

The procedure discussed above has been called a typical test procedure. It appears to be a fairly straightforward procedure. There are, however, possible problems associated with both the two antenna and the three antenna methods, which would both use the setup discussed above.

The first consideration must be a test environment compatible with the measurements to be made. The test range must be of sufficient length to adequately suppress interaction between antennas and to permit the apparent gain to be corrected to the true gain with a negligible error associated with the additive correction factors. Generally, these corrections must be extrapolated between two curves or tables, and corresponding extrapolation errors occur. The greater the test separation, the smaller will be the corresponding correction factors and extrapolation errors between them.

With sufficient length established, sufficient test heights must be obtained to reduce reflections from surrounding objects to an acceptable level. Regardless of the other precautions taken and the precision of the measurements, the results may be worthless if the reflected energy levels are so high as to contribute errors in excess of those allowed in the error budget. Care should also be taken to place the antennas at the same level so as to reduce problems associated with measuring their separation.

The difference in transmitted power level and received power level, $P_o - P_r$, must be measured. The most obvious approach is to measure the input power and received power with a power meter, subtract the two, repeat the measurements many times to get a statistical average of source drift, meter drift, thermal effects, etc., and assume the result to be the actual value of $P_o - P_r$. One is then limited by the accuracy of the power meter.

A more satisfactory procedure is to insert a calibrated directional coupler between the generator and the transmit antenna whose coupling coefficient is such that the output at the auxiliary arm and P_r

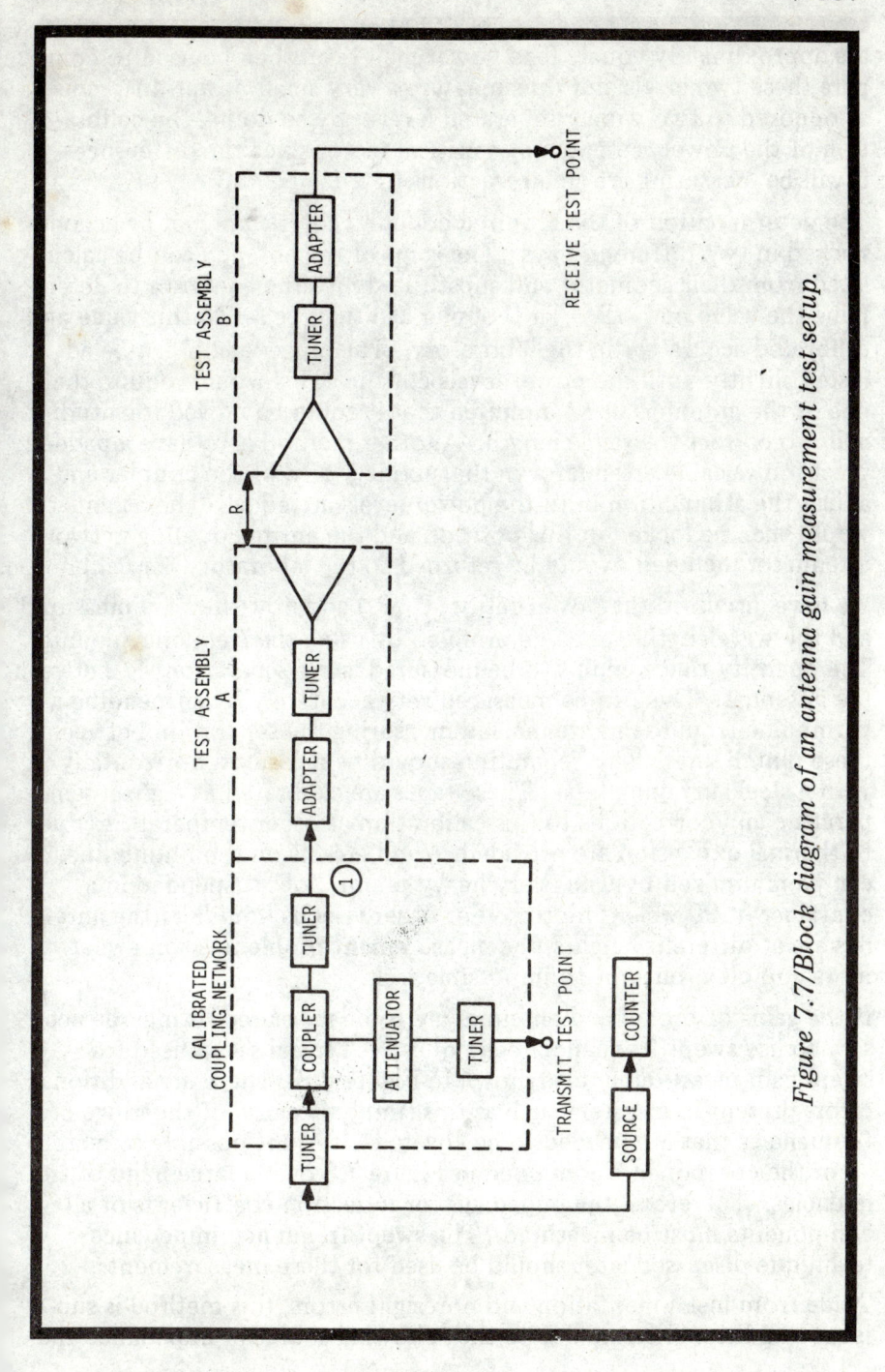

Antenna Measurements

Figure 7.7/Block diagram of an antenna gain measurement test setup.

are approximately equal. The power meter can then be used to compare these two levels and thus measure a very small power difference as opposed to two widely separated levels. By so doing, the calibration of the power meter is not a critical factor since the differences it will be measuring are small fractions of a decibel.

The determination of the required coupling coefficient can be accomplished in two different ways. The gains of the antenna can be calculated from their geometry and substituted into the equation to determine the value of P_o/P_r. The coupler can then be set to this value and calibrated accurately in the laboratory. The range length can be adjusted slightly until the power levels concur. This would require that one of the antennas be so mounted that it could be moved longitudinally to correct the range length. Another method is to have a padded precision variable attenuator in the auxiliary arm of the coupler and adjust the attenuation until the power levels are equal. The attenuator would then be locked in this position and the entire coupling network, attenuator included, would be returned to the laboratory for calibration.

We have discussed the power ratios (P_o/P_r) and how they are measured and the wavelength can be determined by using the frequency counter. The quantity that remains to be measured is the separation, R, between the antennas. This can be measured very accurately by suspending a plumb line from each antenna and measuring the separation between these plumb lines. This separation should be measured horizontally with a steel surveying tape. These tapes are calibrated at a given temperature and corrections to this calibration at other temperatures due to thermal expansion are provided. Wind effects on the plumb line can be minimized by using very heavy plumb bobs suspended in a container of 90 weight motor oil or other viscous liquid. If the antennas are at different heights, the measurement problem is somewhat more complex, but not an impossible task.

If the gains of broadband antennas are to be measured, it may be necessary to use swept frequency techniques. A typical setup used for swept gain measurements is shown in Figure 7.8. There are additional errors present in a system such as this simply because of the range of frequencies that are needed to be covered. It is not possible to match all of the components contained in Figure 7.8 over a large band of frequencies. Therefore, the impedance or reflection coefficients of all components must be measured. The swept frequency impedance technique discussed later should be used for these measurements.

Aside from instrumentation and boresight errors, this method is subject to additional errors due to the following sources: impedance and

Antenna Measurements

Figure 7.8/Swept frequency gain setup.

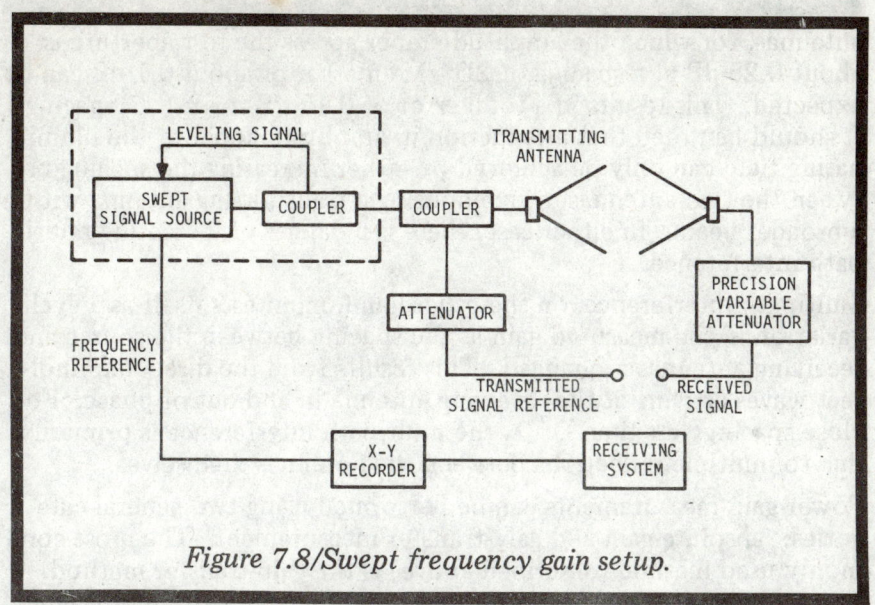

polarization mismatches, proximity effects, and multipath interference. The errors due to impedance and polarization mismatches can be minimized by actually measuring the appropriate complex reflection coefficients and polarization and then correcting the measured power gain accordingly. For example, errors due to impedance mismatches involve not just the antennas, but also the generator and receiver (load).

There is no rigorous method of removing the error due to proximity effects and multipath interference when the absolute gain measurements are performed as outlined above. These effects can only be estimated. Furthermore, a finite spacing between the transmitting and receiving antennas cannot be avoided. The error produced by finite spacing is of a systematic type which results in a measured value of gain lower than the actual far-field gain. This negative-valued systematic error decreases asymptotically with increasing separation between antennas. In order to estimate the magnitude of the correction factor, it is convenient to think of the error as being caused by two effects. One of these effects is the fact that the gain of the test antenna does indeed vary with distance because near-field components are still present even though the spacing may be greater than $2D^2/\lambda$, where D is the maximum dimension of the test antenna and λ is the wavelength. At $2D^2/\lambda$, for a typical dish antenna, this correction factor is about 0.05 dB and it will decrease with increased spacing. The second effect is that of a nonuniform illumination of the test antenna due to the relative sizes of the transmitting and receiving antennas, for which the amplitude taper across the test aperture is about 0.25 dB at a spacing of $2D^2/\lambda$, an error of about 0.1 dB can be expected. This results in a total error of 0.15 dB for $2D^2/\lambda$ spacing. It should be noted that a reduction in amplitude taper of the illuminating field can only be achieved by either increasing the spacing between the two antennas or by choosing a transmitting antenna with a broader beam. In either case, there is a danger of increasing multipath interference.

Multipath interference, on the other hand, manifests itself as a cyclic variation in the measured gain as the spacing between the source and receiving antennas is changed. This results from the direct and indirect waves arriving at the receiving antenna in and out of phase. For close spacing, less than D^2/λ, the multipath interference is primarily due to multiple reflections between the antennas themselves.

Power-gain measurements can be performed using two general categories: absolute-gain and gain-transfer measurement. The most commonly used method (described above) is the gain-transfer method. The measurement can be performed on either a free-space or ground reflection range. This method has been shown for both signal frequency and swept measurements. By following the description and procedure discussed above, you should be able to obtain a very important antenna parameter — *gain*.

7.3 Directivity

Directivity is a measure of the ability of an antenna to concentrate energy in a preferred direction. Antennas with narrow beamwidths have a greater directivity than do those characterized as being omnidirectional. The gain of an antenna is closely linked to its directivity — if the antenna has no heat loss and is perfectly matched to the source or load, then the gain is equal to the directivity. However, the gain decreases proportionally to the losses. If, for example, the antenna is 50% efficient then the gain is 50% (3 dB) less than the directivity.

As mentioned previously, the gain of an antenna is linked to the directivity. It is the product of the directivity times the antenna efficiency. The antenna efficiency is the ratio of the power radiated into free-space to the power accepted at the antenna terminals.

To obtain a meaningful measurement of the radiation characteristics of an antenna, it is essential to duplicate those features of the actual operating environment which can significantly affect the character-

Antenna Measurements

istics in question. Certain types of antennas, especially low-gain antennas mounted on air frames or space frames, are not suited for the direct measurement of radiation characteristics. These antennas and their mounts must be simulated by models of a reasonable size. Proper interpretation of radiation measurements which are made on a correctly scaled model antenna can lead to an accurate evaluation of the characteristics of a full-scale system.

Pattern integration is a convenient method for making the necessary measurements to determine the gain of a model antenna. By performing pattern integration over a sphere surrounding the test model, one may gather sufficient data to compute the directivity of the antenna system. It is then possible to use the measured directivity of the model and the calculated efficiency of the full-scale system to compute the gain of the actual system. This process eliminates the need for compensating for the mismatch of the model antenna and the difference in efficiency between the model and the full-scale system, which are necessary in gain comparison methods.

The mathematics for determination of directivity measurement by pattern integration is beyond this presentation. For our applications we will attempt to explain the procedure involved and leave any further higher calculations to those interested in empirical solutions.

First of all, a directivity figure must be determined for some reference direction (usually where the directivity is at a maximum — a spot where the efficiency is at a maximum point). This will be called D_r or D_{max}.

You can then derive from the antenna under test a signal which bears a direct relationship to the radiation intensity for a particular antenna polarization. This can be done by the use of sampling techniques which simulate a continuous measurement over an imaginary sphere produced by the antenna.

The most commonly used procedure for obtaining the required test data is to position the antenna at successive equally spaced θ angles, and to make a revolution in ϕ for each θ angle, so that the data are taken in ϕ cuts. Cuts are made at equally spaced θ angles until the imaginary sphere surrounding the antenna has been covered, or until all regions of appreciable power flow have received complete coverage. Measurements are made in this manner for a selected antenna polarization and the orthogonal polarization.

As was previously stated, the arrival at a figure for antenna directivity is one which involves a mathematical approach rather than a practical straightforward approach as has been emphasized throughout this book.

Figure 7.9/Tri-plane circular polarized system.

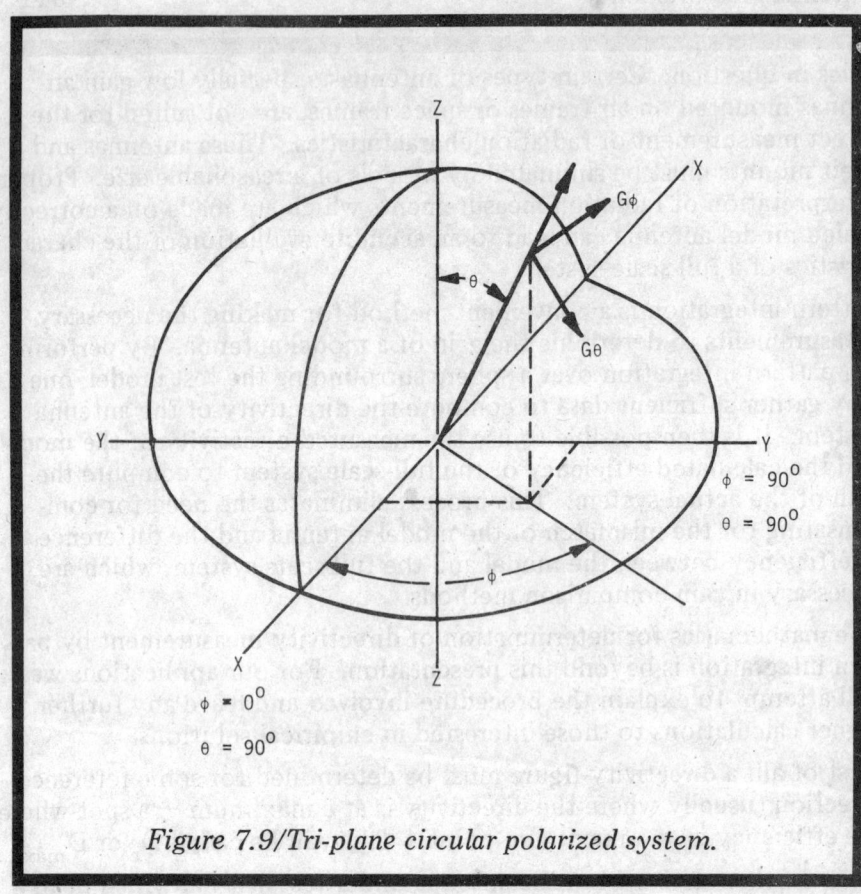

The topic was presented here to point out the importance of the term to antenna systems and to provide a definition for the term in understandable language.

7.4 Polarization

An electromagnetic wave is made up of just what the name says — two components, one electric (E) and one magnetic (M). Polarization refers to the orientation of the E component or E-vector. As an aid to understanding, consider polarization to be the property of an antenna that allows only waves of a certain specified orientation to be transmitted or received, much the same way as a filter passes only certain frequencies. Three types of polarization are commonly considered: linear, circular, and elliptical. The easiest of the trio to understand is *linear*

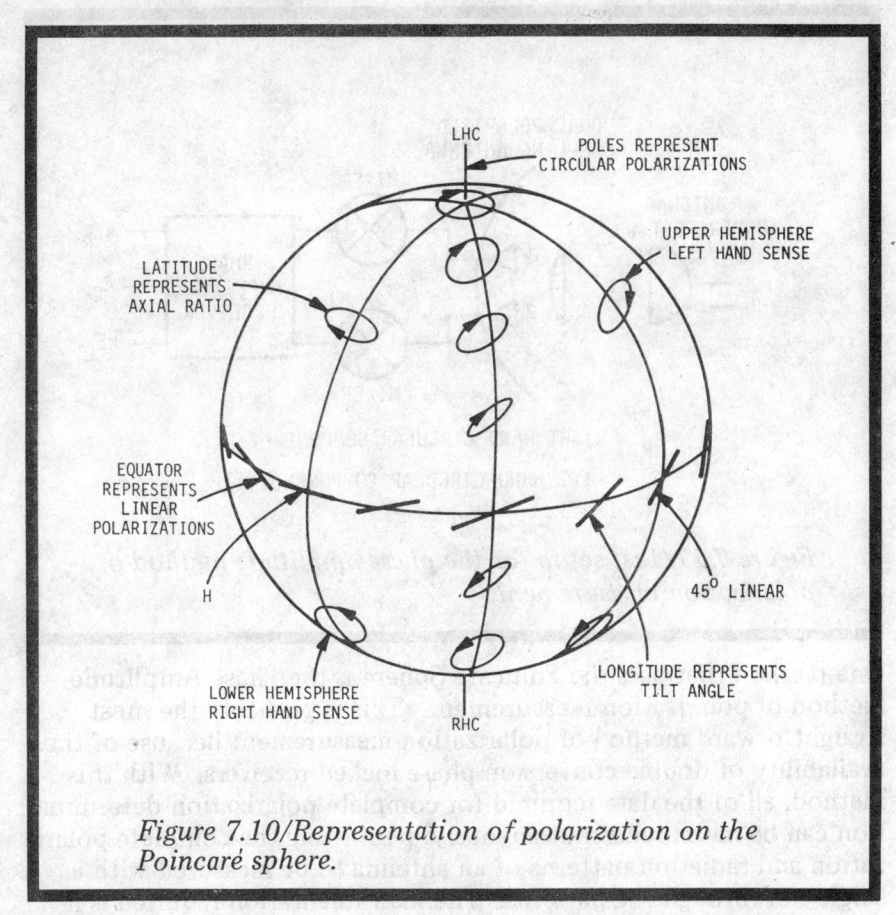

Antenna Measurements

Figure 7.10/Representation of polarization on the Poincaré sphere.

polarization. The E-vector is associated with only a single, either vertical or horizontal, plane; that is, the electric field (E) is in a plane oriented vertically or horizontally.

Circular polarization involves more than one plane and is characterized as being involved with the two-plane atmosphere shown in Figure 7.9. The E-vector rotates in either a clockwise or counterclockwise direction — if it moves clockwise it is said to be right-hand circular; counterclockwise, left-hand circular. For circular polarization, the E-vector must remain at a constant amplitude. If it changes magnitude as it rotates, its trip traces an ellipse and the wave is said to be *elliptically* polarized.

A very useful tool that is used to aid in visualizing the polarization of an antenna is the Poincaré Sphere. Figure 7.10 is a representation of this popular sphere. There is a one-to-one correspondence between all the possible polarizations of a monochromatic wave and the points on the sphere.

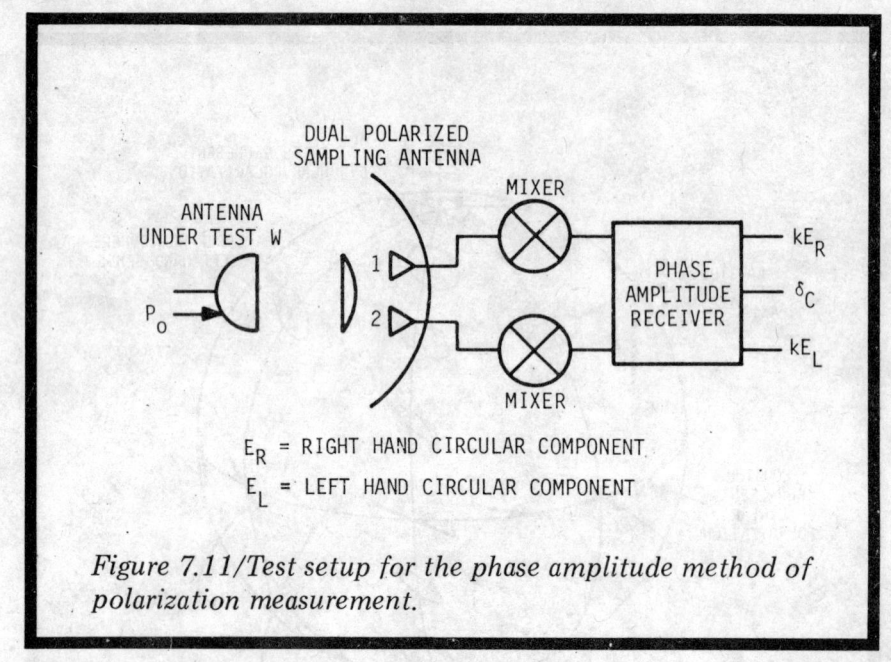

Figure 7.11/Test setup for the phase amplitude method of polarization measurement.

One method that uses the Poincaré Sphere is the Phase-Amplitude method of polarization measurement. This is probably the most straightforward method of polarization measurement because of the availability of double conversion phase-locked receivers. With this method, all of the data required for complete polarization determination can be measured simultaneously, permitting the complete polarization and radiation patterns of an antenna to be measured with a single run through the patterns. The instrumentation required is illustrated in Figure 7.11. A dual-polarized receiving antenna is used to sample the field of the antenna under test, which is, in this case, used as a transmitting antenna. The outputs of the receiver will be the magnitudes of the responses for each of the polarizations of the sampling antenna and their relative phase. If the two polarizations are orthogonal, the complex polarization ratio can be obtained. The polarizations of the sampling antenna must be known and the gains of the two antenna-receiver channels must be made identical.

The measurement of the polarization of the test antenna using circuitry polarized sampling antennas is illustrated by use of the Poincaré sphere in Figure 7.12. The complex circular polarization ratio and angle are obtained from the measurement, and hence the polarization of the wave radiated by the test antenna is uniquely determined.

Antenna Measurements

Figure 7.12/Phase amplitude measurement with circularly polarized sampling antennas.

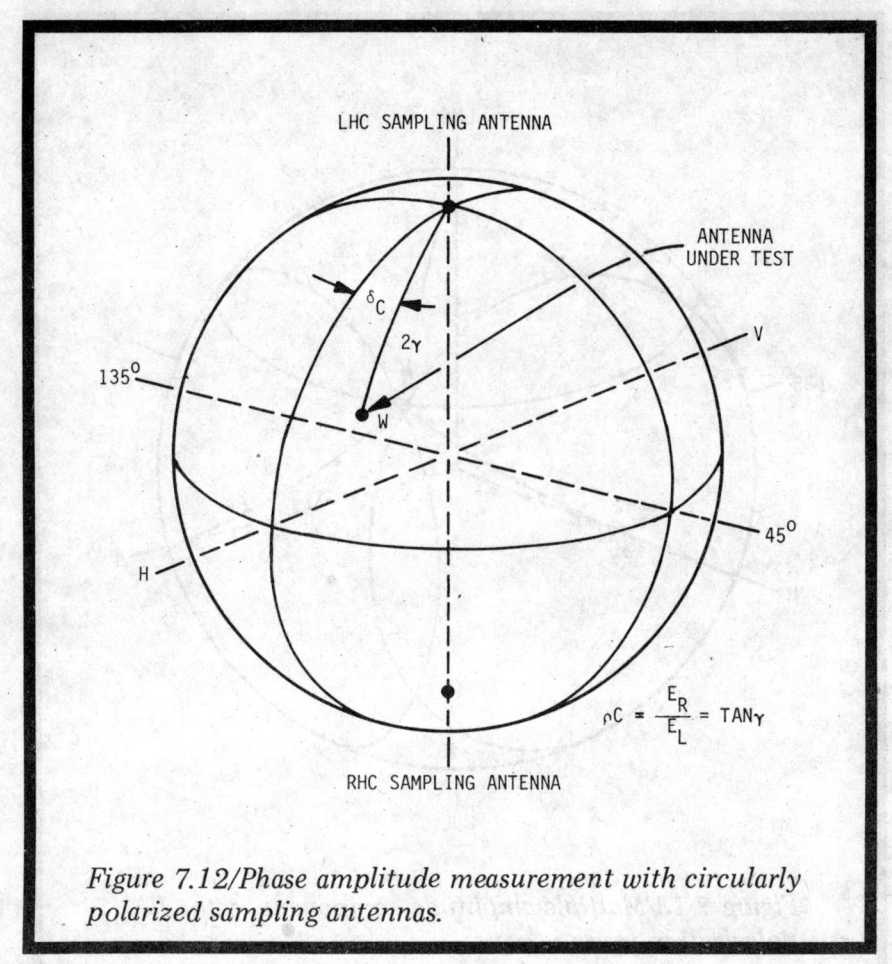

The multiple-amplitude-component methods can be used to determine the polarization completely without the measurement of phase. It has been shown that the polarization of a wave can be determined from the magnitudes of the responses of four antennas having different, but known, polarizations. The most convenient choices of polarizations of the sampling antenna are: horizontal or vertical linear, 45 or 135° linear, right or left hand circular, and any fourth component from this set of six components. These sampling antennas must have known gains and the instrumentation must be suitably calibrated to compensate for gain differences. From these data, the polarization of the wave, and hence that of the test antenna, can be completely determined. Graphical construction or a solution of a system of linear equations may be used to find the Stokes parameters.

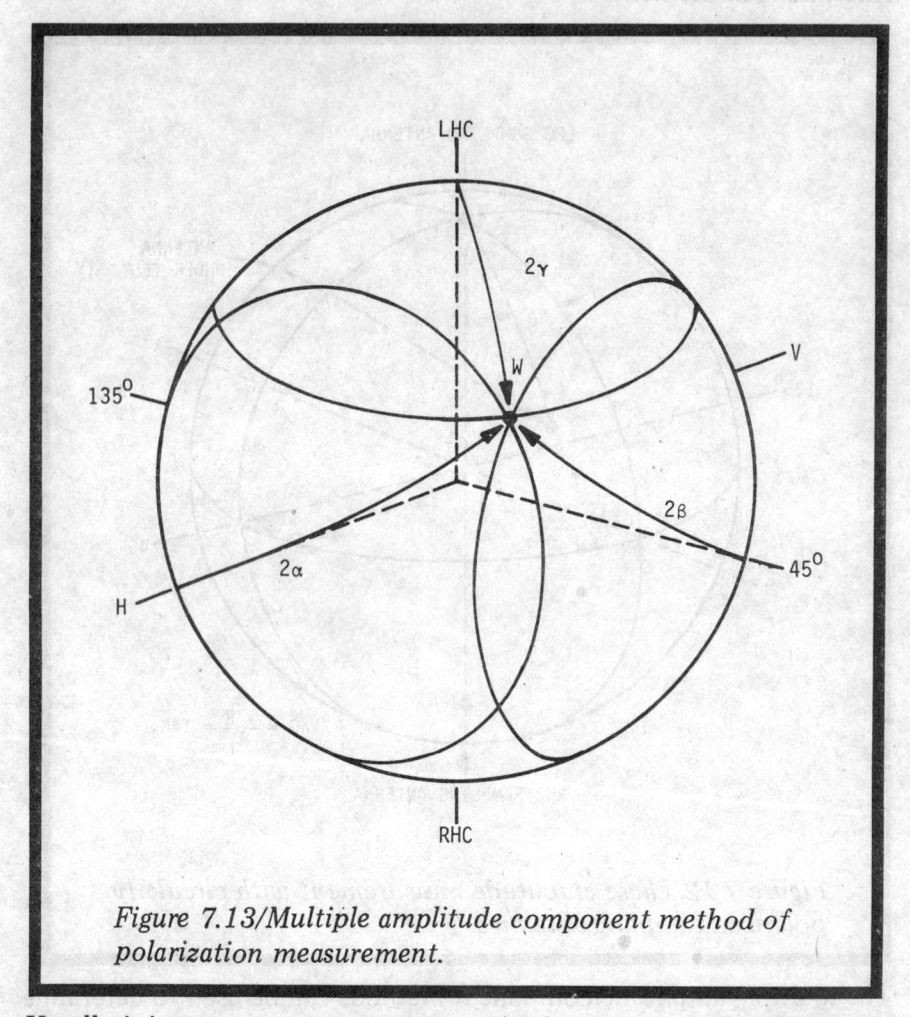

Figure 7.13/Multiple amplitude component method of polarization measurement.

Usually it is more convenient to measure the magnitudes of the polarization ratios, hence, all six components are used. This method is illustrated by use of the Poincaré sphere as shown in Figure 7.13. The linear, diagonal linear, and circular-polarization ratios are measured. From these data, the angles which define the loci of all possible polarizations on the Poincaré sphere and which correspond to the polarization ratios are determined. The common intersection of the three loci determines the polarization of the wave.

A modified version of the multiple-amplitude-component method, which involves measurements with a single linearly polarized antenna, may be employed to determine the axial ratio and tilt angle over the entire radiation pattern of the test antenna, provided that the sense

Antenna Measurements 317

is not required. The pattern of the test antenna is measured with the source antenna oriented at 0° (horizontal), 45 and 90° (vertical) and 135°. From these data, the polarization ratios are determined. With these results, the phase angle can be computed; the value obtained can be used to determine the tilt angle.

To sum up the measurement of an antenna polarization, we can say that the following methods can be used:

(a) Measurement of the ratio of the amplitudes of two orthogonal linear polarization components of the field and their relative phase.

(b) Measurement of the ratio of the amplitudes of two orthogonal circular polarization components of the field and their relative phase.

(c) Measurement of the ratios of the amplitudes of two orthogonal circular polarization components, two orthogonal linear polarization components and a second set of orthogonal linear components which are displaced 45 degrees in tilt angle from the first set. This method does not require measurement of phase.

(d) Measurement of the axial ratio and tilt angle of the major axis of the polarization ellipse by means of the *polarization pattern* and determination of the sense of polarization by means of an auxiliary measurement with at least one antenna which is not linear polarized.

(e) Measurement of the amplitudes of two orthogonal linear components, the tilt angle τ and the sense of polarization. This method breaks down when the two linear components become equal and is not recommended.

7.5 Boresight Measurements

Boresighting is defined as the initial alignment of a directional microwave or radar antenna or system through the use of an optical procedure or a fixed known target. From this definition we can see that boresight measurements are measurements of any variation from this initial alignment.

The antenna types for which boresight calibration is required may be grouped into two broad classifications:

(a) Antennas (and associated signal-processing systems) which produce a null in the sensing function at boresight

(b) Antennas which produce a maximum in the sensing function at boresight

Type (a) systems include con-scan radars, beam-switching radars, amplitude monopulse and phase monopulse tracking and direction sensing systems, etc. Type (b) systems include various fixed-orientation antennas and radars of both electronic-scan and mechanical-scan types, such as height-finders and mortar-locators; important subgroups of this type of system are the family of Doppler radars such as the Janus and beam-intersection systems, and missile-borne fuse-train antennas.

Boresight calibrations are usually directed toward establishing coincidence (or in some cases simply parallelism) of the electrical boresight aixs of an antenna with a mechanical reference axis. The mechanical reference is often called the optical boresight axis, and is fixed with respect to the antenna or its operational positioning structure.

To alleviate the generally significant system-handling problem on the test range, boresight calibrations are typically performed on subsystems which include only the antenna and that portion of the associated circuitry which is necessary to generate the sensing function. The calibration is subsequently transferred to the total operational configuration by optical, mechanical or electrical measurement techniques. Also, for reasons of simplicity and economy, boresight calibrations are most often performed with a CW or amplitude modulated source, as opposed to pulsed sources, regardless of the nature of the using system. Notable exceptions to these convenient approaches are sometimes made for extremely high power systems (although the power handling capabilities of such systems are generally proven in bench tests or microwave darkroom measurements) and for such two-way systems as Doppler radars.

In any case, the test range must provide a source of radiation (or a target) whose location in the test coordinate system is known to an accuracy commensurate with the calibration tolerances. This requirement implies the necessity for calibration or proof-of-performance for the test facility itself. The major sources of error in boresight measurements include:

(a) Positioning system and readout error

Geometric error
- Coordinate axis alignment
- Orthogonality
- Collimation (deviation from alignment)

Antenna Measurements

        Shaft-position error
- Synchros
- Resolvers
- Encoders

        Deflection error
- Load stresses
- Solar heating
- Wind loading

(b) Test instrumentation errors

(c) Improper characteristics of the illuminating field

        Phase curvature
        Amplitude taper, lateral and longitudinal

(d) Parallax

(e) Extraneous signal interference

        On-site reflections and diffraction
        Spurious radiation

(f) Improper test procedures

(g) Data processing errors

With the sources of error identified, let us briefly look at three measurement systems: a beamshift test system, an amplitude boresight error test system, and a phased array test system.

Boresight and beamshift error test ranges normally consist of a transmitting antenna and a receiving antenna or array of antennas. The receiving and transmitting antennas are separated on the test range by a minimum distance of $2D^2/\lambda$, where

D = diameter of the largest antenna aperture
λ = freespace wavelength of the test frequency

The heights of the receiving and transmitting apertures are selected to accommodate the test range terrain. It is advantageous that the radome test range designer include the capability of testing one-way power transmission loss simultaneously with the boresight error.

A typical beamshift error test range is shown in Figure 7.14. A shaped beam test antenna is employed as the transmitting aperture. The three antennas, A, B and C, make up the receiving array. Antennas A and C are the beamshift detection antennas while antenna B is utilized as the one-way power transmission loss detection antenna. The beamshift

Figure 7.14/Beamshift error test system.

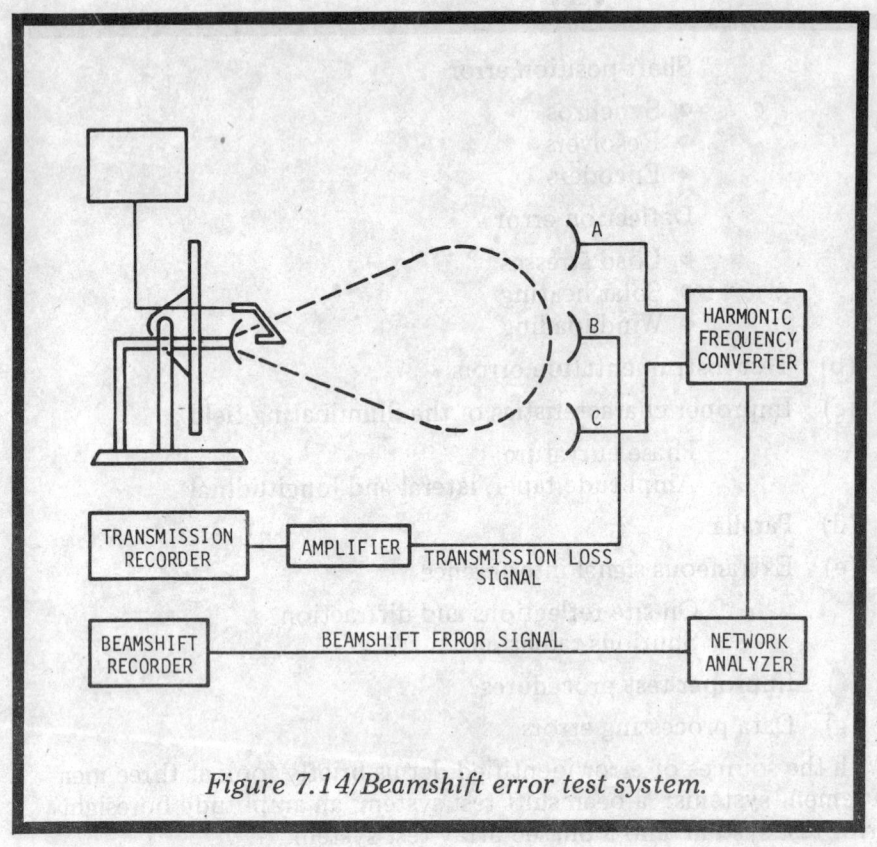

antennas A and C are separated vertically by a distance preferably equal to the 3 dB beamwidth of the test antenna aperture, although it is acceptable to separate them as little as the 1 dB beamwidth of the test antenna. The test antenna is electrically aligned with the array antenna B which is located in the same plane and in the center of antennas A and C. With the antennas aligned in this manner, antennas A and C will receive the radiated signal at equal amplitude and 1 to 3 dB less than antenna B, dependent upon the separation of A and C.

The amplitude decrease caused by losses due to the radome wall is undetected in the beamshift antennas since both antennas will experience the same amplitude decrease. It is the ability to detect the relative amplitude change between antennas A and C that is important in beamshift error measuring. With the test antenna properly aligned, the RF output of antenna A is fed into the test channel of the harmonic frequency converter and the RF output of antenna B is fed into the reference channel of the harmonic frequency converter.

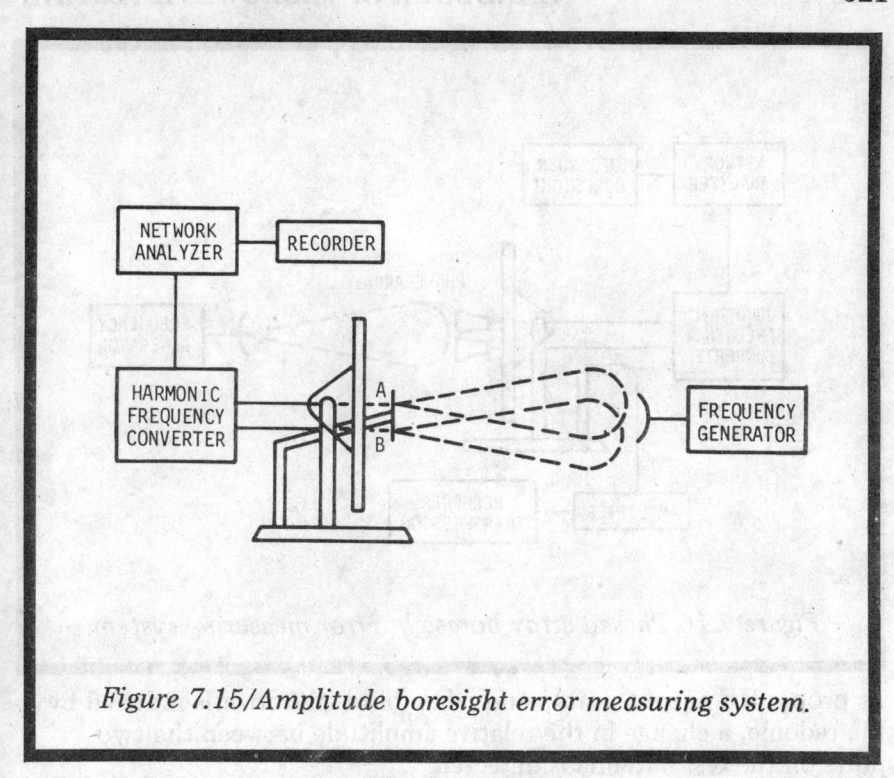

Antenna Measurements

Figure 7.15/Amplitude boresight error measuring system.

A null or 0 dB reference output is established on the phase gain indicator by means of the calibrated step attenuator on the network analyzer. The dc output on the phase gain indicator is coupled to the y-axis of a suitable recorder. The x-axis of the recorder is synchronized with the radome positioning fixture.

When the beam is shifted up, antenna A experiences a small increase in power and antenna C experiences a small decrease in power. A positive dc output will be detected on the y-axis of the recorder proportional to the upward beamshift error. A downward shift in the beam results in a converse reaction.

A typical amplitude boresight (multiple beam antenna) measuring range is shown in Figure 7.15. In this case, it is preferable to conduct one way power transmission and boresight error measurements separately unless a sum output of the multiple beams is provided on the test antenna. The test antenna is utilized as the receiving antenna and consists of two independent apertures which maintain a beam crossover between 1 and 3 dB down from the peak of the beam. The transmitting antenna is aligned electrically to the crossover of the multiple beams. The received signal amplitude is then monitored at

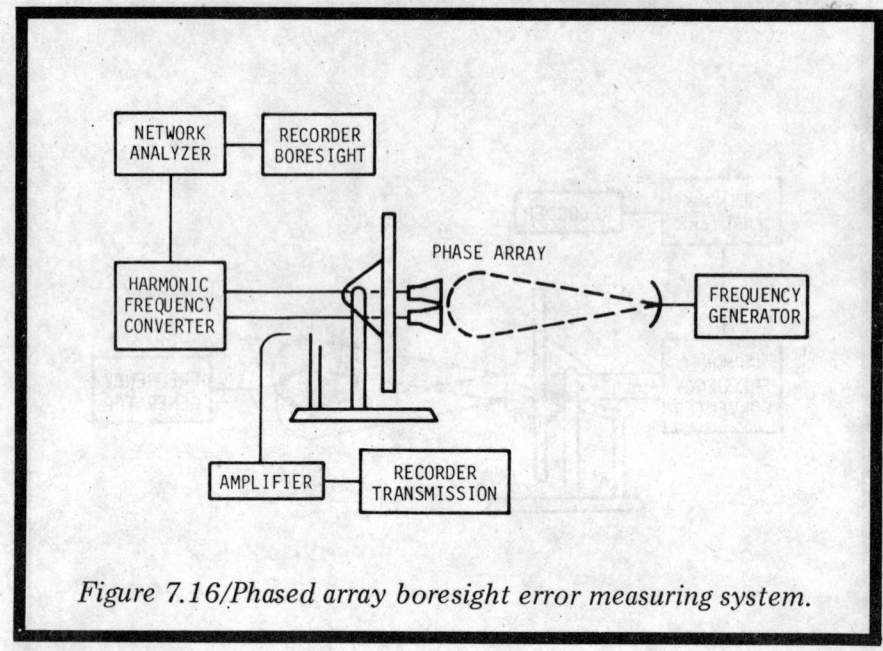

Figure 7.16/Phased array boresight error measuring system.

its proper RF output ports; and when boresight error is induced by the radome, a change in the relative amplitude between the two ports on the test antenna is detected.

The amplitude boresight system operates on the same principle as the beamshift error system. The only changes necessary will be to physically relocate the test equipment to accommodate the antenna-radome system under test.

A typical phased array antenna boresight measuring range is shown in Figure 7.16. In this case the antenna under test is employed as the receiving antenna. Boresight error is detected as the relative phase change between the two arrays of the test antenna and is defined as:

$$\phi = \frac{d \sin \Theta \; 360°}{\lambda}$$

where

ϕ = phase change in RF degrees
Θ = boresight error angle
d = phase center separation between the two arrays
λ = free-space wavelength

The relative phase shift between the two arrays is directly proportional to boresight error. The transmitting antenna is optically aligned with

Antenna Measurements

the phased array to establish electrical boresight. Any boresight error caused by the presence of the radome will be detected as a relative phase change between the two arrays. The RF output of port A on the test array is coupled to the test channel of the harmonic frequency converter. The RF output of port B on the test array is coupled to the reference channel of the harmonic frequency converter. The phase gain indicator is adjusted by means of the calibrated step phase shifter to indicate a null or $0°$ phase reference between array A and B. The dc output on the phase gain indicator is coupled to the y-axis of a suitable recorder. The x-axis of the recorder is synchronized with the radome-antenna positioning fixture.

When an upward boresight error is induced on the test antenna, port A of the phased array experiences a phase lead and port B experiences a phase lag. Thus, a positive dc output proportional to the upward boresight shift will be experienced on the y-axis of the recorder. Conversely, a negative or downward boresight shift can be recorded. Calibration of the system can be confirmed by the self contained phase shifter, by an external phase shifter in one of the receive ports, or by physically displacing the antenna in known angular offsets.

Chapter Summary

In this chapter, we have attempted to introduce you to basic antenna measurements. It is not intended to be a comprehensive text on the topic, since there are outstanding books available which deal with all of the mathematics and intricacies of this area of complex measurements. The main idea, as stated previously, was to introduce you to the major parameters of gain, directivity, polarization, and boresight. With this knowledge we can investigate more complex and involved methods of obtaining data on these parameters.

Automatic

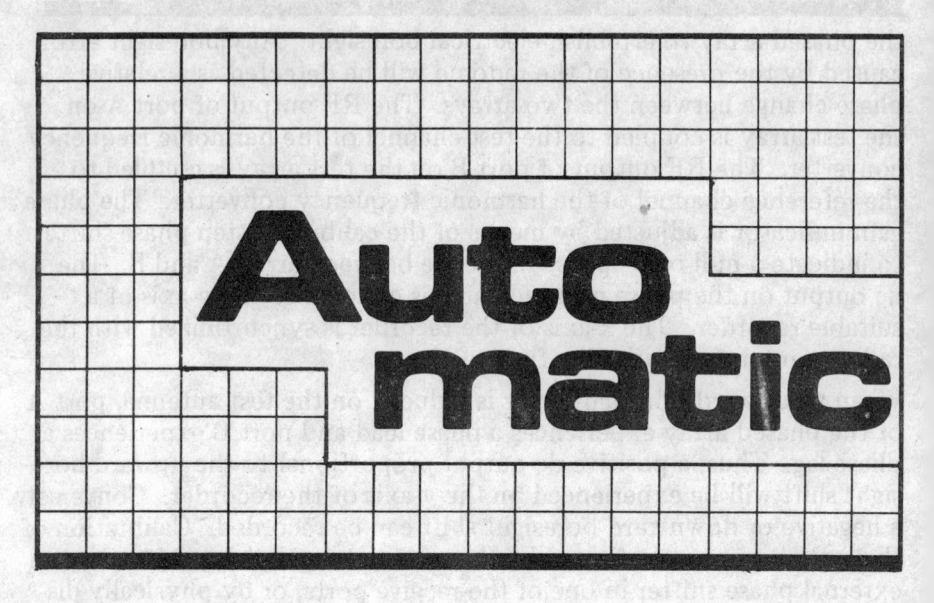

8 Testing

8. Definition

The measurements taken thus far are considered to be basically *manual* measurements. That is, any corrections or adjustments to be made are made by an operator. While they are categorized as *manual*, you could stretch the point on some of them because they involve linking devices together electrically to automate repetitious, routine measurements. We would probably do better to label them *semi-automatic* though, since many functions must still be performed manually by an operator.

The true *automatic* measurement is one that is set up and controlled by a computer or calculator. With an arrangement such as this, there is very little for the operator to do except insert the device to be tested, operate the calculator, and read that data output.

The type of measurement above sounds very desirable and quite simple. However, there is one area that has not been mentioned yet and as such it poses an interesting question: "how do we get from the calculator to the test instruments and back again?" There obviously has to be some sort of interface between these two areas, but it must be much more than a couple of wires. This interface is the heart of all automatic testing and will be the main topic of this chapter. A look at the history and evolution of the present day interface will serve as an introduction to automatic testing.

Historically, the high cost of interfacing related to the lack of interface standardization. Interface methods increased as each engineer designed special links between various instrumentation devices — resulting in different codes, formats, signed levels, and timing factors. Whatever was best for the particular application was used. There were times when even devices from the same manufacturer often had dissimilar control, data and signal lines — to say nothing of differing data exchange rates, codes, and even connectors. It was very obvious that this chaos needed some sort of standardization if users were to enjoy flexible and affordable interface capabilities.

The interest in establishing a suitable standard for interfacing was not limited to United States manufacturers. There was also a growing interest throughout the world for such a device. European organizations, particularly in Germany, started the standardization movement on an international scale.

In mid 1972, American companies (Hewlett-Packard in particular) began participating with various national and international standardization committees. The U.S. Advisory Committee, composed of both users and manufacturers, adopted the interface concept utilized by the then-new digital interface bus from Hewlett-Packard, and this concept was submitted to the International Electrotechnical Commission (IEC) in the autumn of 1972.

By 1974, Hewlett-Packard had developed and marketed a number of different instruments interfaceable via the basic digital techniques contained in the IEC draft document. At this time, the in-house name for this interface was the Hewlett-Packard Interface Bus, or HP-IB.

In April 1975, the Institute of Electrical and Electronics Engineers (IEEE) published IEEE Standard 488-1975, entitled *Digital Interface for Programmable Instrumentation*. In January 1976, the American National Standards Institute published the same document as *ANSI Standard MC1.1*. Also, in 1976, IEC member nations voted on the main interface document, which has now been approved for publication by the IEC.

Today, this interface concept is commonly referred to as the IEEE 488 Bus, the general purpose interface bus (GP-IB), or the Hewlett-Packard Interface Bus (HP-IB). Its widespread acceptance is revealed by the large quantity of bus-compatible instruments being produced today.

In the following sections, we will take the IEEE 488 bus specification and explain each section in simple down-to-earth terminology. Specific test setups will not be considered since there are such a variety of

Automatic Testing

setups that are possible. Instead, we will cover the interface bus and allow you to use your knowledge of the bus to derive a setup suitable for your own application.

8.1 Section 1 (General Information)

This section describes the interface bus and its capabilities. A group of definitions is presented to enable the user to speak the same language as the bus and the instruments that are being connected together.

The IEEE 488 interface does not apply to just any instrument or combination of instruments. There are certain restrictions that must be observed. These restrictions are:

- The data exchange between the interconnected devices must be *digital* (Analog signals will not be accepted).
- The number of devices connected to one bus must not exceed 15.
- The total length the signals travel over the interconnecting cable must not exceed 20m (65.61 feet).
- The data rate on any signal line across the interface must not exceed 1 Mb/S (one million bits per second).

The 488 standard has multiple objectives. These objectives were set up in an attempt to satisfy the needs of the industry by supplying a truly universal interface bus. The objectives of the standard are:

- Definition of a general purpose system for use in limited distance applications.
- Specific mechanical, electrical, and functional interface requirements so that the interconnection between devices can be achieved and communications between devices is possible.
- Specific terminology and definitions related to the system.
- Allow interconnection of equipment from different manufacturers to result in one single system.
- Allow connection of equipment with a wide range of capability — simple or complex — to the system simultaneously.
- Permit direct communications between the apparatus without requiring all messages to be routed to a control or intermediate unit.
- Define a system that would minimize the restrictions on the performance characteristics of any devices connected to that system.
- Define a system that permits simultaneous communication over a wide range of data rates.

- Define a system that is low in cost and also permits interconnection of low cost devices.
- Define a system that is easy to use.

These are the objectives of the IEEE 488 standard and it will be interesting to see just how each of these are met as we go through the remaining sections.

In order to understand the entire IEEE 488 standard, it will be necessary to speak the same language. To aid in this understanding we now present some of the common terms that are used, and a definition of each.

Bi-Directional Bus — A bus used for two-way transmission of message, that is, both input and output.

Bit — An abbreviation for binary digit. A unit of information equal to one binary decision. It is also the designation of one of two possible and equally likely values or states (1 or 0 for example) of anything used to store or convey information. It may also mean "yes" or "no".

Bit Parallel — A set of simultaneous data bits present on a like number of signal lines used to carry information. Bit-parallel data bits may be acted upon as a group or independently as individual data bits.

Bus — A signal line or set of signal lines used by an interface system to which multiple devices can be connected and over which messages are carried.

BYTE — A group of binary bits processed together. It frequently consists of a group of eight bits.

BYTE-Serial — A sequence of bit-parallel data BYTES used to carry information over a common bus.

Handshake Cycle — The process where digital signals effect the transfer of each data BYTE across the interface by means of an interlocked sequence of status and control signals. (Interlocked refers to a fixed sequence of events in which one event in the sequence must occur before the next event.)

High State — The logic state in a two binary logic state system that is the more positive.

Interface — A shared boundary between one system and another, or between parts of systems, through which information is transported.

Interface System — The mechanical, electrical, and functional elements of an interface which are necessary to provide the proper communication between devices or sets of devices. Such areas as cables, connectors, driver and receiver circuits, signal line descriptions, timing and control conventions, and functional logic circuits are elements of an interface system.

Local Control — A method where a device is programmable by means of its front or rear panel controls (referred to as local controls). This is also called manual control.

Low State — The logic state in a two binary logic state system that is the less positive.

Programmable — The characteristic of a device which makes it capable of accepting data. This data alters the state of the circuitry which allows the device to perform specific tasks.

Programmable Measuring Apparatus — A measurement apparatus that either may perform specified operations on command from a system or may transmit the result of the measurements to the system.

Remote Control — A method whereby a device is programmable by way of its electrical interface connection. This is as opposed to local control where the device is programmed at its front (or back) panel.

Signal Line — A signal conductor in an interface system that is used to transfer messages from one device to another.

Terminal Unit — The unit which allows interconnection of one interface system with another.

Unidirectional Bus — The signal line or set of signal lines that a device uses for one-way transmission of message. That is, input only or output only.

The definitions presented above are terms that will be used throughout the following sections. You may find it necessary to refer back to them periodically to assure yourself that you understand what is being explained.

This interface, or any interface system, has basically one purpose to its existence — it must provide an effective and efficient communications link between two devices. This communications link must be capable of accurately carrying messages between the interconnected devices so that each of them is able to perform its assigned functions.

Messages, which are *quantities of information* that are carried by an interface system, belong in either of two categories:

(1) Messages used for operation of the interface system (*interface messages*).

(2) Messages used by the interconnected devices (*device dependent messages*).

Note: The device dependent messages are also carried by the interface system, but are not used or acted on by the system. The system only acts as a medium of transmission for them.

We have stated that the main purpose of the interface system (IEEE-488) is as a communications link. In order for this link to be effective, it must contain three basic functional elements which organize and manage the flow of information. These are:

- A device acting as a listener
- A device acting as a talker
- A device acting as a controller

How do these three elements apply to our interface system? Let us examine them and find out:

- A device with the capability to listen can be addressed by an interface message (message type #1) to receive device dependent messages (message type #2) from another device on the interface system. An example of a listening device is a microwave signal generator.

- A device with the capability to talk can be addressed by an interface message to send device dependent messages to another device on the interface system. An example of a talking device would be a frequency counter.

- A device with the capability to control can address other devices to talk or to listen. In addition, this device can send interface messages to command specified actions within other devices. A controlling device neither sends nor receives device dependent messages.

The part that each of these individual devices plays in the overall interface system is shown in Figure 8.1. You can see devices that are able to talk only; listen only; talk and listen; or talk, listen, and control.

We have discussed the interface objectives, definitions, and device connections to this point. It is now time to get the construction of the interface system itself.

The interface system contains a set of 16 signal lines used to carry all information, interface messages, and device dependent messages between interconnected devices.

The bus structure itself is organized into three sets of signal lines:

(1) Data bus, 8 signal lines
(2) Data BYTE transfer control bus, 3 signal lines
(3) General interface management bus, 5 signal lines

These signal lines are shown in Figure 8.1. We will use Figure 8.1 to describe these and all the functions of the interface system.

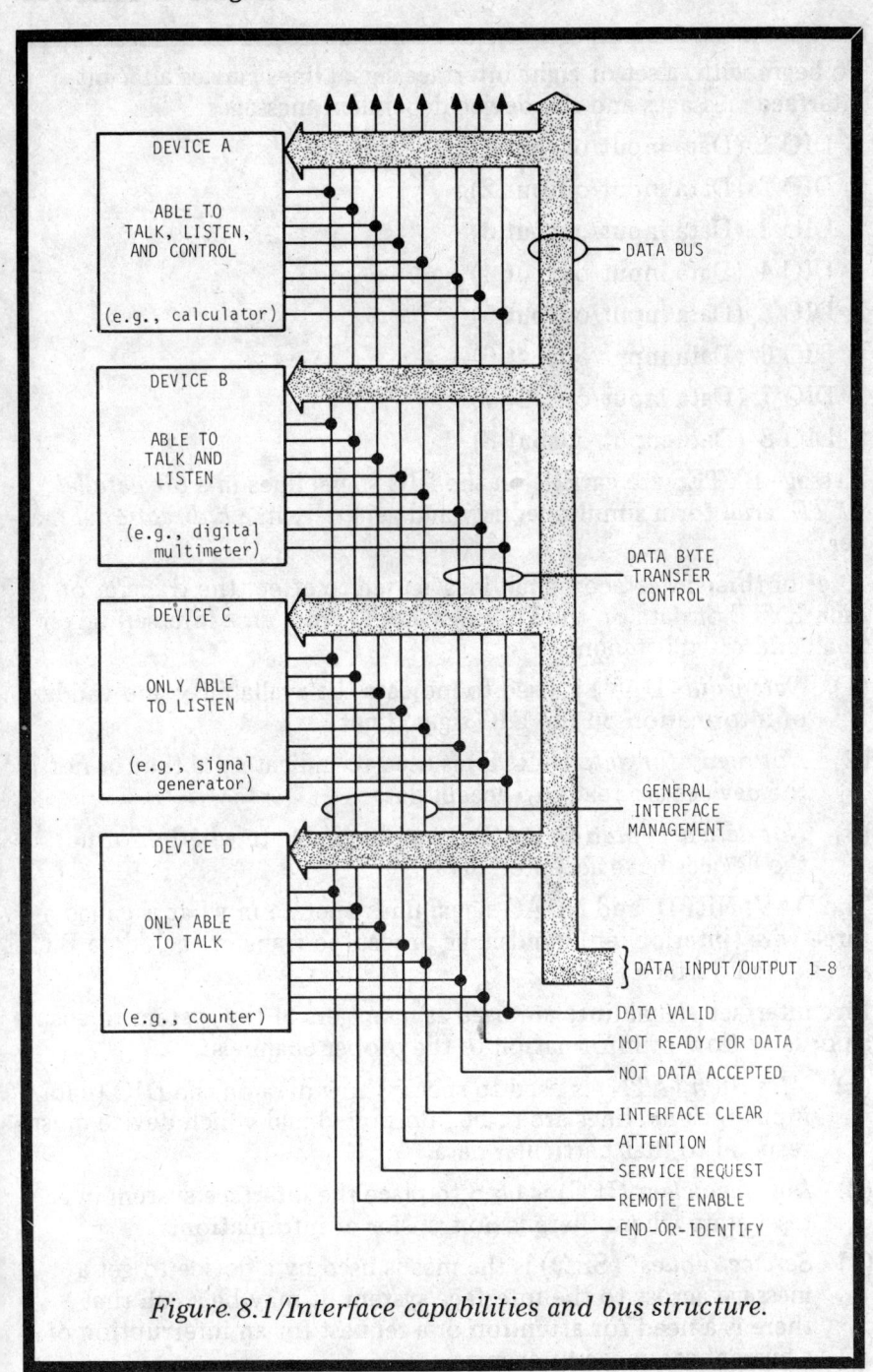

Figure. 8.1/Interface capabilities and bus structure.

To begin with, a set of eight interface signal lines carries all 7-bit interface messages and the device dependent messages:

1. DIO 1 (Data input/output 1)
2. DIO 2 (Data input/output 2)
3. DIO 3 (Data input/output 3)
4. DIO 4 (Data input/output 4)
5. DIO 5 (Data input/output 5)
6. DIO 6 (Data input/output 6)
7. DIO 7 (Data input/output 7)
8. DIO 8 (Data input/output 8)

Message BYTES are carried on the DIO signal lines in a *bit-parallel BYTE serial* form simultaneously and generally in a *bidirectional* manner.

A set of three interface signal lines is used to effect the transfer of each BYTE of data on the DIO signal lines from an addressed talker to all addressed listeners:

(1) *Data valid* (DAV) is used to indicate the availability and validity of information on the DIO signal lines.

(2) *Not ready for data* (NRFD) is used to indicate whether or not the devices are ready to accept data.

(3) *Not data accepted* (NDAC) is used to indicate whether or not the devices have accepted data.

The DAV, NRFD, and NDAC signal lines operate in what is called a three-wire (interlocked) handshake process to transfer each data BYTE across the interface.

Five interface signal lines are used as managers of the system to ensure an orderly flow of information to the proper channels:

(1) *Attention* (ATN) is used to specify how data on the DIO (data input/output) lines are to be interpreted and which device must respond to that particular data.

(2) *Interface clear* (IFC) is used to place the interface system in a reset state where there is no transfer of information.

(3) *Service request* (SRQ) is the means used by a device to get a message across to the interface system. It may be a call that there is a need for attention or a request for an interruption of a current sequence of events.

Automatic Testing 333

(4) *Remote enable* (REN) is used to select between two sources of device data. This signal line is usually used in conjunction with other messages.

(5) *End or identify* (EOI) is used to indicate the end of a multiple BYTE tranasfer sequence or may be used with ATN to execute an identification sequence.

This is a basic description of the interface system: a system which transfers messages between talkers, listeners, and control devices by means of signal lines which are used as input/output lines, data transfer lines, and managerial control lines. With the basic concepts of what the system is intended to do and how it does it, let us continue into the IEEE 488 specification and investigate the primary elements of the interface system:

- Functional Elements
- Electrical Elements
- Mechanical Elements

8.2 Section 2 (Functional Specifications)

The main question that arises when choosing a device, system, or an interface standard is "is it functional?" That is, will it do the job that I have intended it to do? The answer to that very important question is the objective of this section. It will investigate the function that the IEEE 488 standard can perform and allow you to decide if the standard will do the job that you have intended for it.

To put our discussion into its proper perspective, you must realize that a device can be placed in three major functional areas. These areas are:

(1) *Device Functions* — The definition of this function depends on your particular application. It may be a sweep generator, power meter, and counter which make up this function.

(2) *Interface Functions* — These functions are independent of the particular application. It, for example, does not matter if we are taking a frequency reading, power reading, or return loss. It is defined the same, no matter what the application.

(3) *Message Coding Logic* — All communications to and from the interface function are defined in terms of messages and state linkage. All messages are coded according to a coding logic.

The device functions are not really the objective of this chapter. Rather, we are concerned with the interface system and how it can be used to provide measurements on microwave components and systems. There comes a time, however, when the devices must connect to the interface bus. For that reason, we will present a brief discussion of the device function of a typical piece of microwave test equipment. The device we are referring to is an HP 436A power meter. This device has both a talker function and a listener function.

If the basic talker function is selected, the power meter is configured to talk when the controller places the interface bus in the command mode and outputs talk address M. The power meter then remains configured to talk (output data when the interface bus is in the data mode) until it is unaddressed to talk by the controller. To unaddress the power meter, the controller can either generate an interface clear, or it can place the interface bus in the command mode and output a new talk address, or a universal untalk command.

When the power meter functions in the Talk Only Mode, it is automatically configured to *talk* when the interface bus is in the Data Mode and there is at least one listener. Since there can only be one talker per interface bus, this function is normally selected only when there is no controller connected to the system. An example of this condition would be when the power meter is connected to a recorder.

The power meter is configured to listen when the controller places the interface bus in the command mode (ATN and REN lines low; IFC line high) and outputs listen address "-" (minus sign). The power meter then remains configured to listen (accept programing inputs when the interface bus is in the data mode) until it is unaddressed by the controller. To unaddress the power meter, the controller can either set the IFC line low or the REN line high, or it can place the interface bus in the command mode and generate a universal unlisten command.

Note: You will recall from our previous discussions that ATN is the *attention* interface signal line, REN is the *remote enable* line, and IFC is the *interface clear* signal line.

Table 8.1 shows the output data characters for the 436 power meter. Figure 8.2 shows the interface bus output data format.

As we previously stated, our main concern here is the interface function as opposed to the device function. As interface function is the system element which provides the basic operational facility through which a device can receive, process, and send messages. A number of interface functions, each of which act in accordance with specific set

Automatic Testing

	Definition		Character ASC II	Octal	Decimal
S	Measured value valid		P	120	80
T	Watts Mode under Range		Q	121	81
A	Over Range		R	122	82
T	Under Range dBm or dB (REL)				
U	Mode		S	123	83
S	Power Sensor Auto Zero Loop Enabled; Range 1 Under Range (normal for auto zeroing on Range 1)		T	124	84
	Power Sensor Auto Zero Loop Enabled; Not Range 1, Under Range (normal for auto zeroing on Range 2-5)		U	125	85
	Power Sensor Auto Zero Loop Enabled; Over Range (error condition — RF power applied to Power Sensor; should not be)		V	126	86
R	Most Sensitive	1	I	111	73
A		2	J	112	74
N		3	K	113	75
G		4	L	114	76
E	Least Sensitive	5	M	115	77
M	Watt		A	101	65
O	dB REL		B	102	66
D	dB REF (switch pressed)		C	103	67
E	dBm		D	104	68
S	space (+)		SP	40	32
I	— (minus)		—	55	45
G					
N					
D	0		0	60	48
I	1		1	61	49
G	2		2	62	50
I	3		3	63	51
T	4		4	64	52
	5		5	65	53
	6		6	66	54
	7		7	67	55
	8		8	70	56
	9		9	71	57

TABLE 8.1

Interface Bus Output Data Characters

Figure 8.2/Interface bus output data format.

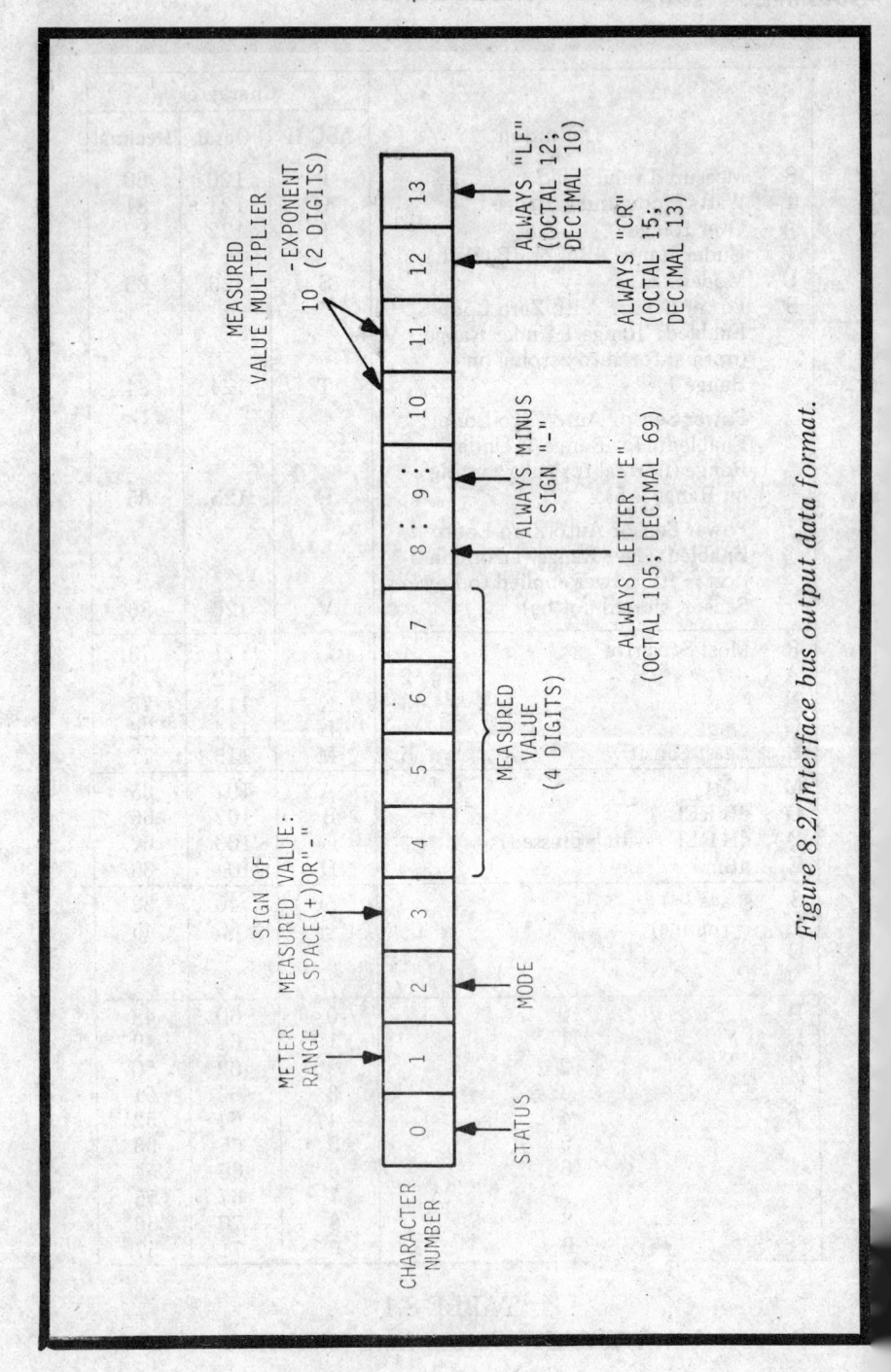

Automatic Testing

requirements, will be covered in the following sections. Each specific interface function may only send or receive a limited set of messages within the particular classes of set messages.

Each of the interface functions is defined in terms of one or more groups of interconnected, mutually exclusive states. For each of these states, a definition will be given for the following:

- Messages that may or must be sent over the interface while that state is active
- Conditions under which the function must leave that state and enter another state

The combination of these messages and the condition above define the processing capability of a particular state.

The state diagrams used to define the interface function that we will be covering are intended to allow the use of a wide variety of logic circuit implementations. Because of this, the designer is free to combine and implement two or more interface functions with one logic design provided all of the conditions for each state of each interface function are met as they are defined in this section. All aspects of the interface functions to be presented should be approached from the device perspective.

The message portion of the interface system is a very important area. The following points can be made concerning these messages:

- Each message constitutes a quantity of information and will be received whether true or false at any specific time.
- All communications between an interface function and its environment are accomplished through messages sent or received.
- Messages sent between a device function and an interface function are called *local messages*.
- The designer is *not* allowed to introduce new local messages to interface functions, but is allowed to introduce a local message derived from any state of any interface function to device functions.
- Local messages sent by device functions must exist for enough time to cause the required change of state.
- Messages sent by way of the interface between interface functions of different devices are called *remote messages*.
- Each remote message is either an interface message or a device dependent message.
- Each interface message is sent to cause a change of state within another interface function. The interface message will not be passed along to the device when received by an interface function.

- Device dependent messages are passed between the device function and the message coding logic by way of specified state functions within the interface function. Device dependent messages are device programing data, device measurement data, and device status data.
- Message coding, to be covered in detail in a later section, is the translating of remote messages to or from interface signal line values.
- A message which is sent over a single signal line is a *uniline message*. Two or more of these messages can be sent at the same time.
- A message which shares a group of signal lines with other messages is called a *multiline message*. Only one message of this nature (also called a message BYTE) can be sent at one time.

With properties of the messages examined, let us look into the connections used for message transfer.

(1) The value (true or false) of all remote messages capable of being sent by a device must at all times be as dictated by active states of its interface functions.

(2) The interface signal line(s) used to send a message value must be set to the levels specified by Appendix H, "Remote Message Coding."

(3) Since normal interface operation allows two devices to simultaneously send opposite values of the remote messages, a technique must be provided for resolving these conflicts. This is accomplished by implementing two types of message transfer over the interface — active transfer and passive transfer. The interface is structured so that in all conflicts between two message values, one of them will be active and the other passive. Messages must be transferred in such a way that the active value overrides the passive value in every conflict that arises.

(4) A remote message can be transferred in one of four ways:

  (a) An active true value being sent is guaranteed to be the value received, and the device need not allow it to be overridden.

  (b) A passive true value being sent is not guaranteed to be the value received, and the device must allow it to be overridden.

  (c) An active false value being sent is guaranteed to be the value received, and the device need not allow it to be overridden.

Automatic Testing   339

(d) A passive false value being sent is not guaranteed to be the value received, and the device must allow it to be overridden.

(5) Throughout the text, the terms "true" and "false," if not qualified, are assumed to mean active true and active false during all discussions of remote message values sent by an interface function.

(6) For two specific remote messages, DCA (Data Accepted) and RFD (Ready for Data), only false values are defined to be sent actively. Thus, an AND operation can be considered to be performed on the interface signal lines.

(7) For one remote message, SRQ (Service Request), only true values are defined to be sent actively. Thus, an OR operation can be considered to be performed on the interface signal lines.

(8) Only the multiline message(s) to be sent true will be specified for an interface function state since multiline messages (sent via the DIO lines) are by their nature mutually exclusive. It should be understood that all unspecified multiline messages are sent passive false while the state is active.

A very important part of the IEEE-488 standard involves an understanding of *state diagrams*. The state diagrams used in the IEEE-488 standard have specific notations attached to them. Each state that an interface function can assume is represented by a circle. A four letter upper case label (which always ends in S as a state designation) is used within the circle to idensity the state. An example is shown below:

$$\boxed{XYZS}$$

Note: If you read the IEEE-488 standard directly, you will note that the terms (or labels as we call them) are termed mnemonics. It seemed much more understandable to the author to designate these letters "labels" and thus that is what they will be called throughout the remainder of this chapter.

The possible transitions between the states are represented by arrows between them. Each of the transitions are qualified by an expression whose value is either true or false. This value is changed or remains the same by a unique set of conditions. To understand this, consider Figure 8.3. If we take a case where expressions 1, 2, 3, and 4 are all false, then ABCS, DEFS, and GHJS will remain in their present state. If, for example, expression 1 becomes true, the interface function ABCS must enter the state of DEFS. Similarly, if expression 3 becomes true, then function DEFS must enter the state of function GHJs. This is true,

Figure 8.3/State diagram example.

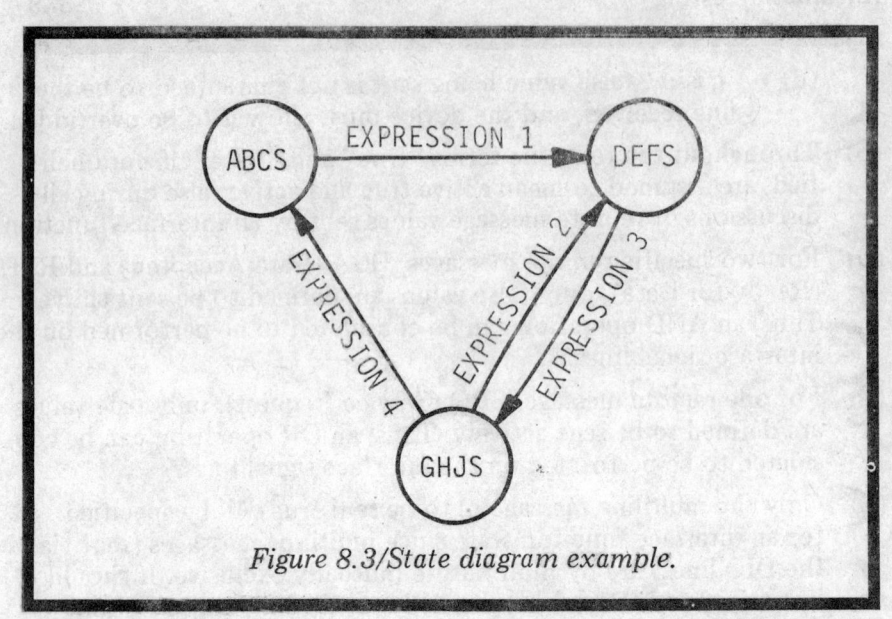

since the standard indicates that for a condition of all expressions being false, the functions remains in their set state; if one expression becomes true, the interface function must enter the state it is pointing to.

The expression we have been referring to above consist of one or more local messages, remote messages, state linkage, or minimum time limits used in conjunction with the digital operators — *and, or,* or *not*.

Local messages are represented by a three-letter lower case label — rdy, for example. The remote messages (received by way of the interface) are represented by a three-letter upper case label such as ATN.

A linkage from another state diagram is represented by a four letter label enclosed in an oval. An example of a stage linkage is (LACS) . State linkages are true if the enclosed state (the LACS state above) is currently active; otherwise, it is false.

A minimum time limit is represented by the symbol T_n. This achieves a true value only after the interface has been in the state that originates the transition for a specified value of time. This time could be anywher from 200 ns to 2 microseconds.

The AND operator is represented by the symbol Λ, while the OR operator is represented by V with the AND operator taking precedence over the OR operator unless otherwise specified. The NOT operator is represented by a horizontal bar placed over the portion of the expression to be negated. The resulting negated expression has a true

Automatic Testing

value if and only if the value of the expression under the bar is false.

If a transition is further qualified by a maximum time limit (within T_n), then the state pointed to must be entered within the specified amount of time after the expression becomes true. Also, if a portion of an expression is optional in that its true value need not be required for the complete expression to be true (at the designer's choice), then it is enclosed within square brackets [. . . .].

If a specific expression causes a transition to a state from all other states of the diagram, a shorthand notation is used instead of all the individual transitions being drawn. An arrow without a state at its origin is used for this condition. This is now assumed to be the origin of all states.

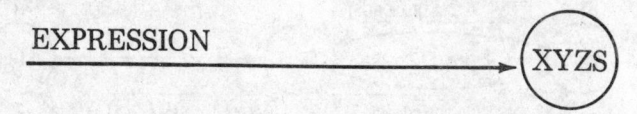

EXPRESSION ──────────────► (XYZS)

Examples of such a case are IFC (Interface Clear) and REN (Not Remote Enable).

A special case where a shorthand notation is used is the POFS (power-off) state. Although it is a valid state in most interface functions and should be shown on the diagram with a transition leading to the state to be entered at power-on time, the notation is still used in many cases. The shorthand notation is as shown below:

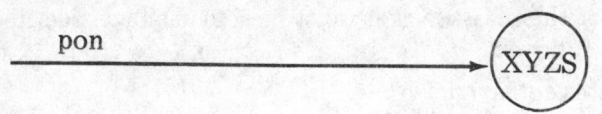

pon ──────────────► (XYZS)

The complete notation would be as follows:

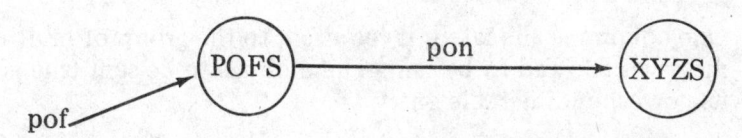

(POFS) ──── pon ────► (XYZS)
   ↖ pof

Each interface function we will cover has a message output table associated with it. This table summarizes only the remote messages that are allowed to be sent during each of the states of the function.

Rows of the table are used to indicate the state of the interface function, while columns are used to indicate the remote messages that are allowed to be sent during at least one state of the interface function.

Figure 8.4/SH (Source Handshake) State Diagram.

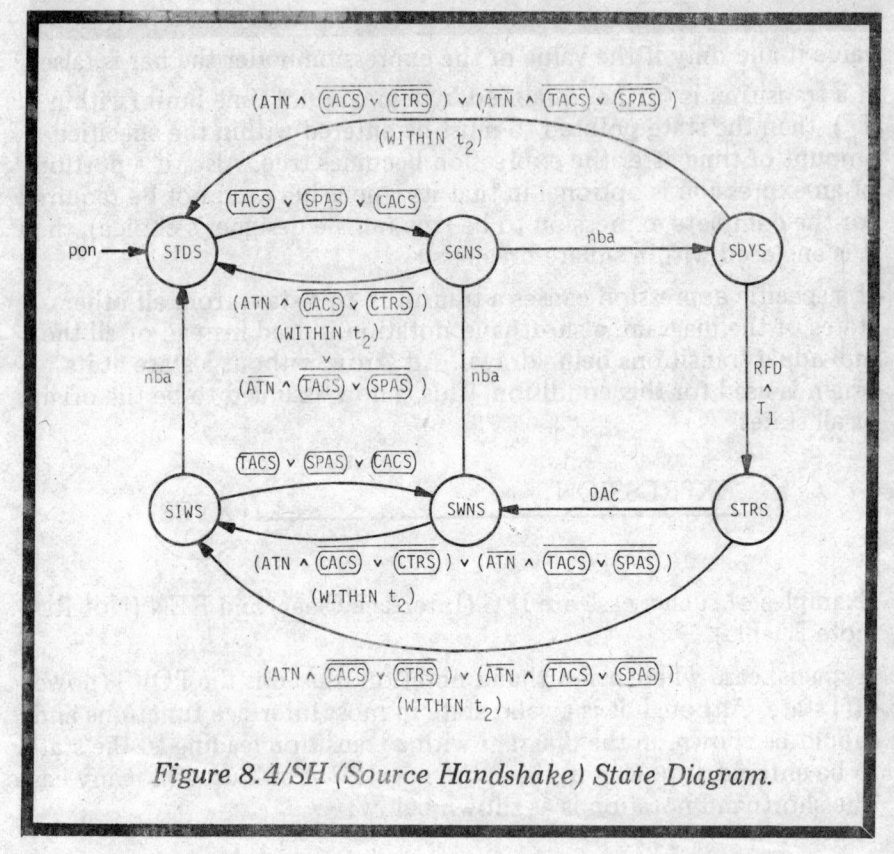

The value of the message that must be sent while a specified state is active is as follows:

- T indicates *active true*
- F indicates *active false*
- (T) indicates *passive true*
- (F) indicates *passive false*

One column is allocated, if required, to the group of multiline remote messages allowed to be sent. The message to be sent true is placed in its corresponding table entry.

A separate column for device function interaction summarizes the types of messages (or resultant action) device functions are allowed to send or receive.

With the basic definitions and function explanations discussed, we are now ready to investigate the interface functions themselves. We will basically cover ten functions:

TABLE 8.2

SH Labels

Messages		Interface States	
pon	= power on	SIDS	= source idle state
nba	= new byte available	SGNS	= source generate state
ATN	= attention	SDYS	= source delay state
RFD	= ready for data	STRS	= source transfer state
DAC	= data accepted	SWNS	= source wait for new cycle state
		SIWS	= source idle wait state
		(TACS)	= talker active state (T function)
		(SPAS)	= serial poll active state (T function)
		(CACS)	= controller active state (C function)
		(CTRS)	= controller transfer state (C function)

- Source Handshake — SH
- Acceptor Handshake — AH
- Talker — T
- Listener — L
- Service Request — SR
- Remote Local — RL
- Parallel Poll — PP
- Device Clear — DC
- Device Trigger — DT
- Controller — C

The first of the interface functions is the SOURCE HANDSHAKE (SH). This function provides a device with the capability to guarantee the proper transfer of multiline messages. This is possible since the SH function controls the initiation and termination of the transfer of a multiline message byte. It utilizes the DAV (Data Valid), RFD (Ready for Data), and DAC (Data Accepted) messages to effect each message byte transfer. These messages, you will recall, are in the set of signal lines used to effect the transfer of byte data from an addressed talker to an addressed listener.

The SH function acts in accordance with the state diagrams shown in Figure 8.4. The two tables shown are what we are referring to as SH

TABLE 8.3

SH Message Outputs

SH State	Remote Message Sent DAV	Device Function (DF) Interaction
SIDS	(F)	DF can change remote multiline messages
SGNS	F	DF can change remote multiline messages
SDYS	F	multiline messages must not change
STRS	T	multiline messages must not change
SWNS	T or F	DF requested to change multiline messages
SIWS	(F)	DF requested to change multiline messages

state diagram labels and the SH message outputs. The first (Table 8.2) table specifies the set of messages and state required to effect transition from one active state to another. The second table (Table 8.3) specifies the messages that must be sent and the device function (DF) interaction required while each state is active.

There are six interface states incorporated within the SH interface function:

(1) SIDS (Source Idle State)
(2) SGNS (Source Generate State)
(3) SDYS (Source Delay State)
(4) STRS (Source Transfer State)
(5) SWNS (Source Wait for New Cycle State)
(6) SINS (Source Idle Wait State)

Let us see where each of these six states fits into the overall interface system.

In the Source Idle State (SIDS), the SH interface function is not engaged in the handshake cycle and does not have a new message byte available. The Source Handshake (SH) function is powered on in SIDS.

The SH function leaves the SIDS state and enters the next state (Source Generate State), SGNS, if one of the following conditions occur:

(1) The TACS (Talker Active State) is active
(2) The SPAS (Serial Poll Active State) is active
(3) The CACS (Controller Active State) is active

You can see these statements on the state diagram in the upper left hand corner. *Note:* In SIDS, the DATA VALID (DAV) messages must be sent passive false.

The second state within the SH interface function is the SGNS (Source Generate State). In this state is waiting for a new message byte to become available from the device. The SH function must send the DAV (Data Valid) message false when the system is in this state.

The SH function must exist SGNS and enter:

(1) The SDYS (Source Delay State) if the nba (new byte available) message is true.

(2) The SIDS (Source Idle State) within t_2 if either:

    (a) The ATN (Attention) message is true and neither CACS (Controller Active State) is active nor CTRS (Controller Transfer State) is active

or

    (b) The ATN (Attention) message is false and neither TACS (Talker Active State) is active nor SPAS (Series Poll Active State) is active.

You will note that these conditions are all outlined in the source diagram across the top. This is the area labeled:

(ATN ∧ $\overline{\text{CACS}}$ V $\overline{\text{CTRS}}$) V (ATN ∧ $\overline{\text{TACS}}$ V $\overline{\text{SPAS}}$)

These are all associated with the circle labeled SGNS.

The third state within the SH interface function is SDYS (Source Delay State). When in the SDYS State, the SH function is waiting for a message byte to settle on the interface signal line after the change during SGNS (Source Generate State) covered previously. It is also waiting for all the acceptor functions to indicate their readiness to accept the message byte. The SDYS is a state which also causes the SH function to send the DAV (Data Valid) message false.

The SH function must exit SDYS and enter:

(1) The STRS (Source Transfer State) only after T_1 if the RFD (Ready for Data) message is true.

(2) The SIWS (Source Idle Wait State) within t_2 if either:

    (a) The ATN (Attention) message is true and neither CACS (Controller Active State) is active nor CTRS (Controller Transfer State) is active

or

(b) The ATN (Attention) message is false and neither TACS (Talker Active State) is active nor SPAS (Serial Poll Active State) is active.

These conditions can be traced on the state diagram on the right side for case #1 above and the upper portion for case #2.

The fourth state within the SH interface is the STRS (Source Transfer State). In this state, the SH function indicates to the AH (Acceptor Handshake) function that it is continuously sending a valid message byte. This state also results in the SH function sending the DAV (Data Valid) message true.

The SH function must exit STRS (Source Transfer State) and enter:

(1) The SWNS (Source Wait for New Cycle State) if the DAC (Data Accepted) message is true.

(2) The SIWS (Source Idle Wait State) within t_2 if either:

    (a) The ATN (Attention) message is true and neither CACS (Controller Active State) nor CTRS (Controller Transfer State) is active

                            or

    (b) The ATN (Attention) message is false and neither TACS (Talker Active State) nor SPAS (Serial Poll Active State) is active.

These conditions for STRS exit can be seen on the state diagram on the lower side.

The fifth state within the SH interface is the SWNS (Source Wait for New Cycle State), mentioned above. In this state, the SH function is waiting for the device to start a new message generation cycle. This state is more versatile than the previous ones since it allows the SH function to send the DAV (Data Valid) message true or false.

The SH function must exit SWNS (Source Wait for New Cycle State) and enter:

(1) The SGNS (Source Generate State), if the mba message is false.

(2) The SIWS (Source Idle Wait State) within t_2 if either

    (a) The ATN (Attention) message is true and neither CACS (Controller Active State) nor CTRS (Controller Transfer State) is active

                            or

    (b) The ATN (Attention) message is false and neither TACS (Talker Active State) nor SPAS (Serial Poll Active State) is active.

The state description above can be seen in the state diagram in Figure 8.4. The area covered by SWNS is the lower center portion.

The final state for the SH function is the SIWS (Source Idle Wait State). In SIWS, the SH function is not active in the external message byte transfer process, but is active in the internal process of waiting for the device to start a new message generation cycle. This state allows a sequence of message byte transfers to be interrupted without loss of data over the interface while at the same time, the device may continue to prepare for the new (next) message byte generation cycle. The DAV (Data Valid) messages for this state must be sent passive false.

The SH function must exit SIWS and enter:

(1) The SIDS (Source Idle State) if the nba (new byte available) message is false.

(2) The SWNS (Source Wait for New Cycle State) if either:

    (a) The TACS (Talker Active State) is active

or

    (b) The SPAS (Serial Poll Active State) is active

or

    (c) The CACS (Controller Active State) is active

These conditions are shown on the lower left side of the state diagram in Figure 8.4.

The nba (new byte available) true message in the SH function indicates that the device has generated a new message byte and has made it available on the interface signal lines.

The nba message must become true only in SIDS (Source Idle State) or SGNS (Source Generate State). The nba may become false in any of the other four SH states.

One additional area in the SH function says that the device must not change the multiline message being sent in the SDYS (Source Delay State) and the STRS (Source Transfer State). The device may change the multiline message being sent in the SWNS (Source Wait for New Cycle State).

The second interface function is the ACCEPTOR HANDSHAKE (AH). This function provides a device with the capability to guarantee proper reception of remote multiline messages. When one or more AH functions are interlock handshaked, a guaranteed transfer of each message byte is obtained. An AH function may delay either the initiation or

348 HANDBOOK OF MICROWAVE TESTING

Figure 8.5/AH (Acceptor Handshake) state diagram.

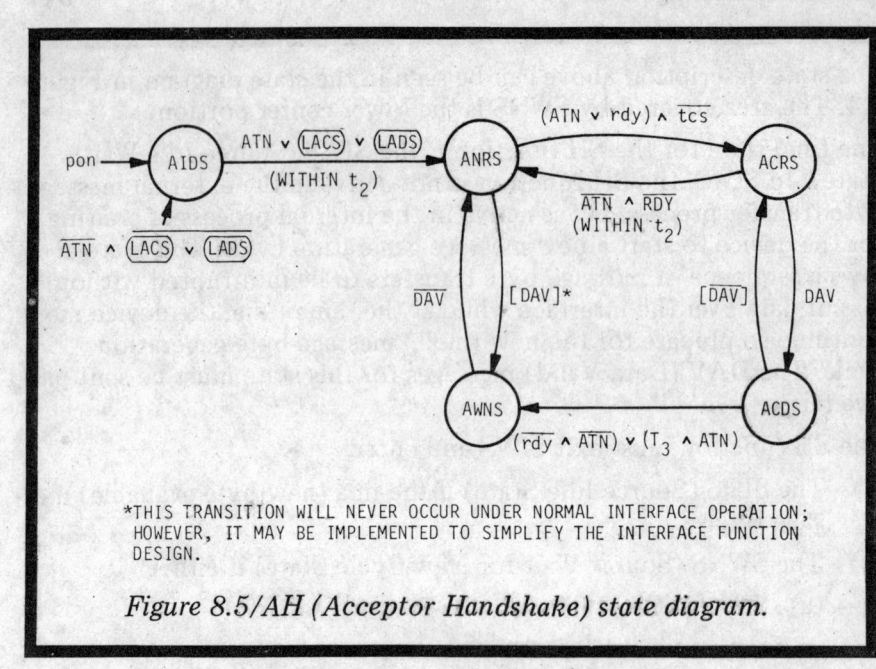

*THIS TRANSITION WILL NEVER OCCUR UNDER NORMAL INTERFACE OPERATION; HOWEVER, IT MAY BE IMPLEMENTED TO SIMPLIFY THE INTERFACE FUNCTION DESIGN.

termination of multiline message transfer until they are prepared to continue with the transfer process. The function uses the DAV (Data Valid), RFD (Ready for Data), and the DAC (Data Accepted) messages to effect each message byte transfer.

TABLE 8.4
AH Labels

Messages		Interface States	
pon	= power on	AIDS	= acceptor idle state
rdy	= ready for next message	ANRS	= acceptor not ready state
tcs	= take control synchronously	ACRS	= acceptor ready state
ATN	= attention	ACDS	= accept data state
DAV	= data valid	AWNS	= acceptor wait for new cycle state
		LADS	= listener addressed state (L function)
		LACS	= listener active state (L function)

TABLE 8.5
AH Message Outputs

STATE	Remote Message Sent RFD	DAC	Device Function (DF) Interaction
AIDS	(T)	(T)	DF cannot receive remote multiline messages
ANRS	F	F	DF cannot receive remote multiline messages
ACRS	(T)	F	DF cannot receive remote multiline messages
AWNS	F	(T)	DF cannot receive remote multiline messages
ACDS	F	F	DF can receive remote multiline messages if LACS is active

The AH function operates in accordance with the state diagram shown in Figure 8.5. Table 8.4 is an explanation of the AH labels used in the state diagram. Table 8.5 is the message outputs for the Acceptor Handshake (AH) function.

Within the AH function there are five distinct and interactive states, as can be seen on the state diagram. They are: AIDS (Acceptor Idle State), ANRS (Acceptor Not Ready State), ACRS (Acceptor Ready State), ACDS (Accept Data State), and AWNS (Acceptor Wait for New Cycle State). We will look at all five states and see how they operate on the state diagram.

The AH function powers on (pon) in the AIDS (Acceptance Idle State) state. In this state, the RFD (Ready for Data) and DAC (Data Accepted) messages must be sent passive true (T).

The AH function must exit AIDS and enter ANRS (Acceptor Not Ready State) within t_2 if either:

(1) The ATN (Attention) message is true

or

(2) LACS (Listener Active State) is active

or

(3) LADS (Listener Addressed State) is active

These conditions can be seen at the left side of the state diagram.

The second state within the AH function is ANRS (Acceptor Not Ready State). This state tells the AH function to indicate to the interfact that it is not yeat prepared internally to continue with the handshake cycle. In contrast to the AIDS state, the ANRS sends the RFD (Ready for Data) and DAC (Data Accepted) messages false.

The AH functions must exist ANRS and enter:

(1) The ACRS (Acceptor Ready State) if the tcs (take control synchronously) message is false and either

   (a) the ATN (Attention) message is true

or

   (b) the rdy (ready for next message) message is true.

(2) The AIDS (Acceptor Idle State) if the ATN (Attention) message is false and neither:

   (a) The LADS (Listener Addressed State) is active

nor

   (b) LACS (Listener Active State) is active.

(3) The AWNS (Acceptor Wait for New Cycle State) if the DAV (Data Valid) message is true (this transition will never occur under normal operation. It may, however, be implemented if simplification of the interface function is desired.)

These states can be seen in Figure 8.5 in the upper center of the diagram.

The third state within the AH interface function is the ACRS (Acceptor Ready State). When in this state, the AH function indicates to the interface that it is ready to receive multiline messages. The DAC (Data Accepted) message is sent false, F, in this state and the RFD (Ready for Data) message sent passive true (T).

The AH function must exit ACRS (Acceptance Ready State) and enter:

(1) The ACDS (Accept Data State) if the DAV (Data Valid) message is true.

(2) The AIDS (Acceptor Idle State) if the ATN (Attention) message is false and neither:

(a) The LADS (Listener Addressed State) is active

nor

(b) LACS (Listener Active State) is active.

(3) The ANRS (Acceptor Not Ready State) within t_2 if both the ATN (Attention) and the rdy (ready for next message) messages are false.

The conditions presented above are illustrated in Figure 8.5, in the upper right corner of the state diagram. You should be able to justify all of the states and changes of states from all the material previously covered.

The fourth state within the AH (Acceptance Handshake) interface function is ACDS (Accept Data State). This state allows the AH function to indicate to the SH (Source Handshake) function that it should maintain a valid message byte. This is the only state in which multiline messages on the DIO (Data Input/Output) signal lines are valid. The ACDS provides indications to both the interface functions and the device functions as follows:

- To the interface function, it indicates that an interface message is present and valid if the ATN (Attention) message is true.
- To the device function, it indicates that a device dependent message is present and valid if LACS (Listener Active State) is active.

As in the case of ANRS (Acceptor Not Ready State), the ACDS state requires that the DAC (Data Accepted) and RFD (Ready for Data) messages be sent false, F.

The AH function must exit the ACDS and enter:

(1) The AWNS (Acceptor Wait for New Cycle State) if either:

(a) The ATN (Attention) message is true and a period of T_3 has elapsed

or

(b) The ATN (Attention and rdy (ready for next message) messages are both false.

(2) The AIDS (Acceptance Idle State) if the ATN (Attention) message is false and neither:

(a) The LADS (Listener Ad dressed State) is active

nor

(b) LACS (Listener Active State) is active.

352 HANDBOOK OF MICROWAVE TESTING

Figure 8.6/T state diagram.

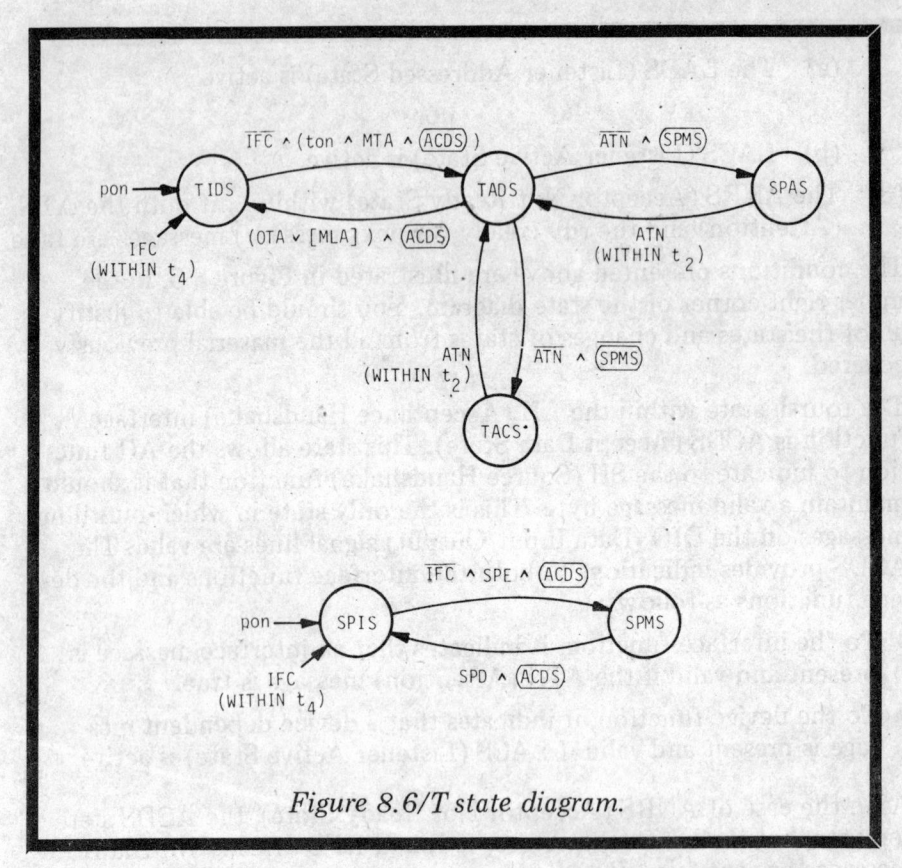

(3) The ACRS (Acceptor Ready State) if the DAV (Data Valid) message is false (This transition can occur only when the controller takes control).

These conditions are illustrated in the lower right corner of the AF function state diagram in Figure 8.5.

The final state within the AH interface function is AWNS (Acceptor Wait for New Cycle State). In this state, the AH function indicates that it has received a multiline message byte. It requires that the RFD (Ready for Data) message be sent false, F, and the DAC (Data Accepted) message be sent positive true, (T).

The AH function must exit the AWNS and enter:

(1) The ANRS (Acceptor Not Ready State) if DAV (Data Valid) is false.

(2) The AIDS (Acceptor Idle State) if the ATN (Attention) message is false and neither:

Automatic Testing

Figure 8.7/TE state diagram.

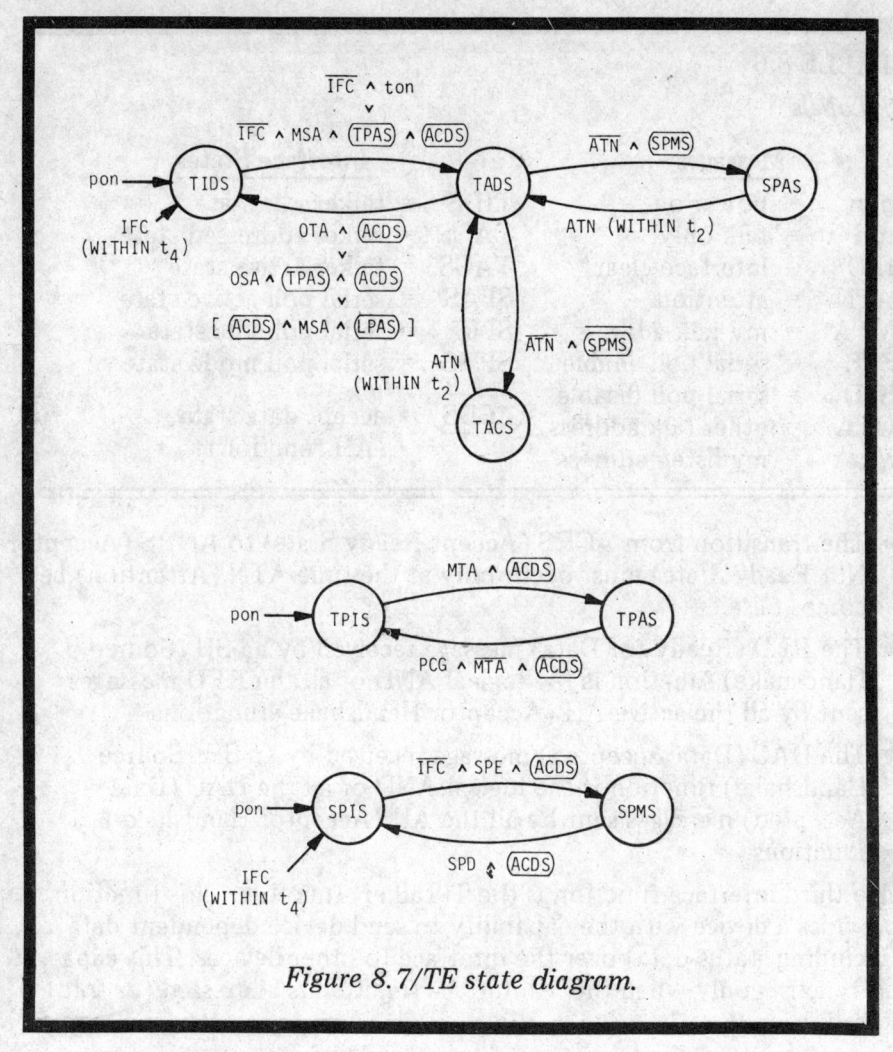

(a) The LADS (Listener Addressed State) is active

nor

(b) LACS (Listener Active State) is active.

These conditions can be seen in the lower center of the state diagram in Figure 8.5.

Additional requirements and guidelines for the AH functions are:

- The local message rdy (ready for next message) must not become false during the ACRS (Accept Ready State).

TABLE 8.6
T Labels

Messages			Interface States		
pon	=	power on	TIDS	=	talker idle state
ton	=	talk only	TADS	=	talker addressed state
IFC	=	interface clear	TACS	=	talker active state
ATN	=	attention	SPAS	=	serial poll active state
MTA	=	my talk address	SPIS	=	serial poll idle state
SPE	=	serial poll enable	SPMS	=	serial poll mode state
SPD	=	serial poll disable	(ACDS)	=	accept data state (AH function)
OTA	=	other talk address			
MLA	=	my listen address			

- The transition from ACRS (Accept Ready State) to ANRS (Acceptor Not Ready State) must occur only at the time ATN (Attention) becomes false.
- The RFD (Ready for Data) message received by an SH (Source Handshake) function is the logical AND of all the RFD messages sent by all the active AH (Acceptor Handshake) functions.
- The DAC (Data Accepted) message received by an SH (Source Handshake) function is the logical AND of all the DAC (Data Accepted) messages sent by all the AH (Acceptor Handshake) functions.

The third interface function is the T(Talker) function. This function provides a device with the capability to send device dependent data (including status data) over the interface to other devices. This capability exists only when the T interface function is addressed (or told) to talk.

There are two versions of the T interface function: the normal T (Talker) function which uses a 1 byte address, and the TE (Extended Talker) function which uses a 2 byte address. In all other respects, the capabilities of both versions are the same and they will be described together throughout this section. *Note:* Although the two functions are very similar, only one of the two T functions need be implemented in a specific device.

The state diagram for the T (Talker) interface function is shown in Figure 8.6. The state diagram for the TE (Extended Talker) function is shown in Figure 8.7. The T labels are shown in Table 8.6; the T and TE message outputs are in Table 8.7; and TE labels in Table 8.8.

TABLE 8.7
T or TE Message Outputs

T STATE	Qualifier	Remote Messages Sent*			Device Function (DF) Interaction
		Multiline	END	RSQ	
TIDS		(NUL)	(F)	(F)	DF not allowed to send messages
TADS		(NUL)	(F)	(F)	DF not allowed to send messages
TACS		DAB† or EOS	T or F	(F)	DF can send DAB or END messages
SPAS	APRS Inactive	STB†	(F)	F	DF can send one STB message
SPAS	APRS Active	STB†	(F)	T	DF can send one STB message

*Messages enabled by the T function originating within the device functions.

TABLE 8.8

TE Labels

Messages		Interface States	
pon = power on		TIDS = talker idle state	
ton = talk only		TADS = talker addressed state	
IFC = interface clear		TACS = talker active state	
ATN = attention		SPAS = serial poll active state	
MTA = my talk address		TPIS = talker primary idle state	
OTA = other talk address		TPAS = talker primary addressed state	
OSA = other secondary address		SPIS = serial poll idle state	
		SPMS = serial poll mode state	
PCG = primary command group		(ACDS) = accept data state (AH function)	
SPE = serial poll enable		(LPAS) = listener primary addressed state (L function)	
SPD = serial poll disable			
MSA = my secondary address			

The first state within the T(Talker) function is TIDS (Talker Idle State). In this state, the T or TE function are inactive. That is, they are not sending data or status bytes. The state powers on the T or TE function.

When the function is in the TIDS state, the END and RQS (Request Service) messages are sent passive false (F), while the NUL message is sent passive true, (T).

The T function must exit the TIDS (Talker Idle State) when the IFC (Interface Clear) message is false and enter the TADS (Talker Addressed State) if either:

(1) The MTA (My Talk Address) message is true and ACDS (Accept Data State) in the AH function is active

or

(2) The ton (Talk Only) message is true.

The TE (Extended Talker) must exit TIDS (Talker Idle State) and enter TADS (Talker Ad dressed State) if the IFC (Interface Clear) message is false and either:

(1) The MSA (My Secondary Address) message is true and ACDS (Accept Data State) is active, and the TPAS (Talker Primary Address State) is active

or

Automatic Testing

(2) The ton (Talk Only) message is true.

You can also see each of these descriptions on the state diagrams in Figures 8.6 and 8.7.

The second state in the T or TE interface function is TADS (Talker Addressed State). In this state, the T function has received its necessary talk address, but has not sent any data or status bytes. It is, however, prepared to send these bytes whenever commanded to. The TE function has received both its primary and secondary talk addresses, but is in the same condition as the T function — prepared to send data or status bytes, but has not sent them.

In TADS (Talker Addressed State), the END or RQS (Request Service) messages must be sent passive false, (F). The NUL message must be sent passive true, (T).

The T (Talker) function must exit TADS (Talker Addressed State) and enter:

(1) The TACS (Talker Active State) if the ATN (Attention) message is false and the SPMS (Serial Pole Mode State) is inactive.

(2) The SPAS (Serial Poll Active State) if the ATN (Attention) message is false and the SPMS (Serial Poll Mode State) is active.

(3) The TIDS (Talker Idle State) if either:

    (a) The OTA (Other Talk Address) message is true and ACDS (Accept Data State) is active

or

    (b) The MLA (My Listen Address) is true and ACDS (Accept Data State) is active

or

    (c) Within t_4 if the IFC (Interface Clear) message is true.

The TE (Extended Talker) function must exit TADS and enter:

(1) The TACS (Talker Active State) if the ATN (Attention) message is false and SPMS (Serial Poll Mode State) is inactive.

(2) The SPAS (Serial Poll Active State) if the ATN (Attention) message is false and SPMS is active.

(3) The TIDS (Talker Idle State) if either:

    (a) The OTA (Other Talk Address) message is true and ACDS (Accept Data State) is active

or

(b) The OSA (Other Secondary Address) message is true and TPAS (Talker Primary Addressed State) and ACDS (Accept Data State) are active

or

(c) The MSA (My Secondary Address) message is true and both the LPAS (Listener Primary Addressed State) and ACDS are active

or

(d) Within t_4 if the IFC (Interface Clear) message is true.

Notes: The MLA message is optional in the T function and the TE function.

The state descriptions presented above can be seen in the center of the state diagrams in Figures 8.6 and 8.7.

The third state in the T or TE functions is TACS (Talker Active State). In this state, the T or TE function is allowed to transfer the DAB (Data Byte) message and END (if used) from the device function to the interface signal lines. The SH (Source Handshake) determines *when* this transfer occurs with the message content being determined solely by the device function.

Messages that can be sent by the device function when the T or TE function is in SACS are DAB (Data Byte), EOS (End of String), or END.

The T and TE function must exit TACS (Talker Active State) and enter:

(1) The TADS (Talker Addressed State) within t_2 if the ATN (Attention message is true.

(2) The TIDS (Talker Idle State) within t_4 if the IFC (Interface Clear) message is true.

The fourth state in the T or TE function is SPAS (Serial Poll Active State). When in the SPAS state, the T (Talker) or TE (Extended Talker) function is able to transfer a single status message from the device function to the interface signal lines. The control of this transfer operation is by means of the SH (Source Handshake) function. If the controller does not assert ATN (Attention) after the first message transfer, the device can repeat the combined message bytes of STB (Status Byte) and RQS (Request Service).

When in SPAS, the END (end) message must be sent passive false (F), and the RQS (Request Service) message is sent true if APRS (Affirmative Poll Response State) is active, or false if APRS is negative.

Automatic Testing

The T function or TE function must exit SPAS (Serial Poll Active State) and enter:

(1) The TADS (Talker Addressed State) within t_2 if the ATN (Attention) message is true.

(2) The TIDS (Talker Idle State) within t_4 if the IFC (Interface Clear) message is true.

These conditions can be seen at the right hand side of state diagrams in Figure 8.6 and 8.7.

The fifth state within the T or TE function is the SPIS (Serial Poll Idle State). In this state, the T or TE function powers on and does not provide a remote sending capability.

The T or TE function must exit the SPIS and enter SPMS (Serial Poll Mode State) if the SPE (Serial Poll Enable) message is true, the ACDS (Accept Data State) is active, and the IFC (Interface Clear) message is false.

These conditions for SPIS exit are shown on the state diagrams in Figures 8.6 and 8.7 at the lower side of the diagram.

The sixth state within the T or TE function is SPMS (Serial Poll Mode State) mentioned above. This state allows the T or TE function to participate in the serial poll. This state (like the SPIS) does not provide a remote message sending capability.

The T or TE function must exit SPMS and return to SPIS (Serial Poll Idle State) if either:

(1) The SPD (Serial Poll Disable) message is true and the ACDS (Accept Data State) is active

or

(2) Within t_4 if the IFC (Interface Clear) message is true.

These conditions can also be seen at the bottom of the state diagrams shown in Figures 8.6 and 8.7.

The seventh state within the TE function only is TPIS (Talker Primary Idle State). In this state, the TE function is prevented from responding to a secondary address and is only allowed to recognize the primary address. TPIS also does not provide a remote message sending capability.

The TE function must exit TPIS and enter TPAS (Talker Primary Addressed State) if the MTA (My Talk Address) message is true and ACDS (Accept Data State) is active. The condition for TPIS exit are illustrated in the state diagram in Figure 8.7.

The eighth, and final, state within the TE function is TPAS (Talker Primary Addressed State). This is the state in which the TE function is able to recognize and respond to its secondary address. As in previous states, the TPAS does not provide a remote message sending capability.

The TE function must exit TPAS and return to TPIS (Talker Primary Idle State) if the PCG (Primary Command Group) message is true, the MTA (My Talk Address) message is false, and ACDS (Accept Data State) is active.

These conditions are also illustrated in Figure 8.7, which is the TE state diagram.

There are many allowable subsets for the T and TE functions. The list of these subsets is presented in the appendix at the end of this book.

Each device which includes a T or TE function must provide some means by which the talk address (or secondary address) can be changed in the field by a user of the device. This address may be recognized as MTA (My Talk Address) or MSA (My Secondary Address).

This interruption of a device sending data by transitions in and out of TACS (Talker Active State) should have no effect on the format of the output data.

Each device which includes the ton (talk only) message must be provided with a manual switch to generate the talk only condition.

The next logical function to follow a T (Talker) function is the fourth function of the interface standard — the L (Listener) function. This function enables a device to receive data over the interface from other devices. That is, it gives a device the capability to be able to listen for data and messages coming from other devices. The device can listen for information only when the L (Listener) function is addressed (or told) to listen.

As in the case of the T (Talker) function, there are two versions of the L (Listener) function — one with an address extension and one without. Also similar to the T function, the normal L function uses a one byte address. The function with the extended address is called LE (Extended Listener) and uses a 2 byte address.

Only one of the two alternative functions need be used on a specific device.

The state diagram for the L (Listener) function is shown in Figure 8.8 with the LE (Extended Listener) function in Figure 8.9. Table 8.9

Automatic Testing

Figure 8.8/L (Listener) state diagram.

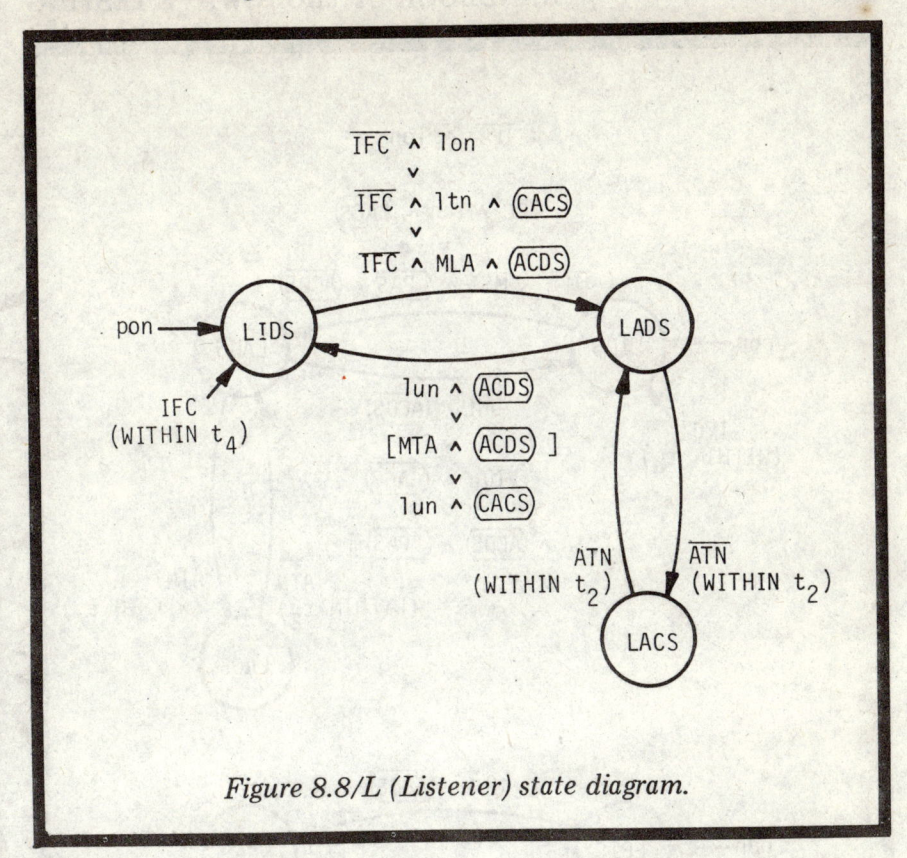

shows all of the labels used in the L function, Table 8.10 lists the L or LE messages, and Table 8.11 lists the LE labels used on the state diagrams.

TABLE 8.9
L Labels

Messages		Interface States	
pon	= power on	LIDS	= listener idle state
ltn	= listen	LADS	= listener addressed state
lun	= local unlisten	LACS	= listener active state
lon	= listen only		
IFC	= interface clear	(ACDS)	= accept data state
ATN	= attention		(AH function)
UNL	= unlisten		
MLA	= my listen address	(CACS)	= controller active state
MTA	= my talk address		(C function)

Figure 8.9/LE (Extended Listener) state diagram.

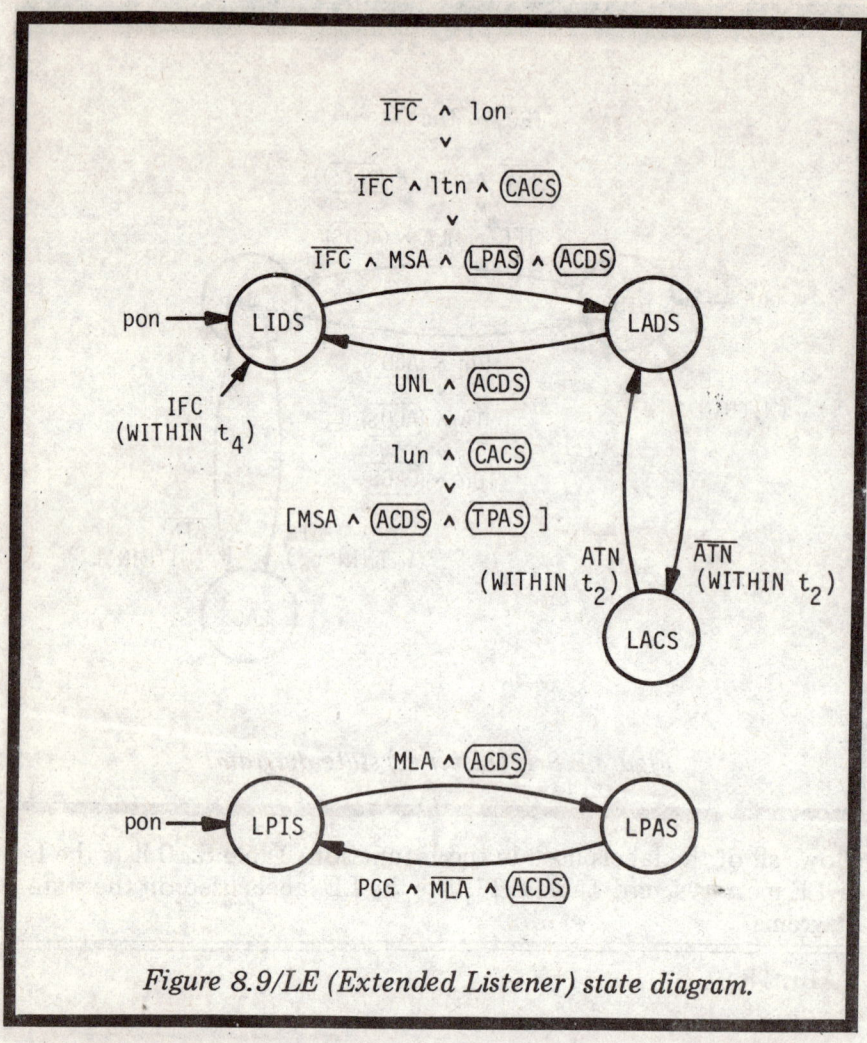

TABLE 8.10

L or LE Message Outputs

L or LE States	Remote Messages Sent	Device Function (DF) Interaction
LIDS	None	device not addressed to receive
LADS	None	device not addressed to receive
LACS	None	DF can receive one device dependent message byte each time ACDS is active

TABLE 8.11
LE Labels

Messages			Interface States		
pon	=	power on	LIDS	=	listener idle state
ltn	=	listen	LACS	=	listener active state
lun	=	local unlisten	LADS	=	listener addressed state
lon	=	listen only	LPIS	=	listener primary idle state
IFC	=	interface clear	LPAS	=	listener primary addressed state
ATN	=	attention			
UNL	=	unlisten	(ACDS)	=	accept data state (AH function)
MLA	=	my listen address			
PCG	=	primary command group	(CACS)	=	controller active state (C function)
MSA	=	my secondary address	(TPAS)	=	talker primary addressed state (T function)

The first state in the L or LE function is LIDS (Listener Idle State). This state powers on the L or LE function. As might be expected, there is no transfer of messages in this state since it does just what the name says — idles.

The L (Listener) function must exit LIDS and enter LADS (Listener Addressed State) if the IFC (Interface Clear) message is false and

(1) The MLA (My Listener Address) message is true and ACDS (Accept Data State) is active

or

(2) The lon (Listen Only) message is true

or

(3) The ltn (Listener) message is true and CACS (Controller Active State) is active.

The LE (Extended Listener) must exit LIDS and enter LADS if the IFC (Interface Clear) message is false and either:

(1) The MSA (My Secondary Address) message is true and the ACDS (Accept Data State) and LPAS (Listener Primary Addressed STate) are both active

or

(2) The lon (Listen Only) message is true

or

(3) The ltn (Listen) message is true and CACS (Controller Active State) is active.

These conditions and state changes can be seen on the state diagrams in Figures 8.8 and 8.9.

The second state in the L or LE function is LADS (Listener Addressed State). When in this state, the L function has received its listen address and is ready to transfer device dependent messages. It has, however, not actively engaged in any transfer in this state. The LE function in this state has received both its primary and secondary listen addresses and is also prepared to transfer device dependent messages. But, just as in the case of the L function, it to is not engaged in any transfer.

The L function must exit LADS and enter:

(1) The LACS (Listener Active State) within t_2 if the ATN (Attention) message is false.

(2) The LIDS (Listener Idle State) if either:

    (a) The UNL (Unlistener) message is true and ACDS (Accept Data State) is active

or

    (b) The lun (Local Unlistener) message is true and ACDS is active

or

    (c) The MTA (My Talker Address) message is true and ACDS is active

or

    (d) Within t_4 if the IFC (Interface Clear) is true.

The LE function must exit LADS and enter:

(1) The LACS (Listener Active State) within t_2 if the ATN (Attention) message is false.

(2) The LIDS (Listener Idle State) if either:

    (a) The UNL (Unlistener) message is true and ACDS (Accept Data State) is active

or

    (b) The lun (Local Unlistener) message is true and CACS (Controller Active State) is active

or

(c) The MSA (My Secondary Address) message is true and both TPAS (Talker Primary Addressed State) and ACDS (Accept Data State) are active

or

(d) Within t_4 if the IFC (Interface Clear) message is true.

Examination of the state diagrams in Figures 8.8 and 8.9 graphically illustrate the condition and state changes discussed above.

The third state within the L or LE function is LACS (Listener Active State). In this state, the L or LE function is able to transfer any device dependent messages to the device function that are received on the interface signal lines. Messages that can be transferred are:

- DAB (Data Byte)
- EOS (End of String)
- STB (Status Byte)
- END (End)
- RQS (Request Service)

This message transfer is controlled by the AH (Acceptor Handshake) interface function.

The L or LE function must exit LACS and enter:

(1) The LADS (Listener Addressed State) within t_2 if the ATN (Attention) message is true.

(2) The LIDS (Listener Idle State) within t_4 if the IFC (Interface Clear) message is true.

These reversals back in state can be seen in the state diagrams in Figures 8.8 and 8.9.

The fourth state within the LE interface function only is LPIS (Listener Primary Idle State). This state allows the LE function to recognize its primary address only. There is no response to a secondary address. This is a power on state for the LE function.

The LE function must exit LPIS and enter LPAS if the MLA (My Listen Address) message is true and ACDS (Accept Data State is active.

A state similar to, and associated with LPIS is LPAS (Listener Primary Addressed State). This state allows the LE function to both recognize and respond to its secondary address. Neither LPIS nor LPAS provide a remote message sending capability.

The LE function must exit LPAS and return to LPIS if the PCG (Primary Command Group) message is true, the MLA (My Listen Address) is false, and ACDS is active.

Figure 8.10/SRQ state diagram.

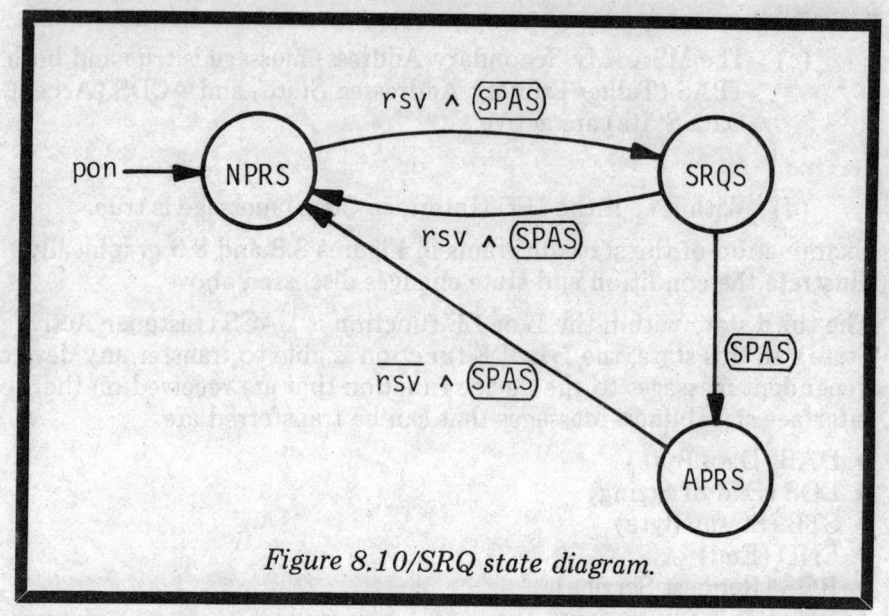

Each device which includes the L (Listen) function or LE (Extended Listener) function must provide a means by which the listen address and secondary address, which are recognized as MLA (My Listen Address) and MSA (My Secondary Address), can be changed in the field by users of the device.

Each device which includes the lon (Listen Only) message must be provided with a manual switch to generate the listen only condition.

The fifth function within the interface system is the SR (Service Request) interface function. This function allows a device to request service from the controller in charge of the interface. It also synchronizes the value of the service request bit of the status byte present during a serial poll so that the SRQ (Service Request) messages can be removed from the interface once this bit is received true by the controller.

TABLE 8.12

SR Labels

Messages		Interface States	
pon	= power on	NPRS	= negative poll response state
rsv	= request service	SRQS	= service request state
		APRS	= affirmative poll response state
		(SPAS)	= serial poll active state (T function)

Automatic Testing

The Service Request state diagram is illustrated in Figure 8.10, with SR labels for the diagram shown in Table 8.12 and SR message outputs in Table 8.13.

TABLE 8.13
SR Message Outputs

SRQ STATE	Remote Message Sent SRQ	Device Function Interaction
SPRS	(F)	None
SRQS	T	None
APRS	(F)	None

The power on state for the SR function is NPRS (Negative Poll Response State). In this state, the SRQ (Service Request) message must be sent passive false, (F).

The SR function function must exit NPRS and enter SRQS (Service Request State) at any time the rsv (Request Service) message is true and SPAS (Serial Poll Active State) is not active.

The second SR function state is SRQS (Service Request State). In this state, the SR function is continually telling the interface that it is requesting service. This state also requires that the SRQ (Serial Request) message be sent true.

The SR function must exit SRQS and enter:

(1) The NPRS (Negative Poll Response State) if the rsv (Request Service) message is false and SPAS (Serial Poll Active State) is not active. That is, $\overline{rsv} \wedge \overline{(SPAS)}$

(2) The APRS (Affirmative Poll Response State) if SPAS is active.

These changes in state (Forward and Reverse) are identified on the SR state diagram in Figure 8.10.

The third state in the SR function is APRS (Affirmative Poll Response State). In this state, the SR function requires service, but does not actively request it over the interface. The state requires that the SRQ (Service Request Message) be sent passive false, (F).

The SR function must exit APRS and return to the NPRS (Negative Poll Response State) at any time the rsv (request service) is false and SPAS (Serial Poll Active State) is not active. These conditions are identical to those required when the SR function exits the SRQS (Service Request State) and enters (or returns to) NPRS (Negataive Poll Re-

Figure 8.11/RL state diagram.

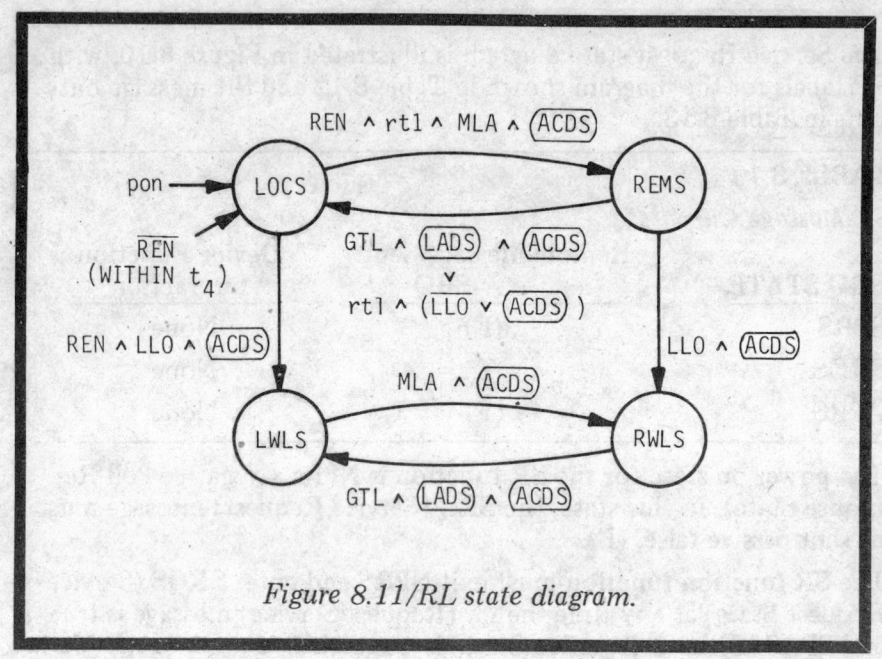

sponse State). The state diagram in Figure 8.10 illustrates this identity and all conditions for the SR function.

The SR function has really only one use — that is, for requesting service from a device. If more than one service request is needed, a separate SR function and rsv (request service) message must be used.

The sixth function within the interface system is RL (Remote Local). With this function, the device has a choice between two sources of input information. This choice is either from the front panel of a device (local) or from the interface (remote). The function indicates which one is to be used.

TABLE 8.14

RL Labels

Messages		Interface States	
pon	= power on	LOCS =	local state
rtl	= return to local	LWLS =	local with lockout state
REN	= remote enable	REMS =	remote state
LLO	= local lockout	RWLS =	remote with lockout state
GTL	= go to local		
MLA	= my listen address	ACDS =	accept data state (AH)
		LADS =	listener addressed state (L)

Automatic Testing

The state diagram for the RL function is shown in Figure 8.11. The labels used in the state diagram and the message outputs referred to are shown in Tables 8.14 and 8.15, respectively.

TABLE 8.15
RL Message Outputs

RL State	Remote Messages Sent	Device Function Interaction
LOCS	None	device is in "local control" mode
LWLS	None	device is in "local control" mode
REMS	None	device is in "remote control" mode
RWLS	None	device is in "remote control" mode

There are four possible states within the RL function (as can be seen in Figure 8.11):

- LOCS (Local State)
- LWLS (Local With Lockout State)
- REMS (Remote State)
- RWLS (Remote With Lockout State)

The LOCS (Local State) is the power on state for the RL interface functions. In this state, all of the local controls of a device function (front or rear panel) are operative. The device being used may store device dependent messages from the interface, but will not respond to them. The state does *not* provide a remote message sending capability.

The RL function must exit LOCS if the REN (Remote Enable) message is true and enter:

(1) The REMS (Remote State) if the rtl (return to local) message is false, the MLA (My Listen Address) message is true, and the ACDS (Accept Data State) is active.

(2) The LWLS (Local With Lockout State) if the universal coded command LLO (Local Lockout) is true and ACDS is active.

The LWLS (Local With Lockout State) permits all device functions to be operative. The device may store, but not respond to device dependent messages from the interface. That is, the rtl (return to local) message is ignored in LWLS. This state also does not provide a remote message sending capability.

The RL function must exit LWLS and enter:

(1) The RWLS (Remote With Lockout State) when MLA (My Listen Address) is true and ACDS (Accept Data State) is active.

(2) The LOCS (Local State) within t_4 if the REN (Remote Enable) message is false.

In REMS (Remote State), the local controls, discussed earlier, which have corresponding remote controls, are inoperative and the device functions are under the control of a remote device (such as a calculator). The REMS, as with all states of the RL function, does not provide a remote message sending capability.

The RL function must exit REMS and enter:

(1) The RWLS (Remote WithLockout State) if the LLO (Local Lockout) message is true and ACDS (Accept Data State) is active.

(2) The LOCS (Local State):

    (a) Within t_4 if the REN (Remote Enable) message is false

        or

    (b) The GTL (Go to Local) message is true and both ACDS and LADS (Listener Addressed States) are active

        or

    (c) The rtl (return to local) message is true and either the LLO (Local Lockout) is false or ACDS is inactive.

In RWLS (Remote With Lockout State), the REMS (Remote State) is duplicated, but the rtl (return to local) message is ignored.

The RL function must exit RWLS and enter:

(1) The LOCS (Local State) within t_4 if the REN (Remote Enable) message is false.

(2) The LWLS (Local With Lockout State) if the GTL (Go To Local) message is true and both LADS and ACDS are active.

You can see from the descriptions of the four states above that the ability of a device to either send device dependent messages over the interface or to receive and utilize device dependent messages not in conflict with locally available data is independent of the state which is active within the function.

When either REMS (Remote State) or RWLS (Remote With Lockout State) is active, the device being used must ignore any local controls and respond only to data on the interface. A device may, however, be designed to either:

(1) Use current local control settings until they are overridden by future input data on the interface

        or

Automatic Testing

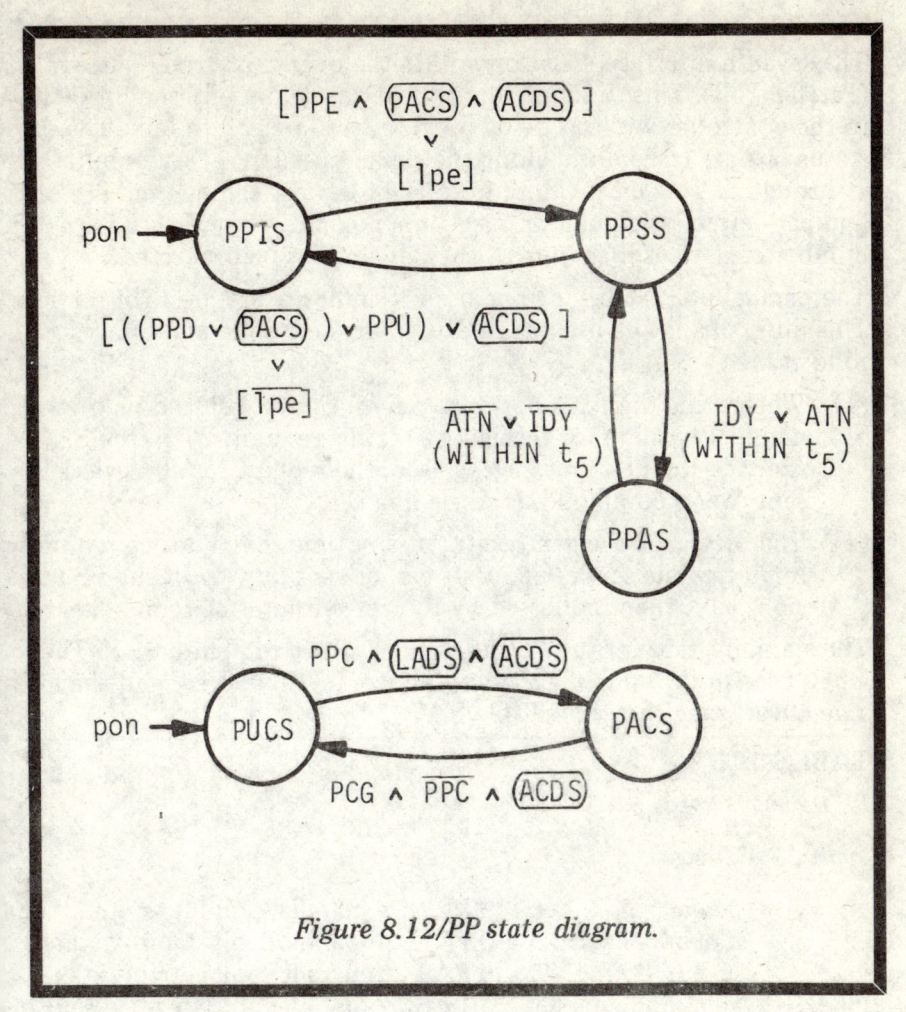

Figure 8.12/PP state diagram.

(2) Use data previously received on the interface.

Also, when REMS or RWLS become inactive, the device must respond to future use of local controls and ignore data received on the interface. In this case, a device may be designed to either:

(1) Use the most recent input data until overridden by local controls

or

(2) Use the current values of all controls at the time that REMS (Remote State) or RWCS (Remote With Lockout State) becomes inactive.

The RL (Remote Local) interface function plays a very important part in automatic microwave testing.

The seventh interface function within the overall interface system is PP (Parallel Poll). This function allows a device to present one bit of status to the controller without being previously addressed to talk. The device status bits are transferred along the signal lines DIO (Data Input/Output 1 through 8. This means that up to eight devices can be used if a one-line-per-device assignment is used. It is possible, however, to have any number of devices if sharing of the DIO lines is incorporated.

The parallel poll facility can be used to indicate a request for service. This differs from the use of the SRQ (Service Request) messages as follows:

(1) When using the PP function, a device can be assigned its own individual bus line for receiving a service request. When the SRQ (Service Request) message is used, the request for all devices is sent over a common bus single line.

(2) The SRQ message can be sent at any time that a device requires service, while service requests can be sent by way of the PP function only when requested by the current interface controller.

The state diagram for the PP function is shown in Figure 8.12. The labels used in the state diagram are shown in Table 8.16, and the message outputs are shown in Table 8.17.

TABLE 8.16
PP Labels

Messages

pon	= power on	PPIS	= parallel poll idle state
ist	= individual status	PPSS	= parallel poll standby state
		PPAS	= parallel poll active state
lpe	= local poll enabled	PUCS	= parallel poll unaddressed to configure state
ATN	= attention		
IDY	= identify	PACS	= parallel poll addressed to configure state
PPE	= parallel poll enable		
PPD	= parallel poll disable		
PPC	= parallel poll configure	(ACDS)	= accept data state (AH function)
PCG	= primary command group	(LADS)	= listener addressed state (L function)
PPU	= parallel poll unconfigure		

Automatic Testing

TABLE 8.17
PP Message Outputs

PP STATE	Qualifier	Remote Message Sent PPRN*	Device Function Interaction
PPIS		(F)	None
PPSS		(F)	None
PPAS	ist $\equiv$ S	T	None
PPAS	ist $\not\equiv$ S	(F)	None

*This column refers only to the specific message assigned by the device.

The first state in the PP (Parallel Poll) function is the same as those of other functions — PPIS (Parallel Poll Idle State). This is the power on state of the PP function. This state allows the function to respond to the parallel poll used by the interface controller. All PPR (Parallel Poll Response) messages in this state must be sent passive false, (F).

The PP function must exit PPIS and enter PPSS (Parallel Poll Standby State) if either:

(1) The PPE (Parallel Poll Enable) message is true and PACS (Parallel Poll Addressed to Configure State) and ACDS (Accept Data State) are active

or

(2) The lpe (local poll enable) message is true.

The second state in the PP function is PPSS (Parallel Poll Standby State). This allows the PP function to respond to parallel polls from the device when they occur. All PPR (Parallel Poll Response) messages must be sent passive false, (F).

The PP function must exit PPSS and enter:

(1) The PPAS (Parallel Poll Active State) within t_5 if the IDY (Identify) and ATN (Attention) messages are true (this means a parallel poll is in progress).

(2) The PPIS (Parallel Poll Idle State) if:

   (a) The lpe (local poll enable) message is false

or

(b) The PPD (Parallel Poll Disable) message is true and both PACS (Parallel Poll Addressed to Configure State) and ACDS (Accept Data State) are active

or

(c) The PPU (Parallel Poll Unconfigure) message is true and ACDS is active.

The third state of the PP function is where the function responds to the parallel poll from the interface controller, PPAS (Parallel Poll Active State).

In this state, one of the response messages must be sent true if, and only if, the value of the ist (individual status) message is equal to the value of the sense (s) bit received as part of the most recent Parallel Poll Enable (PPE) command. The PPR (Parallel Poll Response) message to be sent must be the one specified by the bits P1 through P3 listed in Table 8.18. This table lists the response messages (PPR) which are specified by the combination of P1 through P3. Any other PPR messages must be sent passive false, (F).

TABLE 8.18

PPR Message Specified by Each of the Combinations of Values P1 Through P3

Bits Received With Most Recent PPE Command			PPR Message Specified
P3	P2	P1	
0	0	0	PPR1
0	0	1	PPR2
0	1	0	PPR3
0	1	1	PPR4
1	0	0	PPR5
1	0	1	PPR6
1	1	0	PPR7
1	1	1	PPR8

The PP function must exit PPAS and return to PPSS (Parallel Poll Standby State) within t_5 if either the IDY (Identify) or ATN (Attention) message is false. This indicates that the parallel poll is over.

Automatic Testing

```
(DCL ∨ [SDC ∧ (LADS)]) ∧ (ACDS)
    DCIS  →  DCAS
(DCL ∨ [SCD ∧ (LADS)]) ∨ (ACDS)
```

Figure 8.13/DC state diagram.

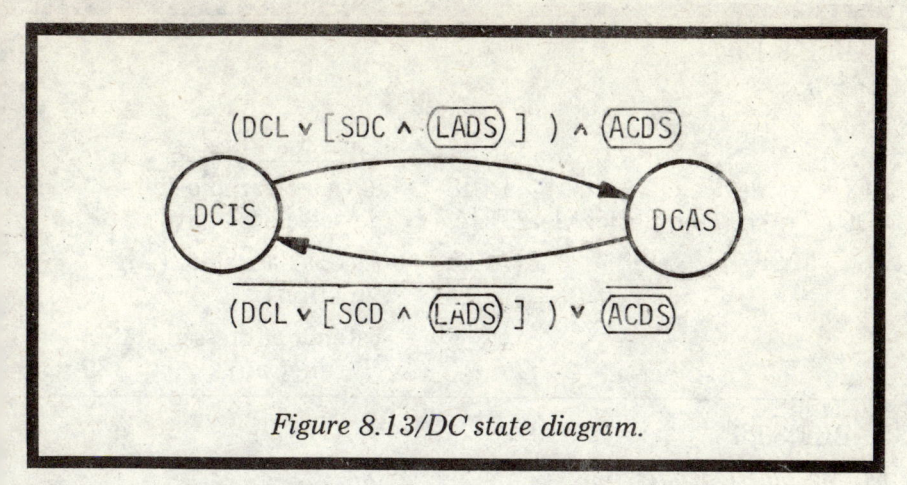

There are two additional states within the PP function: one unaddressed to Configure and one Addressed to Configure.

The first state is the PUCS (Parallel Poll Unaddressed to Configure State) which is also a power on for the PP function. In this state, the PP function ignores any PPE (Parallel Poll Enable) or PPD (Parallel Poll Disable) messages which are received over the interface. This is where the "Unaddressed to Configure" terminology comes into play.

The PUCS state moves on to the PACS (Parallel Poll Addressed to Configure State) if the PPC (Parallel Poll Configure) message is true and both LADS (Listener Addressed State) and ACDS (Accept Data State) are active. As might be expected, the PACS allows the PP function to act on PPE or PPD (Able and Disable) messages over the interfaces. If the enable message is received, the bits S, Pl, P2 and P3 must be preserved by the PP function. (Neither PUCS or PACS provide a remote message sending capability.)

The PP functions return to the PUCS from PACS when the PCG (Primary Command Group) message is true, the PPC (Parallel Poll Configure) message is false, and ACDS (Accept Data State) is active. Figure 8.12 shows the transition from PUCS to PACS and back and the requirements to make the transitions.

The eighth function within the interface system is DC (Device Clear). This function allows a device to be cleared either individually or as part of a group of devices. The group of devices may be either a subset or all addressed devices in one system. Figure 8.13 is the state diagram for the DC function, with Table 8.19 listing the DC labels and Table 8.20 the DC message outputs.

TABLE 8.19
DC Labels

Messages		Interface States
DCL = device clear	DCIS =	device clear idle state
SDC = selected device clear	DCAS =	device clear active state
	(ACDS) =	accept data state (AH function)
	(LADS) =	listener addressed state (L function)

TABLE 8.20
DC Message Outputs

DC State	Remote Message Sent	Device Function (DF) Interaction
DCIS	None	normal device function operation
DCAS	None	DF should return to a known fixed state

The first DC state is DCIS (Device Clear Idle State). In this state, the DC function is inactive and does not provide a remote message sending capability. The DC function leaves this dormant state and enters the active state (DCAS) if the ACDS (Accept Data State) is active, and either:

(1) The DCL (Device Clear) message is true

or

(2) The SDC (Selected Device Clear) message is true and LADS (Listen Addressed State) is active.

The DCAS (Device Clear Active State), as mentioned above, is the state which the DC function enters following the dormant DCIS condition. In this state, the DC function sends an internal message to the device function (or functions) which causes all of the devices to be cleared. (The DCAS also does not provide a remote message sending capability.)

The DC function returns to the DCIS (Device Clear Idle State) if either ACDS (Accept Data State) is inactive or neither:

(1) The DCL (Device Clear) message is true

nor

Automatic Testing

Figure 8.14/DT state diagram.

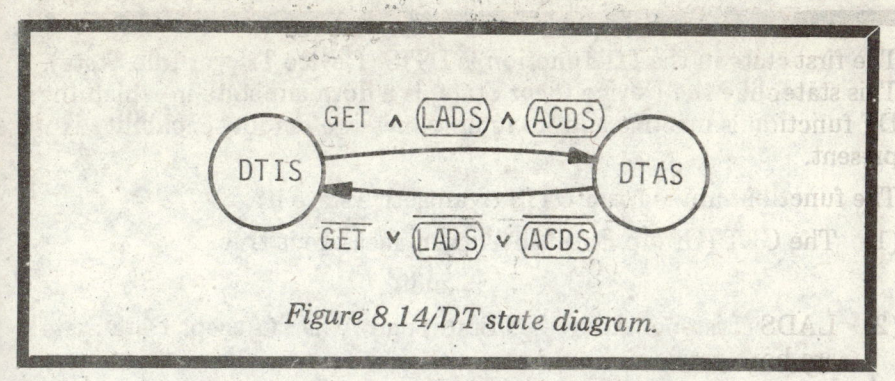

(2) The SDC (Selected Device Clear) message is true and LADS (Listener Addressed State) is active.

The DCAS portion of the DC function effects only device functions along the interface and has no effect on interface functions. Normal usage of this function would be to use it to place device functions in a power on condition.

The ninth function within the interface system is the DT (Device Trigger) function. This function provides a device with the capability to have its basic operation started individually or as part of a group of devices. The state diagram is relatively basic and is shown in Figure 8.14. Table 8.21 lines the DT labels while Table 8.22 provides the DT message outputs.

TABLE 8.21
DT Labels

Messages		Interface States	
GET	= group execute trigger	DTIS	= device trigger idle state
		DTAS	= device trigger active state
		$\overline{ACDS}$	= accept data state (AH function)
		$\overline{LADS}$	= listener addressed state (L function)

TABLE 8.22
DT Message Outputs

DT STATE	Remote Messages Sent	Device Function (DF) Interaction
DTIS	None	normal DF operation
DTAS	None	DF should start performing triggered operation

The first state in the DT function is DTIS (Device Trigger Idle State). This state, like the Device Clear State, is a dormant state in which the DT function is inactive and no remote message sending capability is present.

The function moves from DTIS to an active state if:

(1) The GET (Group Execute Trigger) message is true

and

(2) LADS (Listener Addressed State) and ACDS (Accept Data State) are both active.

The DTAS (Device Trigger Active State) is the state in which the DT function sends an internal message to the device function which causes it to begin its basic operations. DTAS, once again, does not provide a remote message sending capability.

The DT function must return to the DTIS (Device Trigger Idle State) if either:

(1) The GET (Group Execute Trigger) message is false

or

(2) LADS (Listener Addressed State) is inactive

or

(3) ACDS (Accept Data State) is inactive.

The DTAS indicates that a device is starting to perform its operation. This operation should begin immediately after the DTAS becomes active and remains in operation until its assigned tasks are complete. Any state transitions before completion of the operation of a device should be ignored until the operations are complete. When complete, a new operation in response to the next DTAS can begin.

The final function within the interface system is probably the most important — the C (Controller) function. This function provides a device with the capability to send device addresses, universal commands, and addressed commands to other devices over the interface. It also provides the capability to conduct parallel polls to determine which device requires service.

A C (Controller) function can exercise its capabilities only when it is sending the ATN (Attention) message over the interface. This is where the ATN (Attention) message comes from what we have been referring to throughout the description of the individual functions.

If more than one device on the interface has a C interface function,

Automatic Testing

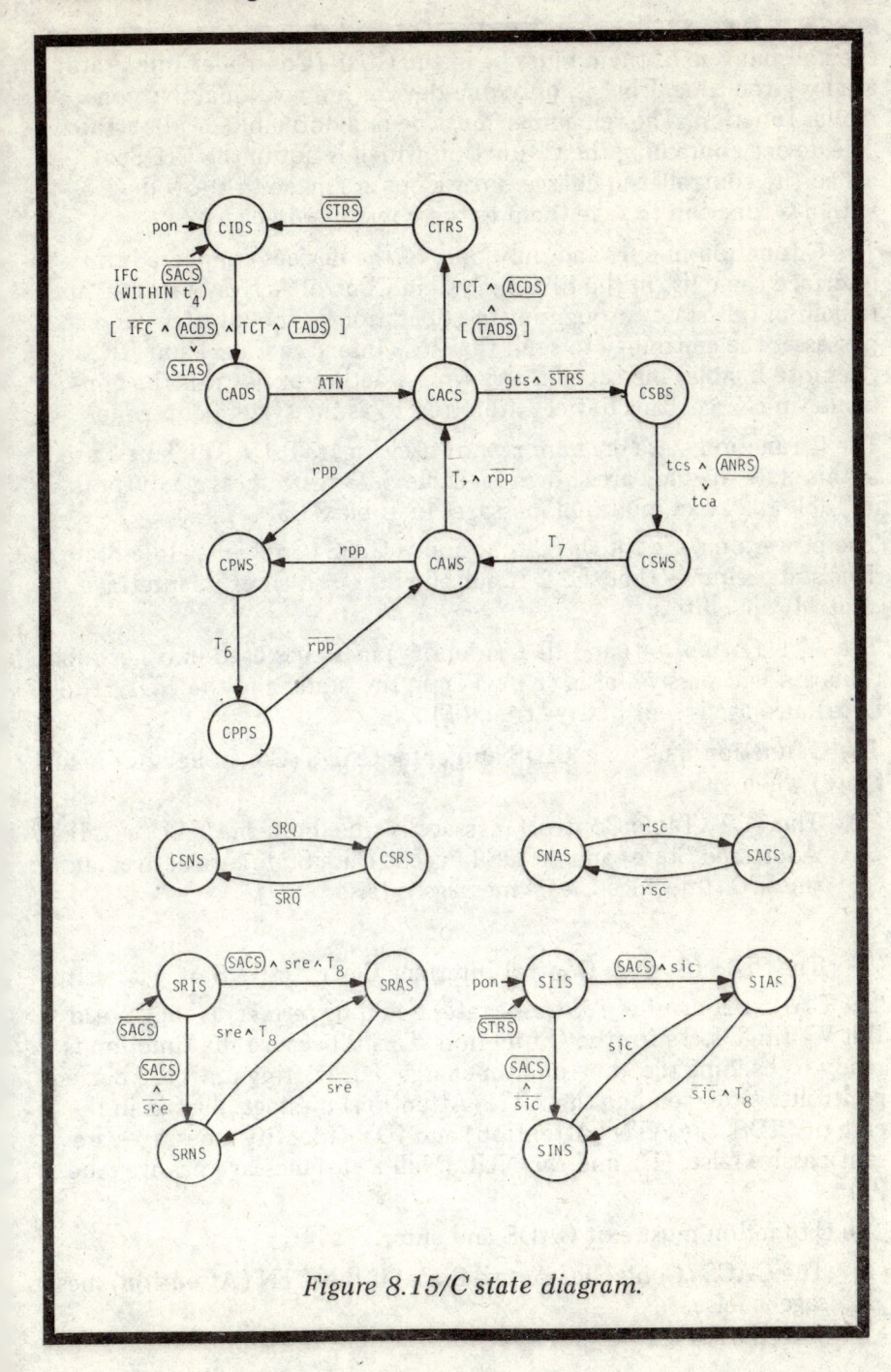

Figure 8.15/C state diagram.

then all but one of them must be in the CIDS (Controller Idle State) at any given time. That is, only one device can have an active controller function. The remainder must be in a dormant (or idle) state. The device containing the C function which is not in the CIDS is called the controller-in-charge. Provisions are made to allow devices with a C function to take them as the controller-in-charge.

The C function in one, and only one, of the devices connected to an interface can exist in the SACS (System Control Active State). It must remain in this state throughout the operation of the interface and so possesses the capability to send the IFC (Interface Clear) and REN (Remote Enable) messages at any time whether or not it is the controller-in-charge. This device is referred to as the system controller.

The C function state diagram is shown in Figure 8.15. The labels used in this state diagram are shown in Table 8.23, with message outputs in Table 8.24 and multiline messages in Table 8.25.

The power on state for the C function is CIDS (Controller Idle State). This state requires that the C function relinquish all of its interface control capabilities.

The ATN (Attention) and IDY (Identify) messages used in other functions are sent passive false, (F), when in this state, and the NUL (Null Byte) message is sent passive true, (T).

The C function must exit CIDS and enter CADS (Controller Addressed State) when either:

(1) The TCT (Table Control) message is true, both the TADS (Talker Addressed State) and ACDS (Accept Data State) are active, and the IFC (Interface Clear) message is false

or

(2) The SIAS (System Control Interface Clear Active State) is active.

The CADS (Controller Address State) could be termed as the "Ready, But Waiting" state for the C function. This is because the function is ready to become the controller-in-charge, but waiting until the current controller stops sending the ATN (Attention) message. Just as in the case of CIDS, the ATN (Attention) and IDY (Identify) messages are sent passive false, (F), and the NUL (Null Byte) message passive true, (T).

The C function must exit CADS and enter:

(1) The CACS (Controller Active State) if the ATN (Attention) message is false.

Automatic Testing

TABLE 8.23

C Labels

Messages			Interface States		
pon	=	power on	CIDS	=	controller idle state
rsc	=	request system control	CADS	=	controller addressed state
			CTRS	=	controller transfer state
rpp	=	request parallel poll	CACS	=	controller active state
			CPWS	=	controller parallel poll wait state
gts	=	go to standby			
tca	=	take control asynchronously	CPPS	=	controller parallel poll state
			CSBS	=	controller standby state
tcs	=	take control synchronously	CAWS	=	controller active wait state
			CSWS	=	controller synchronous wait state
sic	=	send interface clear			
sre	=	send remote enable	CSRS	=	controller service requested state
IFC	=	interface clear			
ATN	=	ATTENTION	CSNS	=	controller service not requested state
TCT	=	take control			
			SNAS	=	system control not active state
			SACS	=	system control active state
			SRIS	=	system control remote enable idle state
			SRNS	=	system control remote enable not active state
			SRAS	=	system control remote enable active state
			SIIS	=	system control interface clear idle state
			SINS	=	system control interface clear not active state
			SIAS	=	system control interface clear active state
			(ACDS)	=	accept data state (AH function)
			(ANRS)	=	acceptor not ready state (AH function)
			(STRS)	=	source transfer state (SH function)
			(TADS)	=	talker addressed state (T function)

TABLE 8.24
C Message Outputs

C STATE	REMOTE MESSAGES SENT ATN	IDY	MULTILINE	DEVICE FUNCTION (DF) INTERACTION
CIDS	(F)	(F)	(NUL)	DF must not send interface messages
CADS	(F)	(F)	(NUL)	DF must not send interface messages
CACS	T	F		DF can send interface messages
CPWS	T	T	(NUL)	DF must not send interface messages
CPPS	T	T	(NUL)	DF can receive PPR message
CSBS	F	(F)	(NUL)	DF must not send interface messages
CSWS	T	T or (F)	(NUL)	DF must not send interface messages
CAWS	T	F	(NUL)	DF must not send interface messages
CTRS	T	F	TCT	DF must finish sending TCT message

C STATE	REMOTE MESSAGES SENT IFC	DEVICE FUNCTION (DF) INTERACTION
SHS	(F)	None
SINS	F	None
SIAS	T	None

C STATE	REMOTE MESSAGES SENT REN	DEVICE FUNCTION (DF) INTERACTION
SRIS	(F)	None
SRNS	F	None
SRAS	T	None

C STATE	REMOTE MESSAGES SENT	DEVICE FUNCTION (DF) INTERACTION
CSNS	None	No service requests exist
CSRS	None	DF notified of request for service

Automatic Testing

TABLE 8.25
Multiline Messages

Universal Commands (Multiline)	Address
LLO	(LAD)
DCL	(TAD)
SPE	UNL
SPD	
PPU	

Addressed Commands	Secondary Commands
GET	(SAD)
GTL	PPD
PPC	PPE
SDC	
TCT	

(2) The CIDS, once again, within t_4 if the IFC (Interface Clear) message is true and SACS (System Control Active State) is not active.

The CACS (Controller Active State) allows the transfer of multiline interface messages from the device function to the interface signal lines. These messages include device addresses, universal commands, or addressed commands. You will recall that these messages were mentioned when explaining the overall C function previously. It is quite easy to see from this that the CACS is the main state of the C interface function.

The SH (Source Handshake) function determines when the device function may change the message content of the multiline message being sent. Message control itself is determined solely by the device function or functions.

The ATN (Attention) message must be sent continuously true and the IDY (Identify) message continuously false when CACS is active. Under these conditions any of the multiline messages in Table 8.25 may be sent by the device function.

The C function must exit CACS and enter:

(1) The CTRS (Controller Transfer State) if the TCT (Take Control) message is true; TADS (Talker Addressed State), which is optional, is inactive; and ACDS (Accept Data State) is active.

(2) The CPWS (Controller Parallel Poll Wait State) if the rrp (request parallel poll) message is true.

(3) Return to CIDS (Controller Idle State) within t_4 if the IFC (Interface Clear) message is true and SACS (System Control Active State) is inactive.

(4) CSBS (Controller Standby State) if the gts (go to standby) message is true and STRS (Source Transfer State) is inactive.

The CPWS (Controller Parallel Poll Wait State) is another "Ready and Waiting" state. It allows the C function to conduct a parallel poll and wait for the DIO (Data Input/Output) lines to settle.

In this state, the ATN (Attention) and IDY (Identify) must be sent true, T, and the NUL (Null Byte) message sent passive true, (T).

The C function must exit CPWS and enter:

(1) The CPPS (Controller Parallel Poll State) after a period of t_6 has elapsed.

(2) The CIDS (Controller Idle State) within t_4 if the IFC (Interface Clear) message is true and SACS (System Control Active State) is not active.

The state in which the C function is actively engaged in transferring PPR (Parallel Poll Response) message values to the device function is CPPS (Controller Parallel Poll State).

Once again, as in the CPWS, the ATN (Attention) and IDY (Identify) messages are sent true, T, and the NUL (Null Byte) message is passive true, (T).

The C function must exit CPPS and enter:

(1) The CAWS (Controller Active Wait State) if the rpp (request parallel poll) message is false.

(2) The CIDS (Controller Idle State) within t_4 if the IFC (Interface Clear) message is true and SACS (System Control Active State) is not active.

When the C function allows two or more devices to transfer device dependent messages, the function is in CSBS (Controller Standby State). When in this state, the ATN (Attention) message must be sent false, F, the IDY (Identify) sent passive false, (F), and the NUL (Null Byte) sent passive true, (T).

The C function must exit CSBS and enter:

(1) The CSWS (Controller Synchronous Wait State) if either:

Automatic Testing

    (a)   The tcs (Take Control Synchronously) message is true and ANRS (Accept Not Ready State) is active

or

    (b)   The tca (Take Control Asynchronously) message is true.

(2) The CIDS (Controller Idle State) within t_4 if the IFC (Interface Clear) message is true and the SACS (System Control Active State) is not active.

A state mentioned above was CSWS (Controller Synchronous Wait State). In this state, the function is preparing to enter the CAWS (Controller Active Wait State), but must wait a specified time, T_7, to be sure that the current active talker recognizes the ATN (Attention) message being sent. If this state were entered by way of the tcs (Take Control Synchronously) message, the device function must continue to send the ATN message true during this state. This causes the AH (Acceptor Handshake) function to continue sending the RFD (Ready for Data) message false. This would keep any data from being used and hold off the next data byte.

The ATN (Attention) message in the CSWS must be sent true, T; the IDY (Identify) message either active or passive false, F or (F); and the NUL (Null Byte) message passive true, (T).

The C function must exit CSWS and enter:

(1) The CAWS (Controller Active Wait State) after a period of T_7.

(2) The CIDS (Controller Idle State) within t_4 if the IFC (Interface Clear) message is true and SACS (System Control Active State) is not active.

The CAWS (Controller Active Wait State) is another "waiting" state of the controller function. In this state, the function is waiting for a period of T_9 before entering CACS (Controller Active State). This guarantees that the EOI (End or Identify) lines have settled and no device is falsely responding to what appears to be a parallel poll.

The ATN (Attention) message in CAWS is sent true, T; the IDY (Identify) message false, F; and the NUL (Null Byte) message sent passive true, (T).

The C function must exit CAWS and enter:

(1) The CACS (Controller Active State) if the rpp (Request Parallel Poll) message is false and a period T_9 has elapsed.

(2) The CPWS (Controller Parallel Poll Wait State) if the rpp (Request Parallel Poll) message is true.

(3) The CIDS (Controller Idle State) within t_4 if the IFC (Interface Clear) message is true and SACS (System Control Active State) is not active.

In the CTRS (Controller Transfer State), the C function is in the process of becoming idle since it is sending the TCT (Take Control) command to another device.

The ATN (Attention) message in CTRS is sent true, T; the IDY (Identify) message false, F; and the TCT (Take Control) message must be sent true, T.

The C function must exit CTRS and enter CIDS (Controller Idle State). We have said previously that this is what the function was in the process of doing, if either:

(1) The STRS (Source Transfer State) becomes inactive

or

(2) Within t_4, if the IFC (Interface Clear) message is true and SACS (System Control Active State) is not active.

In the CSRS (Controller Service Request State), the C function is notifying the device functions that at least one device on the interface is requesting service.

The function exits CSRS and enters CSNS (Controller Service Not Requested State) if the SRQ (Service Request) message is false. The CSNS enables the C function to notify the device functions that no device on the interface is requesting service. Neither CSRS nor CSNS provide a remote message sending capability. All messages are local.

As might be expected, the C function will exit CSNS and enter CSRS if the SRQ (Service Request) message is true.

The following eight states of the C function are all concerned with system control, as opposed to the first eleven, which dealt with the controller states themselves. The SNAS (System Control Not Active State) has the C function relinquishing all system control capabilities while the SACS (System Control Active State) allows the function to exercise system control. Neither state provides a remote message sending capability. Conditions for state exits are as follows:

- SNAS (System Control Not Active State) is exited and SACS (System Control Active State) is entered if the rsc (request system control) message is true.
- SACS is exited and SNRS entered if the rsc message is false.

Automatic Testing

The SIIS (System Control Interface Clear Idle State) allows the C function no capability to clear the interface. It is merely a power on state for the function. The SINS (System Control Interface Clear Not Active State) is simply not engaged in clearing the interface, while the SIAS (System Control Interface Clear Active State) has interface clearing as its only task.

The main message involved in these three states is the IFC (Interface Clear) message. The message should be sent as follows:

- SIIS — Passive False, (F)
- SINS — False, F
- SIAS — True, T

Conditions for state exists can be listed as follows:

- Exit SIIS, Enter SINS (System Control Interface Clear Not Active State) if the sic (Send Interface Clear) message is false.
- Exit SIIS, Enter SIAS (System Control Interface Clear Active State) if the sic message is true.
- Exit SINS, Enter SIAS (System Control Interface Clear Active State) if the local sic message is true.
- Exit SINS, Enter SIIS if SACS (System Control Active State) is not active.
- Exit SIAS, Enter SINS if the sic message is false and SIAS has been active for at least a period of T_8.

The SRIS (System Remote Enable Idle State) has absolutely no remote enable capability. The SRNS (System Control Remote Enable Not Active State) simply is not engaged in enabling remote operation, while the SRAS (System Control Remote Enable Active State) has as its main task the job of enabling remote operation.

The main message involved in these three states is the REN (Remote Enable) message. This message should be sent as follows:

SRIS — Passive False, (F)
SRNS — Passive False, (F)
SRAS — True, T

Conditions for state exits are as listed below:

- Exit SRIS (System Control Remote Enable Idle State), Enter SRNS (System Control Remote Enable Not Active State) if the sre (Send Remote Enable) message is false and SACS (System Control Active State) is active.

- Exit SRIS, Enter SRAS (System Control Remote Enable Active State) if the sre message is true, SACS is active, and SRIS has been active for at least a period of T_8.
- Exit SRNS (System Control Remote Enable Not Active State), Enter SRAS if the sre message is true for at least a period of T_8.
- Exit SRNS, Enter SRIS if SACS (System Control Active State) is not active.
- Exit SRAS, Enter SRNS if the sre message is false.
- Exit SRAS, Enter SRIS if SACS is not active.

For the remote messages covered in all of the functions above, there is a message coding used to distinguish them. These messages, whether sent or received by an interface function, are transferred by means of signal lines. Coding of these signal lines is the topic of discussion which follows.

Messages may be coded into the logical state (1 or 0) of one or more signal lines. The message derived from or sent as the logic state of only one signal line is referred to as a uniline message. The ATN (Attention) message is an example of a uniline message. A uniline message value is considered valid as soon as its corresponding logic state is detected.

A message derived from or sent as a combination of logic states of two or more signal lines is referred to as a multiline message. The DCL (Device Clear) message is an example of a multiline message. A multiline message is valid only within the context of the SH (Source Handshake) and AH (Acceptor Handshake) function. A transmitted multiline message is valid while the SH function is in the STRS (Source Transfer State). A received multiline message is valid while the AH function is in the ACDS (Accept Data State).

A message may be defined as a logical combination (AND, OR, or NOT) of other messages. The OTA (Other TAlk Address) message is an example of a message made up of a combination.

A message is sent by driving one or more specified lines (uniline or multiline) to a logic 1 or logic 0. Lines not specified as part of the message coding must not be driven.

A message is received by sensing one or more specified bus signal lines to determine the logic state of each line as 1 or 0. Lines not specified as part of the message coding are ignored.

All passive message values are transferred as 0 signal line states. This requires only the logic OR of signal line states to be performed on the interface.

Automatic Testing

A table of remote message coding is presented in Appendix H. This Table correlates the message value (true or false) to the bus signal line logic values (1 or 0) and vice-versa.

Each remote message entry in the table specifies both the encoding required to send the message and the decoding required to receive it.

True values of uniline messages are specified by the assignment of a specific logic state to a signal, while the true value of a multiline message is specified by the assignment of a unique set of logic states to the corresponding set of signal lines containing the message.

Note: The false value of a message is any combination of logic states other than the set that specifies the true value.

Each message entry is identified by type: U for uniline and M for multiline. The messages are also identified by class (1 of 7), according to the function it performs within either the interface or device function. The classes are:

AC — Addressed Command
AD — Address (Talk or Listen)
DD — Device Dependent
HS — Handshake
UC — Universal Command
SE — Secondary
ST — Status

The logic state a signal line may have is specified as a 0, 1, or X. These are defined as:

0 — logic zero (High State signal level)
1 — Logic one (Low State signal level)
X — Don't Care (for the coding of a received message)
X — Must not drive (for the coding of a transmitted message)

The codes used in the IEEE-488 standard can be used along with a convenient ISO-7 bit code. For the relationship between ISO codes and IEEE-488, the standard should be consulted directly.

8.3 Section 3 (Electrical Specifications)

The previous section described the functional aspects of the interface system. This section will describe the electrical specifications for systems to be used where the physical distance between devices is short and electrical noise is relatively low. All of the electrical specifications described are based on the use of TTL (transistor-transistor logic) theory.

You will recall from the previous section on message coding that a logic 1 and 0 is used in the coding of signal lines. The question that must be answered is: "What level is a logic 1 or 0?" The answer is shown below:

Logic State	Electrical Signal Level
0	Corresponds to $\geq$ +2.0 volts (Called *High State*)
1	Corresponds to $\leq$ 0.8 volts (Called *Low State*)

These high and low states are based on standard TTL levels for which the power source does not exceed +5.25 volts, referenced to logic ground.

Messages may be sent in either an active or passive manner, as described previously. All passive true (T) message transfer occurs in the high states (0) and must be carried on a signal line using open collector drivers. Messages such as SRQ (Service Request), NRFD (Not Ready For Data), and NDAC (Not Data Accepted) fall into the category of passive true. Open collector drivers or three-state drivers may be used to drive DIO 1-8 (Data Input/Output 1-8), DAV (Data Valid), IFC (Interface Clear), ATN (Attention), REN (Remote Enable), and EOI (End or Identify). The only exception within this group is that DIO 1-8 must use open collector drivers for parallel polling as described in the previous section. (The three-state driver is very useful where higher speed operation is required.)

The specifications for the above listed drivers are:

Low State: Output voltage (three-state or open collector) $<$ +0.4 volts at +48 mA sink current.

High State: Output Voltage (three-state) $\geq$ +2.4 volts at −5.2 mA; Output Voltage (open collector) +5.25 volts at +0.25 mA.

(All voltages are measured at the device connector between the signal line and logic ground).

The standard specifications for receivers with ordinary noise immunity are:

Low State: $\leq$ +0.8 volts
High State: $\geq$ +2.0 volts

For added noise immunity in the receiver, Schmitt-type circuits for the signal lines are used. The specification now becomes:

Hysteresis: V_t pos $-$ V_t neg $\geq$ +0.4 V
Low State: V_t neg $\geq$ +0.6 volts (recommend +0.8 volt)

Automatic Testing

Figure 8.16/DC load boundary specification.

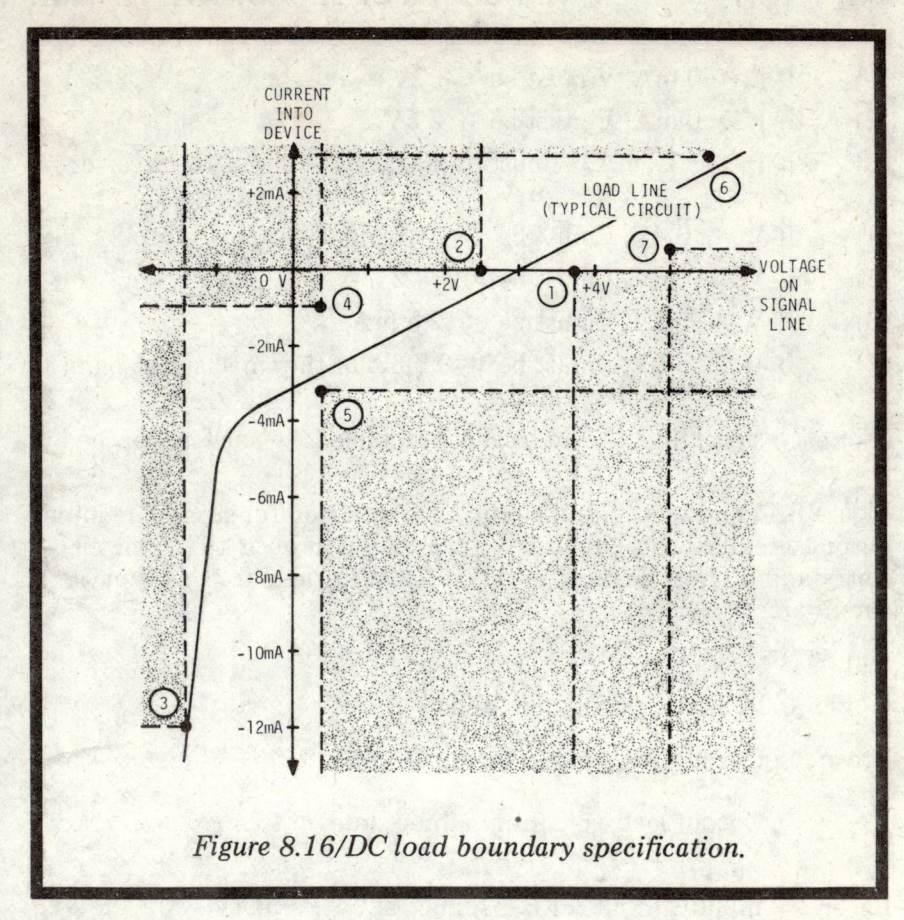

High State: V_t pos $\leq$ +2.0 volts
where V_t neg = Negative Threshold Voltage
                V_t pos = Positive Threshold Voltage

Each signal line (whether or not it is connected to a driver or receiver) must be terminated within the device by a resistive load. This load is used to establish a steady-state voltage when all drivers on the line are in a high impedance state. It is also used to maintain a uniform device impedance on the line and improve noise immunity.

The DC load characteristics of a device are affected by the driver and receiver circuits as well as the resistive termination and voltage clapping circuits used to limit the negative voltage swings on the signal lines. Each signal line interface within a device must have the following DC load characteristics and fall within the shaded areas of Figure 8.16.

(1)   If $I \leq 0$ mA, V must be < 3.7V
(2)   If $I \geq 0$ mA, V must be > 2.5V
(3)   If $I \geq -12$ mA, V must be > -1.5V (only if the receiver exists)
(4)   If $V \leq 0.4$V, I must be < -1.3 mA
(5)   If $V \geq 0.4$V, I must be > -3.2 mA
(6)   If $V \leq 5.5$V, I must be < 2.5 mA
(7)   If $V \geq 5.0$V, I must be > 0.7mA or the small signal Z must be $\leq$ 2k ohm at 1 MHz.

(For each signal line, the internal capacitance load shall not exceed 100 pF.)

Figure 8.17 shows a typical circuit configuration for signal line input/output circuits. This circuit is compatible with both TTL logic circuitry and discrete element devices. Specifications for this typical circuit are:

R_{L1} : 3K Ohm ± 5% (to V_{cc})

R_{L2} : 6.2K Ohm ± 5% (to ground)

Driver: Output leakage current (open collector) +0.25 mA @ Vo = +5.25V
          Output leakage current (three-state) ± 40 microamp max @ Vo = +2.4V

Receiver: Input Current −1.6 mA max @ Vo = +0.4V
          Input Leakage Current +40 microamps max @ Vo = +2.4V
          +1.0 microamp max @ Vo = 5.25V

V_{cc} : +5V ± 5%

The ground requirements for the interface are of great importance to the overall system performance. The overall shield of the interconnecting cable shall be grounded through a contact on the connector to earth (chassis) ground to minimize susceptibility to and generation of external noise.

> Warning: Devices should not be operated at significantly different earth ground potentials. The interface connection system may not be capable of handling the excessive ground currents produced.

Automatic Testing

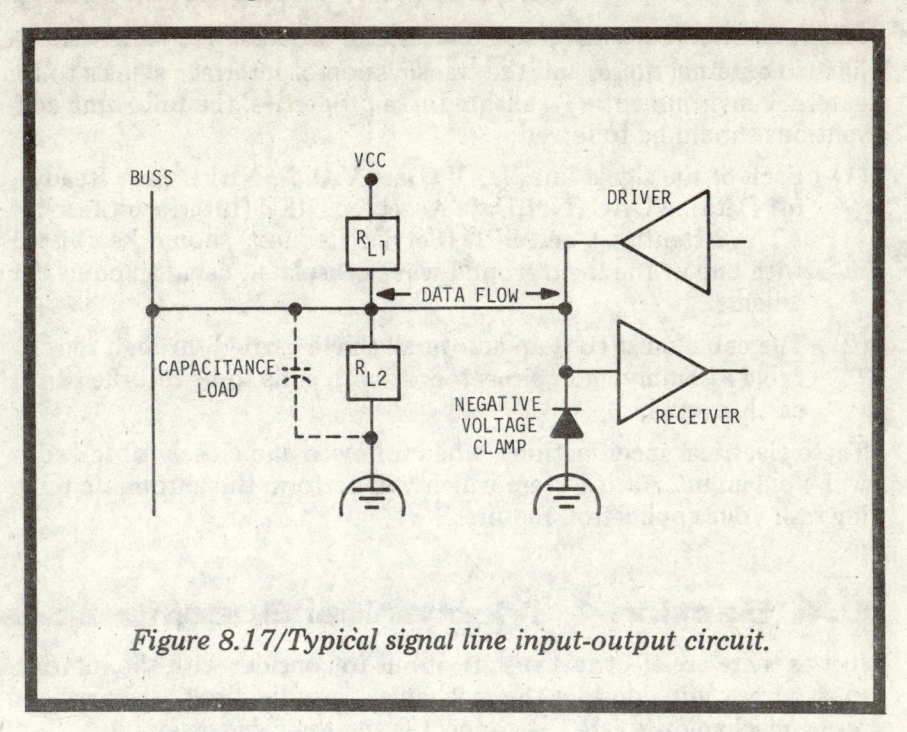

Figure 8.17/Typical signal line input-output circuit.

It is recommended that the ground return of the individual control and status signal lines be returned to logic ground at the logic circuit driver or receiver to minimize cross-talk interference.

The cable that is used for the interface also has specific electrical requirements in order for the system to operate properly. The maximum resistance of the conductor in the cable is required to be as follows:

(1) Each signal line - 0.14 ohm/meter
(2) Each signal line ground return - 0.14 ohm/meter
(3) Common logic ground return - 0.085 ohm/meter
(4) Overall shield - 0.0085 ohm/meter

The cable must contain at least 24 conductors, of which 16 will be for signal lines and the balance used for logic ground returns and overall shield. The capacitance (at 1 kHz) measured between any signal line and all other lines (signal, ground, and shield) connected to ground should be 150 pF/meter.

The shield itself shall contain a braid of 36 AWG wire with at least 85% coverage.

The construction of the cable shall be such that it minimizes the effect of cross-talk between signal lines, the susceptibility of the signal lines to external noise, and the transmission of interface signals to the external environment. To ensure these properties, the following conventions should be followed:

(1) Each of the signal lines DAV (Data VAlid), NRFD (not Ready for Data), NDAC (Not Data Accepted), IFC (Interface Clear), ATN (Attention), and SRQ (Service Request) should be twisted with one of the logic ground wires or isolated using an equivalent scheme.

(2) The cable must contain an overall shield carried through the cable assembly and connectors at both ends to be returned to earth ground.

These electrical specifications, when followed and closely observed, will yield an interface system which will perform the automatic testing task your application requires.

8.4 Section 4 (Mechanical Specifications)

Just as there are electrical specifications to consider with the interface, so also must you consider the mechanical specifications. Primary concerns, mechanically, are the connector and the cable assembly.

The following areas should be considered when selecting a connector for the interface system:

Voltage Rating:	200 Vdc
Current Rating:	5 Amps/Contact
Contact Resistance:	< 10 Ohm
Contact Material:	Gold over Copper
Insulation Resistance:	> 10G ohm
Number of Contacts:	24
Contact Surfaces:	Self-Wiping
Shell Shape:	Trapezoidal Polarization
Shell Material:	Corrosion Resistant Plating
Endurance:	≥ 1000 Insertions
Wire Termination Diameter:	Accept at least 0.35 mm^2 (Standard
External Dimensions:	See Figure 8.18(a)

Automatic Testing

A 46.8mm
B 15.5mm
C 2.16mm
D 4.29mm
E 3.33mm

(a) EXTERNAL DIMENSIONS

A DIMENSION	
36mm	OUTSIDE MOUNT
40.6mm	INSIDE MOUNT

46.8 ± 0.5mm
15.8mm TYP
4.75mm DIAMETER (INSIDE MOUNT) OR ADEQUATE FOR THREAD CLEARANCE (OUTSIDE MOUNT)

(b) CONNECTOR PANEL CUTOUT

14° 45'
3.5mm TYP
CONTACT 12 RECEPTACLE
CONTACT 1
11.94mm TYP
1.14mm TYP
2.16mm TYP
28.40mm TYP
36.12mm TYP

(c) ATTACHMENT

Figure 8.18/Connector dimensions.

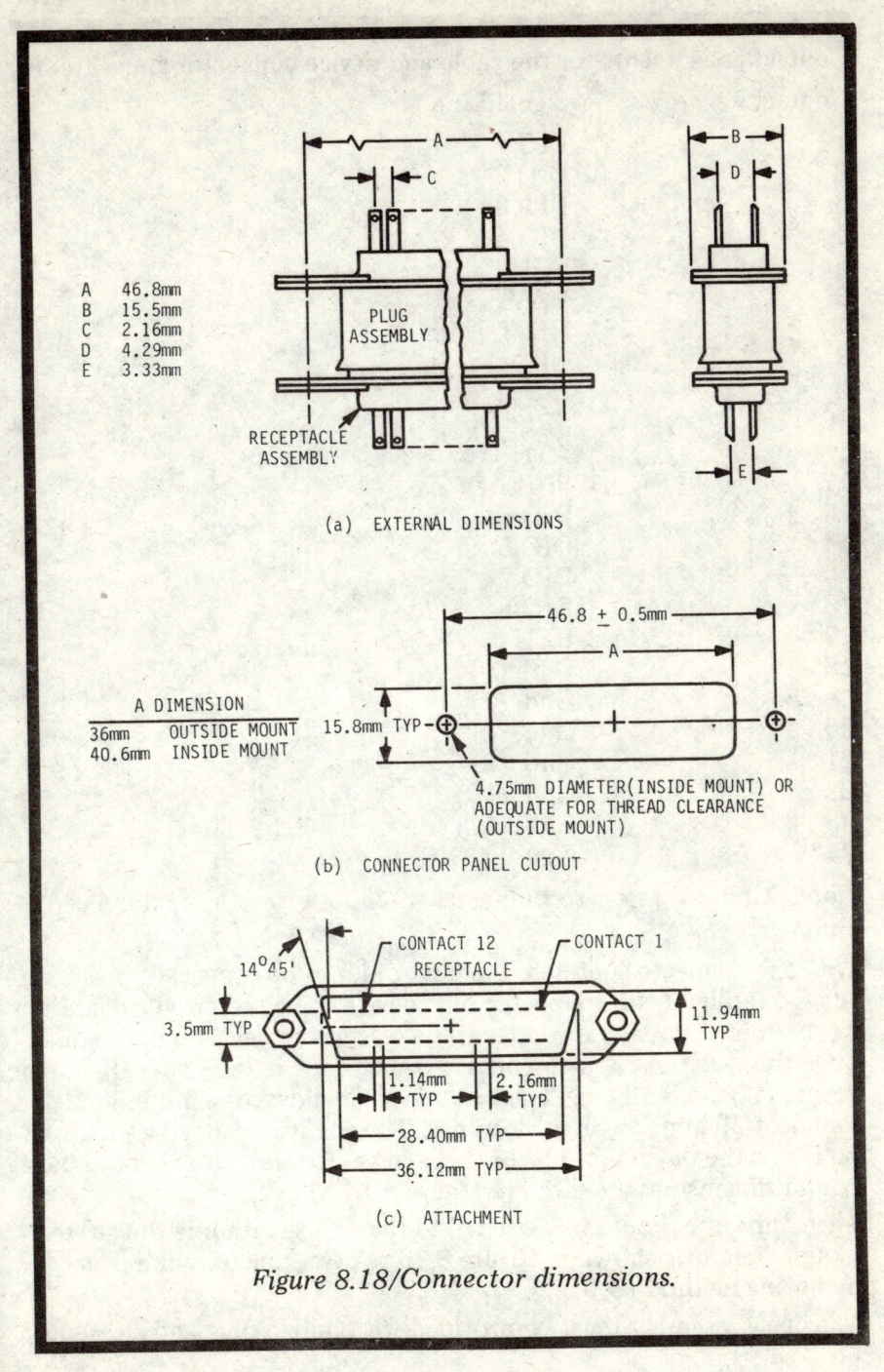

Contact assignments for the cable and device connector are as follows:

Contact	Signal Line
1	DIO 1
2	DIO 2
3	DIO 3
4	DIO 4
5	EOI
6	DAV
7	NRFD
8	NDAC
9	IFC
10	SRQ
11	ATN
12	SHIELD
13	DIO 5
14	DIO 6
15	DIO 7
16	DIO 8
17	REN
18	Gnd. (6)
19	Gnd. (7)
20	Gnd. (8)
21	Gnd. (9)
22	Gnd. (10)
23	Gnd. (11)
24	Gnd. LOGIC

Note: Gnd. (n) refers to the signal ground return of the referenced contact.

With the connector defined, the next order of business is to investigate the mounting of the connector on a device. The preferred orientation of the connector, as viewed from the rear of the device in its normal operating position, is with contact number one in the upper right hand corner (Figure 8.18(c)). The location shall allow for a minimum band radius of 40 mm for cable clearance. The connector may be mounted on either the outside or inside of the panel for which the typical panel cutout dimensions are given in Figure 8.18(b).

The connector should be attached to the device with one of the steel mount standoffs shown in Figure 8.19 as determined by the panel mounting method used.

The cable assembly must be provided with both a plug and a recept-

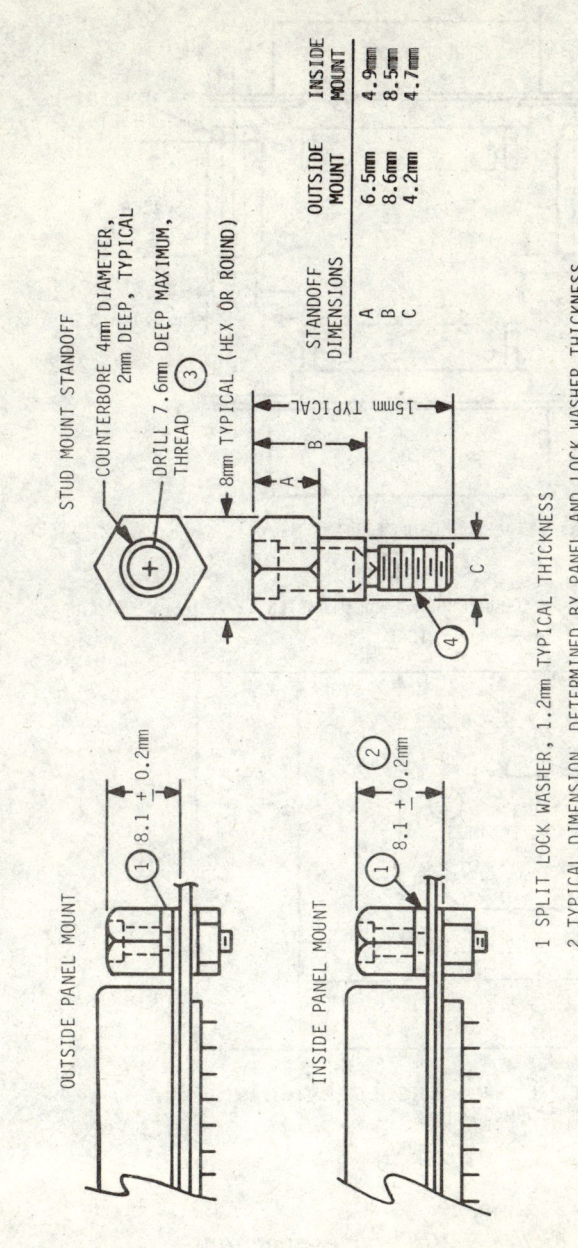

Figure 8.19/Mounting dimensions.

398 HANDBOOK OF MICROWAVE TESTING

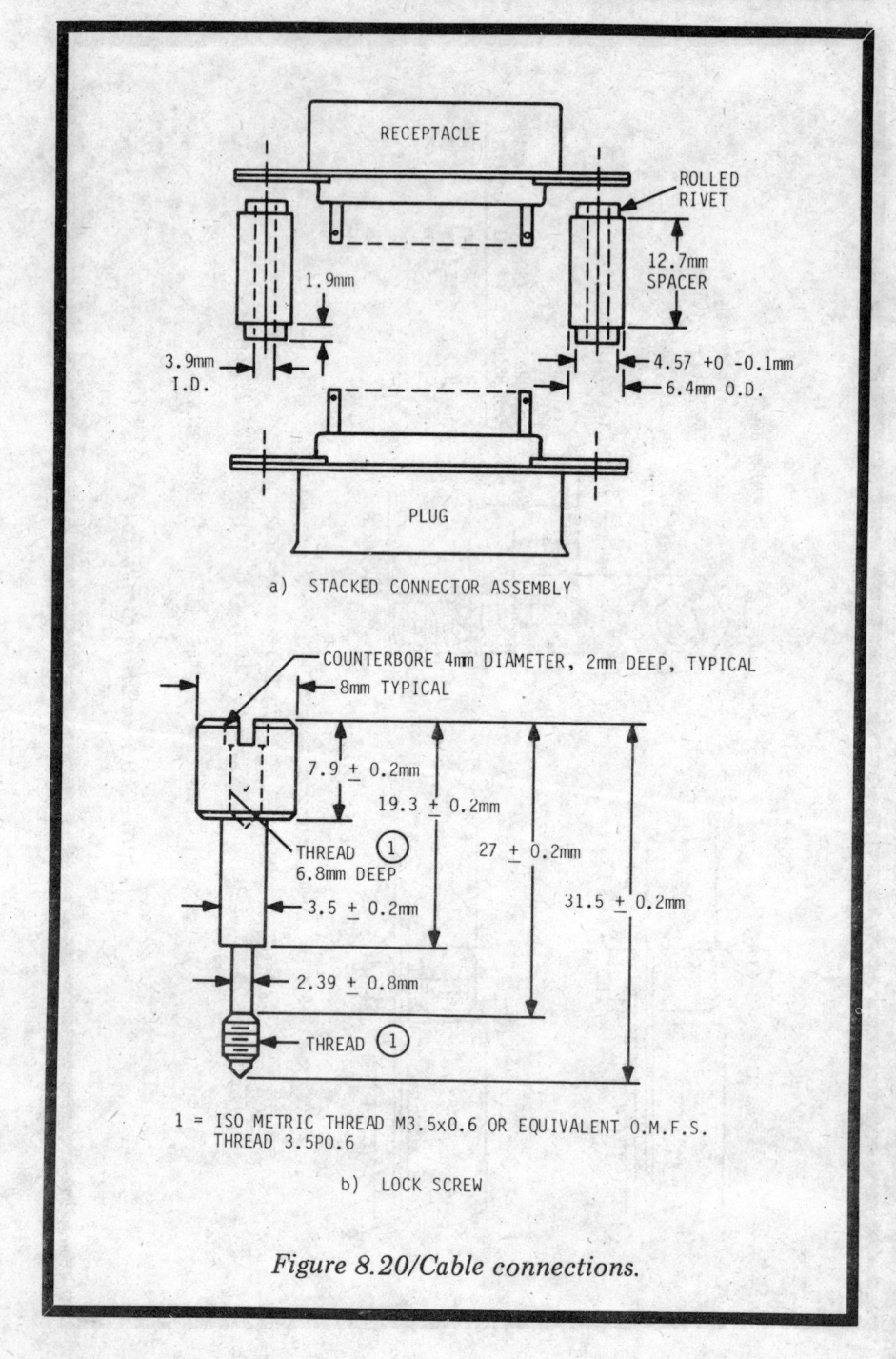

a) STACKED CONNECTOR ASSEMBLY

1 = ISO METRIC THREAD M3.5x0.6 OR EQUIVALENT O.M.F.S. THREAD 3.5P0.6

b) LOCK SCREW

Figure 8.20/Cable connections.

Automatic Testing

Figure 8.21/Cable connector housing.

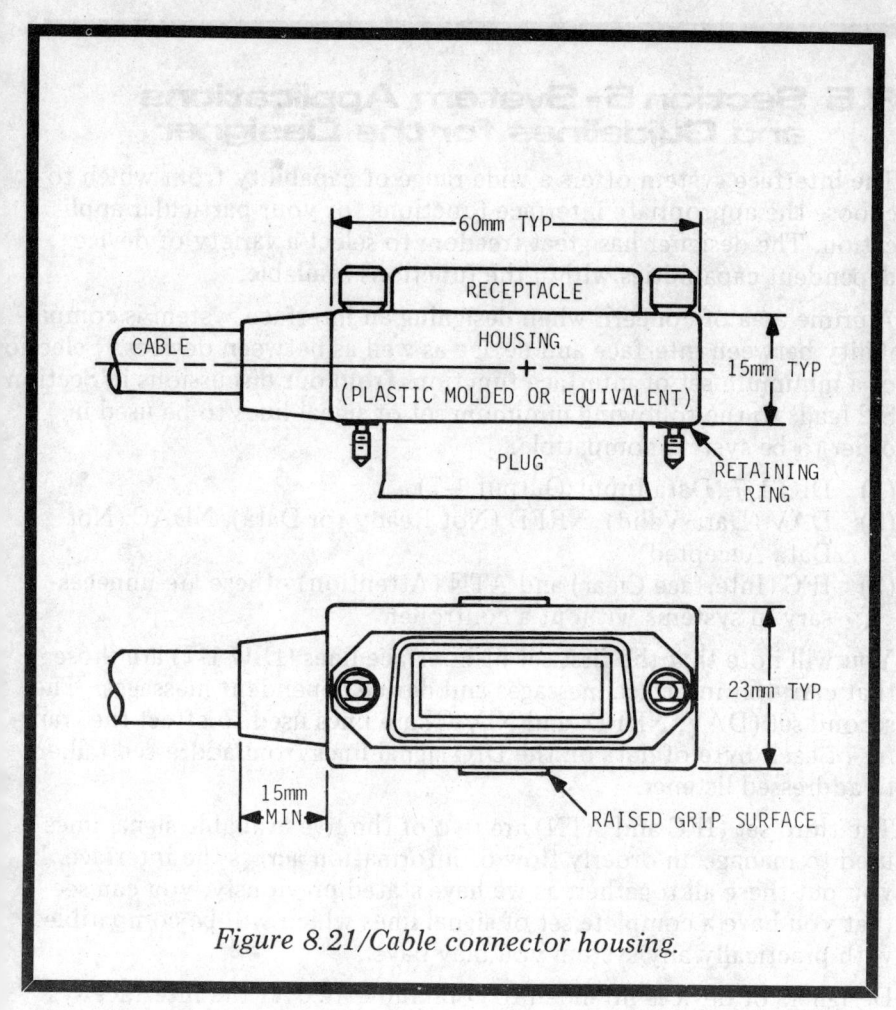

acle connector type at each end of the cable. The preferred method of assembling the stacked connectors contains a rigid structure shown in Figure 8.20(a). This method assures a reliable and positive connection of multiple cable assemblies.

Each connector assembly is fitted with a pair of captive locking screws which conform to the mechanical dimensions in Figure 8.20(b). Retaining rings are used to keep the screws captive. Each pair of connectors is partially enclosed with a housing like the one shown in Figure 8.21.

The individual cable assemblies we have been talking about have one more restriction, mechanically. They are limited in length to no more than four meters (13.123 feet).

8.5 Section 5 - System Applications and Guidelines for the Designer

The interface system offers a wide range of capability from which to choose the appropriate interface functions for your particular application. The designer has great freedom to select a variety of device dependent capabilities within the functions available.

A prime area of concern when designing an interface system is compatibility between interface and device as well as between devices. Selection of a minimum set of interface functions from our discussions in Section 8.2 leads to the following minimum set of signal lines to be used in order to be system compatible:

(1) DIO 1-7 (Data Input/Output 1-7)
(2) DAV (Data Valid), NRFD (Not Ready for Data), NDAC (Not Data Accepted)
(3) IFC (Interface Clear) and ATN (Attention) - these are unnecessary in systems without a controller

You will note that the first set of interface lines (DIO 1-7) are those that carry bit interface messages and device dependent messages. The second set (DAV, NRFD, and NDAC) are lines used to effect the transfer of each byte of data on the DIO signal lines from addressed talker to addressed listener.

The third set (IFC and ATN) are two of the five available signal lines used to manage an orderly flow of information across the interface. If you put these all together, as we have stated previously, you can see that you have a complete set of signal lines which will be compatible with practically any system you may have.

Designers of devices intended to communicate over the interface system bus should consider the relationships between various levels of system performance and the specific device circuits used to provide these levels. The following statements are guides for the designer.

A standard performance bus will operate at distances up to 20 meters, at a maximum of 250,000 bytes/second, with an equivalent standard load for each two meters of cable, using 48 milliamp open collector drivers.

A bus will also operate over 20 meters, 500,000 bytes/second, and with a standard load for each two meters of cable, using 48 milliamp three-state drivers.

Automatic Testing

The bus will similarly operate at 1,000,000 bytes/second using 48 milliamp three-state drivers; however, the 1 megabyte capability cannot be achieved over the 20 meter distance. The approximate maximum distance for 1 megabyte operation is 1 meter/device.

Detailed explanations of BUSY FUNCTION, NRFD, HOLD, RL APPLICATIONS, AND and OR MESSAGE FUNCTIONS, SH and AH MESSAGES, and SRQ MESSAGES are presented in the actual IEEE-488 standard. If the information is needed, you may want to consult the standard. For this text, they are too involved and thus will not be covered here.

Normally, a device will be assigned a single talk and single listen address to perform the necessary tasks. It may, however, be useful to design a device with multiple talk (or listen) to aid in the operation of a specific system. In this case, a device can be assigned two talk addresses (possibly, one to output raw data, the other to output processed data). The use of multiple address should be limited to only those places where they are absolutely needed.

A designer is free to select the particular interface functions required to meet his particular requirement. The selection of certain interface functions demands the inclusion of other functions defined throughout the allowable subset classes described in Appendix J. The list below represents typical combinations of interface functions and should be used as a guide.

Device	Typical Interface Function Used
Signal Generator (Listen Only)	AH, L, RL, DT
Tape Reader (Talk Only)	SH, AH, T, DC
Digital Voltmeter (Talk and Listen)	SH, AH, T, L, SR, RL, PP, DC, DT
Calculator (Talk, Listen, and Control)	SH, AH, T, L, C

where

```
AH  = Acceptor Handshake
SH  = Source Handshake
T   = Talker
L   = Listener
SR  = Service Request
RL  = Remote Local
PP  = Parallel Poll
```

DC = Device Clear
DT = Device Trigger
C = Controller

8.6 Section 6- System Requirements and Guidelines for the User

There are many system requirements that a designer or user must take into consideration when adapting to interface to a particular application:

- The maximum number of devices that can be connected together to form one interface system is 15.
- An interface system must contain one or more devices containing at least one T (Talker) function, one L (Listener) function, and one C (Controller) function.
- All system configurations containing more than one controller must satisfy the following conditions:

    (a) There must not be more than one C function in a system that is in the SACS (System Control Active State).

    (b) Every controller in the system must be able to pass and receive control of the interface.

- A system must be operated with at least one more than half the devices with power on.
- Maximum length of cable which can be used to connect a group of devices within one system is:

    (a) 2 meters times the number of devices

    or

    (b) 20 meters (whichever is less).

Most interface communication tasks require a sequence of coded messages to be sent over the interface. The following sequences are recommended for specified tasks. You will recognize the message and function terms from the discussions presented in Section 8.2.

The end product of all of the explanations and discussions presented thus far is an interface system that will handle the instruments you have to use for your system.

The typical system shown in Figure 8.22 illustrates the capability of the interface system to handle a variety of instrumentation system needs. Two possible event sequences, which can accomplish specific measurement tasks using the interface system, are included as example

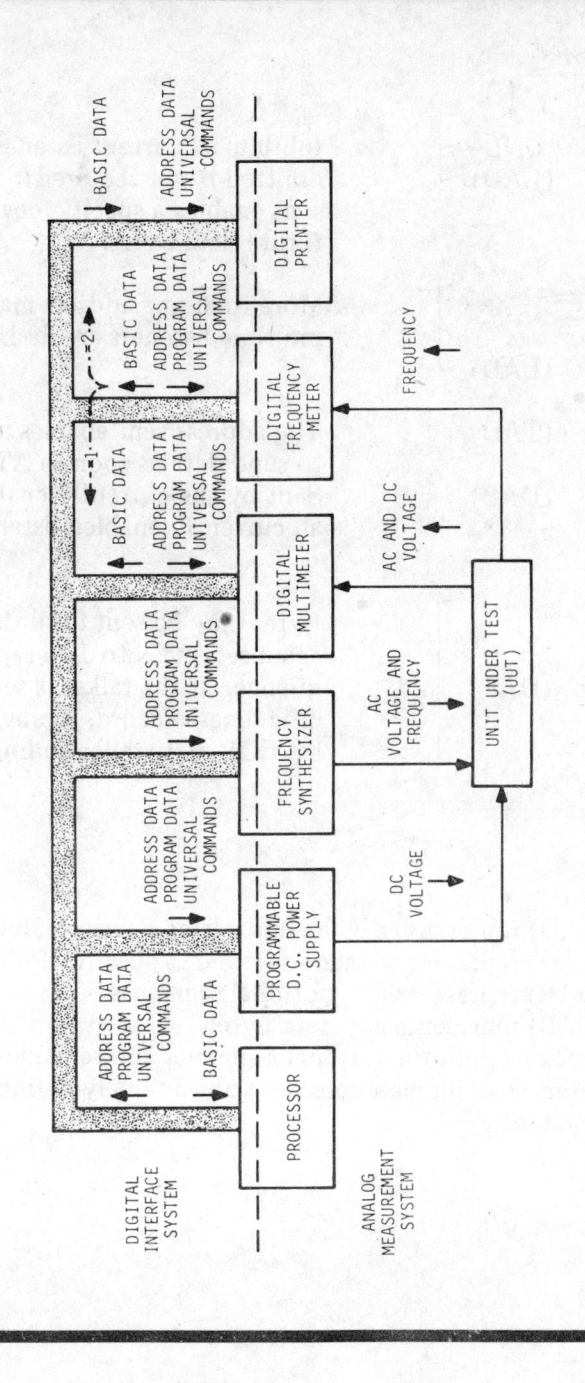

Figure 8.22/Typical instrument system.

Data Transfer

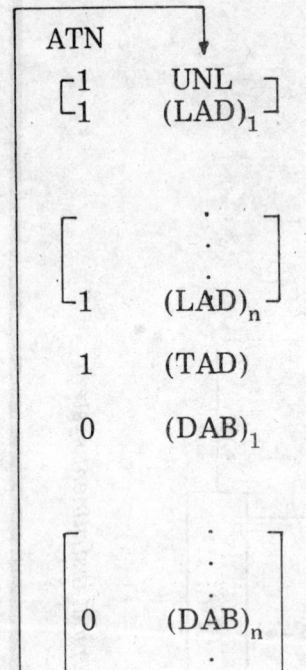

$$\begin{bmatrix} \begin{bmatrix} 1 & \text{UNL} \\ 1 & (\text{LAD})_1 \end{bmatrix} \\ \begin{bmatrix} \vdots & \vdots \\ 1 & (\text{LAD})_n \end{bmatrix} \\ \begin{matrix} 1 & (\text{TAD}) \\ 0 & (\text{DAB})_1 \end{matrix} \\ \begin{bmatrix} \vdots & \vdots \\ 0 & (\text{DAB})_n \\ \vdots & \vdots \end{bmatrix} \end{bmatrix}$$

ATN (arrow pointing down to the top of the bracket)

Inhibits all current listeners (can be omitted if not required). Each address sent enables a specific device to receive future data bytes.

More than one address may be sent if multiple listeners are desired.

The address sent enables a specific device to send data as soon as ATN becomes 0. Sent by the currently enabled talker to all currently enabled listeners.

Bytes may be sent until the controller again sets ATN to 1 to repeat the sequence. If the talker is sending a specific length record, it may optimally set EOI = 1 while sending the last byte.

Note: (LAD) represents a listen address of a specific device.
(TAD) represents a talk address of a specific device.
(DAB) represents any optional segments of a
(DAB) represents any data byte.
Brackets indicate optional segments of a sequence.
Parenthesis indicate messages not uniquely defined in this standard.

Automatic Testing

Serial Poll (issued by controller usually whenever SRQ = 1 on the interface)

ATN		
1	UNL	Prevents other devices from listening to status sent. (Controller continues to listen without being addressed.)
1	SPE	Puts interface into serial poll mode during which all devices send status instead of data when enabled.
1	$(TAD)_n$	Enables a specific device to send status. Within this loop, devices should be sequentially enabled.
0	(SBN) or (SBA)	Status byte sent by enabled device. If SBN was sent, loop should be repeated. If SBA was sent, the enabled device is identified as having sent SRQ over the interface and will automatically remote it.
1	SPD	Removes the interface from serial poll mode.

Note: (TAD) represents a talk address of a specific device.
(SBN) represents a status byte sent by a device in which a request for service is not indicated (bit 7 = 0).
(SBA) represents a status byte sent by a device in which a request for service is indicated (bit 7 = 1).

Control Passing

ATN		
1	(TAD)	The address sent should be that of the device to which control is being passed.
1	TCT	Notifies addressed device to take over control of the interface. New controller-in-charge at this time.
0		

Note: (TAD) represents a talk address of a specific device.

Parallel Poll

ATN	IDY	
1	1	Whenever the bus is in this state, predetermined devices will each place their requests on a specific DIO line. If more than one device is sharing a DIO line, the line value can indicate either an ORing or an ANDing of requests depending on commands previously sent to the device instructing them to use the 0 or 1 value to request service.

Placing Devices in Forced Remote Control

ATN	REN		
1	1	LLO	Disables all devices' "return-to-local" button.
1	1	(LAD)	Each address sent places the addressed device into remote state, disabling all local controls.
1	1	.	
1	1	.	
1	1	.	
		(LAD)	

Note: (LAD) represents a listen address of a specific device. (Devices will all revert back to local state as a group at any time a 0 value of REN is placed on the interface.)

Automatic Testing

A. Event Sequence 1 (Basic Data Returned to Processor)

Processor programs instruments and initiates measurements: resulting basic data is returned to processor.

(1) Processor initializes the interface system by sending the IFC message true.

(2) Processor causes all devices to set their internal conditions to a predefined state by sending the DCL message true.

(3) Processor sends the listen address of the DC power supply followed by program data for that device.

(4) Processor sends the unlisten command then the listen address for the next device, followed by program data for it.

(5) Event (4) is repeated until each device of interest for this specific test has been addressed and programmed, then the unlisten command is sent.

(6) Processor sends listen address of selected measurement device (for example, the digital frequency meter), then that program code required to initiate a measurement.

(7) Processor sends unlisten command, address itself to listen, then sends talk address of the measurement device.

(8) Upon completion of its internal measurement cycle, the digital frequency meter sends (talks) its measurement results (basic data) to the addressed listener, the processor.

B. Event Sequence 2 (Basic Data Directed to Digital Printer)

Processor programs instruments and initiates measurements: resulting basic data is returned to another device.

(1) - (6) Identical to Event Sequence 1.

(7) Processor sends unlisten command, then the listen address of the digital recorder, followed by the talk address of the measurement device.

(8) Upon completion of its measurement, the measurement device again sends its resulting basic data to the addressed listener, the digital recorder.

Note: If the processor were to address both the digital recorder and itself, the resulting basic data would be accepted by both devices, even though the two may have vastly different rates at which data can be accepted.

Chapter Summary

This chapter has examined the IEEE-488 Interface System so that you may be able to incorporate the system into your particular application. You will recall that Section 1 dealt with the basic system, with signal lines being identified and terms defined which were to be used throughout the system.

Section 2 was the heart of the system, with all of the interface functions and their individual states being identified, defined, and characterized. It is interesting to note how practically every function discussed began in an *idle* state in which it powered on. It then went to an addressed (or not ready or standby) state, and finally to an active state. There are times when the function went right from *idle* to *active* (Device Clear and Device Trigger) with no states between.

Generally, a state would revert back to its previous state if ACDS (Accept Data State) was active and a variety of additional messages were true. The Controller Function (C) always moves back to the *idle* state from where it is if no action occurs within a time t_4 (< 100 microseconds).

Many times through the section there were references to a t or T number in which an event would take place. The values for these numbers are contained in a table located in Appendix I, along with all the functions allowable, substates, message explanations, and message coding.

Sections 3 and 4 described electrical and mechanical specifications, respectively, for the interface. General specifications were presented which aid in overall system operation.

Section 5 was system applications and guidelines for designers, and Section 6 was system requirements and guidelines for the user. This section was followed by the presentation of a typical instrument system with setup and event sequences.

With the explanations presented in this chapter and references to the IEEE-488 standard, it should be possible to design a test setup which will provide accurate and reliable automatic data for your particular application.

Misc Measure

9

Miscellaneous Measurements

9. Definition

To this point, all of the measurements have fallen into nicely definable categories. There are, however, some measurements that do not readily fit into a category as such. These are the measurements we will address in this chapter: the *miscellaneous measurements*.

We will cover four measurements that fall into the category of miscellaneous: Phase Noise, Microwave Q, TDR (Time Domain Reflectometry), and swept impedance. You can readily see that none of these measurements fall into any category we have presented in the previous chapters. Each measurement will, therefore, be treated as if it were in a category by itself.

9.1 Phase Noise

To begin our discussion on Phase Noise measurements, we should probably ask two very basic questions. First, "What is phase noise?" and, second, "Why wasn't it covered in Chapter 4 with the rest of the microwave noise measurements?"

When we speak of phase noise, we must also speak of frequency stability. This is because the amount of phase noise at the output of a device affects the total frequency stability of a system. Frequency stability itself consists of three separate parts, of which phase noise is a part of one of them.

The first part to be considered when addressing frequency stability is *long term stability*. This is usually expressed in parts/million of frequency change/hour, day, month or year. This term is reasonably predictable since it is determined by the aging process of the material used in the frequency determining element (oscillator). This can be predicted with very great accuracy.

The second part is *environmentally induced frequency shifts*. These are changes which occur due to temperature, pressure, or in some cases even gravity. With proper designs, these effects can be minimized.

The third area is where phase noise enters the picture. This is *short term frequency fluctuations*. This area involves those elements that cause frequency changes about a center frequency in less than a few seconds. Carrier sidebands can be related to amplitude and phase modulation on a carrier. In our case, we are concerned with the Phase Modulation (PM) signals. In general, there are two categories of these PM signals. The first is *deterministic*. This includes discrete signals which can be traced to known conditions such as power line frequency, vibration frequencies, or ac magnetic fields. These signals show up in the spectral plot as exactly what they are — distinct components. The second category is *random*. This one should probably be called its most common name — *phase noise*. That is, the spectral density of the phase noise sidebands shows a continuous spectrum over a wide range of frequency similar to that of broadband noise. In this case, the signals that are around the carrier are not distinct — they are random. Since the measured level of noise is a function of the detector bandwidth, it is customary to normalize the phase noise measurements to a 1 Hz bandwidth — this results in a consistent standard for measurement. So, at this point, we can probably answer the first question by saying that phase noise is "a phase modulated signal consisting of random frequencies which comprise a continuous spectrum similar to broadband noise." Figure 9.1 shows an RF Sideband Spectrum with the Phase Noise Sideband shown in the insert.

Note: It is interesting to mention that, as can be seen in Figure 9.1, the Phase Noise spectral plot does *not* include the fundamental signal or carrier. Only the sidebands are involved with phase noise.

The second question that we initially asked should have just about answered itself by this point. You can see from our definition of phase noise that it is really not the same type of noise we were speaking of in Chapter 4. In that chapter, we defined microwave noise as

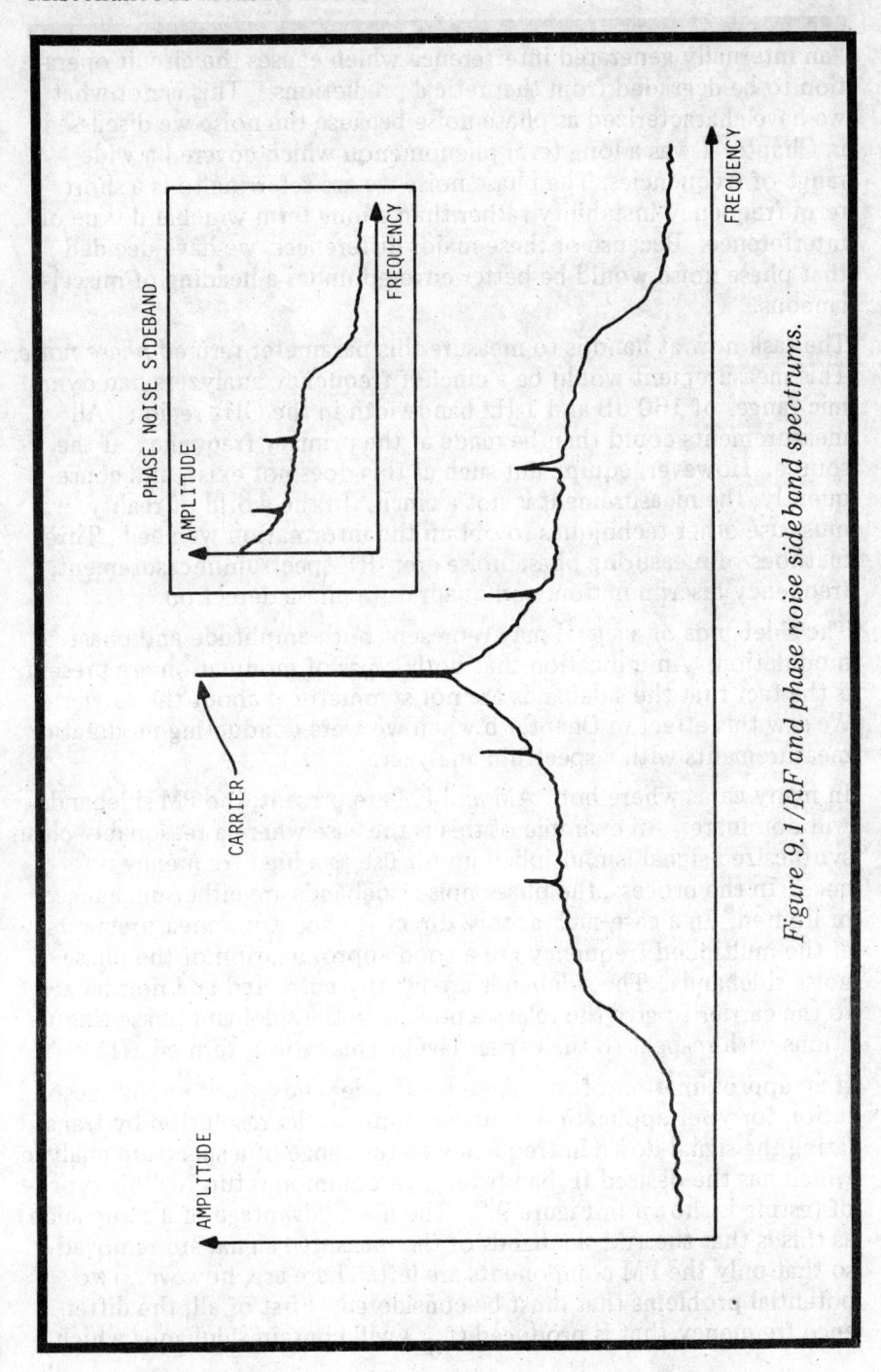

Miscellaneous Measurements

Figure 9.1/RF and phase noise sideband spectrums.

"an internally generated interference which causes the circuit operation to be degraded from theoretical predictions." This is not what we have characterized as phase noise because the noise we discussed in Chapter 4 was a long term phenomenon which covered a wide range of frequencies. The phase noise we are referring to is a short term frequency instability rather than a long term wideband type of interference. Because of these major differences, we have decided that phase noise would be better covered under a heading of miscellaneous.

The task now at hand is to measure this parameter termed *phase noise*. This measurement would be a cinch if frequency analyzers had dynamic ranges of 160 dB and 1 Hz bandwidth in the GHz region. All measurements could then be made at the primary frequency of the source. However, equipment such as this does not exist, and consequently, the measurement is not a cinch. In the world of reality, we must use other techniques to obtain the information we need. Three methods of measuring phase noise are: RF spectrum measurement, frequency discrimination, and quadrature phase detection.

The sidebands of a signal may represent both amplitude and phase modulation. An indication that both types of modulation are present is the fact that the sidebands are not symmetrical about the carrier. We saw this effect in Chapter 5 when we were conducting modulation measurements with a spectrum analyzer.

In many cases where both AM and PM are present, the PM sidebands will dominate. An example of this is the case where a reasonably clean synthesized signal is multiplied up for use as a high frequency reference. In the process, the phase noise sidebands are either unchanged or limited. In a case such as this, direct RF spectrum measurements at the multiplied frequency are a good approximation of the phase noise sidebands. The sidebands are usually corrected and normalized to the carrier to give the relative powers in the sideband phase fluctuations with respect to the carrier level. This ratio is termed $\mathcal{L}(f)$.

If an approximation of the phase noise sidebands is not enough resolution for your application, you can improve the resolution by translating the signal down in frequency to the range of a spectrum analyzer which has the desired IF bandwidth. A common setup for this type of testing is shown in Figure 9.2. The main advantage of a setup such as this is that the AM sidebands of the measured signal are removed so that only the PM components are left. There are, however, two potential problems that must be considered. First of all, the difference frequency that is produced (f_{IF}) will contain sidebands which

Miscellaneous Measurements 415

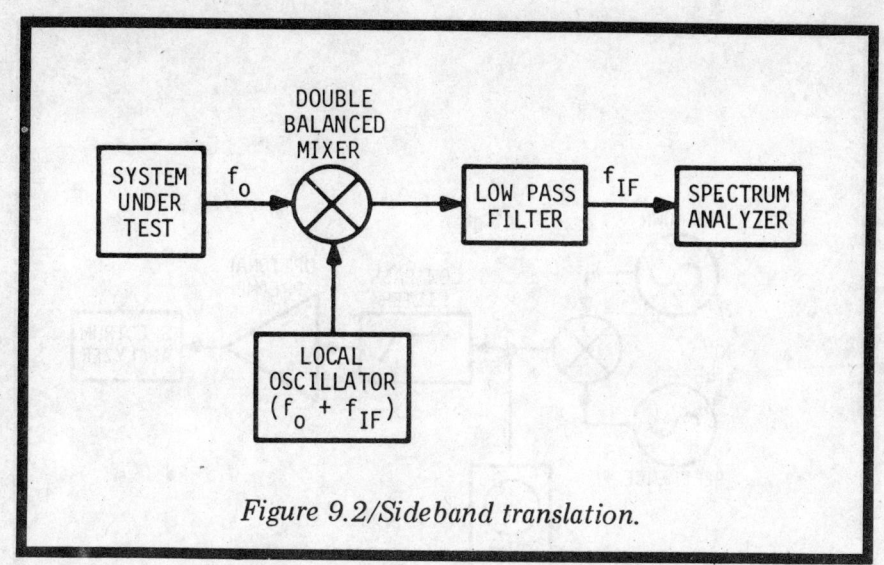

Figure 9.2/Sideband translation.

are below zero frequency. This may or may not cause a problem, depending on what type of system is being measured. You should, however, be aware of the existence of the additional sidebands.

The second potential problem is that the phase noise sidebands from the local oscillator will also be translated down at the mixer. This problem can be avoided by using a source (local oscillator) with better phase noise specifications than the system you are testing. (Phase noise is expressed in dBc/Hz; dBc is the number of dB below the carrier level). When using the RF spectrum measurement technique, the entire spectrum as shown in Figure 9.1 is presented. This is why we can refer to a level with respect to the carrier and sidebands set up by the difference frequency. For some applications, the full spectrum is acceptable; for others it is a hindrance.

One case in which the carrier must be eliminated is in measuring sidebands that are beyond the dynamic range of the analyzer. With the carrier removed, you can extend your viewing to a greater extent and characterize the sidebands. This elimination can be accomplished by using a frequency discriminator. To ensure that the calibration factor for the system is constant, you should check the linearity of the discriminator over the frequency range of interest. That is, the volt/Hz characteristic of the discriminator must be linear or else you should obtain a good calibration curve to tell you what the discriminator will do when put into the test setup. For microwave frequencies, the cavity discriminator is very useful. The high Q of the cavity results in a linear device at the higher microwave frequencies.

Figure 9.3/Quadrature phase detection.

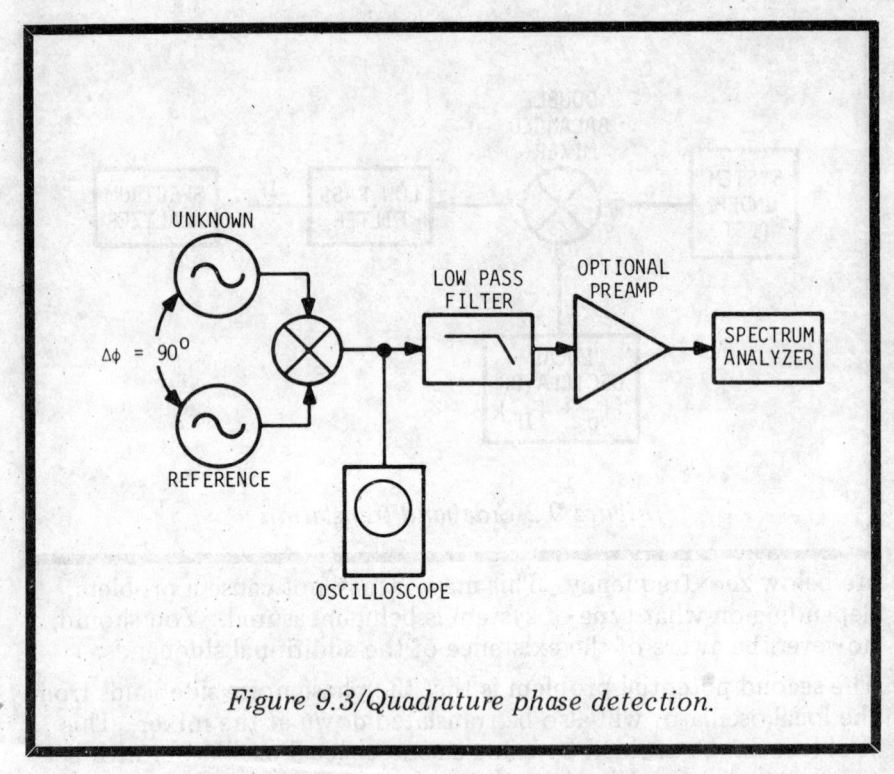

Probably the most versatile, and the most widely used, setup for measuring phase noise is a double balanced mixer with the unknown and the reference sources separated from each other in phase by 90° (in phase *quadrature*). The setup used for this method is shown in Figure 9.3. When the quadrature condition exists at the input ($\Delta\phi$ = 90°), the difference frequency between the sources is 0 Hz. One assumption that is made in this setup is that the reference oscillator is perfect (no phase noise) and that it can be adjusted in frequency. We must also assume that both oscillators (sources) are extremely stable so that the phase quadrature (90°) can be maintained without the use of a phase lock loop. The double balanced mixer acts as a phase sensitive detector so that when two signals are identical in frequency and are nominally in phase quadrature, the mixer output is a small fluctuating voltage centered on approximately 0 volts. The small fluctuating voltage represents the phase modulation PM sideband component of the signal.

There are times when the sources do not remain stable, as assumed in the above discussion, long enough for a quadrature phase relationship to be held during the measurement period. For cases like this, the

Miscellaneous Measurements

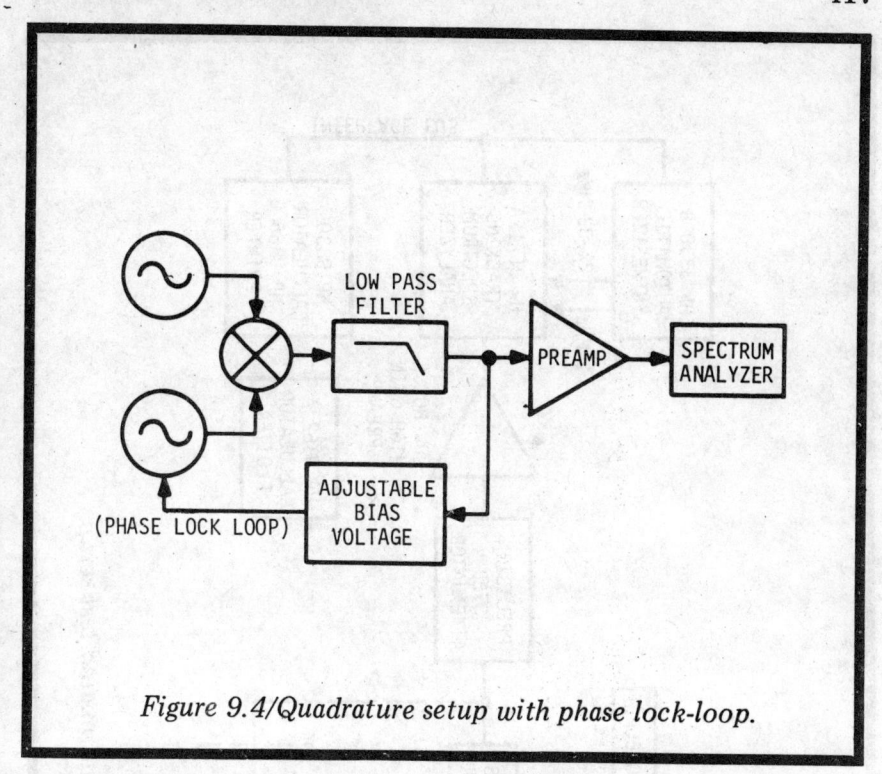

Figure 9.4/Quadrature setup with phase lock-loop.

sources must either be adjusted periodically or a phase lock loop is used if either source, or both, have a voltage control with which to make small frequency adjustments. This setup is shown in Figure 9.4.

To retain a constant relationship between phase and voltage fluctuations, the low frequency cutoff of the phase locked loop must be below the lowest frequency to be analyzed. If the breakpoint is moved out, by adding gain in the loop, the voltage fluctuations at frequencies below the breakpoint will represent frequency fluctuations. To calibrate with the phase locked setup, disconnect the feedback voltage and observe the beat signal as if there were a straight quadrature detection setup.

The phase noise measurement system shown in Figure 9.5 is program controlled by the Hewlett-Packard 9830 programmable calculator. Each step of the calibration and measurement sequence is included in the program. The measurements are not completely automated since the calibration sequence requires several manual operations. The baseband measurements are fully automated. The software program controls frequency selection, bandwidth settings, settling time, amplitude ranging, measurements, calculations, and graphics and data plotting.

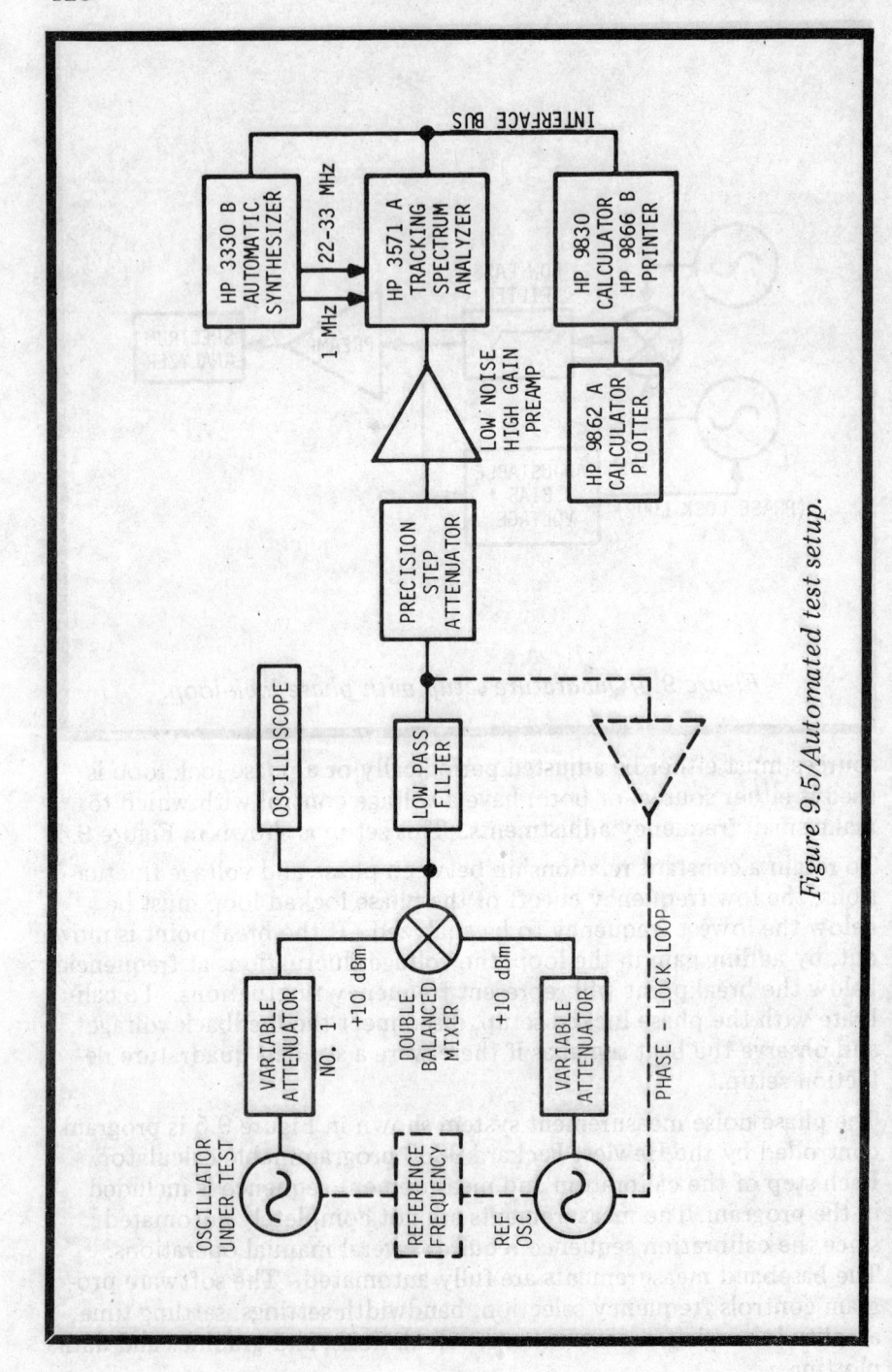

Figure 9.5/Automated test setup.

Miscellaneous Measurements 419

The system will be described as it is used to obtain a direct plot of $\mathcal{L}(f)$. The direct measurement of $\mathcal{L}(f)$ is represented by the following equation:

$$\mathcal{L}(f) = \frac{(\text{Noise Power Level})}{(\text{Carrier Power Level})} \text{ in dB}$$

$$-6 \text{ dB} + 2.5 \text{ dB} - 10\log(B)$$

$$-3 \text{ dB} \qquad\qquad (\text{dBc/Hz})$$

The noise power is measured relative to the carrier power level, and the remaining terms of the equation represent corrections that must be applied due to the type of measurement and the characteristics of the measurement equipment as follows:

- The basic measurement of noise sidebands with the signals in phase quadrature requires the −6 dB correction.
- The nonlinearity of the spectrum analyzer logarithmic IF amplifier results in compression of the noise peaks which, when average detected, require the +2.5 dB correction for the HP 3571A tracking spectrum analyzer.
- The bandwidth correction is required because the spectrum analyzer measurements of random noise are a function of the particular bandwidth used in the measurement.
- The −3 dB correction is required since this is a measurement of $\mathcal{L}(f)$ using two oscillators, assuming that the oscillators are of a similar type and that the noise contribution is the same for each oscillator. If one oscillator is sufficiently superior to the other, this correction is not used.

The calibration and measurement sequence is as follows:

(1) Measure the noise power bandwidth of each IF bandwidth setting on the tracking spectrum analyzer.

(2) Obtain a carrier power reference level (referenced to the output ot the mixer).

(3) Adjust for quadrature of the two signals applied to the mixer. *Note:* If the noise floor of the system has not been established, it is advisable to include a short program to check the phase noise versus system noise floor. This noise floor scan can be at a small number of points at selected Fourier frequencies over the measurement range.

(4) Noise power is measured at the selected Fourier frequencies, the calculations are performed, and the data is plotted (or stored) using calculator and program control (fully automated).

(5) Measure and plot the system noise floor characteristics if desired.

Instruments are available that measure phase noise directly. Phase noise measurements are performed by comparing the phase of the IF signal produced from the interaction between the RF signal and the local oscillator, and the phase of a VXCO. The IF reference oscillator is phase locked to the IF signal through an adjustable loop bandwidth filter. The VXCO responds only to the phase noise components which are passed by this filter.

Typical specifications for a Phase Noise Analyzer are:

Loop Bandwidth	5 Hz to 500 Hz
Video Bandwidth	300 Hz to 300 kHz
Frequency Range	1 - 18 GHz
Signal Source Level	−20 dBm to +10 dBm
Maximum Residual Error	< 1.0 degrees

You can see that there are a variety of ways of checking the level of pahse modulated signals, consisting of random frequencies which comprise a continuous spectrum similar to broadband noise — that is, *phase noise*.

9.2 Q Measurements

The most widely used method of measuring the Q (figure of merit) of a resonant cavity is by measuring the bandwidth between the 3 dB (or half-power) points. The term "Q" is defined as the ratio of the resonan frequency (fo) to the frequency difference between the half-power (3 dB) frequencies ($f_1 - f_2$).

$$Q = \frac{f_o}{f_1 - f_2}$$

One method that utilizes the half-power points within the measurement is the *impedance measurement technique*. This technique uses a slotted line section with an SWR meter and the tuned cavity with a match termination on it.

When tuned off resonance (detuned), the transmission or absorption type cavities present a short circuit to incoming signals. The slotted section probe can be adjusted to locate the minimum of the standing-

Miscellaneous Measurements

wave pattern, which represents the detuned short position. The source must present a perfect match to the reflected signal.

If the cavity is tuned to exact resonance, the cavity absorbs power and the VSWR in the line decreases. Therefore, the output power level at the standing-wave indicator increases. If the cavity is tuned to exact resonance and if there are no reflections from the source, the slotted section probe will be at a voltage maximum or voltage minimum on the standing-wave pattern. The cavity presents a pure resistance to the line. If the slotted section probe, at the detuned short position, is at a voltage maximum of the standing-wave pattern, the cavity impedance is $Z_c = Z_o$ (VSWR). The cavity-coupling parameter β is defined as Z_c/Z_o and is greater than unity. The cavity impedance is a maximum, greater than Z_o, and the cavity is overcoupled. Alternately, if the slotted section is at a minimum of the standing-wave pattern, $Z_c = Z_o/\text{VSWR}$, and the cavity is undercoupled.

Measurement Procedure:

(1) Tune the cavity off resonance and locate the detuned short position by tuning the slotted section probe to a minimum of the standing-wave pattern.

(2) Tune the cavity to resonance as indicated by a maximum output reading on the standing-wave indicator.

(3) Set up a reference level on the standing-wave indicator at a power level within the square-law-response region of the probe-detecting element.

(4) Tune the cavity to each side of resonance and record the cavity frequencies (f_1 and f_2) at which the standing-wave indicator level drops by 3 dB.

(5) Tune the cavity back to resonance. Move the slotted section probe and note whether the probe (at the detuned short position) was at a minimum or at a maximum of the standing-wave pattern. If at a maximum, $Z_c/Z_o = \text{VSWR}$, if at a minimum, $Z_c/Z_o = 1/\text{VSWR}$.

(6) Measure the VSWR and calculate the parameters as follows:

$$Q_L = \frac{f_o}{\Delta_f}$$

$$Q_0 = \left(1 + \frac{Z_c}{Z_o}\right) Q_L$$

where:

Δ_f = bandwidth at 3 dB points $(f_1 - f_2)$.

Z_o = waveguide characteristic wave impedance

Z_c = cavity impedance

Q_L = loaded Q of the cavity

Q_0 = unloaded Q of the cavity

f_1 = high-frequency 3 dB point

f_2 = low-frequency 3 dB point

f_0 = cavity resonant frequency

The Q of reaction type frequency meters can be measured using the same technique preceded by the following: Tune the cavity near resonance, then alternately vary the cavity and short circuit to obtain a minimum VSWR. Repeat the previous measurement procedure. If the probe was not at exact minimum or maximum as noted in Step (5), the original adjustments of the cavity and short circuit were not properly tuned for minimum VSWR. The complete procedure must be repeated.

A less widely used method of measuring cavity Q is one based on the comparison of the measured device with a reference resonator (cavity) or with a reference resonance circuit with known parameters. Comparison methods are convenient in those cases where accuracy in measuring the absolute value of Q is of a secondary nature. The primary consideration is rather to indicate and measure the small relative change in the resonator parameters being measured.

The method presented here is based on the comparison of the power being transmitted through the measured and reference resonators. The powers are displayed on an oscilloscope screen with the X-axis representing the power of the reference resonator and the Y-axis displaying the power difference between the reference and tested resonator. The shape of the curve displayed depends on the difference in the parameters of both resonators, and expressively varies with small changes in the parameters. The tuning of the measured resonator to the resonance frequency of the reference resonator is indicated very markedly. Also, it is possible to read from the graph of the resultant curve all the data which are necessary for the calculation of the cavity quality factor.

Miscellaneous Measurements

Figure 9.6/Q measurement test setup.

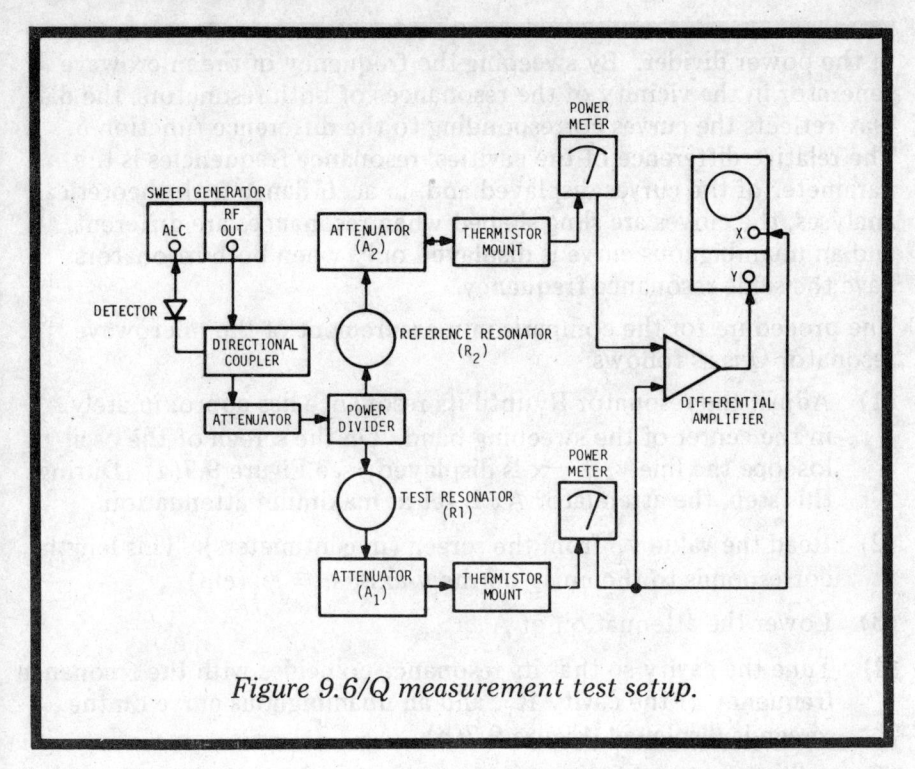

The method is very convenient for the measurement of small changes in the quality factor and resonant frequency of the microwave resonator.

The test setup for measuring the cavity quality factor by the comparison method is shown in Figure 9.6. The test setup consists of an amplitude-leveled microwave sweep generator connected through a level attenuator to the hybrid power divider. The power divider must have a balanced power division, constant insertion loss, low and constant SWR, and high isolation between the output arms. At each output arm of the power divider, a transmission-type resonator is located, each terminated with the matched power sensor of a microwave power meter. One is the test resonator, and its resonance frequency and quality are unknown; the second is a reference resonator with a calibrated dial and a known loaded quality factor. The characteristics of the applied power meters must be linear. The output signals of both power meters are fed into the horizontal amplifier of the oscilloscope and differential amplifier. The output of the differential amplifier is connected to the vertical amplifier of the oscilloscope. The setup is completed with two level attenuators located at the two output arms of the power divider. By sweeping the frequency of the microwave generator in the vicinity of the resonances of both resonators, the display reflects the curves corresponding to the difference function δ. The relative difference of the cavities' resonance frequencies is the parameter of the curves displayed and, in accordance with theoretical analyses, the curves are sling shaped when resonances are different, and an unambiguous curve is displayed only when both resonators have the same resonance frequency.

The procedure for the comparison measurement of the microwave resonator Q is as follows:

(1) Adjust the resonator R_1 until its resonance lies approximately in the center of the sweeping band. On the screen of the oscilloscope the line $y = -x$ is displayed; see Figure 9.7(a). During this step, the attenuator A_2 is set at maximum attenuation.

(2) Read the value y_1 from the screen (in centimeters). This length corresponds to the unity on the y-axis, $1 \equiv y_1 (cm)$.

(3) Lower the attenuation of A_2.

(4) Tune the cavity so that its resonance coincides with the resonance frequency of the cavity R_1, and an unambiguous curve on the screen is displayed, Figure 9.7(b).

(5) Adjust the variable attenuator A_2 so that the end point A of the curve coincides with the point (1.0) on the x-axis, see Figure 9.7(c).

(6) Adjust the gain of the vertical amplifier of the oscilloscope until the displayed curve covers the whole vertical side of the screen. This step is convenient in increasing the sensitivity of the method. The unity of the y-axis is 1/K now, where K is the ratio of gain increase.

(7) Repeat steps (4) and (5), in order to correct inaccuracy of previous adjustments.

(8) Read the value of the curve height y_2 from the display. The value of the extreme of the differential frequency is $\delta_1 = y_2/K$.

(9) Compute the ratio of the loaded quality factors Q_{L2}/Q_{L1} and, finally, the quality factor of the unknown resonator.

Note: The ratio of the loaded Q factor is found by applying the following formula:
$$\frac{Q_{L2}}{Q_{L1}} = \frac{1 - \delta_1}{1 + \delta_1}$$

Miscellaneous Measurements

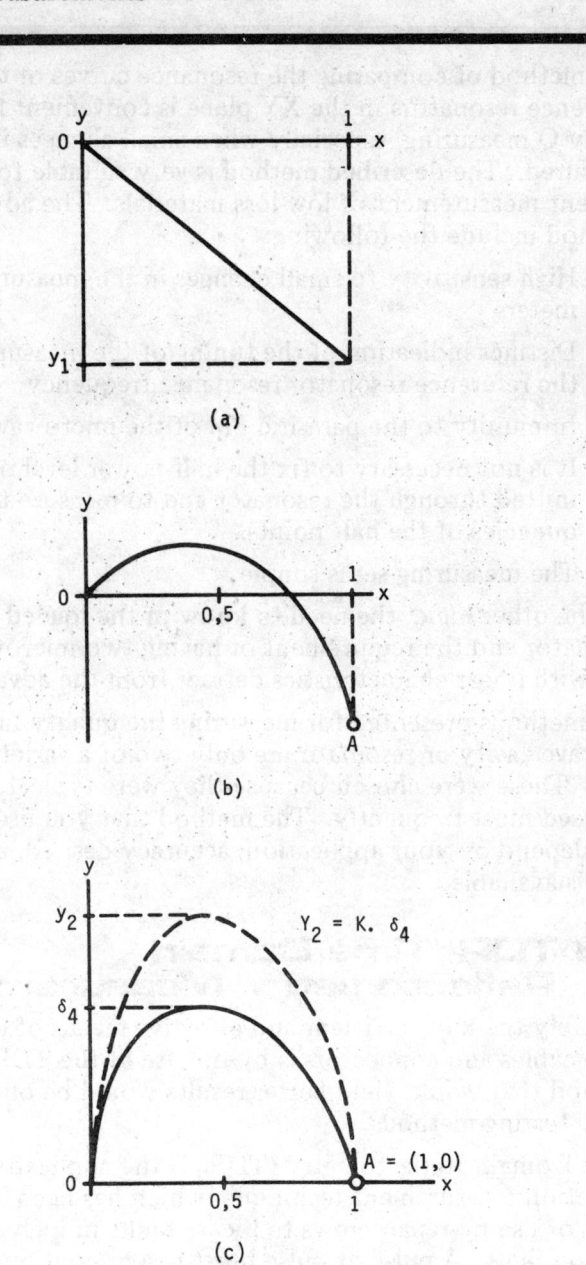

Figure 9.7/Oscilloscope displays of Q measurement.

The method of comparing the resonance curves of the measured and reference resonators in the XY plane is convenient for microwave cavity Q measuring, especially when small changes in the Q are to be measured. The described method is very suitable for microwave loss tangent measurement of low-loss materials. The advantages of the method include the following:

(1) High sensitivity to small changes in the measured resonator parameters.

(2) Distinct indication of the tuning of the measured resonator on the reference-resonator resonance frequency.

(3) Immunity to the parasitic FM of the microwave signal.

(4) It is not necessary to fix the half-power level of the signal transmitted through the resonator and to measure precisely the frequencies of the half points.

(5) The measuring set is simple.

On the other hand, the need to know of the loaded Q of the reference resonator and the requirement of having two microwave milliwattmeters with linear characteristics detract from the advantages listed.

The methods presented for measuring the quality factor (Q) of a microwave cavity or resonator are only two of a variety that are available. These were chosen because they were typical of the type that are used most frequently. The method that you use for your testing will depend on your application, accuracy desired, and equipment that is available.

9.3 TDR (Time Domain Reflectometry) Measurement

Probably the most efficient and effective means of checking microwave cables and connectors is by the use of the TDR. The only method that would yield better results would be one utilizing automatic testing methods.

Time Domain Reflectometry (TDR) is the application of the pulse reflection measurement technique, which has been proven through years of use by repair crews to locate faults in high voltage transmission lines. A pulse or pulse burst is sent continuously down the line. If the pulse encounters a short or open circuit, the reflection travels back to the sending point where it is compared in phase, time, and amplitude with the original. This likening determines the distance to the fault and provides an indication of its nature (short, open, etc.).

Miscellaneous Measurements

The Time Domain Reflectometer is thus a type of closed-loop, one-dimensional radar system, in which the transmitted signal is a very fast step function and the reflected signals are monitored on an oscilloscope screen. The faster the step, the greater the distance resolution, since distance is related to time in this technique. The amplitude of the reflected step is associated directly with impedance; any slight deviation from the 50 ohm output of the TDR can be recognized easily and measured. Connectors, baluns, striplines, tapered sections, and a host of other devices also can be analyzed with a TDR.

Time-domain reflectometry has many advantages over conventional CW reflectometry. Discontinuities in a system are clearly separated in time on the CRT. Therefore, it is easy to see the mismatch caused by a connector, even in the presence of a load discontinuity somewhere else in the system. It is even possible to determine which connector is troublesome and in what way. Having determined that a discontinuity appears in a waveform, it is simple to locate it in the system, suing the equation:

$$d = \frac{ct}{2\sqrt{\epsilon_t}}$$

where c is the speed of light (3×10^{10} cm/sec), t is the time factor, and ϵ_t is the relative dielectric constant of the transmission system. A timesharing procedure is to calibrate the system such that 1 cm on the horizontal axis is the equivalent of the certain number of centimeters for the coaxial line. Again the limiting factor is the system rise time.

The finite rise time also places a lower limit on the size of the reactive impedance than can be distinguished. For example, a small shunt capability in a 50 ohm system will cause the waveform to depart from the ideal response.

The idea of an "ideal" response brings up an interesting question. That obvious question is: "What is an ideal response?". To answer this question, we will present the responses for both the resistive and reactive loads on a TDR display.

If a pure resistive load is placed on the output of the reflectometer and a step signal is applied, a step signal is observed on the CRT whose height is a function of the resistance (Figure 9.8). It is a good practice to separate the system under test from the TDR unit by at least 8 inches of 50 ohm cable. This moves the reflections away from the leading edge of the step, such that overshoot and ringing are not superimposed on the observed signal.

Figure 9.8/Resistive loads on the TDR.

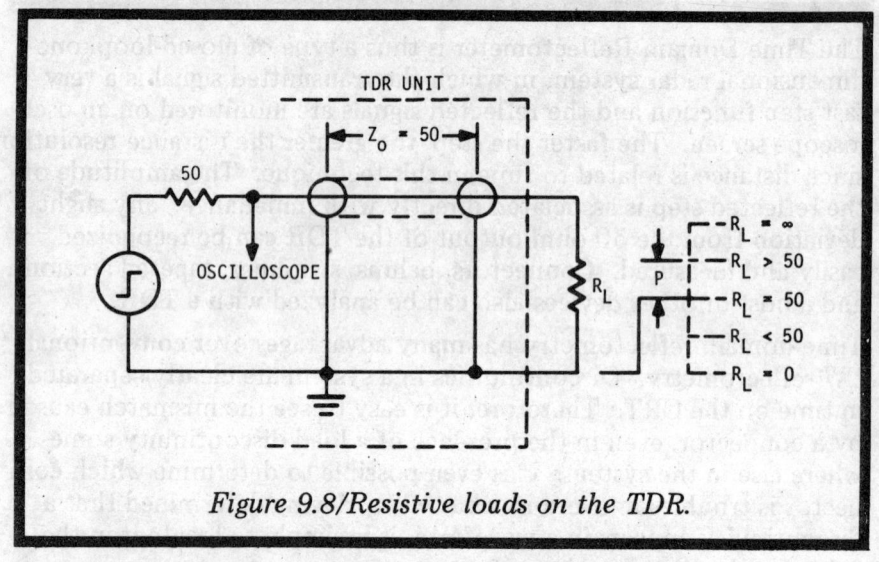

The magnitude of the step caused by the resistive load may be expressed as a fraction of the input signal as given by

$$\rho = \frac{R_L - 50}{R_L + 50}$$

For reactive loads, the observed waveform depends upon the time constant formed by the load and the 50 ohm source (Figure 9.9).

Consider the case of a coaxial cable terminated in a resistive load. Because of the lead length in the load, there is excess inductance. By use of TDR (Figure 9.10), we can determine quite simply:

(a) The characteristic impedance of the cable
(b) The degree of resistive mismatch in the termination
(c) The amount of excess inductance in the termination
(d) The length of the cable

An important part of using the TDR technique is the calibration of the unit. With proper calibration you have one of the most accurate instruments available for checking cables, connectors, and transmission lines.

The generator is connected to the sampler channel and an accurate 50 ohm termination is connected to the output. For the initial setup, however, this termination can be a 1 percent film resistor with short lead lengths. The vertical gain is adjusted to give a 10 cm step, and the 50 ohm termination is removed and test system is connected.

Miscellaneous Measurements

Figure 9.9/Reactive load on TDR.

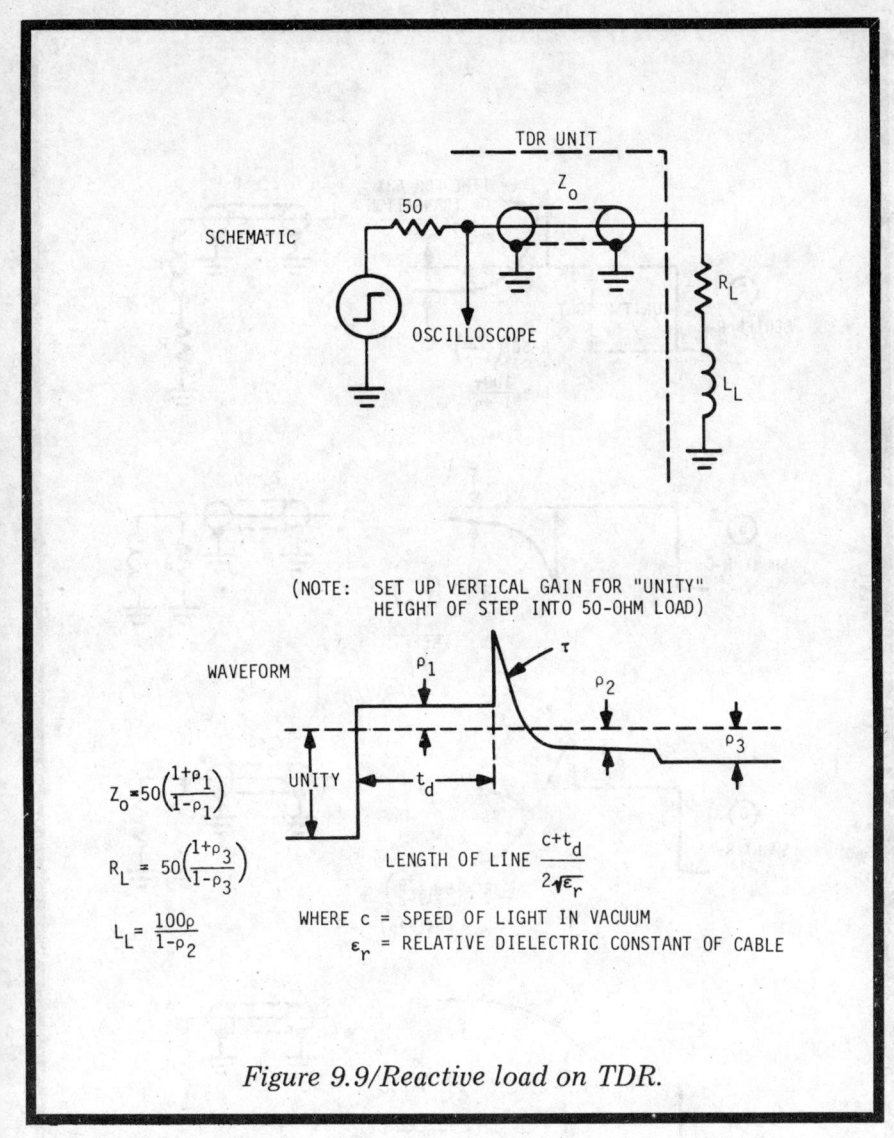

The magnitude and location of relections can be viewed more closely by choosing an appropriate time scale and vertical magnification. Calibration of the horizontal deflection allows rapid physical location of points of interest. Calibration of the vertical axis in terms of reflection coefficient allows interpretation of results directly in terms familiar to engineers who have been using the conventional VSWR measurements.

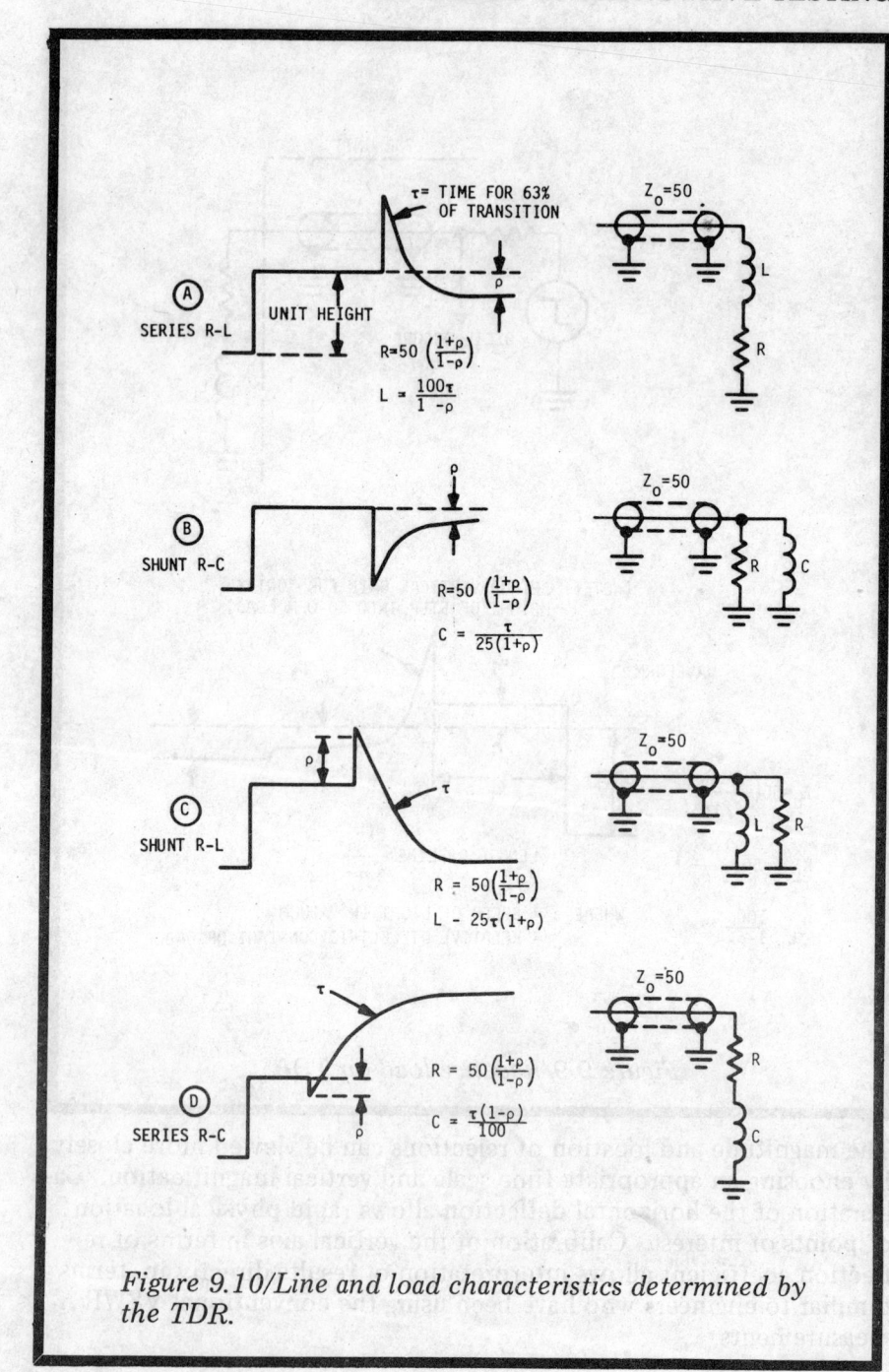

Figure 9.10/Line and load characteristics determined by the TDR.

Miscellaneous Measurements

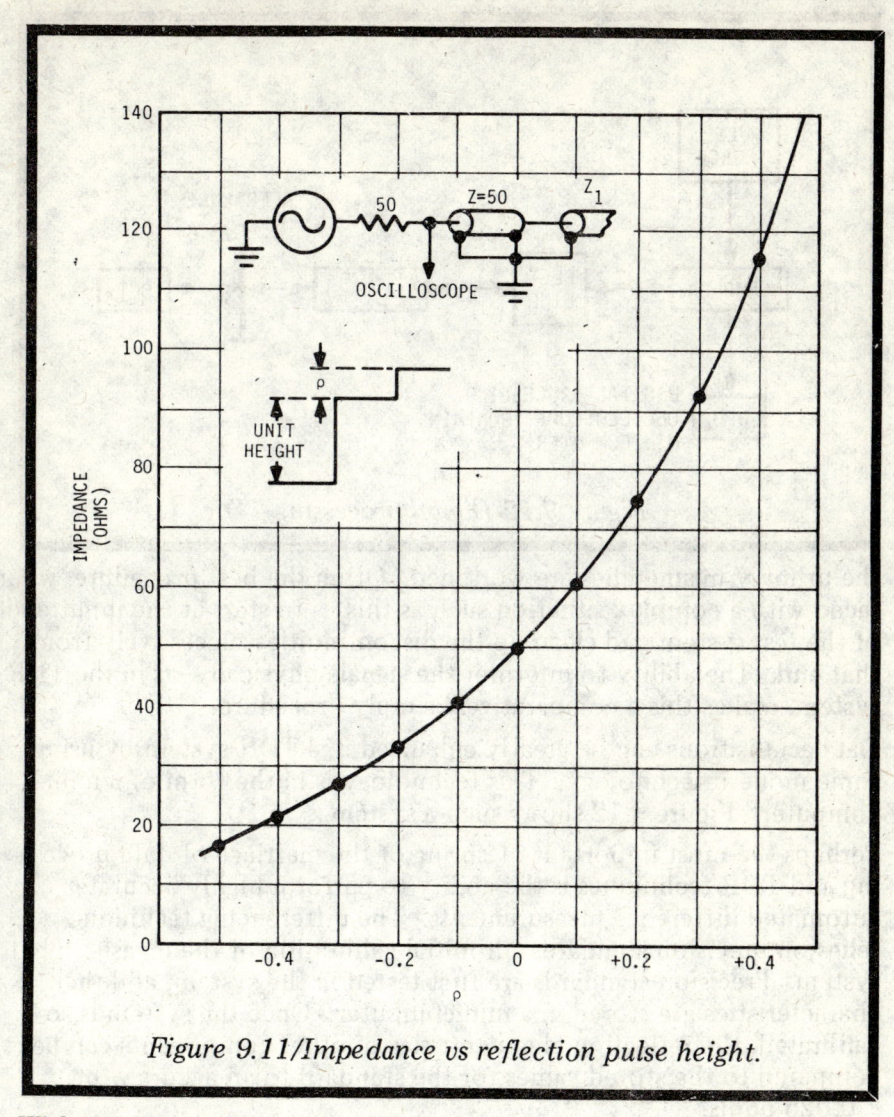

Figure 9.11/Impedance vs reflection pulse height.

With the instrument calibrated, cable impedance can be determined from the height of the reflected step. The graph shown in Figure 9.11 may be used to translate step heights into impedance.

For a system with a large number of bad mismatches, great care must be taken to avoid confusion because of rereflections. These rereflections usually can be separated and identified by deliberately worsening each known mismatch in turn and observing the change in reflected waveform. Disturbances caused by rereflections will increase when

Figure 9.12/Digital processing TDR.

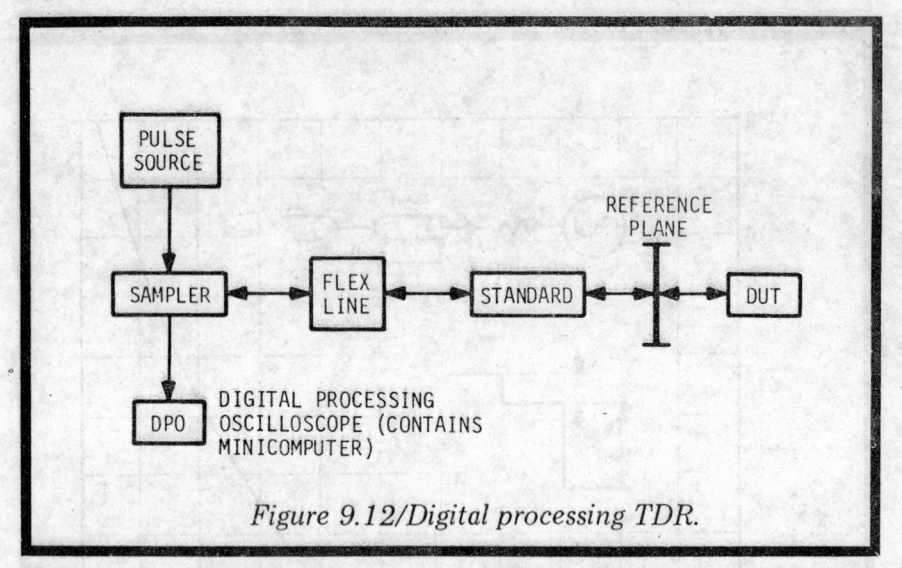

the primary mismatches are worsened. Often the best procedure, when faced with a complex situation such as this, is to start at the input end of the test system and clean up the discontinuities successively from that end. The ability to interpret the signals physically, as in the TDR system, makes this a comparatively simply procedure.

Data acquisitions can be greatly enhanced in a TDR system by using some modern technology. This technology is in the form of a minicomputer. Figure 9.12 shows such a system.

Perhaps the most important offspring of the marriage of data processing and TDR techniques is the ability to perform highly accurate, automated difference measurements. The differencing technique relies on precision standards for initial calibration of the measurement system. Precision standards are first tested in the system, and their characteristics are stored in a minicomputer. Once the system is so calibrated, the reflection characteristics of other components can be compared to the stored values for the standard to an accuracy of ± 0.025 ohms.

But the minicomputer can do much more than merely store data points in a TDR system. One of the major drawbacks of a non-assisted TDR system is a limited signal-to-noise ratio (SNR), typically about 42 dB for a sampler with a bandwidth from dc to 14 GHz. With digital processing techniques, however, the limiting SNR can be improved to approximately 66 dB for more exacting, repeatable measurements.

Techniques such as these have greatly enhanced the TDR so that its use is not as limited as it might have been.

9.4 Swept Impedance

When impedance measurements are needed, they usually are obtained at a number of specific frequencies. This, however, is not the only way to obtain impedances across a band of frequencies. The following sections will show swept setups and procedures using slotted lines, reflectometers, VSWR bridges, and the HP8755.

9.4.1 Swept Slotted Line

In the past, the general consensus has been that the slotted line is seldom used for measurements other than single frequencies because the procedure is so time consuming. Until the storage oscilloscope came into existence, this statement was very true. However, the storage scope has introduced the slotted line to the world of swept measurements, making this highly accurate instrument available for a wide range of applications. There are various methods of sweeping a slotted line — one is to modify the cw setup by replacing the cw generator with a sweep generator and the SWR meter with an oscilloscope with storage capability. The sweep generator is set to sweep the frequency band of interest. Multiple sweeps are made while the carriage of the slotted line is moved manually. The measurement is displayed as an envelope of SWR (the scope calibrated in dB). At any given frequency, the ratio of the maximum and minimum amplitude of the envelope is related to SWR.

This swept slotted line technique eliminates the loss of measurement continuity described previously and reduces measurement time (though it still requires considerably more time than other methods and requires an expensive storage scope). Other disadvantages with regard to single frequency setups are still present; in addition, the display is not a direct readout.

A variation of the above procedure is available using an instrument known as an RF ratio meter. A block diagram of a commercially available ratio meter is shown in **Figure 9.13**. The instrument does virtually all of the necessary calculations within the device. It is basically a two channel device that takes the reference (incident wave) and test (reflected) signals, compares them in a log ratio detector, and produces a dc voltage proportional to their ratio. This voltage is displayed on the main meter and is available for use with an X-Y recorder or an oscilloscope. Both the signal channel and the reference channel usually require 1.0 KHz demodulated input signals from **barretters**,

Figure 9.13/Ratio meter block diagram.

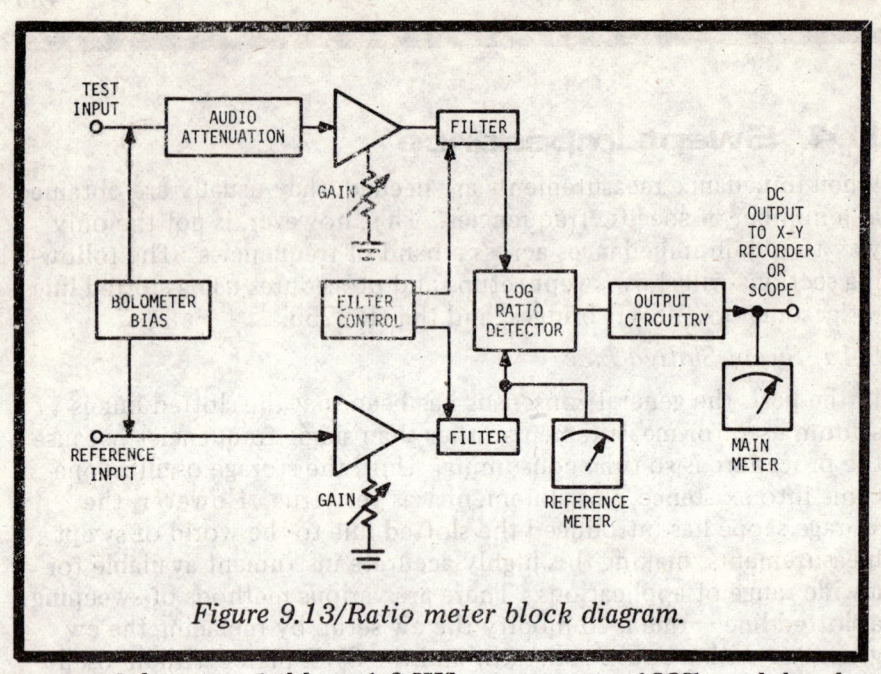

or crystal detectors fed by a 1.0 KHz square wave 100% modulated source. Barretter bias current is supplied through the unit so that the operator is not limited to the use of crystal detectors. The audio attenuator shown in the test channel is an adjustable precision attenuator which results in an overall RF attenuation range of the system of 50 dB and resolution in the order of 0.005 dB.

The ratio meter can perform swept and fixed frequency attenuation measurements accurately over wide frequency ranges since the only limitation on frequency is that of the signal source and crystal detectors or barretters. A typical frequency range of microwave measurements is 50 MHz to 18 GHz.

A test setup using the ratio meter is shown in Figure 9.14. The 1 KHz swept source is divided to provide a *test* and *reference* channel for the setup. An attenuator is placed between both the slotted line and the power divider and the reference detector and the power divider. This is to ensure that the input to the slotted line and the input to the detector both see a good impedance match. The average 2-way power divider has a VSWR of from 1.4 to 1.6:1. A slotted line exhibits a very nice VSWR of 1.06:1 and an average detector of the type used in this setup will read 1.3:1. Taken by themselves, the components look pretty good. However, when you start to connect them together

Miscellaneous Measurements

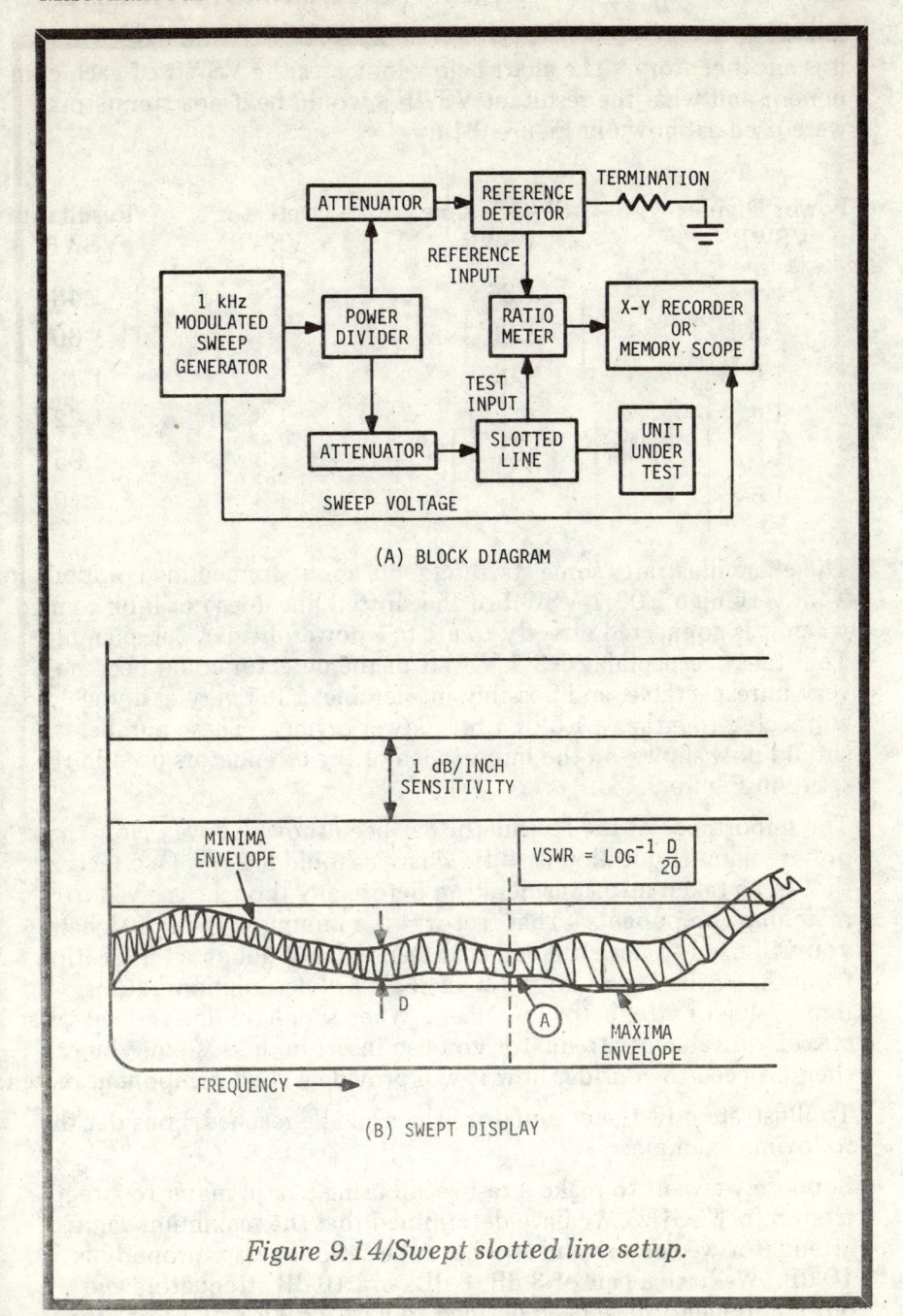

Figure 9.14/Swept slotted line setup.

(A) BLOCK DIAGRAM

(B) SWEPT DISPLAY

$$VSWR = \log^{-1} \frac{D}{20}$$

it is another story. The chart below indicates the VSWR of each component and what the resultant VSWR's would be if noa ttenuators were used as shown in Figure 9.14.

Power Divider VSWR	Slotted Line VSWR	Detector VSWR	Resultant VSWR
1.4	1.06	-	1.48
1.5	1.06	-	1.60
1.6	1.06	-	1.70
1.4	-	1.3	1.82
1.5	-	1.3	1.95
1.6	-	1.3	2.10

The chart illustrates some startling facts about connecting components. That very nice 1.06:1 VSWR of the slotted line does not look so nice when it is connected directly to a 1.5:1 power divider, for example. That fairly respectable 1.3:1 VSWR of the detector could become a very unrespectable, and possibly intolerable, 2.0:1 very easily when connected to either a 1.5 or 1.6:1 power divider. These numbers should now show you the importance of the attenuators used in the setup in Figure 9.14.

The importance of the attenuator has been shown above. Now the only remaining question is: "How large should it be?" One factor should be taken into consideration before any thought is given to matching components. That factor is the minimum level of signal you will need to have the detector respond and obtain an indication from the probe of the slotted line. This will determine what maximum value of attenuator you need. When you have determined what *maximum* value of attenuator you can insert in the system, you can then proceed to consider how it will provide a good component match.

To illustrate how the attenuator values can be reached, consider the following example:

Suppose we want to make a test setup using a ratio meter to sweep from 8 to 12 GHz. We have determined that the maximum value of attenuator we can use and still have the setup operate properly is 10 dB. We have a pair of 3 dB, 6 dB, and 10 dB attenuators and our components have been defined to have VSWR's of 1.06:1 for the slotted line; 1.3:1 for the detector; and 1.5:1 for the power

Miscellaneous Measurements 437

divider. With this information available we then set up the charts below. (All of these values assume that the attenuators have negligible VSWR's themselves.)

Power Divider VSWR	VSWR with Attenuator 3 dB	6 dB	10 dB	Slotted Line VSWR	Resultant VSWR 3 dB	6 dB	10 dB
1.5	1.28	1.13	1.05	1.06	1.36	1.2	1.11

Power Divider VSWR	VSWR with Attenuator 3 dB	6 dB	10 dB	Detector VSWR	Resultant VSWR 3 dB	6 dB	10 dB
1.5	1.28	1.13	1.05	1.3	1.66	1.46	1.36

It is very obvious that the best VSWR will be realized by using the 10 dB attenuator in the setup. By using the 10 dB attenuator, the power divider will show a 1.05:1 VSWR to the detector to go with the 1.3:1 VSWR of the detector, and it will show the same 1.05:1 VSWR to the slotted line which has a 1.06:1. These will result in VSWR's of 1.36:1 between the detector and the power divider and 1.11:1 between the slotted line and the power divider.

We stated that the best performance would be obtained by using the 10 dB attenuator. However, we also said that 10 dB was the *maximum* amount of attenuation we could use and still have the circuit operate. Therefore, if you tend to be on the conservative side and really don't care to push your setup to a maximum point, you could use the 6 dB attenuators and receive very adequate performance. Actually, the best arrangement would be a variable attenuator in each side but these are usually not too plentiful around a lab and the typical fixed attenuators which were covered in Chapter 2 are readily available and will do a more than adequate job.

With the attenuators determined, the remaining components in the setup are pretty straightforward. The detector is any device which can make the conversion of the incoming RF signal to a dc voltage to be used as a reference for the ratio meter. The slotted line should be one which operates over the required frequency range and has the proper means of connection to the other components in the setup without the use of adapters. The last point may seem trivial, but we have just gone through a very long and involved explanation of how we provide a good match between the power divider and the slotted line. Any adapters that are used will also have a VSWR and will degrade the match we have so carefully created by the use of an attenuator.

The indicating device you will use depends on what type of data you desire. If a permanent copy is needed, you should use the X-Y recorder. If you need only certain points with which to calculate VSWR, a storage scope will do just fine. (If a scope camera is available you also can have a copy of your data by taking a picture of the storage scope display and making points and calibration scales on it.)

With a test setup explained and ready to work, let us see how we obtain a swept measurement of VSWR from the method. For initial considerations we will use the X-Y recorder as an indicating device and explain the measurement procedure using it. VSWR measurements employing this technique are made in basically three steps:

(1) The X-Y recorder is adjusted for the desired sensitivity over the selected frequency range (a sensitivity of 1 dB/inch is used in Figure 9.14(b)).

(2) Repeated swept recordings are made to produce an envelope of the VSWR of the full frequency range as shown in the figure.

(3) VSWR is read in dB at any frequency in the swept range.

The audio attenuator of the ratio meter is used to calibrate the vertical displacement on the X-Y recorder to the desired resolution in dB. Then, with the slotted-line probe in a set position, the frequency band of interest is swept repeatedly, each time with the probe reset to a slightly different position. This action eventually results in clearly visible maxima and minima envelopes. The VSWR, in dB, at any given frequency is equal to the vertical distance between the maxima and minima envelopes, as illustrated in Figure 9.14(b). The envelopes are clearly visible, with a variety of plots between them.

To illustrate how you convert the display in (b) to a VSWR, consider an example. Suppose we have chosen the point in the figure marked A. The distance, D, between the maxima and minima envelopes appears to be on the order of 0.5 dB. Our formula states that VSWR = $\log^{-1} d/20$. Therefore:

VSWR = $\log^{-1} d/20$
VSWR = $\log^{-1} 0.5/20$
VSWR = $\log^{-1} (.025)$

The VSWR at A is equal to the number whose log is .025; if you consult a slide rule, calculator, or a book of tables, that number is 1.05. Thus, the VSWR at point A is 1.05:1.

Miscellaneous Measurements

Figure 9.15/Swept slotted line system.

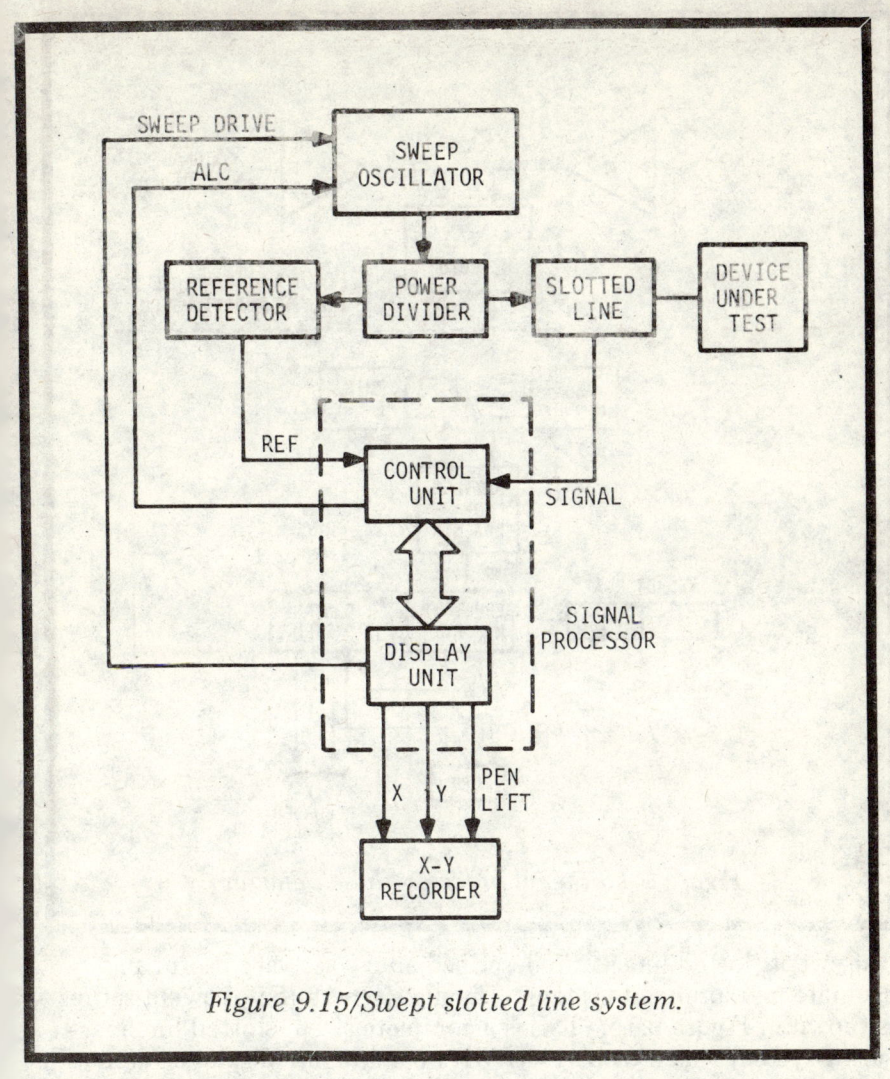

The actual measuring time can be greatly reduced by using a memory scope that can store each of the many sweeps necessary for a good measurement. This process saves the time otherwise taken as the X-Y recorder repeatedly moves mechanically from one side to the other.

Fixed frequency VSWR measurements using this technique can be made by moving the probe between maxima and minima along the slotted line at the selected frequency and reading directly from the VSWR scale of the ratio meter.

Figure 9.16/Signal processor block diagram.

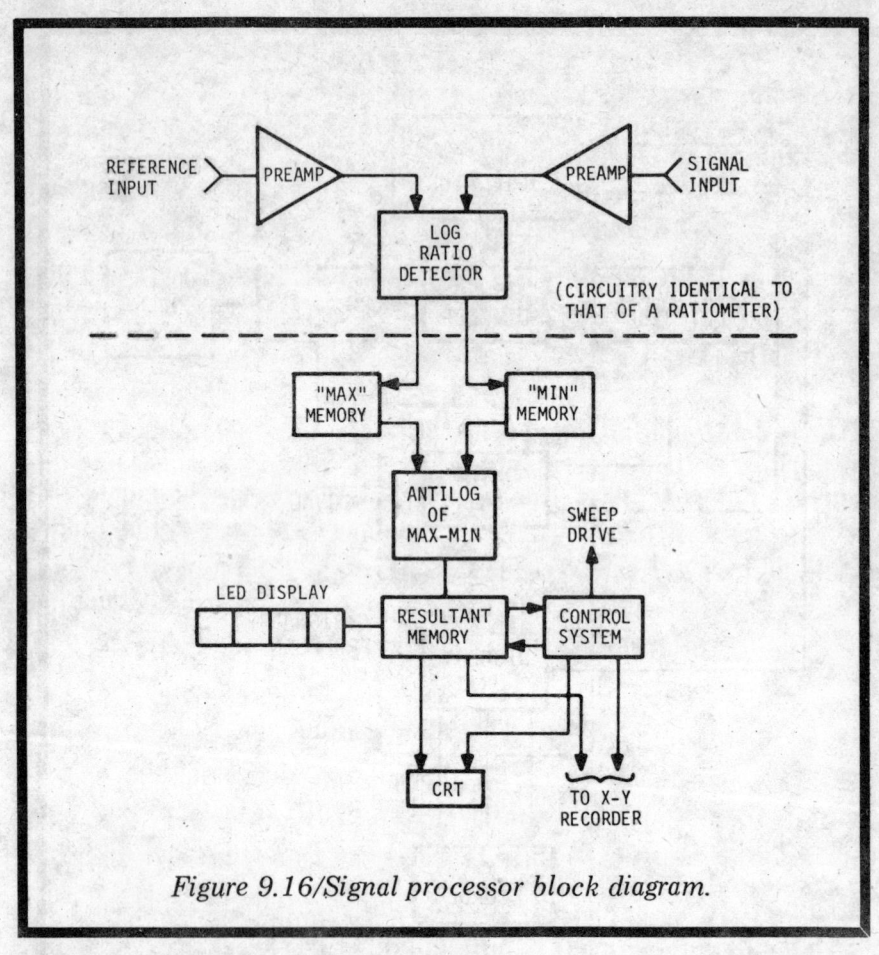

The swept slotted line setup described above is a very useful and accurate measurement system. The next generation of swept setups is shown in Figure 9.15. This is a complete swept slotted line measuring system which displays VSWR on a numeric display as well as providing an X-Y recorder presentation. Notice the similarity between the setup we previously discussed and the setup in Figure 9.15. Both setups have a sweep generator whose output is sent to a power divider with one output going to a reference detector and the other to a slotted line. The output of the slotted and the reference detector go to a ratio meter in one case and to a signal processor in the other. This is where the difference in setups occurs.

Figure 9.16 shows a block diagram of a signal processor used for swept slotted line measurements (Weinschel 1840). Once again, there

Miscellaneous Measurements 441

Figure 9.17/Swept reflectometer setup.

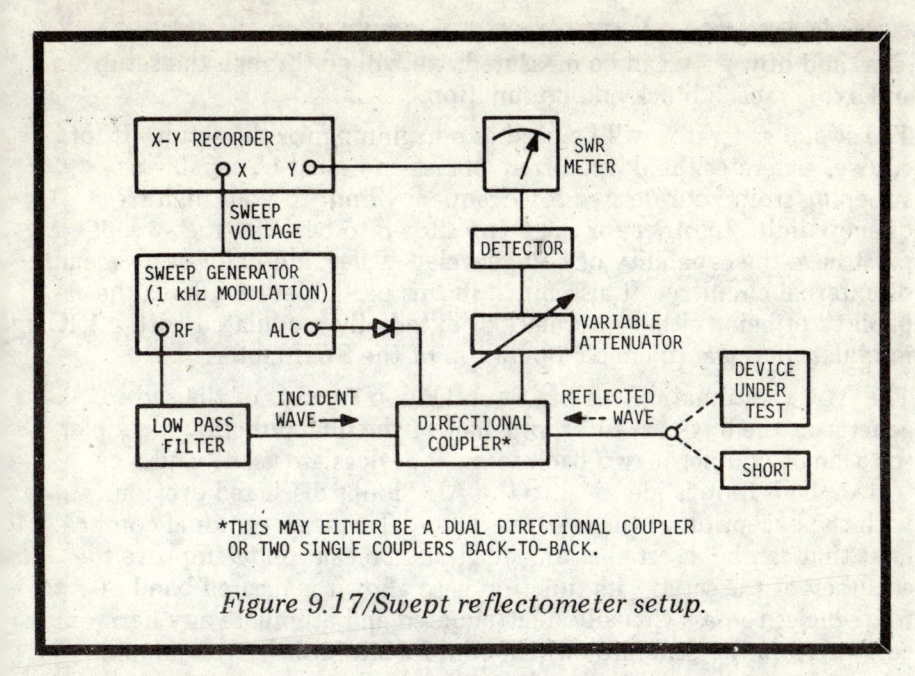

are similarities between the signal processor of Figure 9.16 and the ratio meter of Figure 9.13. Each has a reference input and a signal (or test) input with pre-amplification at each port. Each channel is fed to a log ratio detector for both devices and this is where the similarity ends. All of the circuitry below the dashed line in Figure 9.16 is the new generation which eliminates calculation to obtain VSWR. Memories to store minimum and maximum readings (previously recorded on an X-Y recorder or a storage scope and then substituted into a formula), anti-log circuitry of the max-min values, and a resultant memory produces a numerical readout on an LED display of VSWR readings ranging from 1.01:1 to 2.6:1. It is very easy to see how the signal processor (Weinschel 1840 in this case) is the absolute heart of the new generation of swept slotted line systems.

9.4.2 Swept Reflectometer

Nowhere do the terms "reflection coefficient" and "return loss" appear more understandable than in a reflectometer setup. All of the methods used for measuring impedance use reflection coefficient and return loss to do so, but they seem so much more obvious using the reflectometer.

Figure 9.17 is a typical swept reflectometer setup and illustrates the ideas of reflection coefficient and return loss. To understand these ideas and how they can be measured, we will go through the setup and explain each block and its function.

The sweep generator will be used as a beginning for the setup. It, of course, generates the desired frequencies and must be capable of sweeping from your desired low frequency limit to your high frequency limit. In order for the setup shown to be valid, the sweeper must have the capability of being leveled either internally or by means of external circuitry. It also must, in this particular case, have the capability of being either internally or externally modulated with a 1 KHz modulation signal to enable operation of the SWR meter.

The ALC (Automatic Leveling Circuit) loop consists of the sweep generator, the low-pass filter, one half of the dual directional coupler (or a single coupler if two back-to-back devices are used), and a detector. This loop is identical to the ALC loops discussed previously, with the exception of the low-pass filter. This is an optional component that can be inserted as an additional precaution to improve the accuracy of the setup. Its function is to allow the desired band of frequencies to pass virtually unattenuated and attenuate any harmonics coming from the generator which could work into the system and cause false readings. This also helps the ALC loop to do a better job, since it is leveling a much cleaner spectrum.

The use of an ALC loop in the setup eliminates some of the components that are necessary in the CW measurement systems. For example, there is a need for only one detector and SWR meter in the swept setup since the incident power level is held constant by the ALC loop. The only level that needs to be metered is the reflected level caused by the mismatches of the device under test. This reduces setup complexity and measuring time.

The heart of a reflectometer setup is the directional coupler. If a mediocre coupler is used, the accuracy of the measurements will suffer greatly. It pays to have a top quality directional coupler for the best possible measurements. The directional coupler is of such importance to a good reflectometer setup that the explanation of couplers presented previously is repeated here. A thorough understanding of directional couplers will make your understanding of the reflectometer setup presented here, and any you may use later, much more meaningful.

We called the directional coupler the heart of the measurement system. This is because it determines, to a great extent, the overall system accuracy. A term called *directivity* is the prime parameter which does this. This parameter can be explained by referring to

Miscellaneous Measurements

Figure 9.18/Directional couplers.

(a) Basic Directional Coupler

(b) Dual-Directional Coupler

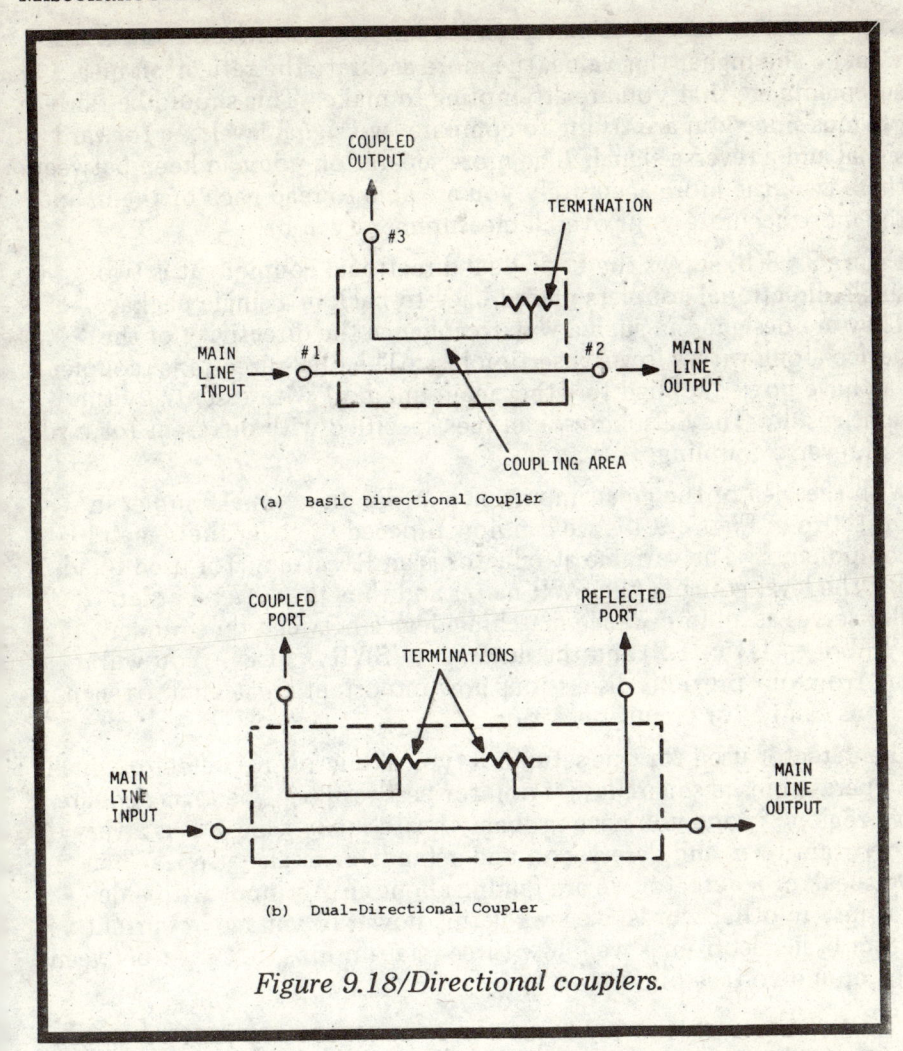

Figure 9.18(a). This is the basic directional coupler. Under normal operations, power is applied at port 1; the same level of power, minus the internal loss of the device (insertion loss), is seen at port 2; and the coupled port 3 has a reduced power level determined by the coupling region of the device. If power were to be applied at port 2, such as the case when reflections are present at a device, the power would be seen at port 1 (once again minus the insertion loss) and port 3 would be termed an *isolated* port. A measure of this isolation minus the value of normal coupling is what the device exhibits as *directivity*. In plain simple terms it is the value of desired power minus the value of undesired power (what we want minus what we don't

want). The higher this value, the more accurate the reflection measurements are that you are attempting to make. This should be fairly obvious since you are trying to compare two signal levels: a forward signal and a reverse signal. The more separation you can keep between these two, the more accurately you are able to read each of them, and the more accurate your overall measurement will be.

Figure 9.18(b) shows the basic dual-directional coupler. It is two single directional couplers placed back-to-back in a single package. They are designed in such a way to enhance the directivity of the device along with a lower insertion loss. The dual-directional coupler can have power applied to either main line port with essentially the same result. They also are sometimes specified with different forward and reverse coupling.

With the idea of the great importance of the directional coupler in the setup in Figure 9.17, we can now proceed to cover the remaining components. The variable attenuator is an RF attenuator used to adjust the level going to the SWR meter and thus the X-Y recorder. It also serves as an impedance matching device between the coupler (VSWR = 1.15 - 1.2) and the detector (VSWR = 1.3). You will recall from our previous discussions how important an attenuator such as this can be for circuit operation.

The detector used for this setup is a typical microwave detector which is operated in its square law region for best results. The term "square law region" is one you have probably heard many times. It is a very important term and deserves an understandable explanation. When we speak of a detector we are talking about an RF input with a dc output; in other words, we are relating power to voltage or current. From basic electronics we know three relationships that exist between power and voltage or current.

1. $P = EI$
2. $P = E^2/R$
3. $P = I^2 R$

The one that is of importance to us is the second one, ($P = E^2/R$).

A power applied to the input of a detector produces a voltage proportional to the square root of the power times the system impedance (R). This fact can be seen by multiplying expression #2 by R and taking the square root ($E = \sqrt{PR}$). For another power level we would achieve a different voltage, and so on. One term has remained a constant over this range — the impedance (R). This value stays constant over a certain input power range, within which the relation of $P = E^2$

Miscellaneous Measurements 445

is valid; thus the term "Square-Law Region". At either extreme of this area the R-term appears, eliminating the linear relationship of $P = E^2$. By keeping the power at the proper level, you can stay in the square law region of the detector and obtain accurate readings. So when we speak of keeping a detector in its square-law region you now know why it is important to do so.

The SWR meter and X-Y recorder are standard devices which are available from a variety of manufacturers. The SWR meter, as mentioned earlier, requires a 1 KHz modulated signal for operation. That is why the sweep generator is marked to have that capability.

With a setup defined and examined, let us go through a typical measurement procedure to see how we can obtain information from the setup. As usually, we will divide the procedure into two sections: Calibration and Measurement.

Calibration

(1) Connect the test setup as shown in Figure 9.17.

(2) Turn on the AC power to the sweep generator, X-Y recorder, and SWR meter. Keep the sweep generator in a *standby* or *RF off* position initially.

(3) Place the short on the output port of the directional coupler.

(4) Set the variable attenuator to its maximum attenuation setting.

(5) Set the SWR meter to a 0 dB setting.

(6) Set the *start* and *stop* frequencies on the sweeper which coincides with the lowest and highest frequencies you will be testing at. Then turn the sweeper to the CW position.

(7) Turn the *RF power on* to the sweep generator. Adjust the RF level until the ALC loop is locked. This may be indicated by a light, meter, or some other means that indicates when the RF output is leveled.

(8) Adjust the variable attenuator for an indication on the SWR meter. Set a convenient level that will allow you enough range on the meter to make your measurement. (If a VSWR of 1.1:1 is expected, for example, you would need at least 26 dB of range on the meter and X-Y recorder. You probably would adjust for 30 dB). Note the settings on the meter and the attenuator.

(9) Set the end limits on the X-Y recorder by putting the sweep generator on a *manual* position and running back and forth through the range of frequencies.

(10) Set the X-Y recorder at the top of the paper; set the sweep generator to a *trigger* mode; set the sweep speed for a medium speed run across the paper; and run the recorder across the paper by dropping the pen and pressing the *trigger* button. This will represent a 0 dB return loss, or all energy reflected back (as is the case for a short circuit). If you do not have an automatic pen lift on your recorder, be sure to watch your plot and manually lift the pen to prevent a retrace from being marked on the plot.

(11) Set in various values of attenuation and run them on the X-Y recorder as outlined above. Typical values are 1, 2, 3, 5, 10, 20, and 30 dB. This covers a wide range of return losses, and thus, VSWR's. You now have a set of calibration time.

(12) Set the SWR meter and attenuator back to their original setting as outlined in step 8.

(13) Put the sweep generator back into a *standby* or *RF off* position and disconnect the short from the coupler.

You now have a calibrated test setup and are ready to take your measurements.

Measurement

(1) Place the device to be tested at the directional coupler output as shown in Figure 9.17.

(2) Turn the *RF power on* and note the reading on the SWR meter. Turn the *servo* control *on* on the X-Y recorder and be sure that this reading corresponds to that on the SWR meter. Change pens on the recorder to make the data recorded a different color from that of the calibration line.

(3) Put the sweeper in the *trigger* mode; drop the pen on the paper; and press the *trigger* button. If you do not have an automatic pen lift, once again, watch the pen and lift it as soon as it gets to the end of the page and stops.

(4) Turn the *servo* on the X-Y recorder *off*; put the sweep generator in *standby* or *RF off*; remove the data from the X-Y recorder; place a clean sheet on the recorder; remove the device you have tested and you are now ready to perform a measurement on another component.

The swept reflectometer setup is a very widely used setup because the components involved in making up the setup are usually readily available. There is, generally, nothing in the setup that is not available in most labs. Although it is a straightforward setup, there still are precautions to be taken to ensure the most accurate readings possible.

The major precaution, and also the number one source of inaccuracies, is the choice of directional couplers. We have said many times how important this component is to the reflectometer setup, both CW and swept. The prime characteristic that this coupler should have is high directivity. The higher the directivity, the higher the accuracy because the incident and reflected signals are separated to a greater extent and the SWR meter will measure only the reflected levels.

Another precaution to be taken is to be sure that the sweep generator is leveled over the band of frequencies you will be testing. This is of utmost importance since the only signals you will be monitoring are the reflected ones. The leveling loop is relied on to keep the incident signals constant. Any wild variations in the incident wave signals will completely invalidate any calibration that has been done.

A third, and final, precaution to be taken is in your choice of a variable attenuator. As was mentioned earlier, this attenuator is an RF attenuator and must hold calibration (or at least have a set calibration) over the entire RF range of frequencies you will be testing. One problem with this type of attenuator is the presence of a "hole" in its calibration. When these attenuators are calibrated they are usually only checked at specific frequencies. They are not swept but only checked at certain CW frequency points. It is entirely possible that in between these frequencies there are areas where they are not within calibration and the attenuation increases greatly. These are called "holes" in the attenuation. This type of problem can cause great difficulty when you are trying to make an accurate measurement.

9.4.3 Swept VSWR Bridge

The VSWR bridge has many benefits applicable to the microwave industry. The previous section has glorified the reflectometer as the test setup to use because of the high directivity achievable from directional couplers. Although there are differences between the bridge and the directional couplers, the bridge available today also has many areas in common with the directional coupler. The bridge performs the same function as a high directivity coupler, and the two devices may be used virtually interchangeably in reflectometer and network analyzer applications. The only significant difference between the two is that the bridge has a 6 dB insertion loss in both the input-to-test and test-to-output paths, while the coupler has a very low (0.1 to 0.5 dB) loss in the main line direction and a loss equal to the coup-

Figure 9.19/VSWR bridge test setup.

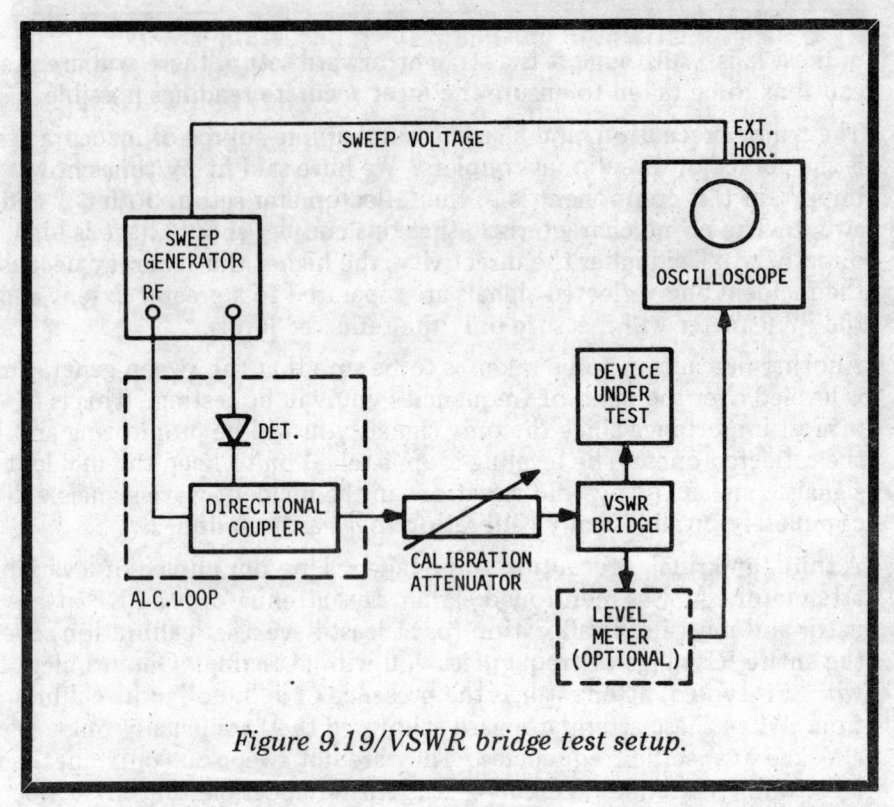

ling value (10, 20, 30 dB) in the coupled line. If the coupling value is in the order of 12 dB, the two devices should be interchangeable.

With a comparison setup between the bridge and directional coupler, let us investigate the swept VSWR bridge method of impedance measurement. Figure 9.19 shows a typical swept VSWR bridge setup. The sweep generator is one that will cover the frequency range required at a power output level sufficient to enable readings to be taken at the output, while being low enough to preserve the components along the way. Most sweep generators have an output of 100 mw or less and will not cause any power problems. However, there are some narrow band sweep generators that are capable of up to 5 watts output when a solid-state power amplifier is used. This could definitely cause problems to devices such as the calibration attenuator which probably will only handle 2 watts of power.

The ALC loop is the standard configuration used previously in other swept setups. Usually a 20 dB coupler is used since the detector does not need a very high level to operate in its square-law region. This low

Miscellaneous Measurements

RF input level to the detector produces a simultaneous low level dc voltage which is used to drive the ALC circuitry within the sweeper. This dc level is low to keep the input circuitry from being overdriven. There are times when a 30 dB coupler is needed to ensure proper operation. However, generally a 20 dB coupler will do the job very nicely.

The calibration attenuator used in the setup is, once again, an RF attenuator. This variable attenuator must hold calibration over the entire frequency range that you are testing. This should be checked very carefully so that the overall system performance can be predicted from one end of the frequency band to the other. Another parameter to watch is the input VSWR of the attenuator. It is not as important to check it at a single setting as it is when the attenuator is varied over a range of attenuation. Serious mismatches can occur throughout the system if the VSWR varies drastically at the calibration attenuator. The attenuator, of course, could be a series of two or three attenuators, if necessary. One would be in 10 dB steps; one in 1.0 dB steps; and the last in 0.1 dB steps. This would be necessary only if accuracies of this magnitude are needed.

The VSWR bridge is the device which will register the reflections from the device under test. This, along with the calibration attenuator, is one of the key components in the setup. The oscilloscope is an indicating device which could very well be replaced by an X-Y recorder if a permanent hard copy of the data is required.

The level meter converts the dc voltage from an external detector (in the bridge, in this case) to an indication of RF power (calibrated in dB per division). The device is used to increase the sensitivity of the measuring system; it permits accurate low level RF measurements to −40 dBm. It is marked "optional" on the setup because some bridges do not have the detector built in. Their output is an RF power level which must be detected before it can be displayed. The ideal setup, however, is using the bridge with an internal detector and then using a level meter to aid in the system sensitivity.

With the test setup described and explained, the one remaining job is to outline the measurement procedure.

Calibration

(1) Connect the test setup as shown in Figure 9.19.

(2) Adjust the calibration attenuator for its maximum attenuation and apply AC power to the sweep generator, oscilloscope, and level meter (if used). The sweep generator should be in a *standby*

or *RF off* position. Set the *start* and *stop* frequencies on the sweeper to correspond to your low and high frequency limits — set the sweeper to the *sweep position*.

(3) When the sweeper and scope have power (after any time delay) turn the sweeper to *RF on*. Be sure that the RF output is leveled. This point usually can be checked by observing the "RF unleveled" light on the front of most sweepers — when the light is on, the output is unleveled.

(4) The *test* port of the bridge should now be left *open*. This corresponds to an infinite VSWR (a 0 dB return loss). Set the sweeper *sweep speed* to obtain a solid sweep on the scope with no flickering.

(5) Using the gain of the oscilloscope and the calibration attenuator, set the 0 dB return loss (VSWR = ∞) line to the bottom of the scope face.

(6) Turn the sweep generator to the *RF off* or *standby* position and prepare for the measurement.

Measurement

Connect the device to be tested to the test port of the bridge; the display will move to some level on the scope corresponding to the return loss, and thus VSWR, of the device. Overlays for the oscilloscope relating return loss to VSWR are available and are shown in Figure 9.20. They are transparent so that the value of the reflection can be read on the face of the scope directly in either VSWR or dB of return loss. The overlays include four lower value VSWR scales in addition to the one for infinity so that intermediate and low VSWR values can be measured with higher resolution. The lowest scale is such that a VSWR as small as 1.066 occupies the full scale of the scope. To illustrate how that scale would work, consider an example. Suppose we calibrated the setup with 10 dB of attenuation set in at the calibration attenuator. When our device is attached the trace on the scope sweeps from 5 dB at the low frequencies down to the 3 dB level at the high end. Thus, the VSWR is approximately 1.42 at the low end (15 dB return loss) and 1.58 at the high (13 dB return loss). As mentioned previously, for a permanent record of the VSWR of a device, an X-Y recorder is used in place of the oscilloscope and overlay. Values of attenuation ranging from 0-30 dB are inserted by way of the calibration attenuator and a swept calibration line is recorded for each. The attenuator is then returned to 0 dB and the device is attached and swept over the calibrated frequency range. The values of return loss indicated can

Miscellaneous Measurements

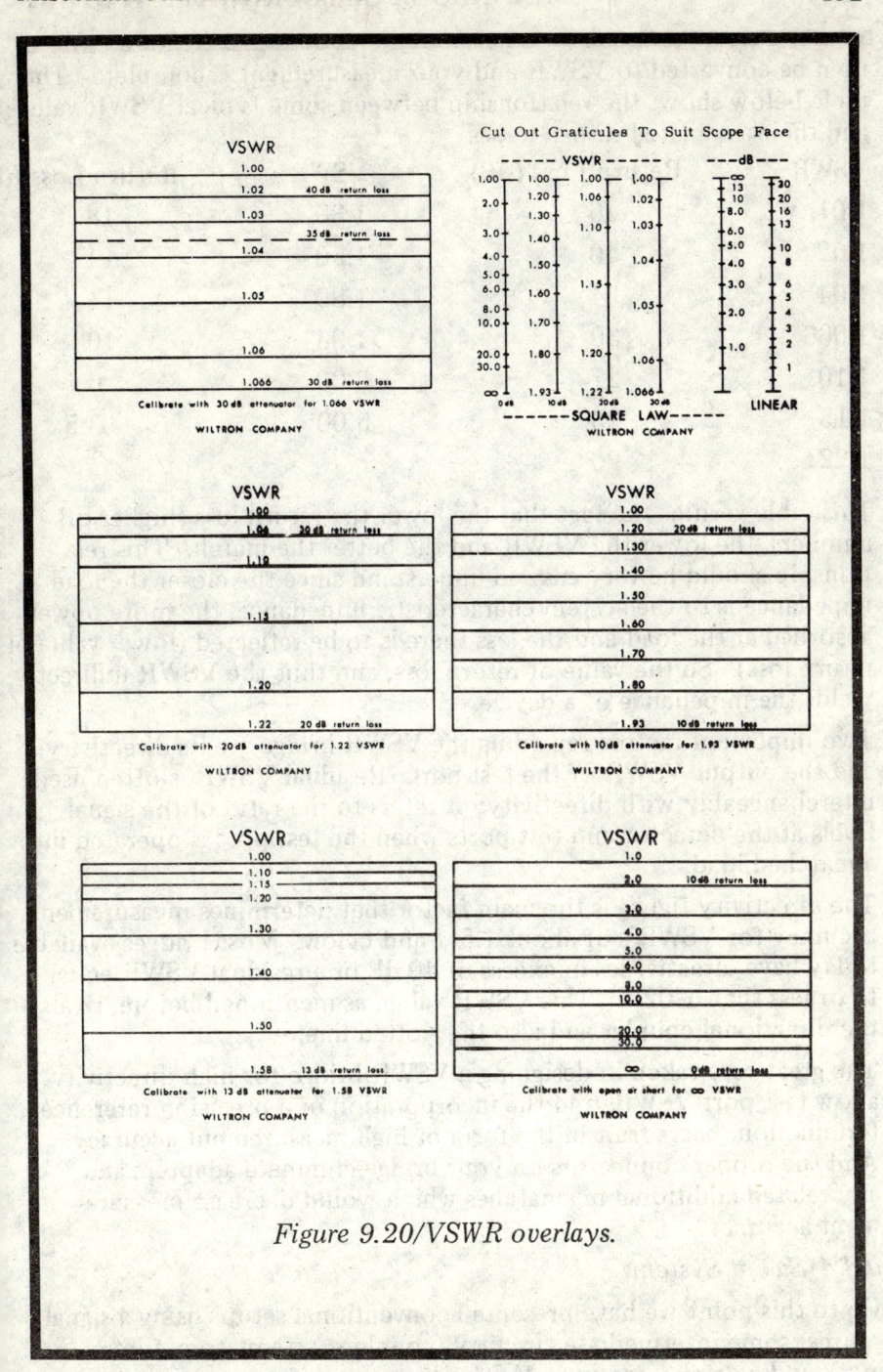

Figure 9.20/VSWR overlays.

then be converted to VSWR and your measurement is complete. The table below shows the relationship between some typical VSWR values and their associated return losses.

VSWR	Return Loss (dB)	VSWR	Return Loss (dB)
1.01	46	1.30	18
1.02	40	1.50	14
1.04	34	1.80	11
1.066	30	1.93	10
1.10	27	3.00	6
1.15	23	5.00	3.5
1.22	20		

This table verifies the fact that the lower the return loss (higher dB number), the lower the VSWR and the better the match. This relationship should be very easy to understand since the closer the load impedance is to the system characteristic impedance, the more power absorbed at the load and the less there is to be reflected (lower value of return loss). So the value of return loss, and thus the VSWR indirectly yields the impedance of a device.

Two important factors regarding the VSWR bridge are its directivity and the output VSWR of the test port. Residual VSWR is often used interchangeably with directivity; it refers to the ratio of the signal levels at the detector and test ports when the test port is operated into a matched load.

The directivity figure is the main factor that determines measurement accuracy for VSWR's of about 1.5:1 and below. Most bridges available today have directivities in excess of 40 dB or a residual VSWR equal to or less than 1.02:1. This VSWR value, as mentioned before, rivals the directional coupler and also the slotted line.

The great care taken in designing a VSWR bridge for high directivity, a low test port VSWR, and the incorporation of a precision reference termination, bears fruit in the form of high measurement accuracy. And the proper connectors on your bridge eliminate adapters and any related additional mismatches which would decrease measurement accuracy.

9.4.4 Swept Systems

Up to this point we have presented conventional setups using a signal source, some intermediate circuitry (couplers, attenuators, filters, etc.), and a display section. Many times your reference was set up;

Miscellaneous Measurements 453

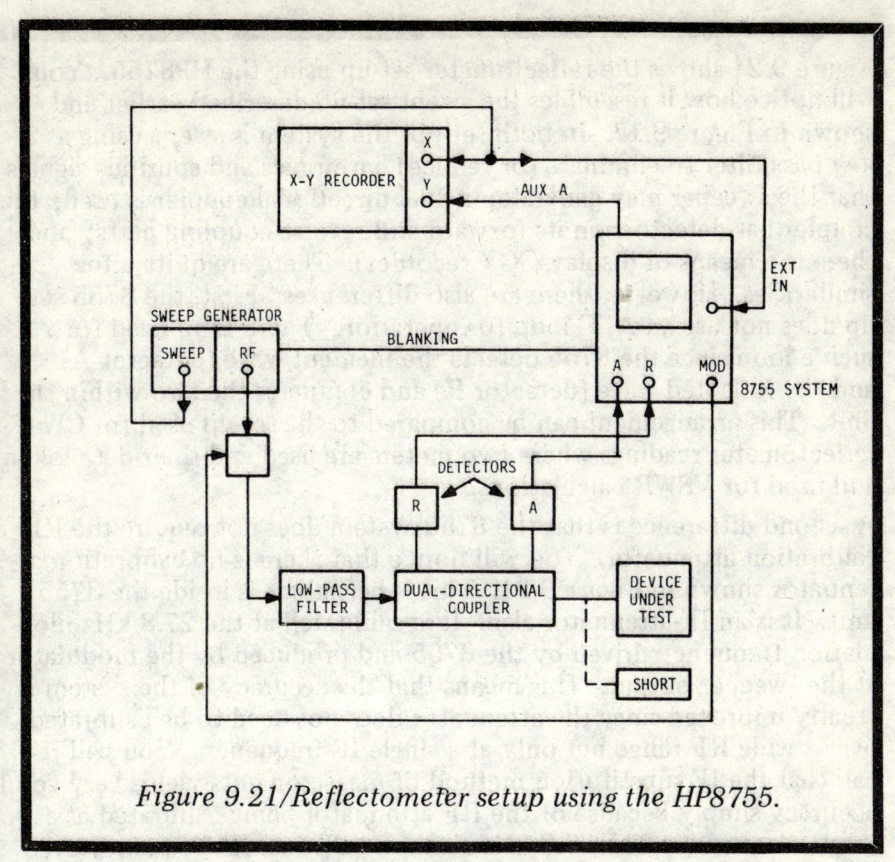

Figure 9.21/Reflectometer setup using the HP8755.

the system disconnected; and the device under test was then connected. It is during this disconnect process or when you are marking a scope with a grease pencil that you wish that someone would provide you with a better and more efficient way of measuring an impedance. Well, it just so happens that Hewlett-Packard has done just that. We are speaking of the HP8755 frequency response test set. This system has been mentioned before in Chapter 2 on test equipment. The intent here is to discuss its applications for impedance (VSWR) measurements.

The 8755 measurement system is capable of making impedance measurements in a variety of ways. It can function as a reflectometer set-up by using a technique similar to that described earlier with a dual-directional coupler; it is capable of swept measurements on a slotted line, and it becomes one of the simplest set-ups available when connected with a reflectometer bridge. Let us investigate each of these and make comparisons with the swept set-ups covered earlier.

Figure 9.21 shows the reflectometer set-up using the HP8755. You will notice how it resembles the swept set-up described earlier and shown in Figure 9.17. In both set-ups the system is swept using a low pass filter to eliminate (or reduce) harmonics and spurious signals that the sweeper may generate; a dual directional coupler is used; the coupler has detectors on its forward and reverse coupling parts; and there is a means of display (X-Y recorder). There are quite a few similarities. However, there are also differences. First, the 8755 set-up does not use an ALC loop for operation. There is no need for such a loop since the 8755 detects the incident wave (detector A) and the reflected wave (detector R) and compares the two within the unit. This arrangement can be compared to the set-up used for CW reflectometer readings where two meters are used and their data taken and used for VSWR calculations.

A second difference is that the 8755 system does not require the RF calibration attenuator. You will notice that there is no calibration attenuator shown in Figure 9.21. This is because it is inside the 8755 unit. It is an IF attenuator also. It is calibrated at the 27.8 kHz modulation frequency driven by the 8755 and produced by the modulator at the sweeper output. This means that the accuracy of the system is greatly improved since the attenuator does not need to be calibrated over a wide RF range but only at a single IF frequency. You will recall that the IF substitution method of measurements yields very good accuracy simply because of the RF attenuator being calibrated at a single frequency. The 27.8 kHz signal is called an IF frequency even though it is not obtained by a mixing process. It is obtained by a demodulation of an RF signal. We refer to it as an IF to illustrate it is a single frequency resultant.

A measurement procedure for the reflectometer set-up would be as follows:

Calibration

1. Connect the set-up as shown in Figure 9.21. Turn on the AC to the sweep generator, X-Y recorder, and 8755 test set. Leave the sweeper in the *standby* or *RF off* position and the X-Y recorder set at *servo off*.

2. Adjust the *start* indicator on the sweeper to your lowest frequency of interest and the *stop* indicator to the highest frequency. Adjust the sweep speed for a solid line on the scope of the 8755.

3. Connect a short to the output port of the directional coupler, as shown.

Miscellaneous Measurements

4. Turn the RF on to the sweep generator. By means of the RF level control on the sweeper and the *offset* dB control on the 8755, adjust the sweep trace to the top of the scope. This corresponds to 0 dB return loss since the short reflects all energy back toward the source. This is the main calibration level.
5. Turn the *servo* on the X-Y recorder to the *on* position and place the sweep generator in the *manual* position. Run the *manual* control from one end to the other and set the left and right limits on the recorder.
6. Set the recorder so that the 0 dB line is at the top of the paper on the recorder. Using the *offset* dB control on the 8755, put in the specified amount of attenuation that you will need for your measurement at the bottom of the paper (30 dB, for example, will allow you to read a 1.055:1 VSWR). Set the final amount of attenuation at the bottom of the page. Recheck to be sure the 0 dB point is still the same. Repeat the procedure until the 0 dB and final attenuation settings are where they should be.
7. Set the plotter pen back to the 0 dB point. Set the sweeper on *trigger* and run the 0 dB plot. Repeat for all desired values of attenuation (such as 1, 2, 3, 5, 10, etc.).
8. Place the pen back on the 0 dB point and turn the *servo* off on the X-Y recorder. Turn the sweep generator to *standby* or *RF off*. Remove the short from the coupler output. Your system is now calibrated and is ready for measurements to be taken on it.

Measurement

1. Connect the device to be tested to the output of the dual-directional coupler. Turn the sweep generator to *RF on*. There should be a trace on the CRT which lies somewhere between the 0 dB line and the maximum attenuation you set for your measurement. (If the trace goes below the maximum attenuation figure you will have to go back and calibrate the system for a wider range of operation [40-45 dB may do the trick]).
2. If the trace is in a space between 0 dB and maximum attenuation, turn the X-Y recorder *servo on*. Place the sweep generator in the *trigger* mode; drop the pen on the X-Y recorder on the paper; and press the *trigger* button. (If an automatic penlift is not used, be sure to watch the recorder and manually lift the pen at the end of the sweep to prevent a sweep retrace from being printed across your data.)

This method of swept reflectometer measurements depends on the square law reflection characteristics of the R detector (we have discussed square law responses earlier in this chapter). This method is generally quite satisfactory since the range of return loss measured in a set-up of this type is usually a maximum of 30 to 34 dB.

The second set-up to be discussed is the swept slotted line. You will recall that two set-ups were presented in Section 9.4.1. One used the ratio meter for signal processing; the second set-up used the Weinschel 1840 as a signal processor; and the set-up to be presented here uses the 8755 as a processor. Figure 9.22 shows the swept slotted line set-up using the 8755. Various displays are also presented in this figure. You will note that the first two displays are for CW readings. The first is for a single frequency and the second is for six different CW frequencies. This shows the versatility of the overall system. We are, however, concerned with swept measurements here and will discuss that aspect of the set-up.

The set-up shown in Figure 9.22 contains many of the standard components of a 8755 set-up: the modulator, which places a 27.8 kHz modulation to aid in processing; reference (R) and measurement (A) detectors; and, in this case, a standard slotted line which will operate over the frequency range you desire. An additional component, the sweep adapter, is added to the set-up to help provide a good source match for the slotted line over the entire swept frequency range. This device can be likened to our previous discussion of swept slotted lines when we pointed out the importance of an attenuator between the input power divider and the slotted line to provide a good match. This unit, of course, is a much more sophisticated component than the attenuator but it accomplishes the same purpose. It also provides access to the incident wave which can be used as a reference signal for the 8755.

There is no need for a short for calibration of the set-up since this is a ratio measurement. The calibration is performed on the 8755 itself. The display is calibrated in dB/division since a total variation is measured and VSWR is calculated from that number. The display at the bottom of Figure 9.22 is achieved by setting the sweeper to the proper frequency limits for your application, letting the sweeper sweep that range, and moving the probe carriage on the slotted line up and down the line. The sweeper creates the display from left to right and the probe movement records the amplitude variations. A storage display on the 8755 stores these variations and they can be measured directly from the CRT.

Miscellaneous Measurements

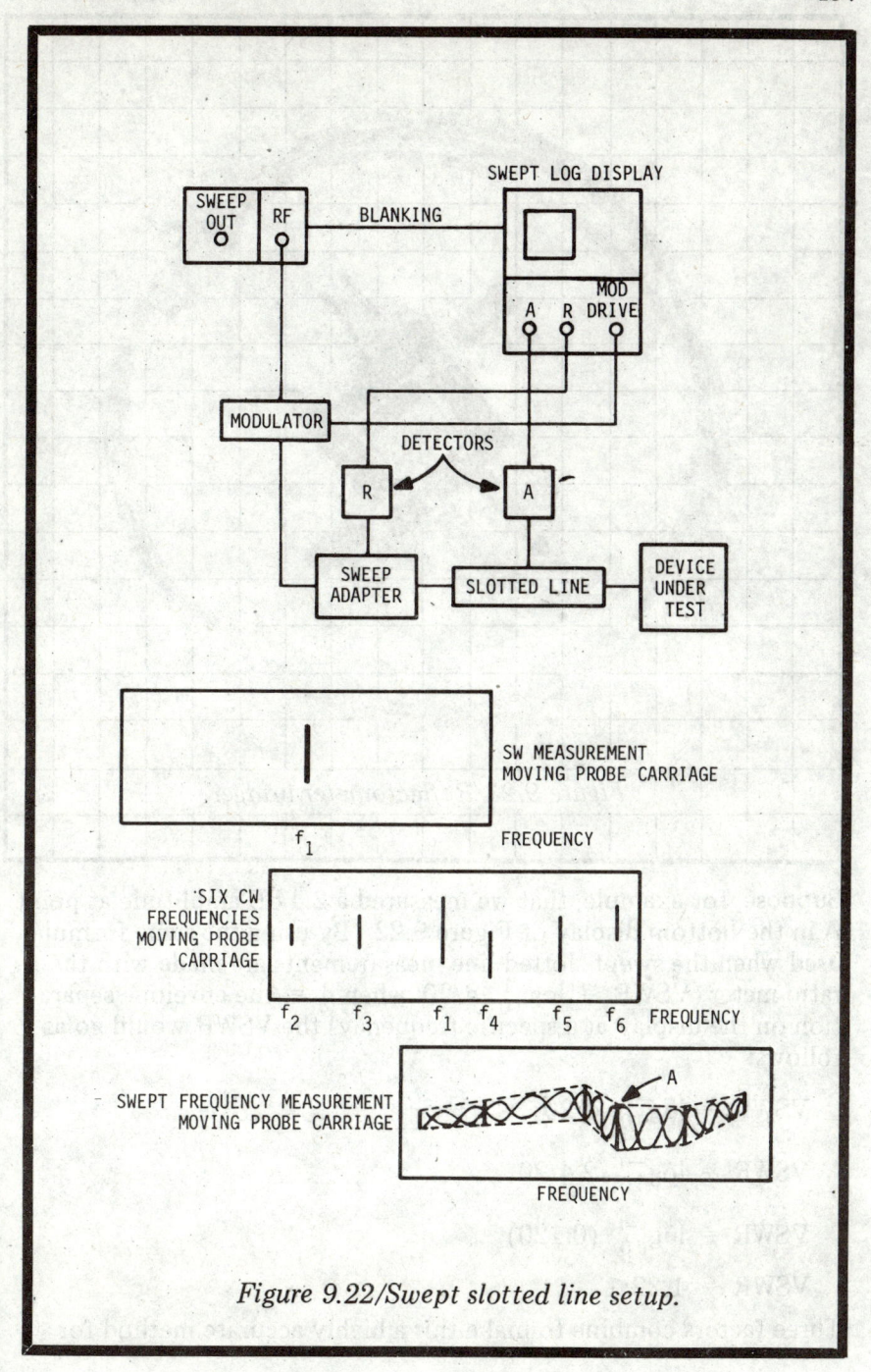

Figure 9.22/Swept slotted line setup.

Figure 9.23/Reflectometer bridge.

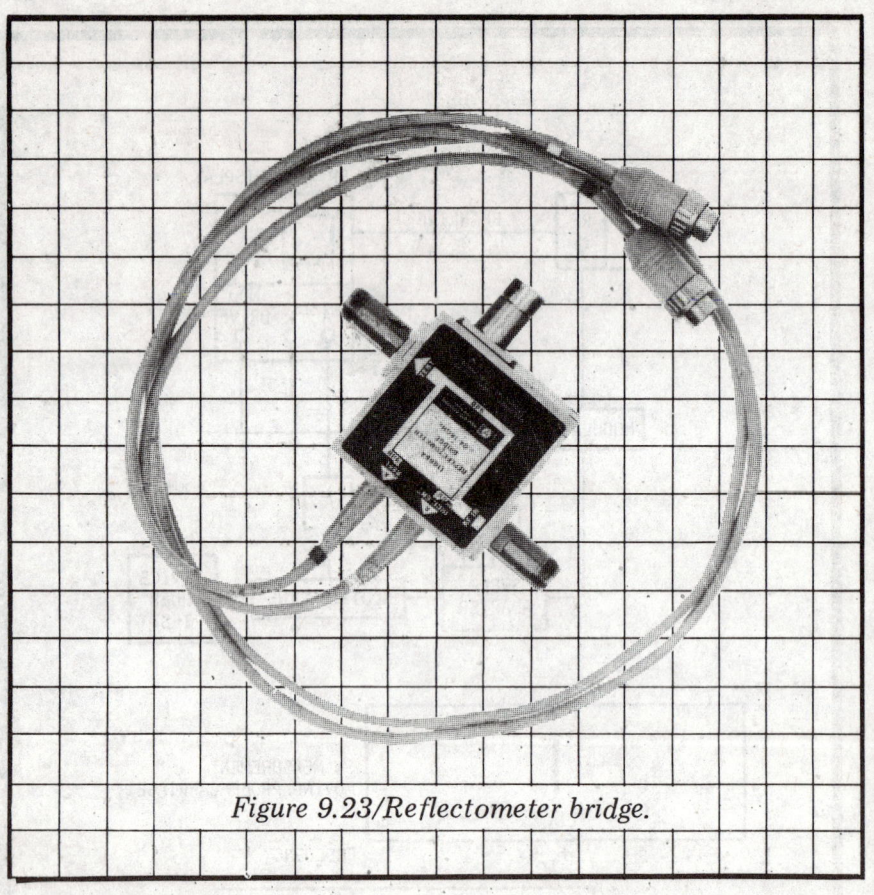

Suppose, for example, that we measured a 2.4 dB amplitude at point A in the bottom display of Figure 9.22. By using the same formula used when the swept slotted line measurements are made with the ratio meter (VSWR = $\log_{10}^{-1}$ d/20, when d = the envelope separation on the display at a specific frequency) the VSWR would go as follows:

VSWR = $\log_{10}^{-1}$ d/20

VSWR = $\log_{10}^{-1}$ 2.4/20

VSWR = $\log_{10}^{-1}$ (0.120)

VSWR = 1.32:1

Three factors combine to make this a highly accurate method for

Miscellaneous Measurements

Figure 9.24/Reflectometer bridge setup.

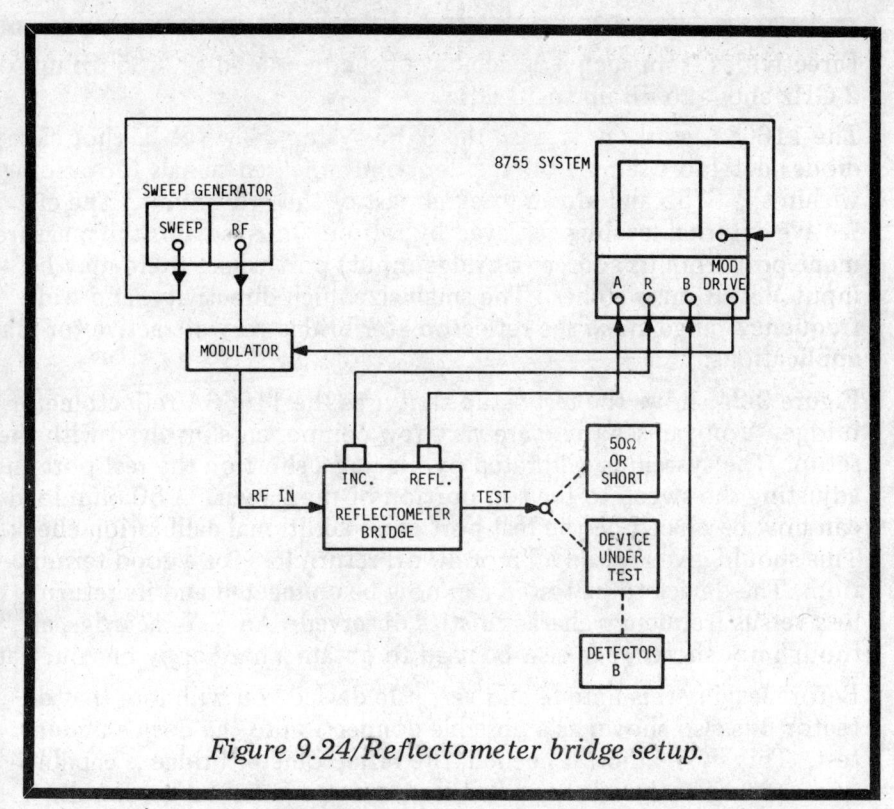

making swept slotted line measurements. First, since only the width of the resulting envelope is measured, frequency response is not a factor; second, residual VSWR (equivalent to directivity) is greater than 34 dB (1.03:1 VSWR) over the full range of 1.8 to 18 GHz; and third, the 8755 instrument uncertainty becomes negligible for low values of measured SWR.

The third setup using the 8755 system is one that uses a *reflectometer bridge*. The reflectometer bridge we are referring to is the HP11666A shown in Figure 9.23. With the bridge it is possible to make reflection measurements from 40 MHz to 18 GHz with one coupling device. Operation of this type of device is based on principles of the resistive Wheatstone Bridge extended to microwave frequencies. When three bridge arms are 50 ohms, the voltage across corners of the bridge is directly proportional to the reflection coefficient of the device connected to the fourth port (test port in our case). Directivity is then a measure of how well the bridge circuit is balanced with a 50 ohm termination connected (this would be a good calibration check for the birdge to check its internal balance with an ideal load at the test port).

Directivities from such a device have been measured as > 35 dB up to 2 GHz and >25 dB up to 18 GHz.

The 11666A works only with the 8755 system. Two HCD (hot carrier diode) detectors sample the incident and reflected signals for ratioing within the 8755 and are an integral part of the bridge unit. The effective external leveling achieved by ratioing thus isolates the measurement port from its source (a bridge input) port where there may be input VSWR mismatches. The small size, high directivity, and wide frequency range make the reflectometer bridge very attractive for many applications.

Figure 9.24 shows the test setup that uses the 11666A reflectometer bridge. You can see there are very few components involved with the setup. The system is calibrated by placing a short on the test port and adjusting the sweep to the top portion of the screen. A 50 ohm load can now be placed on the test port as an additional calibration check. This should give around a 35 or 40 dB return loss for a good termination. The device to be tested can now be connected and its return loss versus frequency characteristics observed. An X-Y recorder, although not shown, can also be used to obtain a hard copy of your data

Before leaving this unique and versatile device you will note that detector B is also shown as a possible connection to the device under test. This shows that, if needed, the reflectometer bridge is capable of providing both reflection and transmission information simultaneously.

The HP8755 and its associated components have been shown to be very useful and accurate for swept reflection (VSWR) measurements. All of its uses have not been covered in this chapter — it would surely take too much time and space to cover them all. Your individual application may use a setup similar to one we have shown, or you may derive your own arrangement that will be new and unique.

Chapter Summary

Usually when you see a chapter called "Miscellaneous," you think that it is a catch-all chapter of loose ends. This was not the case in this chapter. We have presented four measurements that are very important to the microwave induatry.

The section on Phase Noise is of utmost importance in characterizing a microwave source for short-term stability. The section on Q Measurements is one that is absolutely necessary if you are using high

Miscellaneous Measurements 461

frequency cavity resonators; measurements with the TDR aid greatly in characterizing the lifelines for microwaves — microwave cables and connectors (to say nothing of characterizing some microwave devices also); and a means of providing swept impedance measurements adds a much needed capability to the microwave industry. So you can see that these measurements were not loose-ends. They are very important, necessary procedures which aid in allowing the world of microwaves to operate at its peak.

Appendix A

Microwave Bands

OFFICIAL JCS BAND DESIGNATION

A	Band	0	MHz	to	250	MHz
B	Band	250	MHz	to	500	MHz
C	Band	500	MHz	to	1000	MHz
D	Band	1.0	GHz	to	2.0	GHz
E	Band	2.0	GHz	to	3.0	GHz
F	Band	3.0	GHz	to	4.0	GHz
G	Band	4.0	GHz	to	6.0	GHz
H	Band	6.0	GHz	to	8.0	GHz
I	Band	8.0	GHz	to	10.0	GHz
J	Band	10.0	GHz	to	20.0	GHz
K	Band	20.0	GHz	to	40.0	GHz
L	Band	40.0	GHz	to	60.0	GHz
M	Band	60.0	GHz	to	100.0	GHz

OLD JCS BAND DESIGNATION

VHF	Band	100	MHz	to	300	MHz
UHF	Band	300	MHz	to	1.0	GHz
L	Band	1.0	GHz	to	2.0	GHz
S	Band	2.0	GHz	to	4.0	GHz
C	Band	4.0	GHz	to	8.0	GHz
X	Band	8.0	GHz	to	12.4	GHz
Ku	Band	12.4	GHz	to	18.0	GHz
K	Band	18.0	GHz	to	26.0	GHz
Ka	Band	26.0	GHz	to	40.0	GHz
Millimeter		40.0	GHz	to	100.0	GHz

US MICROWAVE BAND

P	Band	225	MHz	to	390	MHz
L	Band	390	MHz	to	1.55	GHz
S	Band	1.55	GHz	to	5.20	GHz
C	Band	3.90	GHz	to	6.20	GHz
X	Band	5.20	GHz	to	10.9	GHz
K	Band	10.9	GHz	to	36.0	GHz
Ku	Band	15.25	GHz	to	17.25	GHz
K_1	Band	15.35	GHz	to	24.5	GHz
Q	Band	36.0	GHz	to	46.0	GHz
V	Band	46.0	GHz	to	56.0	GHz
W	Band	56.0	GHz	to	100.0	GHz

Appendix B

VSWR Versus Return Loss (R)

VSWR	R dB	VSWR	R dB	VSWR	R dB
1.001	66.025	1.041	33.941	1.081	28.196
1.002	60.009	1.042	33.763	1.082	28.093
1.003	56.491	1.043	33.536	1.083	27.992
1.004	53.997	1.044	33.341	1.084	27.892
1.005	52.063	1.045	33.150	1.085	27.794
1.006	50.484	1.046	32.963	1.086	27.696
1.007	49.149	1.047	32.780	1.087	27.600
1.008	47.993	1.048	32.602	1.088	27.505
1.009	46.975	1.049	32.427	1.089	27.411
1.010	46.064	1.050	32.256	1.090	27.318
1.011	45.240	1.051	32.088	1.091	27.226
1.012	44.489	1.052	31.923	1.092	27.135
1.013	43.798	1.053	31.762	1.093	27.046
1.014	43.159	1.054	31.604	1.094	26.957
1.015	42.564	1.055	31.449	1.095	26.869
1.016	42.007	1.056	31.297	1.096	26.782
1.017	41.485	1.057	31.147	1.097	26.697
1.018	40.993	1.058	31.000	1.098	26.612
1.019	40.528	1.059	30.856	1.099	26.528
1.020	40.086	1.060	30.714	1.100	26.444

VSWR Versus Return Loss (R)

VSWR	R dB	VSWR	R dB	VSWR	R dB
1.021	39.667	1.061	30.575	1.102	26.281
1.022	39.867	1.062	30.438	1.104	26.120
1.023	38.885	1.063	30.303	1.106	25.963
1.024	38.520	1.064	30.171	1.108	25.809
1.025	38.170	1.065	30.040	1.110	25.658
1.026	37.833	1.066	29.912	1.112	25.510
1.027	37.510	1.067	29.785	1.114	25.364
1.028	37.198	1.068	29.661	1.116	25.221
1.029	36.898	1.069	29.538	1.118	25.081
1.030	36.607	1.070	29.417	1.120	24.943
1.031	36.327	1.071	29.298	1.122	24.808
1.032	36.055	1.072	29.181	1.124	24.675
1.033	35.792	1.073	29.066	1.126	24.544
1.034	35.537	1.074	28.952	1.128	24.415
1.035	35.290	1.075	28.839	1.130	24.289
1.036	35.049	1.076	28.728	1.132	24.164
1.037	34.816	1.077	28.619	1.134	24.042
1.038	34.588	1.078	28.511	1.136	23.921
1.039	34.367	1.079	28.405	1.138	23.803
1.040	34.151	1.080	28.299	1.140	23.686
1.142	23.571	1.410	15.385	3.100	5.811
1.144	23.457	1.420	15.211	3.200	5.617
1.146	23.346	1.430	15.043	3.300	5.435
1.148	23.235	1.440	14.879	3.400	5.265
1.150	23.127	1.450	14.719	3.500	5.105
1.152	23.020	1.460	14.564	3.600	4.956
1.154	22.914	1.470	14.412	3.700	4.815
1.156	22.810	1.480	14.264	3.800	4.682
1.158	22.708	1.490	14.120	3.900	4.556
1.160	22.607	1.500	13.979	4.000	4.437
1.162	22.507	1.520	13.708	4.100	4.324
1.164	22.408	1.540	13.449	4.200	4.217
1.166	22.311	1.560	13.201	4.300	4.115
1.168	22.215	1.580	12.964	4.400	4.018
1.170	22.120	1.600	12.736	4.500	3.926

Appendix B: VSWR Versus Return Loss (R)

continued

VSWR	R dB	VSWR	R dB	VSWR	R dB
1.172	22.027			4.600	3.838
1.174	21.934	1.620	12.518	4.700	3.753
1.176	21.843	1.640	12.308	4.800	3.673
1.178	21.753	1.660	12.107	4.900	3.596
1.180	21.664	1.680	11.913	5.000	3.522
		1.700	11.725		
1.182	21.576			5.200	3.383
1.184	21.489	1.720	11.545	5.400	3.255
1.186	21.403	1.740	11.370	5.600	3.136
1.188	21.318	1.760	11.202	5.800	3.025
1.190	21.234	1.780	11.039	6.000	2.923
1.192	21.151	1.800	10.881		
1.194	21.069			6.200	2.827
1.196	20.988	1.820	10.729	6.400	2.737
1.198	20.907	1.840	10.581	6.600	2.653
1.200	20.828	1.860	10.437	6.800	2.573
		1.880	10.298	7.000	2.499
1.210	20.443	1.900	10.163		
1.220	20.079			7.200	2.428
1.230	19.732	1.920	10.032	7.400	2.362
1.240	19.401	1.940	9.904	7.600	2.299
1.250	19.085	1.960	9.780	7.800	2.239
1.260	18.783	1.980	9.660	8.000	2.183
1.270	18.493	2.000	9.542		
1.280	18.216			8.200	2.129
1.290	17.949	2.100	8.999	8.400	2.078
1.300	17.692	2.200	8.519	8.600	2.029
		2.300	8.091	8.800	1.983
1.310	17.445	2.400	7.707	9.000	1.938
1.320	17.207	2.500	7.360		
1.330	16.977	2.600	7.044	9.200	1.896
1.340	16.755	2.700	6.755	9.400	1.855
1.350	16.540	2.800	6.490	9.600	1.816
1.360	16.332	2.900	6.246	9.800	1.779
1.370	16.131	3.000	6.021	10.000	1.743
1.380	15.936				
1.390	15.747				
1.400	15.563				

VSWR Versus Return Loss (R) continued

VSWR	R dB	VSWR	R dB
11.000	1.584	42.000	0.414
12.000	1.451	44.000	0.395
13.000	1.339	46.000	0.378
14.000	1.243	48.000	0.362
15.000	1.160	50.000	0.347
16.000	1.087		
17.000	1.023	55.000	0.316
18.000	0.966	60.000	0.290
19.000	0.915	65.000	0.267
20.000	0.869	70.000	0.248
		75.000	0.232
22.000	0.790	80.000	0.217
24.000	0.724	85.000	0.204
26.000	0.668	90.000	0.193
28.000	0.621	95.000	0.183
30.000	0.579	100.000	0.174
32.000	0.543		
34.000	0.511		
36.000	0.483		
38.000	0.457		
40.000	0.434		

Appendix C

Maximum and Minimum Resultant VSWR from Two mismatches

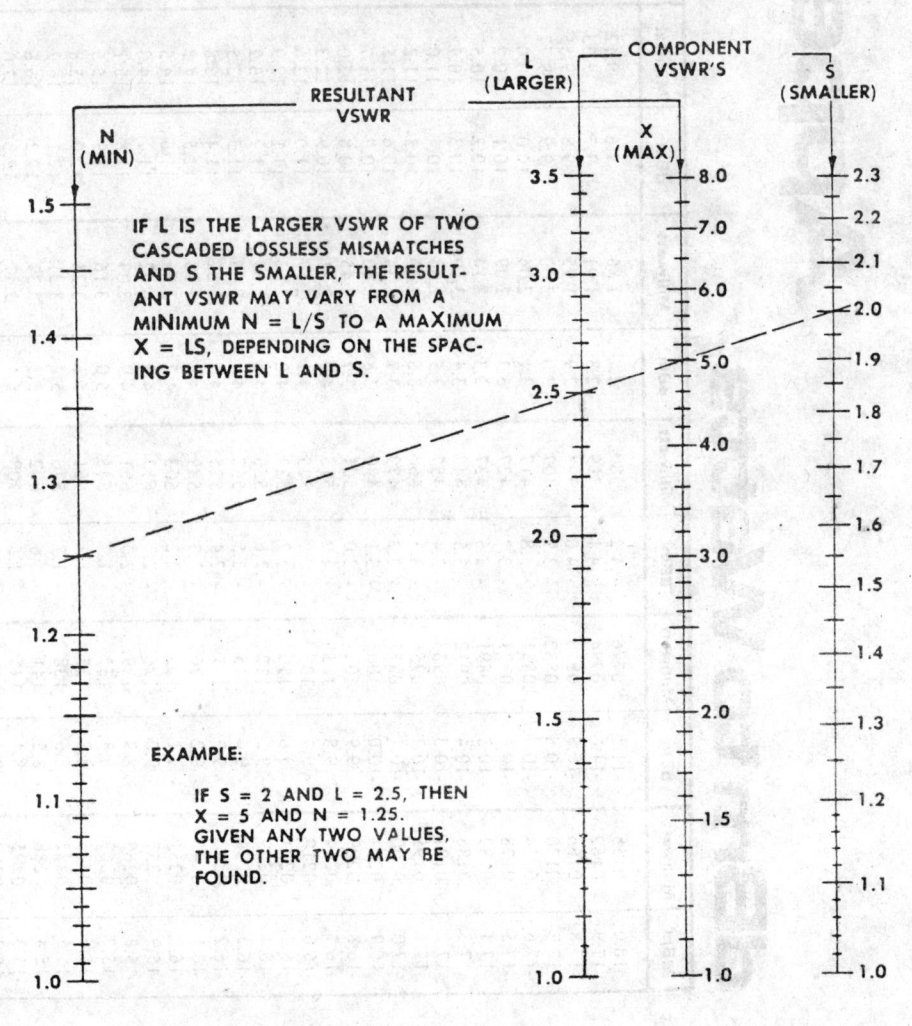

IF L IS THE LARGER VSWR OF TWO CASCADED LOSSLESS MISMATCHES AND S THE SMALLER, THE RESULTANT VSWR MAY VARY FROM A MINIMUM N = L/S TO A MAXIMUM X = LS, DEPENDING ON THE SPACING BETWEEN L AND S.

EXAMPLE:
IF S = 2 AND L = 2.5, THEN X = 5 AND N = 1.25.
GIVEN ANY TWO VALUES, THE OTHER TWO MAY BE FOUND.

Appendix D

dBm to Watts

dBm	Milliwatt	dBm	Milliwatt	dBm	Milliwatt	dBm	Milliwatt	dBm	Milliwatt
-18.0	.0158	-11.1	.0776	-4.2	.380	2.7	1.86	9.6	9.12
-17.9	.0162	-11.0	.0794	-4.1	.389	2.8	1.91	9.7	9.33
-17.8	.0166	-10.9	.0813	-4.0	.398	2.9	1.95	9.8	9.55
-17.7	.0170	-10.8	.0832	-3.9	.407	3.0	2.00	9.9	9.77
-17.6	.0174	-10.7	.0851	-3.8	.417	3.1	2.04	10.0	10.0
-17.5	.0178	-10.6	.0871	-3.7	.427	3.2	2.09	10.1	10.2
-17.4	.0182	-10.5	.0891	-3.6	.437	3.3	2.14	10.2	10.5
-17.3	.0186	-10.4	.0912	-3.5	.447	3.4	2.19	10.3	10.7
-17.2	.0191	-10.3	.0933	-3.4	.457	3.5	2.24	10.4	11.0
-17.1	.0195	-10.2	.0955	-3.3	.468	3.6	2.29	10.5	11.2
-17.0	.0200	-10.1	.0977	-3.2	.479	3.7	2.34	10.6	11.5
-16.9	.0204	-10.0	.100	-3.1	.490	3.8	2.40	10.7	11.7
-16.8	.0209	-9.9	.102	-3.0	.501	3.9	2.45	10.8	12.0
-16.7	.0214	-9.8	.105	-2.9	.513	4.0	2.51	10.9	12.3
-16.6	.0219	-9.7	.107	-2.8	.525	4.1	2.57	11.0	12.6
-16.5	.0224	-9.6	.110	-2.7	.537	4.2	2.63	11.1	12.9
-16.4	.0229	-9.5	.112	-2.6	.550	4.3	2.69	11.2	13.2
-16.3	.0234	-9.4	.115	-2.5	.562	4.4	2.75	11.3	13.5
-16.2	.0240	-9.3	.117	-2.4	.575	4.5	2.82	11.4	13.8
-16.1	.0245	-9.2	.120	-2.3	.589	4.6	2.88	11.5	14.1
-16.0	.0251	-9.1	.123	-2.2	.603	4.7	2.95	11.6	14.5
-15.9	.0257	-9.0	.126	-2.1	.617	4.8	3.02	11.7	14.8
-15.8	.0263	-8.9	.129	-2.0	.631	4.9	3.09	11.8	15.1
-15.7	.0269	-8.8	.132	-1.9	.646	5.0	3.16	11.9	15.5
-15.6	.0275	-8.7	.135	-1.8	.661	5.1	3.24	12.0	15.8
-15.5	.0282	-8.6	.138	-1.7	.676	5.2	3.31	12.1	16.2
-15.4	.0288	-8.5	.141	-1.6	.692	5.3	3.39	12.2	16.6
-15.3	.0295	-8.4	.145	-1.5	.708	5.4	3.47	12.3	17.0
-15.2	.0302	-8.3	.148	-1.4	.724	5.5	3.55	12.4	17.4

dBm	Milliwatt	dBm	Milliwatt
16.5	44.7	23.4	219
16.6	45.7	23.5	224
16.7	46.8	23.6	229
16.8	47.9	23.7	234
16.9	49.0	23.8	240
17.0	50.1	23.9	245
17.1	51.3	24.0	251
17.2	52.5	24.1	257
17.3	53.7	24.2	263
17.4	55.0	24.3	269
17.5	56.2	24.4	275
17.6	57.5	24.5	282
17.7	58.9	24.6	288
17.8	60.3	24.7	295
17.9	61.7	24.8	302
18.0	63.1	24.9	309
18.1	64.6	25.0	316
18.2	66.1	25.1	324
18.3	67.6	25.2	331
18.4	69.2	25.3	339
18.5	70.8	25.4	347
18.6	72.4	25.5	355
18.7	74.1	25.6	363
18.8	75.9	25.7	372
18.9	77.6	25.8	380
19.0	79.4	25.9	389
19.1	81.3	26.0	398
19.2	83.2	26.1	407
19.3	85.1	26.2	417

Appendix D: dBM to Watts

dBm	W	dBm	W	dBm	W	dBm	W	dBm	W	dBm	W
-15.1	.0309	-8.2	.151	-1.3	.741	5.6	3.63	12.5	17.8	19.4	87.1	26.3	427
-15.0	.0316	-8.1	.155	-1.2	.759	5.7	3.72	12.6	18.2	19.5	89.1	26.4	437
-14.9	.0324	-8.0	.158	-1.1	.776	5.8	3.80	12.7	18.6	19.6	91.2	26.5	447
-14.8	.0331	-7.9	.162	-1.0	.794	5.9	3.89	12.8	19.1	19.7	93.3	26.6	457
-14.7	.0339	-7.8	.166	-0.9	.813	6.0	3.98	12.9	19.5	19.8	95.5	26.7	468
-14.6	.0347	-7.7	.170	-0.8	.832	6.1	4.07	13.0	20.0	19.9	97.7	26.8	479
-14.5	.0355	-7.6	.174	-0.7	.851	6.2	4.17	13.1	20.4	20.0	100	26.9	490
-14.4	.0363	-7.5	.178	-0.6	.871	6.3	4.27	13.2	20.9	20.1	102	27.0	501
-14.3	.0372	-7.4	.182	-0.5	.891	6.4	4.37	13.3	21.4	20.2	105	27.1	513
-14.2	.0380	-7.3	.186	-0.4	.912	6.5	4.47	13.4	21.9	20.3	207	27.2	525
-14.1	.0389	-7.2	.191	-0.3	.933	6.6	4.57	13.5	22.4	20.4	110	27.3	537
-14.0	.0398	-7.1	.195	-0.2	.955	6.7	4.68	13.6	22.9	20.5	112	27.4	550
-13.9	.0407	-7.0	.200	-0.1	.977	6.8	4.79	13.7	23.4	20.6	115	27.5	562
-13.8	.0417	-6.9	.204	0.0	1.00	6.9	4.90	13.8	24.0	20.7	117	27.6	575
-13.7	.0427	-6.8	.209	0.1	1.02	7.0	5.01	13.9	24.5	20.8	120	27.7	589
-13.6	.0437	-6.7	.214	0.2	1.05	7.1	5.13	14.0	25.1	20.9	123	27.8	603
-13.5	.0447	-6.6	.219	0.3	1.07	7.2	5.25	14.1	25.7	21.0	126	27.9	617
-13.4	.0457	-6.5	.224	0.4	1.10	7.3	5.37	14.2	26.3	21.1	129	28.0	631
-13.3	.0468	-6.4	.229	0.5	1.12	7.4	5.50	14.3	26.9	21.2	132	28.1	646
-13.2	.0479	-6.3	.234	0.6	1.15	7.5	5.62	14.4	27.5	21.3	135	28.2	661
-13.1	.0490	-6.2	.240	0.7	1.17	7.6	5.75	14.5	28.2	21.4	138	28.3	676
-13.0	.0501	-6.1	.245	0.8	1.20	7.7	5.89	14.6	28.8	21.5	141	28.4	692
-12.9	.0513	-6.0	.251	0.9	1.23	7.8	6.03	14.7	29.5	21.6	145	28.5	708
-12.8	.0525	-5.9	.257	1.0	1.26	7.9	6.17	14.8	30.2	21.7	148	28.6	724
-12.7	.0537	-5.8	.263	1.1	1.29	8.0	6.31	14.9	30.9	21.8	151	28.7	741
-12.6	.0550	-5.7	.269	1.2	1.32	8.1	6.46	15.0	31.6	21.9	155	28.8	759
-12.5	.0562	-5.6	.275	1.3	1.35	8.2	6.61	15.1	32.4	22.0	158	28.9	776
-12.4	.0575	-5.5	.282	1.4	1.38	8.3	6.76	15.2	33.1	22.1	162	29.0	794
-12.3	.0589	-5.4	.288	1.5	1.41	8.4	6.92	15.3	33.9	22.2	166	29.1	813
-12.2	.0603	-5.3	.295	1.6	1.45	8.5	7.08	15.4	34.7	22.3	170	29.2	832
-12.1	.0617	-5.2	.302	1.7	1.48	8.6	7.24	15.5	35.5	22.4	174	29.3	852
-12.0	.0631	-5.1	.309	1.8	1.51	8.7	7.41	15.6	36.3	22.5	178	29.4	871
-11.9	.0646	-5.0	.316	1.9	1.55	8.8	7.59	15.7	37.2	22.6	182	29.5	891
-11.8	.0661	-4.9	.324	2.0	1.58	8.9	7.76	15.8	38.0	22.7	186	29.6	912
-11.7	.0676	-4.8	.331	2.1	1.62	9.0	7.94	15.9	38.9	22.8	191	29.7	933
-11.6	.0692	-4.7	.339	2.2	1.66	9.1	8.13	16.0	39.8	22.9	195	29.8	955
-11.5	.0708	-4.6	.347	2.3	1.70	9.2	8.32	16.1	40.7	23.0	200	29.9	977
-11.4	.0724	-4.5	.355	2.4	1.74	9.3	8.51	16.2	41.7	23.1	204	30.0	1000
-11.3	.0741	-4.4	.363	2.5	1.78	9.4	8.71	16.3	42.7	23.2	209		
-11.2	.0759	-4.3	.372	2.6	1.82	9.5	8.91	16.4	43.7	23.3	214		

dBm to Watts continued

dBm	Watts	dBm	Watts	dBm	Watts	dBm	Watts	dBm	Watts
30.1	1.02	36.8	4.79	43.5	22.40	50.2	105.00	56.9	490.00	63.6	2290.00
30.2	1.05	36.9	4.90	43.6	22.90	50.3	107.00	57.0	501.00	63.7	2340.00
30.3	1.07	37.0	5.01	43.7	23.40	50.4	110.00	57.1	513.00	63.8	2400.00
30.4	1.10	37.1	5.13	43.8	24.00	50.5	112.00	57.2	525.00	63.9	2450.00
30.5	1.12	37.2	5.25	43.9	24.50	50.6	115.00	57.3	537.00	64.0	2510.00
30.6	1.15	37.3	5.37	44.0	25.10	50.7	117.00	57.4	550.00	64.1	2570.00
30.7	1.17	37.4	5.50	44.1	25.7	50.8	120.00	57.5	562.00	64.2	2630.00
30.8	1.20	37.5	5.62	44.2	26.30	50.9	123.00	57.6	575.00	64.3	2690.00
30.9	1.23	37.6	5.75	44.3	26.90	51.0	126.00	57.7	589.00	64.4	2750.00
31.0	1.26	37.7	5.89	44.4	27.50	51.1	129.00	57.8	603.00	64.5	2820.00
31.1	1.29	37.8	6.03	44.5	28.20	51.2	132.00	57.9	617.00	64.6	2880.00
31.2	1.32	37.9	6.17	44.6	28.80	51.3	135.00	58.0	631.00	64.7	2950.00
31.3	1.35	38.0	6.31	44.7	29.50	51.4	138.00	58.1	646.00	64.8	3020.00
31.4	1.38	38.1	6.46	44.8	30.20	51.5	141.00	58.2	661.00	64.9	3090.00
31.5	1.41	38.2	6.61	44.9	30.90	51.6	145.00	58.3	676.00	65.0	3160.00
31.6	1.45	38.3	6.76	45.0	31.60	51.7	148.00	58.4	692.00	65.1	3240.00
31.7	1.48	38.4	6.92	45.1	32.40	51.8	151.00	58.5	708.00	65.2	3310.00
31.8	1.51	38.5	7.08	45.2	33.10	51.9	155.00	58.6	724.00	65.3	3390.00
31.9	1.55	38.6	7.24	45.3	33.90	52.0	158.00	58.7	741.00	65.4	3470.00
32.0	1.58	38.7	7.41	45.4	34.70	52.1	162.00	58.8	759.00	65.5	3550.00
32.1	1.62	38.8	7.59	45.5	35.50	52.2	166.00	58.9	776.00	65.6	3630.00
32.2	1.66	38.9	7.76	45.6	36.30	52.3	170.00	59.0	794.00	65.7	3720.00
32.3	1.70	39.0	7.94	45.7	37.20	52.4	174.00	59.1	813.00	65.8	3800.00
32.4	1.74	39.1	8.13	45.8	38.00	52.5	178.00	59.2	832.00	65.9	3890.00
32.5	1.78	39.2	8.32	45.9	38.90	52.6	182.00	59.3	851.00	66.0	3980.00
32.6	1.82	39.3	8.51	46.0	39.80	52.7	186.00	59.4	871.00	66.1	4070.00
32.7	1.86	39.4	8.71	46.1	40.70	52.8	191.00	59.5	891.00	66.2	4170.00
32.8	1.91	39.5	8.91	46.2	41.70	52.9	195.00	59.6	912.00	66.3	4270.00
32.9	1.95	39.6	9.12	46.3	42.70	53.0	200.00	59.7	933.00	66.4	4370.00
33.0	2.00	39.7	9.33	46.4	43.70	53.1	204.00	59.8	955.00	66.5	4470.00

Appendix D: dBM to Watts

dBm	Watts	dBm	Watts	dBm	Watts	dBm	Watts	dBm	Watts
33.1	2.04	39.8	9.55	46.5	44.70	53.2	209.00	59.9	977.00	66.6	4570.00
33.2	2.09	39.9	9.77	46.6	45.70	53.3	214.00	60.0	1000.00	66.7	4680.00
33.3	2.14	40.0	10.00	46.7	46.80	53.4	219.00	60.1	1020.00	66.8	4790.00
33.4	2.19	40.1	10.20	46.8	47.90	53.5	224.00	60.2	1050.00	66.9	4900.00
33.5	2.24	40.2	10.50	46.9	49.00	53.6	229.00	60.3	1070.00	67.0	5010.00
33.6	2.29	40.3	10.70	47.0	51.10	53.7	234.00	60.4	1100.00	67.1	5130.00
33.7	2.34	40.4	11.00	47.1	51.30	53.8	240.00	60.5	1120.00	67.2	5250.00
33.8	2.40	40.5	11.20	47.2	52.50	53.9	245.00	60.6	1150.00	67.3	5370.00
33.9	2.45	40.6	11.50	47.3	53.70	54.0	251.00	60.7	1170.00	67.4	5500.00
34.0	2.51	40.7	11.70	47.4	55.00	54.1	257.00	60.8	1200.00	67.5	5620.00
34.1	2.57	40.8	12.00	47.5	56.20	54.2	263.00	60.9	1230.00	67.6	5750.00
34.2	2.63	40.9	12.30	47.6	57.50	54.3	269.00	61.0	1260.00	67.7	5890.00
34.3	2.69	41.0	12.60	47.7	58.90	54.4	275.00	61.1	1290.00	67.8	6030.00
34.4	2.75	41.1	12.90	47.8	60.30	54.5	282.00	61.2	1320.00	67.9	6170.00
34.5	2.82	41.2	13.20	47.9	61.70	54.6	288.00	61.3	1350.00	68.0	6310.00
34.6	2.88	41.3	13.50	48.0	63.10	54.7	295.00	61.4	1380.00	68.1	6460.00
34.7	2.95	41.4	13.80	48.1	64.60	54.8	302.00	61.5	1410.00	68.2	6610.00
34.8	3.02	41.5	14.10	48.2	66.10	54.9	309.00	61.6	1450.00	68.3	6760.00
34.9	3.09	41.6	14.50	48.3	67.60	55.0	316.00	61.7	1480.00	68.4	6920.00
35.0	3.16	41.7	14.80	48.4	69.20	55.1	324.00	61.8	1510.00	68.5	7080.00
35.1	3.24	41.8	15.10	48.5	70.80	55.2	331.00	61.9	1550.00	68.6	7240.00
35.2	3.31	41.9	15.50	48.6	72.40	55.3	339.00	62.0	1580.00	68.7	7410.00
35.3	3.39	42.0	15.80	48.7	74.10	55.4	347.00	62.1	1620.00	68.8	7590.00
35.4	3.47	42.1	16.20	48.8	75.90	55.5	355.00	62.2	1660.00	68.9	7760.00
35.5	3.55	42.2	16.60	48.9	77.60	55.6	363.00	62.3	1700.00	69.0	7940.00
35.6	3.63	42.3	17.00	49.0	79.40	55.7	372.00	62.4	1740.00	69.1	8130.00
35.7	3.72	42.4	17.40	49.1	81.30	55.8	380.00	62.5	1780.00	69.2	8320.00
35.8	3.80	42.5	17.80	49.2	83.20	55.9	389.00	62.6	1820.00	69.3	8510.00
35.9	3.89	42.6	18.20	49.3	85.10	56.0	398.00	62.7	1860.00	69.4	8710.00
36.0	3.98	42.7	18.60	49.4	87.10	56.1	407.00	62.8	1910.00	69.5	8910.00
36.1	4.07	42.8	19.10	49.5	89.10	56.2	417.00	62.9	1950.00	69.6	9120.00
36.2	4.17	42.9	19.50	49.6	91.20	56.3	427.00	63.0	2000.00	69.7	9330.00
36.3	4.27	43.0	20.00	49.7	93.30	56.4	437.00	63.1	2040.00	69.8	9550.00
36.4	4.37	43.1	20.40	49.8	95.50	56.5	447.00	63.2	2090.00	69.9	9770.00
36.5	4.47	43.2	20.90	49.9	97.70	56.6	457.00	63.3	2140.00	70.0	10000.00
36.6	4.57	43.3	21.40	50.0	100.00	56.7	468.00	63.4	2190.00		
36.7	4.68	43.4	21.90	50.1	102.00	56.8	479.00	63.5	2240.00		

Appendix E

Decibels vs. Voltage & Power

Voltage Ratio	Power Ratio	-db +	Voltage Ratio	Power Ratio	Voltage Ratio	Power Ratio	-db +	Voltage Ratio	Power Ratio	Voltage Ratio	Power Ratio	-db +	Voltage Ratio	Power Ratio
1.0000	1.0000	0	1.000	1.000	.4467	.1995	7.0	2.239	5.012	.1995	.03981	14.0	5.012	25.12
.9886	.9772	.1	1.012	1.023	.4416	.1950	7.1	2.265	5.129	.1972	.03890	14.1	5.070	25.70
.9772	.9550	2	1.023	1.047	.4365	.1905	7.2	2.291	5.248	.1950	.03802	14.2	5.129	26.30
.9661	.9333	3	1.035	1.072	.4315	.1862	7.3	2.317	5.370	.1928	.03715	14.3	5.188	26.92
.9550	.9120	4	1.047	1.096	.4266	.1820	7.4	2.344	5.495	.1905	.03631	14.4	5.248	27.54
.9441	.8913	5	1.059	1.122	.4217	.1778	7.5	2.371	5.623	.1884	.03548	14.5	5.309	28.18
.9333	.8710	6	1.072	1.148	.4169	.1738	7.6	2.399	5.754	.1862	.03467	14.6	5.370	28.84
.9226	.8511	7	1.084	1.175	.4121	.1698	7.7	2.427	5.888	.1841	.03388	14.7	5.433	29.51
.9120	.8318	8	1.096	1.202	.4074	.1660	7.8	2.455	6.026	.1820	.03311	14.8	5.495	30.20
.9016	.8128	9	1.109	1.230	.4027	.1622	7.9	2.483	6.166	.1799	.03236	14.9	5.559	30.90
.8913	.7943	1.0	1.122	1.259	.3981	.1585	8.0	2.512	6.310	.1778	.03162	15.0	5.623	31.62
.8810	.7762	1.1	1.135	1.288	.3936	.1549	8.1	2.541	6.457	.1758	.03090	15.1	5.689	32.36
.8710	.7586	1.2	1.148	1.318	.3890	.1514	8.2	2.570	6.607	.1738	.03020	15.2	5.754	33.11
.8610	.7413	1.3	1.161	1.349	.3846	.1479	8.3	2.600	6.761	.1718	.02951	15.3	5.821	33.88
.8511	.7244	1.4	1.175	1.380	.3802	.1445	8.4	2.630	6.918	.1698	.02884	15.4	5.888	34.67
.8414	.7079	1.5	1.189	1.413	.3758	.1413	8.5	2.661	7.079	.1679	.02818	15.5	5.957	35.48
.8318	.6918	1.6	1.202	1.445	.3715	.1380	8.6	2.692	7.244	.1660	.02754	15.6	6.026	36.31
.8222	.6761	1.7	1.216	1.479	.3673	.1349	8.7	2.723	7.413	.1641	.02692	15.7	6.095	37.15
.8128	.6607	1.8	1.230	1.514	.3631	.1318	8.8	2.754	7.586	.1622	.02630	15.8	6.166	38.02
.8035	.6457	1.9	1.245	1.549	.3589	.1288	8.9	2.786	7.762	.1603	.02570	15.9	6.237	38.90
.7943	.6310	2.0	1.259	1.585	.3548	.1259	9.0	2.818	7.943	.1585	.02512	16.0	6.310	39.81
.7852	.6166	2.1	1.274	1.622	.3508	.1230	9.1	2.851	8.128	.1567	.02455	16.1	6.383	40.74
.7762	.6026	2.2	1.288	1.660	.3467	.1202	9.2	2.884	8.318	.1549	.02399	16.2	6.457	41.69
.7674	.5888	2.3	1.303	1.698	.3428	.1175	9.3	2.917	8.511	.1531	.02344	16.3	6.531	42.66
.7586	.5754	2.4	1.318	1.738	.3388	.1148	9.4	2.951	8.710	.1514	.02291	16.4	6.607	43.65

Appendix E: Decibels Versus Voltage and Power

.7499	.5623	2.5	1.334	1.778	.3350	1122	9.5	2.985	8.913	.1496	.02239	16.5	6.683	44.67
.7413	.5495	2.6	1.349	1.820	.3311	.1096	9.6	3.020	9.120	.1479	.02188	16.6	6.761	45.71
.7328	.5370	2.7	1.365	1.862	.3273	.1072	9.7	3.055	9.333	.1462	.02138	16.7	6.839	46.77
.7244	.5248	2.8	1.380	1.905	.3236	.1047	9.8	3.090	9.550	.1445	.02089	16.8	6.918	47.86
.7161	.5129	2.9	1.396	1.950	.3199	.1023	9.9	3.126	9.772	.1429	.02042	16.9	6.998	48.98
.7079	.5012	3.0	1.413	1.995	.3162	.1000	10.0	3.162	10.000	.1413	.01995	17.0	7.079	50.12
.6998	.4898	3.1	1.429	2.042	.3126	.09772	10.1	3.199	10.23	.1396	.01950	17.1	7.161	51.29
.6918	.4786	3.2	1.445	2.089	.3090	.09550	10.2	3.236	10.47	.1380	.01905	17.2	7.244	52.48
.6839	.4677	3.3	1.462	2.138	.3055	.09333	10.3	3.273	10.72	.1365	.01862	17.3	7.328	53.70
.6761	.4571	3.4	1.479	2.188	.3020	.09120	10.4	3.311	10.96	.1349	.01820	17.4	7.413	54.95
.6683	.4467	3.5	1.496	2.239	.2985	.08913	10.5	3.350	11.22	.1334	.01778	17.5	7.499	56.23
.6607	.4365	3.6	1.514	2.291	.2951	.08710	10.6	3.388	11.48	.1318	.01738	17.6	7.586	57.54
.6531	.4266	3.7	1.531	2.344	.2917	.08511	10.7	3.428	11.75	.1303	.01698	17.7	7.674	58.88
.6457	.4169	3.8	1.549	2.399	.2884	.08318	10.8	3.467	12.02	.1288	.01660	17.8	7.762	60.26
.6383	.4074	3.9	1.567	2.455	.2851	.08128	10.9	3.508	12.30	.1274	.01622	17.9	7.852	61.66
.6310	.3981	4.0	1.585	2.512	.2818	.07943	11.0	3.548	12.59	.1259	.01585	18.0	7.943	63.10
.6237	.3890	4.1	1.603	2.570	.2786	.07762	11.1	3.589	12.88	.1245	.01549	18.1	8.035	64.57
.6166	.3802	4.2	1.622	2.630	.2754	.07586	11.2	3.631	13.18	.1230	.01514	18.2	8.128	66.07
.6095	.3715	4.3	1.641	2.692	.2723	.07413	11.3	3.673	13.49	.1216	.01479	18.3	8.222	67.61
.6026	.3631	4.4	1.660	2.754	.2692	.07244	11.4	3.715	13.80	.1202	.01445	18.4	8.318	69.18
.5957	.3548	4.5	1.679	2.818	.2661	.07079	11.5	3.758	14.13	.1189	.01413	18.5	8.414	70.79
.5888	.3467	4.6	1.698	2.884	.2630	.06918	11.6	3.802	14.45	.1175	.01380	18.6	8.511	72.44
.5821	.3388	4.7	1.718	2.951	.2600	.06761	11.7	3.846	14.79	.1161	.01349	18.7	8.610	74.13
.5754	.3311	4.8	1.738	3.020	.2570	.06607	11.8	3.890	15.14	.1148	.01318	18.8	8.710	75.86
.5689	.3236	4.9	1.758	3.090	.2541	.06457	11.9	3.936	15.49	.1135	.01288	18.9	8.811	77.62
.5623	.3162	5.0	1.778	3.162	.2512	.06310	12.0	3.981	15.85	.1122	.01259	19.0	8.913	79.43
.5559	.3090	5.1	1.799	3.236	.2483	.06166	12.1	4.027	16.22	.1109	.01230	19.1	9.016	81.28
.5495	.3020	5.2	1.820	3.311	.2455	.06026	12.2	4.074	16.60	.1096	.01202	19.2	9.120	83.18
.5433	.2951	5.3	1.841	3.388	.2427	.05888	12.3	4.121	16.98	.1054	.01175	19.3	9.226	85.11
.5370	.2884	5.4	1.862	3.467	.2399	.05754	12.4	4.169	17.38	.1072	.01148	19.4	9.333	87.10
.5309	.2818	5.5	1.884	3.548	.2371	.05623	12.5	4.217	17.78	.1059	.01122	19.5	9.441	89.13
.5248	.2754	5.6	1.905	3.631	.2344	.05495	12.6	4.266	18.20	.1047	.01096	19.6	9.550	91.20
.5188	.2692	5.7	1.928	3.715	.2317	.05370	12.7	4.315	18.62	.1035	.01072	19.7	9.661	93.33
.5129	.2630	5.8	1.950	3.802	.2291	.05248	12.8	4.365	19.05	.1023	.01047	19.8	9.772	95.50
.5070	.2570	5.9	1.972	3.890	.2265	.05129	12.9	4.416	19.50	.1012	.01023	19.9	9.886	97.72
.5012	.2512	6.0	1.995	3.981	.2239	.05012	13.0	4.467	19.95	.1000	.01000	20.0	10.000	100.00
.4955	.2455	6.1	2.018	4.074	.2213	.04898	13.1	4.519	20.42		10^{-2}	30		10^3
.4898	.2399	6.2	2.042	4.169	.2188	.04786	13.2	4.571	20.89			40		10^3
.4842	.2344	6.3	2.065	4.266	.2163	.04677	13.3	4.624	21.38	10^{-2}	10^{-3}	50	10^2	10^4
.4786	.2291	6.4	2.089	4.365	.2138	.04571	13.4	4.677	21.88			60		10^6
.4732	.2239	6.5	2.113	4.467	.2113	.04467	13.5	4.732	22.39	10^{-3}	10^{-6}	70	10^3	10^6
.4677	.2188	6.6	2.138	4.571	.2089	.04365	13.6	4.786	22.91		10^{-7}	80		10^8
.4624	.2138	6.7	2.163	4.677	.2065	.04266	13.7	4.842	23.44	10^{-4}	10^{-8}	90	10^4	10^8
.4571	.2089	6.8	2.188	4.786	.2042	.04169	13.8	4.898	23.99		10^{-9}			10^9
.4519	.2042	6.9	2.213	4.898	.2018	.04074	13.9	4.955	24.55	10^{-5}	10^{-10}	100	10^5	10^{10}

Appendix F

Microwave Formulas

Wavelength (λ)

$$\lambda(\text{centimeters}) = \frac{3 \times 10^{10}}{f}$$

$$\lambda(\text{meters}) = \frac{3 \times 10^{8}}{f}$$

where f = frequency (hertz)

dB (Power and Voltage)

$$dB_{(power)} = 10 \log_{10} \frac{P1}{P2}$$

$$dB_{(voltage)} = 20 \log_{10} \frac{E1}{E2}$$

where P1 & P2 = system powers
E1 & E2 = system voltages

Characteristic Impedance (Z_0) of RF Cable

$$Z_0 = \frac{138}{\epsilon} \log_{10} \frac{D}{d}$$

where ϵ = dielectric constant
D = inside diameter of outer conductor
d = outside diameter of inner conductor

% Velocity of Propagation (v)

$$v = \frac{1}{\epsilon} \times 100$$

where ϵ = dielectric constant

Noise Figure (NF_{dB})

$$NF_{dB} = 10 \log_{10} \frac{S_i/N_i}{S_o/N_o}$$

where NF_{dB} = noise figure (dB)
S_i/N_i = input signal-to-noise ratio
S_o/N_o = output signal-to-noise ratio

Reflection Coefficient (ρ)

$$\rho = \frac{SWR-1}{SWR+1}$$

where SWR = Standing Wave Ratio

Return Loss in dB

$$dB = -20 \log_{10} |\rho|$$

where ρ = reflection coefficient

VSWR

$$VSWR = \frac{1 + \rho}{1 - \rho}$$

where ρ = reflection coefficient

Appendix G

Decimal-to-Metric Conversion

To Convert	To Obtain	Multiply by
Feet	Centimeters	3.048×10^1
Feet	Meters	3.048×10^{-1}
Inches	Centimeters	2.54
Inches	Meters	2.54×10^{-2}
Yards	Meters	9.144×10^{-1}
Miles (Statute)	Kilometers	1.609
Pounds	Kilograms	4.536×10^{-1}
Pounds	Grams	4.535×10^2
Centimeters	Inches	3.937×10^{-2}
Centimeters	Feet	3.281×10^{-1}
Centimeters	Yards	1.094×10^{-2}
Centimeters	Millimeters	10
Meters	Inches	39.37
Meters	Feet	3.281
Meters	Yards	1.094
Meters	Centimeters	100
Kilometers	Miles (Statute)	6.214×10^{-1}
Kilograms	Pounds	2.2046

Appendix H

Remote Message Coding

			T Y P E	C L A S S	D I O 8	7	6	5	4	3	2	D I O 1	NN D I O AFA VDCN	DRD ATO NI	A T Q	E R C	S F	I E N	R
LABEL	MESSAGE NAME																		
ACG	addressed command group	(Note 10)	M	AC	X	0	0	0	X	X	X	X	XXX	X	X	X	X		
ATN	attention		U	UC	X	X	X	X	X	X	X	X	XXX	1	X	X	X		
DAB	data byte	(Notes 1, 9)	M	DD	D$_8$	D$_7$	D$_6$	D$_5$	D$_4$	D$_3$	D$_2$	D$_1$	XXX	X	X	X	X		
DAC	data accepted		U	HS	X	X	X	X	X	X	X	X	XX0	X	X	X	X		
DAV	data valid		U	HS	X	X	X	X	X	X	X	X	1XX	X	X	X	X		
DCL	device clear	(Note 10)	M	UC	X	0	0	1	0	1	0	0	XXX	X	X	X	X		
END	end	(Notes 9, 11)	U	ST	X	X	X	X	X	X	X	X	XXX	X	1	X	X	X	
EOS	end of string	(Notes 2, 9)	M	DD	E$_8$	E$_7$	E$_6$	E$_5$	E$_4$	E$_3$	E$_2$	E$_1$	XXX	X	X	X	X		
GET	group execute trigger	(Note 10)	M	AC	X	0	0	0	1	0	0	0	XXX	X	X	X	X		
GTL	go to local	(Note 10)	M	AC	X	0	0	0	0	0	0	1	XXX	X	X	X	X		
IDY	identify	(Notes 10, 11)	U	UC	X	X	X	X	X	X	X	X	XXX	X	1	X	X	X	
IFC	interface clear		U	UC	X	X	X	X	X	X	X	X	XXX	X	X	X	1	X	
LAG	listen address group	(Note 10)	M	AD	X	0	1	X	X	X	X	X	XXX	X	X	X	X		
LLO	local lock out	(Note 10)	M	UC	X	0	0	1	0	0	0	1	XXX	X	X	X	X		
MLA	my listen address	(Notes 3, 10)	M	AD	X	0	1	L$_5$	L$_4$	L$_3$	L$_2$	L$_1$	XXX	X	X	X	X		
MTA	my talk address	(Notes 4, 10)	M	AD	X	1	0	T$_5$	T$_4$	T$_3$	T$_2$	T$_1$	XXX	X	X	X	X		
MSA	my secondary address	(Notes 5, 10)	M	SE	X	1	1	S$_5$	S$_4$	S$_3$	S$_2$	S$_1$	XXX	X	X	X	X		
NUL	null byte		M	DD	0	0	0	0	0	0	0	0	XXX	X	X	X	X		
OSA	other secondary address	(Note 10)	M	SE	(OSA = SCG ∧ $\overline{\text{MSA}}$)
OTA	other talk address	(Note 10)	M	AD	(OTA = TAG ∧ $\overline{\text{MTA}}$)

Remote Message Coding continued

PCG	primary command group	(Note 10)	M	— (PCG = ACG ∨		UCG ∨ LAG ∨ TAG)
PPC	parallel poll configure	(Note 10)	M	AC X 0 0 0 0 1 0 1		XXX X X X X X
PPE	parallel poll enable	(Notes 6, 10)	M	SE X 1 1 0 S P P P		XXX X X X X X
				3 2 1		
PPD	parallel poll disable	(Notes 7, 10)	M	SE X 1 1 1 D D D D		XXX X X X X X
				4 3 2 1		
PPR1	parallel poll response 1		U	ST X X X X X X X 1		XXX X X X X X
PPR2	parallel poll response 2		U	ST X X X X X X 1 X		XXX X X X X X
PPR3	parallel poll response 3		U	ST X X X X X 1 X X		XXX X X X X X
PPR4	parallel poll response 4		U	ST X X X X 1 X X X		XXX X X X X X
PPR5	parallel poll response 5		U	ST X X X 1 X X X X		XXX X X X X X
PPR6	parallel poll response 6		U	ST X X 1 X X X X X		XXX X X X X X
PPR7	parallel poll response 7		U	ST X 1 X X X X X X		XXX X X X X X
PPR8	parallel poll response 8		U	ST 1 X X X X X X X		XXX X X X X X
PPU	parallel poll unconfigure	(Note 10)	M	UC X 0 0 1 0 1 0 1		XXX X X X X X
REN	remote enable		U	UC X X X X X X X X		XXX X X X X 1
RFD	ready for data		U	HS X X X X X X X X		X0X X X X X X
RQS	request service	(Note 9)	U	ST X 1 X X X X X X		XXX X X X X X
SCG	secondary command group	(Note 10)	M	SE X 1 1 X X X X X		XXX X X X X X
SDC	selected device clear	(Note 10)	M	AC X 0 0 0 0 1 0 0		XXX X X X X X
SPD	serial poll disable	(Note 10)	M	UC X 0 0 1 1 0 0 1		XXX X X X X X
SPE	serial poll enable	(Note 10)	M	UC X 0 0 1 1 0 0 0		XXX X X X X X
SRQ	service request		U	ST X X X X X X X X		XXX X X 1 X X
STB	status byte	(Notes 8, 9)	M	ST S X S S S S S S		XXX X X X X X
				8 6 5 4 3 2 1		
TCT	take control	(Note 10)	M	AC X 0 0 0 1 0 0 1		XXX X X X X X
TAG	talk address group	(Note 10)	M	AD X 1 0 X X X X X		XXX X X X X X
UCG	universal command group	(Note 10)	M	UC X 0 0 1 X X X X		XXX X X X X X
UNL	unlisten	(Note 10)	M	AD X 0 1 1 1 1 1 1		XXX X X X X X
UNT	untalk	(Note 10)	M	AD X 1 0 1 1 1 1 1		XXX X X X X X

Appendix H: Remote Message Coding

Remote Message Coding
continued

Symbols:

Level Assignment:

0 = High state signal level
1 = Low state signal level

Type	U	=	Uniline message
	M	=	Multiline message
Class	AC	=	Addressed command
	AD	=	Address (talk or listen)
	DD	=	Device dependent
	HS	=	Handshake
	UC	=	Universal Command
	SE	=	Secondary
	ST	=	Status

NOTES:
(1) D1-D8 specify the device dependent data bits.
(2) E1-E8 specify the device dependent code used to indicate the EOS message.
(3) L1-L5 specify the device dependent bits of the device's listen address.
(4) T1-T5 specify the device dependent bits of the device's talk address.
(5) S1-S5 specify the device dependent bits of the device's secondary address.
(6) S specifies the sense of the PPR.

S	Response
0	0
1	1

P1-P3 specify the PPR message to be sent when a parallel poll is executed.

P3	P2	P1	PPR Message
0	0	0	PPR1
.	.	.	
.	.	.	
.	.	.	
1	1	1	PPR8

(7) D1-D4 specify don't-care bits that must be sent all zeroes, but do not need to be decoded by the receiving device.

(8) S1-S6, S8 specify the device dependent status. (DIO7 is used for the RQS message.)

(9) The true message value must be ignored when received if the LACS is inactive.

(10) The true message value must be ignored when received if the ATN message is false.

(11) Interface protocol specifies that the IDY message is sent true only when the ATN message is sent true, whereas the END message is sent true only when the ATN message is sent false.

Appendix I

Time Values

Time Value Identifier*	Function (applies to)	Description	Value
T_1	SH	settling time for multiline messages	$\geq 2\ \mu s$†
t_2	SH, AH, T, L	response to ATN	≤ 200 ns
T_3	AH	interface message accept time‡	> 0 §
t_4	T, TE, L, LE, C	response to IFC or REN false	$< 100\ \mu s$
t_5	PP	response to ATN $\wedge$ EOI	≤ 200 ns
T_6	C	parallel poll execution time	$\geq 2\ \mu s$
T_7	C	controller delay to allow current talker to see ATN message	≥ 500 ns
T_8	C	length of IFC or REN false	$> 100\ \mu s$
T_9	C	delay for EOI**	$\geq 1.5\ \mu s$††

*Time values specified by a lower case t indicate the maximum time allowed to make a state transition. Time values specified by an upper case T indicate the minimum time that a function must remain in a state before exiting.

†If three-state drivers are used on the DIO, DAV, and EOI lines, T_1 may be:
  (1) ≥ 1100 ns.
  (2) Or ≥ 700 ns if it is known that within the controller ATN is driven by a three-state driver
  (3) Or ≥ 500 ns for all subsequent bytes following the first sent after each false transition of ATN (the first byte must be sent in accordance with (1) or (2))

‡Time required for interface functions to accept, not necessarily respond to interface messages.

§Implementation dependent.

**Delay required for EOI, NDAC, and NRFD signal lines to indicate valid states.

††≥ 600 ns for three-state drivers.

Appendix J
Allowable Subsets

J.1 SH Function Allowable Subsets

Identification	Description	States Omitted	Other Requirements	Other Function Subsets Required
SH0	no capability	all	none	none
SH1	complete capability	none	none	none

J.2 AH Function Allowable Subsets

Identification	Description	States Omitted	Other Requirements	Other Function Subsets Required
AH0	no capability	all	none	none
AH1	complete capability	none	none	none

Appendix J: Allowable Subsets

J.3 T Function Allowable Subsets

Identification	Description Capabilities Basic Talker	Serial Poll	Talk Only Mode	Unaddress If MLA	States Omitted	Other Requirements	Other Function Subsets Required
T0	N	N	N	N	all	none	none
T1	Y	Y	Y	N	none	omit [MLA ∧ ACDS]	SH1 and AH1
T2	Y	Y	N	N	none	omit [MLA ∧ ACDS]	SH1 and AH1
T3	Y	N	Y	N	SPIS, SPMS, SPAS	ton always false	SH1 and AH1
T4	Y	N	N	N	SPIS, SPMS, SPAS	omit [MLA ∧ ACDS]	SH1 and AH1
T5	Y	Y	Y	Y	none	include [MLA ∧ ACDS]	SH1 and L1-L4 or LE1-LE4
T6	Y	Y	N	Y	none	include [MLA ∧ ACDS]	SH1 and L1-L4 or LE1-LE4
T7	Y	N	Y	Y	SPIS, SPMS, SPAS	include [MLA ∧ ACDS]	SH1 and L1-L4 or LE1-LE4
T8	Y	N	N	Y	SPIS, SPMS, SPAS	ton always false	SH1 and L1-L4 or LE1-LE4

Allowable Subsets continued

J.4 T Function (With Address Extension) Allowable Subsets

Identi-fication	Description — Capabilities — Basic Extended Talker	Serial Poll	Talk Only Mode	Unaddress If MLA ∧ MSA	States Omitted	Other Requirements	Other Function Subsets Required
TE0	N	N	N	N	all	none	none
TE1	Y	Y	Y	N	none	omit [MSA ∧ (LPAS ∨ ACDS)]	SH1 and AH1
TE2	Y	Y	N	N	none	omit [MSA ∧ (LPAS ∨ ACDS)]	SH1 and AH1
TE3	Y	N	Y	N	SPIS, SPMS, SPAS	ton always false omit [MSA ∧ (LPAS ∨ ACDS)]	SH1 and AH1
TE4	Y	N	N	N	SPIS, SPMS, SPAS	omit [MSA ∧ (LPAS ∨ ACDS)]	SH1 and AH1
TE5	Y	Y	Y	Y	none	ton always false include [MSA ∧ (LPAS ∨ ACDS)]	SH1 and L1-L4 or LE1-LE4
TE6	Y	Y	N	Y	none	include [MSA ∧ (LPAS ∨ ACDS)] ton always false	SH1 and L1-L4 or LE1-LE4

Appendix J: Allowable Subsets

	Basic Listener	Listen Only Mode			States Omitted	Other Requirements	Other Function Subsets Required
TE7	Y	N		Y	SPIS, SPMS, SPAS	include [MSA ∧ (LPAS)/(ACDS)]	SH1 and L1-L4 or LE1-LE4
TE8	Y	N		Y	SPIS, SPMS, SPAS	include [MSA ∧ (LPAS)/(ACDS)] ∧ on always false	SH1 and L1-L4 or LE1-LE4

J.5 L Function Allowable Subsets

Identification	Description				States Omitted	Other Requirements	Other Function Subsets Required
	Capabilities						
	Basic Listener	Listen Only Mode	Unaddress If MTA				
L0	N	N	N		all	none	none
L1	Y	Y	N		none	omit [MTA ∧ (ACDS)]	AH1
L2	Y	N	N		none	omit [MTA ∧ (ACDS)]	AH1
						on always false	
L3	Y	Y	Y		none	include [MTA ∧ (ACDS)]	AH1 and T1-T8 or TE1-TE8
L4	Y	N	Y		none	include [MTA ∧ (ACDS)] ∧ on always false	AH1 and T1-T8 or TE1-TE8

Allowable Subsets *continued*

J 6 L Function (With Address Extension) Allowable Subsets

Identi-fication	Description			States Omitted	Other Requirements	Other Function Subsets Required
	Capabilities					
	Basic Extended Listener	Listen Only Mode	Unaddress If MSA ∧ (TPAS)			
LE0	N	N	N	all	none	none
LE1	Y	Y	N	none	omit [MSA ∧ (TPAS) ∧ (ACDS)]	AH1
LE2	Y	N	N	none	omit [MSA ∧ (TPAS) ∧ (ACDS)]	AH1
					lon always false	
LE3	Y	Y	Y	none	include [MSA ∧ (TPAS) ∧ (ACDS)]	AH1 and T1-T8 or TE1-TE8
LE4	Y	N	Y	none	include [MSA ∧ (TPAS) ∧ (ACDS)]	AH1 and T1-T8 or TE1-TE8
					lon always false	

Appendix J: Allowable Subsets

J 7 SR Function Allowable Subsets

Identification	Description	States Omitted	Other Requirements	Other Function Subsets Required
SR0	no capability	all	none	none
SR1	complete capability	none	none	T1, T2, T5, T6 TE1, TE2, TE5 or TE6

J 8 RL Function Allowable Subsets

Identification	Description	States Omitted	Other Requirements	Other Function Subsets Required
RL0	no capability	all	none	none
RL1	complete capability	none	none	L1-L4 or LE1-LE4
RL2	no local lock out	LWLS and RWLS	rtl always false	L1-L4 or LE1-LE4

Allowable Subsets continued

J 9 PP Function Allowable Subsets

Identi-fication	Description	States Omitted	Other Requirements	Other Function Subsets Required
PP0	no capability	all	none	none
PP1	complete capability	none	include [((PPD ∧ PACS) ∨ PPU) ∧ (ACDS)] include [PPE ∧ PACS ∧ ACDS)] exclude lpe	L1-L4 or LE1-LE4
PP2	omit capability of being con-figured by controller	PUCS, PACS	include lpe exclude [((PPD ∧ PACS)) ∨ PPU) ∧ (ACDS)] exclude [PPE ∧ PACS ∧ ACDS)] local messages must be substituted for S, P1, P2, P3	none

Appendix J: Allowable Subsets

J 10 DC Function Allowable Subsets

Identi-fication	Description	States Omitted	Other Requirements	Other Function Subsets Required
DC0	no capability	all	none	none
DC1	complete capability	none	none	L1-L4 or LE1-LE4
DC2	omit selective device clear	none	omit (SDC ∧ LADS)	AH1

J 11 DT Function Allowable Subsets

Identi-fication	Description	States Omitted	Other Requirements	Other Function Subsets Required
DT0	no capability	all	none	none
DT1	complete capability	none	none	L1-L4 or LE1-LE4

Allowable Subsets

J 12 C Function Allowable Subsets

Identification*	\multicolumn{9}{c}{Capabilities}	Notes
	System Controller	Send IFC and Take Charge	Send REN	Respond to SRQ	Send I.F. Messages	Receive Control	Pass Control	Pass Control to Self	Parallel Poll	Take Control Synchronously	
C0	N	N	N	N	N	N	N	N	N	N	
C1	Y	—	—	—	—	—	—	—	—	—	(1)
C2	—	Y	—	—	—	—	—	—	—	—	(1)
C3	—	—	Y	—	—	—	—	—	—	—	(1)
C4	—	—	—	Y	—	—	—	—	—	—	(1)
C5	—	—	—	—	Y	Y	Y	Y	Y	Y	(2),(3)
C6	—	—	—	—	Y	Y	Y	Y	Y	N	(2),(3)
C7	—	—	—	—	Y	Y	Y	Y	N	Y	(2),(3)
C8	—	—	—	—	Y	Y	Y	Y	N	N	(2),(3)
C9	—	—	—	—	Y	Y	Y	N	Y	Y	(2),(3)
C10	—	—	—	—	Y	Y	Y	N	Y	N	(2),(3)
C11	—	—	—	—	Y	Y	Y	N	N	Y	(2),(3)
C12	—	—	—	—	Y	Y	Y	N	N	N	(2),(3)
C13	—	—	—	—	Y	Y	N	N	Y	Y	(2)
C14	—	—	—	—	Y	Y	N	N	Y	N	(2)
C15	—	—	—	—	Y	Y	N	N	N	Y	(2)
C16	—	—	—	—	Y	Y	N	N	N	N	(2)
C17	—	—	—	—	Y	N	Y	Y	Y	Y	(2),(3),(4)
C18	—	—	—	—	Y	N	Y	Y	Y	N	(2),(3),(4)
C19	—	—	—	—	Y	N	Y	Y	N	Y	(2),(3),(4)
C20	—	—	—	—	Y	N	Y	Y	N	N	(2),(3),(4)
C21	—	—	—	—	Y	N	Y	N	Y	Y	(2),(3),(4)
C22	—	—	—	—	Y	N	Y	N	Y	N	(2),(3),(4)
C23	—	—	—	—	Y	N	Y	N	N	Y	(2),(3),(4)
C24	—	—	—	—	Y	N	Y	N	N	N	(2),(3),(4)
C25	—	—	—	—	Y	N	N	N	Y	Y	(2)
C26	—	—	—	—	Y	N	N	N	Y	N	(2)
C27	—	—	—	—	Y	N	N	N	N	Y	(2)
C28	—	—	—	—	Y	N	N	N	N	N	(2)

*Typical notation to describe a controller consists of the letter C followed by one or more of the numbers indicating the subsets selected. For example; C1, 2, 3, 4, 8.

†This is part of the CIDS to CADS transitional expression.

‡This is part of the CACS to CTRS transitional expression.

Appendix J: Allowable Subsets   493

continued

States Required								Other Requirements			Other Function Subsets Required				
SNAS, SACS	SIIS, SIAS, SINS	SRIS, SRAS, SRNS	CSNS, CSRS	CACS, CSBS, CSWS, CAWS	CADS	CIDS	CTRS	CPWS, CPPS	[TCT ∧ (ACDS) ∧ (TADS)]†	[(TADS)]‡	tcs not always false	C1	C2	AH1	SH1	T1-T8, TE1-TE8
O	O	O	O	O	O	O	O	O	O	O	O	O	—	—	—	—
R	—	—	—	—	—	—	—	—	—	—	—	—	—	—	—	—
—	R	—	—	—	—	—	—	—	—	—	—	R	—	—	—	—
—	—	R	—	—	—	—	—	—	—	—	—	R	—	—	—	—
—	—	—	R	—	—	—	—	—	—	—	—	—	—	—	—	—
—	—	—	—	R	R	R	R	R	R	R	O	—	—	R	R	R
—	—	—	—	R	R	R	R	R	O	R	R	O	—	—	R	R	R
—	—	—	—	R	R	R	R	O	R	R	O	—	—	R	R	R
—	—	—	—	R	R	R	R	O	R	R	O	—	—	—	R	R
—	—	—	+	R	R	R	R	R	O	R	O	—	—	R	R	R
—	—	—	—	R	R	R	R	R	O	R	O	—	—	—	R	R
—	—	—	—	R	R	R	R	O	R	R	O	—	—	R	R	R
—	—	—	—	R	R	R	R	O	R	R	O	—	—	—	R	R
—	—	—	—	R	R	R	O	R	O	O	R	—	R	R	R	—
—	—	—	—	R	R	R	O	R	O	O	R	—	R	—	R	—
—	—	—	—	R	R	R	O	O	O	O	R	—	R	R	R	—
—	—	—	—	R	R	R	O	O	O	O	R	—	R	—	R	—
—	—	—	—	R	O	R	R	R	R	R	R	—	—	R	R	R
—	—	—	—	R	O	R	R	R	R	R	O	—	—	—	R	R
—	—	—	—	R	O	R	R	R	R	R	R	—	—	R	R	R
—	—	—	—	R	O	R	R	O	R	R	O	—	—	—	R	R
—	—	—	—	R	O	R	R	R	O	R	R	—	—	R	R	R
—	—	—	—	R	O	R	R	R	O	O	R	—	—	—	R	R
—	—	—	—	R	O	R	R	O	O	R	R	—	—	R	R	R
—	—	—	—	R	O	R	R	O	O	O	R	—	—	—	R	R
—	—	—	—	R	O	O	O	R	O	O	R	—	—	R	R	—
—	—	—	—	R	O	O	O	R	O	O	R	—	—	—	R	—
—	—	—	—	R	O	O	O	O	O	O	R	—	—	R	R	—
—	—	—	—	R	O	O	O	O	O	O	R	—	—	—	R	—

NOTES:
(1) One or more of subsets C1 through C4 may be chosen in any combination with any one of C5 through
(2) Only one subset may be chosen from C5 through C28.
(3) The CTRS state must be included in devices which are to be operated in multicontroller systems.
(4) These subsets are not allowed unless C2 is included.
O = omit, R = required, hyphen = not applicable or not required, Y = yes, N = no.

Appendix K

Interface Message Reference List

LABEL	Message	Interface Function(s)
LOCAL MESSAGES RECEIVED (By interface functions)		
gts	go to standby	C
isr	individual service request (qual)	PP
lon	listen only	L, LE
lpe	local poll enable	PP
ltn	listen	L, LE
lun	local unlisten	L, LE
nba	new byte available	SH
pon	power on	SH, AH, T, TE, L, LE, SR, RL, PP, C
rdy	ready	AH
rpp	request parallel poll	C
rsc	request system control	C
rsv	request service	SR
rtl	return to local	RL
sic	send interface clear	C
sre	send remote enable	C
tca	take control asynchronously	C
tcs	take control synchronously	C
ton	talk only	AH, C
		T, TE

LOCAL MESSAGES SENT (To interface functions)

None defined; see Message Output tables in Section 2 for description of Device Function Interaction which provides guidelines as to the appropriate states from which local messages may be sent to the device functions.

Appendix K: Interface Message Reference List

REMOTE MESSAGES RECEIVED

ATN	attention	SH, AH, T, TE, L, LE, PP, C
DAB	data byte	(via L, LE)
DAC	data accepted	SH
DAV	data valid	AH
DCL	device clear	DC
END	end	(via L, LE)
GET	group execute trigger	DT
GTL	go to local	L, LE, PP
IDY	identify	T, TE, L, LE, C
IFC	interface clear	RL
LLO	local lockout	L, LE, RL
MLA	my listen address	T
[MLA]	my listen address	TE, LE
MSA or [MSA]	my secondary address	T, TE
MTA	my talk address	L
[MTA]	my talk address	TE
OSA	other secondary address	T, TE
OTA	other talk address	TE, LE, PP
PCG	primary command group	PP
PPC	parallel poll configure	PP
[PPD]	parallel poll disable	PP
[PPE]	parallel poll enable	(via C)
PPRn	parallel poll response n	PP
PPU	parallel poll unconfigure	RL
REN	remote enable	SH
RFD	ready for data	(via L, LE)
RQS	request service	DC
[SDC]	selected device clear	T, TE
SPD	serial poll disable	T, TE
SPE	serial poll enable	(via C)
SQR	service request	(via L, LE)
STB	status byte	C
TCT or [TCT]	take control	L, LE
UNL	unlisten	

Interface Message Reference List continued

REMOTE MESSAGES SENT

ATN	attention	C
DAB	data byte	(via T, TE)
DAC	data accepted	AH
DAV	data valid	SH
DCL	device clear	(via C)
END	end	(via T)
GET	group execute trigger	(via C)
GTL	go to local	(via C)
IDY	identify	C
IFC	interface clear	C
LLO	local lockout	(via C)
MLA or [MLA]	my listen address	(via C)
MSA or [MSA]	my secondary address	(via C)
MTA or [MTA]	my talk address	(via C)
OSA	other secondary address	(via C)
OTA	other talk address	(via C)
PCG	primary command group	(via C)
PPC	parallel poll configure	(via C)
[PPD]	parallel poll disable	(via C)
[PPE]	parallel poll enable	(via C)
PPRn	parallel poll response n	PP
PPU	parallel poll unconfigure	(via C)
REN	remote enable	C
RFD	ready for data	AH
RQS	request service	T, TE
[SDC]	selected device clear	(via C)
SPD	serial poll disable	(via C)
SPE	serial poll enable	(via C)
SRQ	service request	SR
STB	status byte	(via T, TE)
TCT	take control	(via C)
UNL	unlisten	(via C)

Appendix L

Frequency Counters

There are methods available today that allow active frequency measurements to 40 GHz and beyond. There are three basic methods employed by counters to provide frequency measurements. They are:

- *prescaling* — measurements as high as 1.3 GHz.
- *heterodyne down conversion* — measurements as high as 20 GHz.
- *transfer oscillator down conversion* — measurements to 40 GHz and beyond.

Let us investigate each of these methods and see how they can provide accurate and reliable frequency measurements.

Prescaling

Prescaling involves a simple division of the input frequency which results in a lower frequency signal that can easily be counted by digital circuitry. The frequency that the actual counter section measures is related to the input frequency by an integer N (usually ranging from 2 to 16). A display of the input is accomplished either by multiplying the content of the counter by N or by lengthening the counter's gate time (which is the measured time) by the same integer, a prescale factor, N. The accuracy of such a measurement system is subject to plus or minus one count in the least significant digit and the time base error.

Figure L.1 shows a block diagram of a microwave counter using prescaling as its downconversion technique. The input signal is amplified and shaped (input signal conditioner) to properly interact with the prescaling circuitry. It is then divided by the integer N before entering the main gate. The lower portion of the block diagram pretty much resembles any conventional counter. The main gate is opened and closed (by the main gate flip flop) in timing very precisely determined by the crystal time base of the counter.

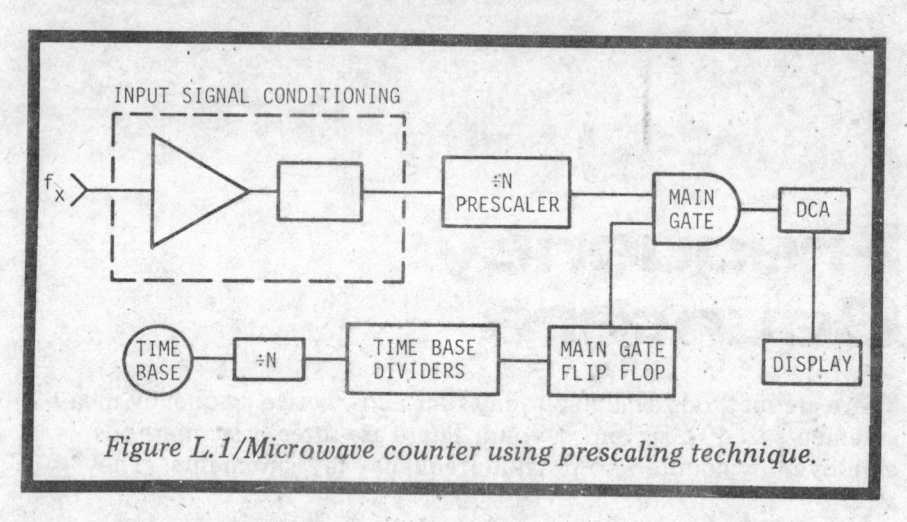

Figure L.1/Microwave counter using prescaling technique.

The DCA (Decade Counting Assembly) accumulates any frequencies under 500 MHz, multiplies them by N, and then transmits the end result to the display.

Frequencies of up to 1.5 GHz can be read with such a conversion technique. Other methods must be used for high readings. There are, however, many applications for frequency counters to work up to 1.5 GHz.

Heterodyne Down-conversion

Heterodyne down-conversion is a considerably more involved technique which allows frequency measurements to about 18 GHz. The key to this technique is a mixer which beats the incoming microwave frequency against a high-stability local oscillator signal, resulting in a difference frequency which is within the conventional counter's 500 MHz bandwidth.

Figure L.2 is the block diagram of an automatic microwave counter using the heterodyne down-conversion technique. The down-converter section is enclosed by the dotted line. Outside the dotted line is the block diagram of a conventional counter, with the addition of

Appendix L: Frequency Counters

a new block called the processor. The decision-making capability of a processor is necessary here in order to lead the counter through its measurement algorithm. The high stability local oscillator of Figure L.2 is generated by first digitally multiplying the frequency of the instrument's time base to a convenient fundamental frequency (designated f_{in}), typically 100 to 500 MHz. This f_{in} is directed to a harmonic generator which produces a "comb line" of frequencies spaced at f_{in} extending to the full frequency range of the counter. One line of this comb is then selected by the microwave filter and directed to the mixer. Since this frequency line is an integral multiple of f_{in}, it is designated Kf_{in}. The down-conversion process now occurs as follows: emerging from the mixer is a video frequency equal to $f_x - Kf_{in}$. This video frequency is amplified and sent to the counter. The display contains the sum of the video frequency and Kf_{in}, which is provided by the processor. (The processor stores the value of K, since it is in control of the microwave filter.)

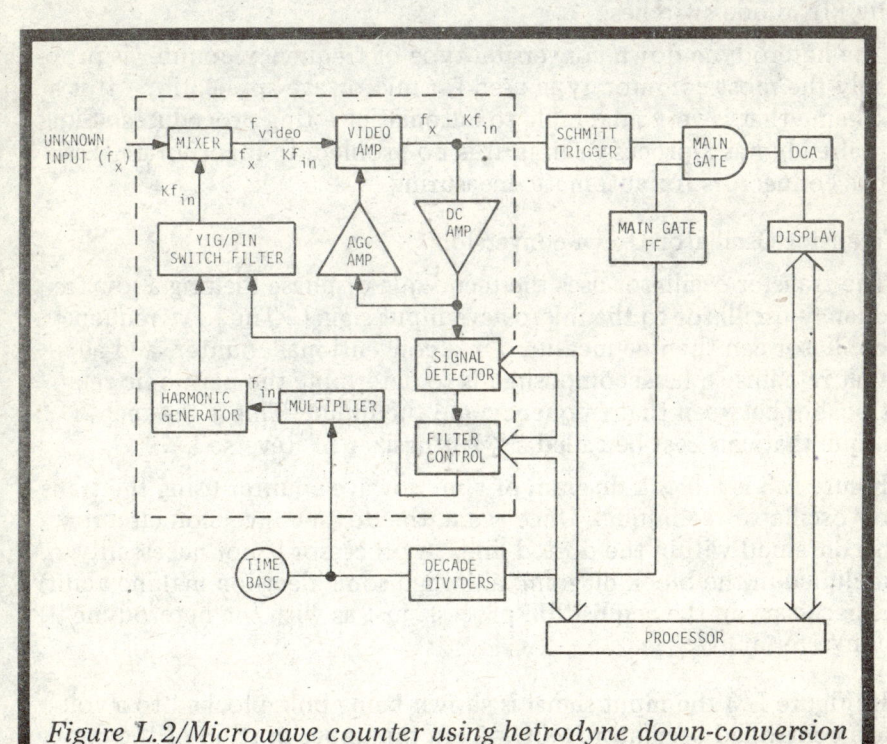

Figure L.2/Microwave counter using hetrodyne down-conversion technique.

The signal detector block in Figure L.2 is necessary for determining the correct K value. In practice, the processor will begin with K = 1 and will "walk" the value of K through the comb line until the signal detector determines that a video frequency is present. At this point the routine is terminated and measurement can begin.

The remaining block in Figure L.2 which has not been discussed is the automatic gain control (AGC) circuit. This circuit provides a degree of noise immunity by desensitizing the video amplifier such that only the strongest frequency components of the video signal will enter the Schmitt trigger and be counted.

A key ingredient in automating the heterodyne down-conversion process is the microwave filter. Years ago this filter consisted of a resonant cavity which was tuned by the operator's turning a crank. Until quite recently this manual technique had a cost advantage over automated counters. Today, the process is automated through use of either a YIG filter or an array of thin film filters which are selected by PIN diode switches.

The heterodyne down-conversion type of frequency counter is probably the most common type used for microwave application. It is a scheme that is very adaptable to automatic testing procedures. Since it already has a processor in it, it is no problem to incorporate IEEE bus connectors for automatic measuring.

Transfer Oscillator Down-conversion

The transfer oscillator uses the technique of phase locking a low frequency oscillator to the microwave input signal. The low frequency oscillator can then be measured in a conventional counter, and all that remains to be accomplished is to determine the harmonic relationship between that frequency and the input. This is the technique that can best be called a "Synthesizer In Reverse".

Figure L.3 is a block diagram of a microwave counter using the transfer oscillator technique. Once again, the down-conversion circuitry is contained within the dotted line. A processor is not necessarily included in the block diagram, although some decision-making ability is necessary in the acquisition process, just as with the heterodyne converter above.

In Figure L.3 the input signal is shown being phase locked to a voltage controlled oscillator (VCO #1) in the upper portion of the converter section. Once phase lock is achieved, the relationship between the input and the VCO frequency is $f_x = Nf_1$, where N is an integer.

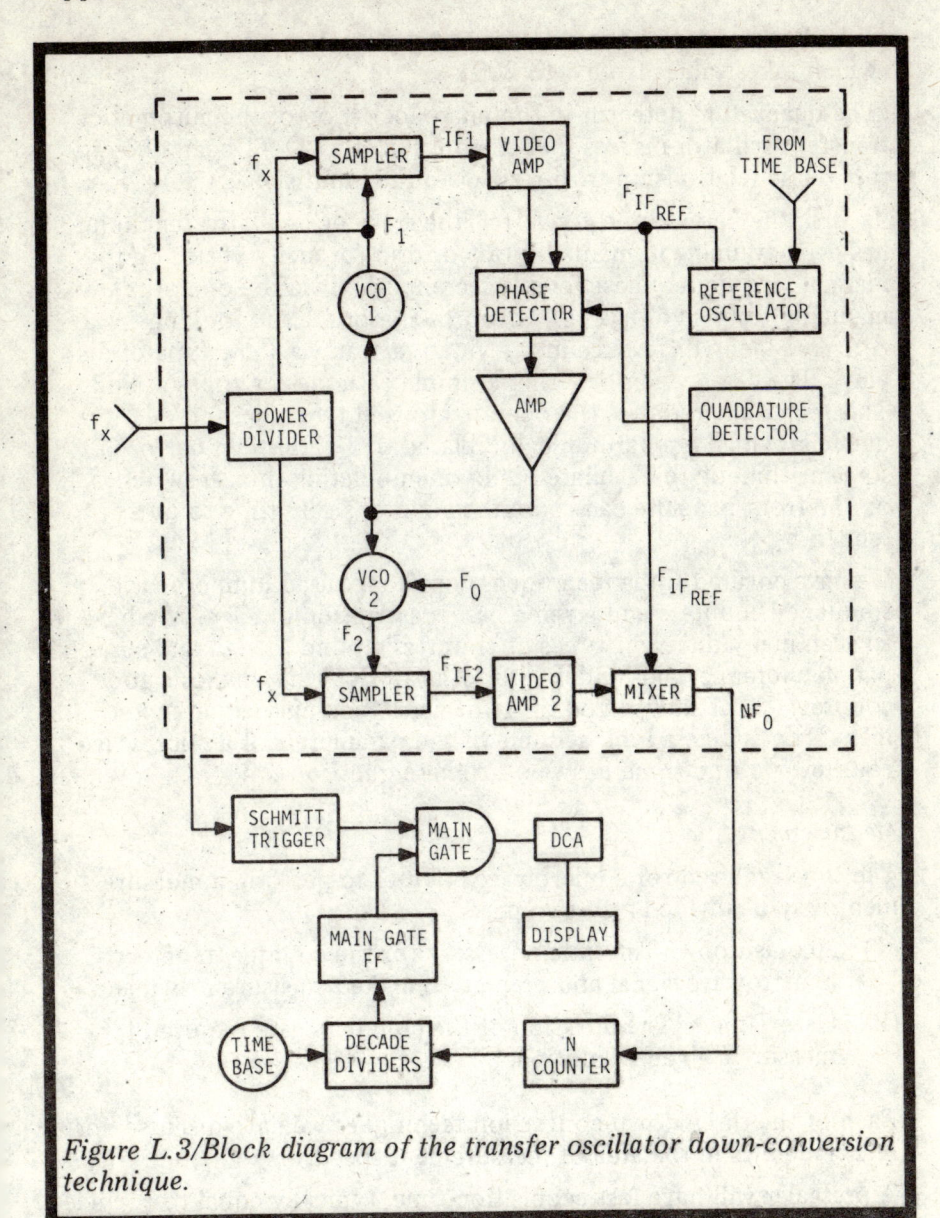

Appendix L: Frequency Counters 501

Figure L.3/Block diagram of the transfer oscillator down-conversion technique.

The remainder of the down-converter circuitry is devoted to the task of determining N. The counter can now measure f_1, (typically 100-200 MHz) and multiply by N for a display of the microwave frequency. As in the case of prescaling above, this multiplication is usually accomplished by extending the gate time of the counter by a factor of N (which takes values from 1 to 200).

(The quadrature detector in the phase lock loop of the automatic transfer oscillator insures that the output of VCO #1 bears the correct phase relationship with respect to the input signal.)

As with the heterodyne converter, the transfer oscillator technique has been available in manually tuned form for many years. In the manual technique, phase lock is accomplished via the operator's manually tuning voltage controlled oscillator #1 and looking for a zero frequency difference in the video section via some type of display. The display can be either a simple cathode-ray tube or a dial. This technique requires the operator to find two adjacent VCO frequencies which are harmonically related to f_x. He then performs some arithmetic to calculate N. By manipulating thumb switches on the front panel he can cause the counter's gate time to be extended by N.

We have covered three techniques commonly used in microwave frequency counters and given a basic description of each. We have all seen many different types of counters around the lab and some people prefer certain models over others. But what makes a good counter? What should you look for when making such an investment? Let's take a look at some of the parameters of a microwave counter and find some answers for these questions.

Measurement Speed

The time required for a microwave counter to perform a measurement may be divided into two parts:

(1) Acquisition — The time necessary for the counter to detect a microwave signal and prepare to make a measurement; and

(2) Gate Time — The duration of the counter's gate required to measure a given resolution.

Each of the three down-conversion techniques we have discussed offers trade-offs in the area of measurement speed.

A prescaler will have fast acquisition time, typically equal to N cycles of the microwave input. The gate time for a prescaling counter will be equal to N/R, where R is the desired resolution in Hz and N is the prescale factor. As mentioned above, this extended gate time is required in order to effectively multiply the counter's contents by N.

Appendix L: Frequency Counters

The heterodyne converter, using the YIG filter, has an acquisition time ranging from 40 milliseconds to over 200 milliseconds. A design using thin film filters has an impressively short acquisition time of less than 1 millisecond. The gate time for heterodyne converter counters is 1/R, unless the conventional portion of the counter uses prescaling (which is sometimes done to reduce cost).

A microwave counter using the transfer oscillator technique will typically have an acquisition time of about 150 milliseconds, which is comparable to the heterodyne converter. Gate times for the transfer oscillator are longer since, as with the prescaler, they must be set to N/R. This factor of N can cause the transfer oscillator counter to measure much more slowly than the heterodyne converter for high resolution (100 Hz or less) measurements of microwave frequencies. For typical measurements with resolution 1 kHz or greater, the difference in measurement speed between the two techniques will not be noticed by the operator.

Accuracy

The accuracy of microwave counter measurements is limited by two factors:

(1) The plus/minus one-count quantization error; and
(2) Time base errors.

Time base errors may further be looked at in two different ways: short-term stability, which generally limits the repeatability from one measurement to the next; and long-term stability, which limits the absolute accuracy of a measurement.

Figure L.4(a) graphs the short-term stability (repeatability) of microwave measurements for each of two down-conversion techniques. If gate time is limited to one second, it is clear that the transfer oscillator is limited to about 1×10^{-8} resolution. The heterodyne converter is limited to about 1×10^{-9}, where short-term instabilities of the crystal oscillator become the limiting factor. With the high stability of an oven oscillator, the heterodyne converter is capable of resolving one part in 10^{10} at microwave frequencies.

Of considerably more importance to the user, however, are the long-term effects which limit the accuracy of microwave counter measurements. Figure L.4(b) graphs the combined effects of inaccuracies due to time base againg and to the resolution of the counting technique. In this figure it is assumed that the time base was calibrated to a high degree of accuracy one month ago. Clearly, even with the best time

504   HANDBOOK OF MICROWAVE TESTING

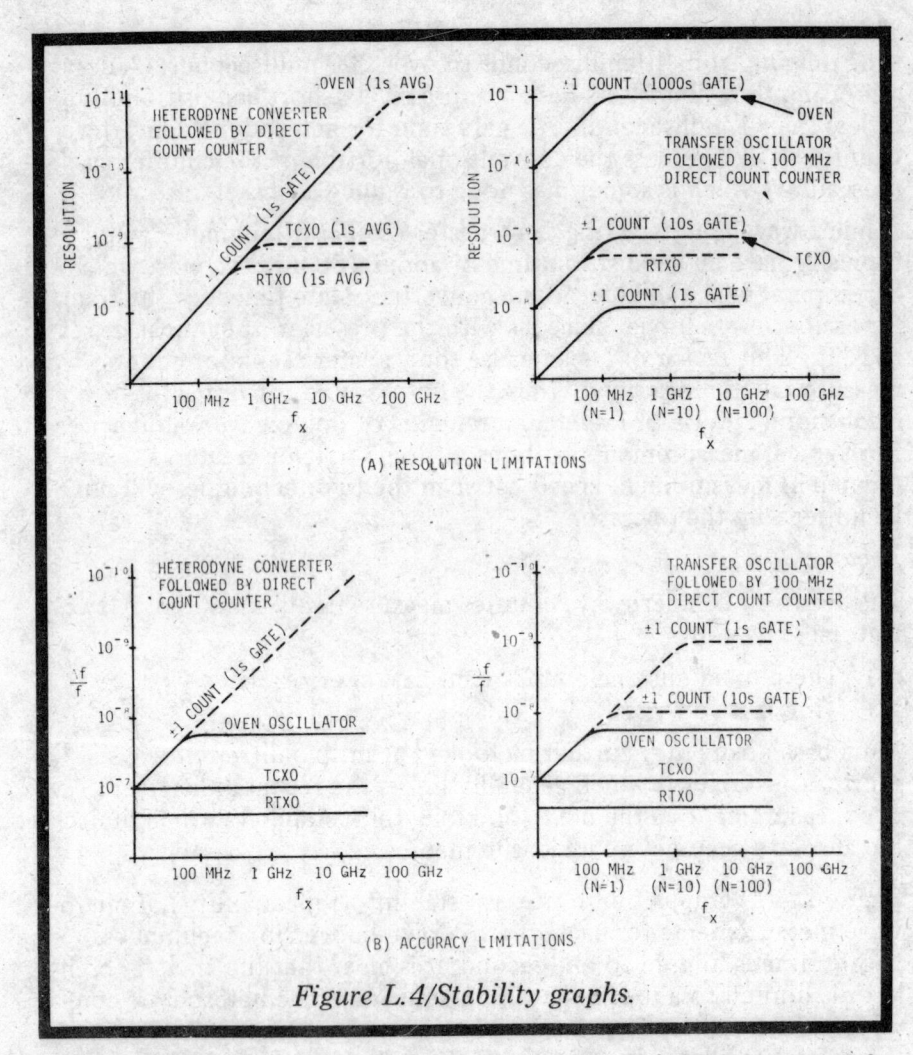

Figure L.4/Stability graphs.

bases available, the long-term instability of the time base becomes the accuracy limitation, no matter which down-conversion technique is used. It may, therefore, be concluded that accuracy is not a consideration in choosing between microwave down-conversion techniques for a particular application.

The plus/minus one-count quantization error is present in a counter due to the fact that the signal to be counted is not synchronized to the opening and closing of the main gate within the counter. The non-synchronization case introduces a ±1 count error in your readings. This error is usually more significant at lower frequencies and when

Appendix L: Frequency Counters

measuring pulse frequencies. For high frequency measurements, the ±1 count gating error is generally insignificant.

Sensitivity and Dynamic Range

Sensitivity and dynamic range are two more parameters to be considered when choosing a microwave counter. The sensitivity of a heterodyne converter must be chosen with great care. On the other hand, since the effective input bandwidth of the heterodyne converter is so wide, a counter which is designed to be too sensitive might register false readings due to broadband low level noise. Typical sensitivity specifications for heterodyne down-converter microwave counters range from −20 dBm to −25 dBm.

The transfer oscillator offers an advantage in the area of sensitivity. Since the input signal into this down-converter enters a narrowband (about 200 kHz) phase lock loop, the problem of noise triggering is greatly diminished. Consequently, these counters can be constructed with exceptional sensitivity specification and the highest level input signal which can be counted reliably. A typical value for the upper limit is +7 dBm. It has been found that the technique with the greater sensitivity, that is, the transfer oscillator, will also have the greater dynamic range.

It should be noted that some microwave counters allow measurements of input to +20 dBm and beyond; this specification is generally accompanied by lesser sensitivity. As an example, there are counters which have a rated maximum input of +20 dBm, but are specified at a sensitivity of only −15 dBm. This results in a 35 dB dynamic range.

Signal-to-Noise Ratio

An important consideration in choosing a microwave counter is the signal-to-noise environment of the measurement. As mentioned in the above paragraph, the apparent amplifier bandwidth at the counter's input limits the amount of noise which the counter can tolerate on the measured signal.

Consider a microwave frequency to be measured which has a good deal of noise surrounding the carrier. A transfer oscillator counter will be capable of measuring the signal if the peak carrier exceeds the noise floor by 20 dB. A typical heterodyne converter counter, however, will require 40 dB or greater separation to allow accurate measurement.

A common situation wherein broadband noise surrounds a signal to be measured is in the monitoring of solid state microwave sources. With this type of spectrum to be measured, a transfer oscillator will provide reliable readings up to the maximum sweeper output power (about +10 dBm). The typical heterodyne converter counter will encounter noise interference at sweeper output levels near −10 dBm. So you can see a 20 dB difference in measurement capability between the heterodyne converter counter and the transfer oscillator simply because of the signal-to-noise environment.

FM Tolerance

All modern microwave counters are capable of measuring today's microwave sources with their inherent incidental frequency modulation. There are applications, however, in which it is desired to measure a microwave communications carrier with frequency modulation present. In these cases, the FM tolerance of microwave counters becomes a consideration for choosing the appropriate instrument.

A heterodyne converter may be thought of as dividing microwave frequency space into distinct bands, of a width equal to the comb line spacing. The design of these instruments is such that the video counting capability of the conventional counter is somewhat greater than the comb line spacing. It is this resulting overlap between adjacent bands that is the measure of the FM tolerance of the counter. An example of this would be a counter with comb line spacing of 500 MHz and a video bandwidth of 530 MHz. The FM tolerance of the particular counter is over 500 MHz with the carrier located mid-band, and diminishes to 30 MHz at band edge. These are typical values for the heterodyne converter technique.

The transfer oscillator's tolerance of frequency modulation is more complex. The maximum allowable peak-to-peak deviation is a function of modulating frequency and carrier frequency. In general, this tolerance is at a minimum at the point where modulating frequency is equal to the bandwidth of the input phase lock loop. If more than one tone modulates the carrier simultaneously (as in the case of the multi-channel communication modulation), this analysis is no longer applicable. A typical response to multichannel FM is shown in Figure L.5. On this chart, tolerance to FM is indicated by the number of voice channels which are modulated onto the carrier. It can be seen that the FM tolerance of the transfer oscillator is in this case dependent upon carrier frequency and the per-channel loading of

Appendix L: Frequency Counters

the radio. Since most microwave radios operate between 86 kHz RMS and 140 kHz RMS per-channel loading, it can be seen that a transfer oscillator counter is capable of measuring just about all fully loaded microwave communications carriers in use today.

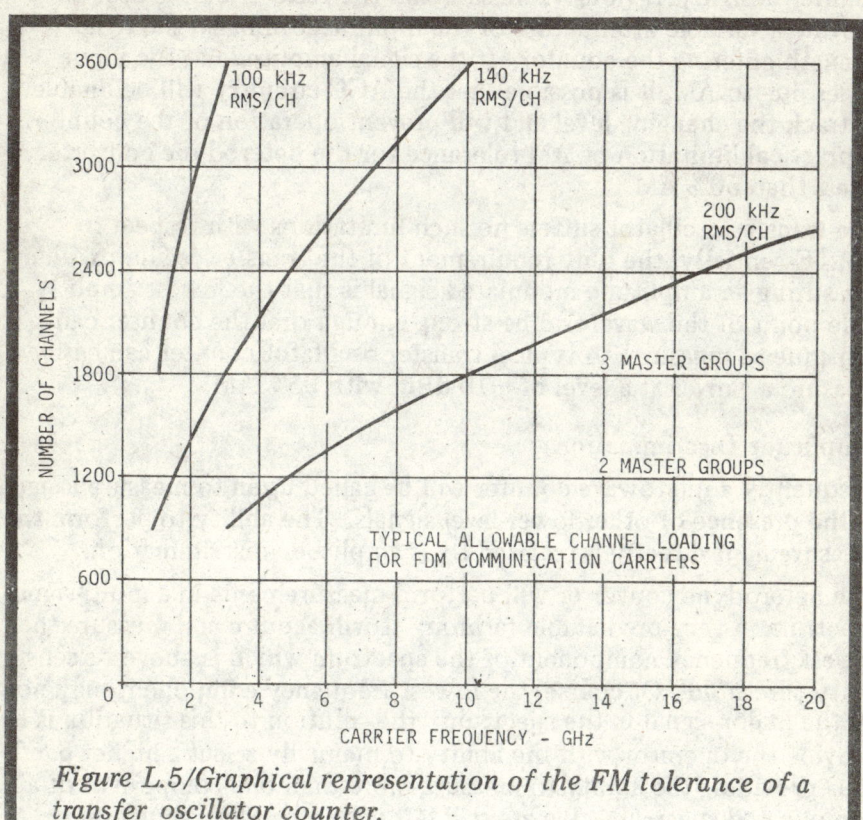

Figure L.5/Graphical representation of the FM tolerance of a transfer oscillator counter.

In summary, although the transfer oscillator is capable of measuring microwave frequencies with all common forms of FM modulation, the heterodyne converter has a clear advantage in the area of FM tolerance.

AM Tolerance

A second form of modulation frequency encountered in the microwave world is amplitude modulation. Few microwave radios use AM for communications transmissions, but nearly all microwave sources provide signals with incidental AM. Also, in many R&D and maintenance environments a time-varying attenuation of the signal is commonly encountered.

The heterodyne converter's tolerance to amplitude modulation is limited by its AGC circuitry when such a circuit is employed in the counter design. Previously, we saw that the AGC circuit is used to provide a variable attenuation of the input according to the signal strength entering the counter. If the signal amplitude at the input varies due to AM, it is possible that the AGC circuitry will be unable to track the changing level and will prevent operation of the counter. A practical limitation of AM tolerance for the heterodyne converter is less than 50% AM.

The transfer oscillator suffers no such limitations with respect to AM. Essentially, the only requirement of the transfer oscillator when measuring an amplitude modulated signal is that the lowest amplitude point of the waveform be strong enough that the counter can continue to measure. A typical transfer oscillator counter can easily measure a carrier at a level of −10 dBm with 95% AM.

Amplitude Discrimination

Frequently a microwave counter will be called upon to measure a signal in the presence of other lower level signals. The ability to perform the measurement directly is referred to as amplitude discrimination.

The heterodyne converter will perform measurements in a multi-signal spectrum in very predictable fashion: it will acquire and measure the lowest frequency component of the spectrum which is above its sensitivity threshold. Of course, the lowest frequency component may not be the major signal in the spectrum; the solution to this situation is to provide the operator with the ability to manually select a higher band, thus providing the ability to measure more than one component of a complicated spectrum. We are still left with a problem, however: what happens when two significant frequency components are within the same band? The answer to this situation is that the AGC circuitry must be able to differentiate between the two signals. Typical AGC circuits found in heterodyne converters today provide discrimination between signals which lie from 4 dB to 30 dB apart, located in the same band.

Automatic amplitude discrimination is one of the key features of the transfer oscillator design. These counters are typically capable of always finding the most prominent component of the spectrum, provided that it is at least 2 dB above nearby signals and at least 10 dB above signals at the far end of the counter's frequency range.

Appendix L: Frequency Counters

We now have covered some of the performance trade-offs between the heterodyne converter and the transfer oscillator down-conversion techniques for microwave counters. (You will recall that the prescaling method was omitted due to its low operating frequencies.) The results of our findings are shown below in Table L.1. It should be noted that these comparisons are made on the basis of typical specifications; a comparison of individual instruments may result in different results in some areas.

TABLE L.1
Summary of the Performance of the Two Principal Microwave Down-Conversion Techniques

Characteristic	Heterodyne	Transfer Oscillator
Frequency Range	20 GHz	40 GHz
Measurement Speed	150 ms acquisition 1/R Gate	150 ms acquisition N/R Gate
Accuracy	Time base limited	Time base limited
Sensitivity/Dynamic Range	−25 dBm/30 dB	−35 dBm/40 dB
Signal-to-Noise Ratio	40 dB	20 dB
FM Tolerance	30-40 MHz peak-peak	1-10 MHz peak-peak
AM Tolerance	Less than 50%	Greater than 90%
Amplitude Discrimination	Displays lowest frequency	Displays strongest frequency

The characteristics covered in Table L.1 are the major items to look for in the microwave counter. There are, however, other areas of importance that are not shown above. Two such areas are *signal inputs* and *systems interface*.

Signal Inputs

The first check of the inputs to a microwave counter is to insure that the frequency ranges covered by the various input connectors satisfy the requirements of the application. At times, it can be burdensome to be continually changing the input connector from one spigot to another. Of course, the ideal situation in a system application is for one connector to cover the full frequency range of the counter. Also, are the connector types appropriate for the frequency range involved? For example, a Precision Type N connector may begin to display high SWR with input very much over 18 GHz.

SWR, of course, is quite important in selecting test equipment; good SWR specifications should always be required of a microwave counter.

High impedance inputs are also necessary for some applications to avoid loading the circuit under test. A question to be asked here: does the high impedance input of the microwave counter satisfactorily cover the full range of applications for high impedance probing? For instance, it is often necessary to use a high impedance input for measurements at the IF of microwave radios.

A consideration of great importance in microwave counters is that of damage level limitations on the input signal level. Well-designed microwave counters today can tolerate up to +30 dBm inputs without damage. This area has been covered many times before and cannot be emphasized too much.

Systems Interface

A consideration of growing importance today in the test equipment world is that of systems compatibility of instruments. For years, specifying this type of compatibility has been impossible due to the lack of a recognized standard. Since the IEEE adopted standard #488 in 1975, however, an elegant solution to systems compatibility for all forms of test instruments has been a reality. Many companies have made their test equipment compatible with the IEEE bus (Hewlett Packard has implemented the IEEE 488 standard by what they term the Hewlett Packard Interface Bus, or HP-IB). The fact that interfacing is possible allows programming and data output of microwave counters using the ASCII code via an 8-bit bidirectional bus. This method of interface is covered in Chapter 8 on *automatic testing*.

References

Chapter 2

EIP, Inc., *The Automatic Microwave Pulse Counter*, Application Note 101, Santa Clara, California, June 1976.

Graff, Rudoff F., *Modern Dictionary of Electronics*, Howard Sams and Co., Indianapolis, Indiana, 4th Edition, 2nd Printing, 1975.

Hewlett-Packard Co., *Signal Generator Seminar*, Palo Alto, California, September, 1974.

Hewlett-Packard Co., *Broadband Sweep Oscillator 8620A/86290A*, Technical Data (5952-812300), Palo Alto, California, June 1975.

Hewlett-Packard Co., *Noise Figure Primes*, Application Note 57, Palo Alto, California, January 1965.

Hewlett-Packard Co., *Understanding Microwave Frequency Measurements*, Application Note 144, Palo Alto, CAlifornia.

Hewlett-Packard Co., *Application and Performance of the 8671A and 8672A Microwave Synthesizer*, Application Note 218-1, Palo Alto, California, February 1977.

Lance, A.L., *Introduction to Microwave Theory and Measurements*, McGraw-Hill, New York, 1964.

Laverghetta, Thomas, S., *Microwave Measurements and Techniques*, Artech House, Dedham, Massachusetts, July 1976.

Watkins-Johnson Co., *The MJ1250 Frequency Synthesizer System*, Application Note 100072-076A, Palo Alto, California, March 1976.

Catalogs

AIL Tech, *Noise Figure Instrumentation*, #NF1175/30MNP.

Hewlett-Packard, *Electronic Instruments and Systems*, 1977.

Polarad Electronics, *Microwave Instrumentation*, Short Form P7, 1976.

Systron Donner Instrument Catalog, 1975/1976.

Systron Donner, *Frequency Counters*, September 1977.

Systron Donner, *Spectrum Analyzers*, 1977.

Chapter 3

Graff, Rudoff F., *Modern Dictionary of Electronics*, Howard Sams and Co., Indianapolis, Indiana, 4th Edition, 2nd Printing, 1975.

Hewlett-Packard Co., *Microwave Power Measurements*, Application Note 64, Palo Alto, California, October 1965.

Hewlett-Packard Co., *Fundamentals of RF and Microwave Power Measurements*, Application Note 64-1, Palo Alto, CAlifornia, August 1977.

ITT, *Reference Data for Radio Engineers*, Howard Sams and Co., Indianapolis, Indiana, 6th Edition, 1975.

Laverghetta, Thomas S., *Microwave Measurements and Techniques*, Artech House, Dedham, Massachusetts, July 1976.

Chapter 4

Hewlett-Packard Co., *Noise Figure Primes*, Application Note 57, Palo Alto, California, January 1965.

Hewlett-Packard Co., *Spectrum Analysis. . . Noise Figure Measurements*, Application Note 150-9, Palo Alto, California, April 1976.

Laverghetta, Thomas S., *Microwave Measurements and Techniques*, Artech House, Dedham, Massachusetts, July 1976.

Microwave Semiconductor Corp., *System Noise Figure Monitoring*, Application Note NM-201, Somerset, New Jersey, September 1976.

Catalogs

AIL Tech, *Noise Figure Instrumentation*, #NF1175/30 MWP.

Hewlett-Packard, *Coaxial and Waveguide Catalog and Microwave Measurement Handbook*, Palo Alto, California, 1977-1978.

Manual

Instruction Manual for Type 75 Precision Automatic Noise Figure Indicator, AIL 75, Revision C.

Chapter 5

EDN Editorial Staff, *Spectrum Analyzers-RF and Digital Technologies Team up to Produce a Winner*, Cahners Publishing Co., Boston, Massachusetts, April 20, 1978.

Engelson, Morris and Telewski, Fred, *Spectrum Analyzer Theory and Applications*, Artech House, Dedham, Massachusetts, 1974.

Graff, Rudoff F., *Modern Dictionary of Electronics*, Howard Sams and Co., Indianapolis, Indiana, 4th Edition, 2nd Printing, 1975.

References

Hewlett-Packard Co., *The Spectrum Analyzer for Design Engineers — Expand Your Measurement Capability*, Number 5952-0932, Palo Alto, California.

Hewlett-Packard Co., *The Spectrum Analyzer Could Be the Most Important Test Instrument on Your Bench*, Palo Alto, California, January 1975.

Hewlett-Packard Co., *Spectrum Analyzer Measurement Seminar*, Palo Alto, California, September 1975.

Hewlett-Packard Co., *Spectrum Analysis . . . Spectrum Analyzers Basics*, Application Note 150, Palo Alto, California, April 1974.

Hewlett-Packard Co., *Spectrum Analysis . . . Amplitude and Frequency Modulation*, Application Note 150-1, Palo Alto, CAlifornia, November 1971.

Hewlett-Packard Co., *Spectrum Analysis . . . Pulsed RF*, Application Note 150-2, Palo Alto, California, November 1971.

Hewlett-Packard Co., *Spectrum Analysis . . . Accuracy Improvements*, Application Note 150-8, Palo Alto, California, March 1976.

Hewlett-Packard Co., *Spectrum Analysis . . . Noise Figure Measurement*, Application Note 150-9, Palo Alto, California, April 1976.

Hewlett-Packard Co., *Spectrum Analysis . . . Distortion Measurement*, Application Note 150-11, Palo Alto, California, October 1976.

IEEE Standard Dictionary of Electrical and Electronic Terms, Wiley Interscience Division of John Wiley and Sons, 2nd Edition New York, 1977 (IEEE STD-100-1977).

Linden, Peter, "Tracking Generators Stretch Your Spectrum Analyzer Capability," *EDN*, November 5, 1976.

Pacific Measurements, *Technical Data-Model 1038-J10 Spectral Power Analyzer*, Palo Alto, California, October 1975.

Catalog

Systron Donner Spectrum Analyzer Catalog 6-77.

Chapter 6

Brinkoetter, Thomas R., "Swept Measurements Speed Gain Compression Tests," *Microwaves*, April 1977.

Hewlett-Packard Co., *The Spectrum Analyzer for Design Engineers*, Publication Number 5952-0932, Palo Alto, California, November 1973.

Hewlett-Packard Co., *Spectrum Analysis . . . Noise Measurements*, Application Note 150-4, Palo Alto, California, January 1973.

Hewlett-Packard Co., *Spectrum Analysis . . . Distortion Measurements*, Application Note 150-11, Palo Alto, California, October 1976.

Hewlett-Packard Co., *Active Device Measurements with the HP 8755 Frequency Response Test Set*, Application Note 155-1, Palo Alto, California, July 1976.

Hewlett-Packard Co., *High Frequency Swept Measurements*, Application Note 183, Palo Alto, Co., May 1975.

McVay, Franz C., "Don't Guess at Spurious Levels," *Electronic Design 3*, February 1, 1967.

Chapter 7

Hewlett-Packard Co., *Antenna/Random Boresight Error Measurements*, Application Note 110, Palo Alto, California, October 1968.

Hollis, J.S. Lynn, T.O., Clayton, L., Jr., *Microwave Antenna Measurements*, Scientific Atlanta Inc., Atlanta, Georgia, 1969.

Kummer, Wolfgang H., Gillespie, Edmond S., "Antenna Measurements — 1978," *Proceedings of the IEEE*, Vol. 66, No. 4, April 1978.

Laverghetta, Thomas S., *Microwave Measurements and Techniques*, Artech House, Dedham, Massachusetts, 1976.

Chapter 8

Hewlett-Packard Co., *Hewlett-Packard Interface Bus-HP IB*, #5952-0050, Palo Alto, California, March 1977.

IEEE-488-1975, *IEEE Standard Digital Interface for Programmable Instrumentation*, IEEE, New York, April 1975.

Kimball, Jim, "IEEE 488 Bus Going Your Way?" *Tekscope*, Vol. 10, No. 2, 1978.

Ableidinger, Bruce, "Unraveling the Mystery on the GPIB," *Tekscope*, Vol. 10, No. 2, 1978.

Manual

HP436A Power Meter Instruction Manual.

Chapter 9

Cowan, Gene, "Digital Processing Adds Accuracy to TDR," *Microwaves*, December 1975.

References

Frequency Engineering Labs, *Phase Noise Measurements at 1-18 GHz Using the Model 800 Series Phase Noise Analyzer*, Application Note 771, Farmingdale, New Jersey, April 1977.

Hewlett-Packard Co., *Selected Articles on TDR Applications*, Application Note 75, Palo Alto, California, March 1966.

Hewlett-Packard Co., *Understanding and Measuring Phase Noise*, Application Note 207, Palo Alto, California, October 1976.

Lance, A.L., Seal, W.D., Mendoza, F.G., and Hudson, N.W., "Automated Phase Noise Measurements," *Microwave Journal*, June 1977.

Lance, A.L., *Introduction to Microwave Theory and Measurements*, McGraw-Hill, New York, 1964.

Laverghetta, Thomas S., *Microwave Measurements and Techniques*, Artech House, Dedham, Massachusetts, July 1976.

Reynolds, Chuck, "Measure Phase Noise," *Electronics Design 4*, February 15, 1977.

Hewlett-Packard Co., *Frequency Response Test Sets 8755L/M*, Technical Data, Palo Alto, CAlifornia, May 1975.

Narda Microwave, *Narda Probe*, Plainsview, LI, New York, October 1977, Vol. 9, #2.

Catalogs

Hewlett-Packard Co., *Coaxial Component Catalog*, Palo Alto, California, 1977.

Weinschel Engineering, *Microwave Measurements and Generation Instruments*, Gaithersburg, Maryland, 1974.

INDEX

Acceptor handshake, 347
Active testing, 257-289
ALC, 16, 33, 109, 262, 448
Anechoic chamber, 296
    Rectangular, 296
    Tapered, 299
Antenna measurement, 291-323
Attenuator,
    Fixed, 53
    Step, 60
Automatic testing, 325-408
Auxiliary components, 53-78

Barretter, 34-96
Beamshift error, 319
Bessel function, 226
Bi-directional bus, 328
Bit, 328
Bit parallel, 328
Bolometer, 34, 96
Boresight, 317
Bus, 328
Byte, 328
Byte-serial, 328

Calibration, 7
Carrier, 215
Compact range, 293
Continuously variable attenuator, 56-59
    Waveguide below cut-off, 56
    Variable card, 58
    Tee, 58
    Variable coupler, 58
    Low-loss cut-off, 59
    Lossy wall, 59
Controller, 378-387

Data bus, 330
Data byte transfer control bus, 330
DB, 94
DBM, 95
DC load, 391
Detector, 30-34
Device clear, 375-377
Device trigger, 377
Directional coupler, 63-71

Directivity, 67
Display, 87-90
    Phase-magnitude, 87
    Phase-gain indicator, 87
    Polar, 88
Distortion, 179
Dynamic range, 188

Elevated range, 292

Frequency accuracy, 187
Frequency counter, 50-52, 499-512
Frequency domain, 178
Frequency measurement, 201-206
    Absolute, 204
    Relative, 202
Frequency stability, 185

Gain, 259, 301
Gain compression, 268
General interface management bus, 330
Ground reflection range, 298

Handshake cycle, 328
Harmonic frequency converter, 83
Hetrodyne down-conversion, 50, 500
High state, 328, 390

IF substitution, 85
Indicating device, 29
Indirect synthesis, 26
Insertion loss, 68
Interface, 328

Linearizing, 21
Listener, 360
Local control, 329
Local message, 337
Low state, 329, 390

Miscellaneous measurements, 411-461
Modulated wave, 216
Modulating wave, 215
Modulation, 16, 215-234
    AM, 16, 216
    FM, 16, 225
Modulator, 73-77

PIN diode, 73
Pulse, 75
Multiline message, 338

Network analyzer, 83
Noise factor, 138
Noise figure, 139
Noise figure meter, 44, 146-151
Noise measurement, 151-169
    Automatic, 162-169
    Manual, 151-161
Noise source, 46, 142-146
    Diode, 143
    Gas discharge, 46, 144
    Hot/cold, 46, 142
    Solid state, 46, 145
Noise temperature, 140
Normalizer, 77

Parallel poll, 372
Peak power measurements, 119-134
    Average power-duty cycle, 119
    Barretter integration differentiation, 128
    DC pulse power comparison, 125
    Direct pulse, 122
    Direct reading, 131
    Notch wattmeter, 124
    Sample and hold, 130
Phase lock, 26
Phase modulation, 286
Phase noise, 411
Poincare sphere, 313
Polarization, 312
    Circular, 313
    Elliptical, 313
    Linear, 312
Power, 93-135
    Low, 97
    Medium, 103
    High, 114
Power detector, 34
Power handling capability, 3
Power meter, 36
Prescaling, 50, 499
Pulsed RF, 234

Q-Measurement, 420

Receiver measurements, 215

Reflectometer, 441
Remote control, 329
Remote local, 368
Residual AM, 283
RF ratio meter, 433

Sensitivity, 32
Service request, 366
Signal generator, 14
Signal Oscillator, 14
Start range, 292
Slotted line, 71
Source handshake, 343
Spectrum analyzer, 40, 177
State diagram, 339
Swept impedance measurements, 433-460
Sweep oscillator, 20
Synthesized signal generator, 24

Talker, 354
TDR, 426
Test equipment, 13
Testing rules, 10
Test parameters, 2
Test range, 292-301
    Free space, 292
    Reflection, 298
Thermistor, 34, 96
Thermocoupler, 34, 96
Third order intercept, 279
Third order intermodulation, 277
Time domain, 178
Tracking generator, 252
Tracking preselector, 250
Transducer, 85
Transfer oscillator down-conversion, 50, 502
Twice-power noise measurement, 151
Two-tone intermodulation, 225

Uni-directional bus, 329

VSWR bridge, 447
VTO, 21

Watt, 94
Work, 94

Y-factor noise measurement, 154
YTO, 21

CHEMISTRY-CHEMICAL ENGINEERING

A-1-1 Alder :	MECHANISM IN ORGANIC CHEMISTRY	$100
A-1-2 Adamson :	PHYSICAL CHEMISTRY OF SURFACES 3/e (1976)	$200
#A-1-3 ACS :	TRANSITION METAL CHEMISTRY (Vol. 1)	$200
A-1-4 ACS :	TRANSITION METAL CHEMISTRY (Vol. 2)	$200
A-1-5 Anderson :	STRUCTURE OF METALLIC CATALYSTS	$150
A-1-6 Aiba :	BIOCHEMICAL ENGINEERING (2nd ed)	$300
*A-1-7 Andrews :	AN INTRODUCTION TO CHROMATOGRAPHY	$ 45
A-1-8 Atkinson :	BIOCHEMICAL REACTORS	$110
A-1-9 Adams :	ELECTROCHEMISTRY AT SOLID ELECTRODES	$150
A-1-10 Abragam :	THE PRINCIPLES OF NUCLEAR MAGNETISM	$300
A-1-11 Acs :	ORBITAL SYMMETRY PAPERS	$220
A-1-12 Anderson :	EXPERIMENTAL METHODS IN CATALYTIC RESEARCH	
	1. KINETICS OF CATALYTIC REACTIONS	$150
	2. PREPARATION AND EXAMINATION OF PRACTICAL CATALYSTS	$150
	3. CHARACTERIZATION OF SURFACE AND ADSORBED SPECIES	$150
*A-1-13 Albery :	ELECTRODE KINETICS	$ 60
A-1-14 Atkins :	QUANTA-A HANDBOOK OF CONCEPTS	$200
ΔA-1-16 Abraham :	PROTON AND CARBON-13 N.M.R. SPECTROSCOPY 1979 (076)	$150
B-1-1 Baumann :	BASIC LIQUID CHROMATOGRAPHY	$120
B-1-2 Butler :	PROBLEMS IN PHYSICAL ORGANIC CHEMISTRY	$ 40
B-1-3 Brownstein :	TRENDS IN PETROCHEMICAL TECHNOLOGY	$150
B-1-4 Brownstein :	US PETROCHEMICALS	$200
B-1-5 Bamford :	COMPREHENSIVE CHEMICAL KINETICS (3 Vols)	$450
	1. THE PRACTICE OF KINETICS	
	2. THE THEORY OF KINETICS	
	3. THE FORMATION AND DECAY OF EXCITED SPECIES	
B-1-6 Baker :	ORGANIC CHEMISTRY PROBLEMS AND SOLUTIONS 1978	$120
B-1-7 Balescu :	EQUILIBRIUM AND NONEQUILIBRIUM STATISTICAL MECHANICS	$220
B-1-8 Braterman :	METAL CARBONYL SPECTRA	$160
B-1-9 Becker :	HIGH RESOLUTION NMR (2/e) 1980	$160
*B-1-10 Brapley :	FAST REACTION	$ 50
B-1-11 Bruins :	UNSATURATED POLYESTER TECHNOLOGY	$300
/B-1-12 Billot :	PERFUMERY TECHNOLOGY	$250
/B-1-13 Bennett :	INDUSTRIAL WAXES	
	Vol. I. NATURAL & SYNTHETIC WAXES	$320
	Vol. II. COMPOUNDED WAXES & TECHNOLOGY	$280
ΔB-1-14 Bent :	THE SECOND LAW (060)	$190
ΔB-1-15 Benson :	THERMOCHEMICAL KINETICS (062)	$130
ΔB-1-16 Blatt :	ORGANIC SYNTHESIS VOLUME (069) (5Vols)	$2000
C-1-1 Coxon :	ORGANIC PHOTOCHEMISTRY	$110
C-1-2 Cundall :	PHOTOCHEMISTRY	$ 85
#C-1-3 Clarke :	ISOLATION AND IDENTIFICATION OF DRUGS (Vol. 1)	$300
#C-1-4 Clarke :	ISOLATION AND IDENTIFICATION OF DRUGS (Vol. 2)	$180
*C-1-5 Cremlyn :	NAMED AND MISCELLANEOUS REACTIONS IN PRACTICAL ORGANIC CHEMISTRY**	$100
C-1-6 Carrington :	INTRODUCTION TO MAGNETIC RESONANCE	$230
C-1-7 Cooney :	BIOMEDICAL ENGINEERING PRINCIPLES	$250
C-1-8 Chang :	BASIC PRINCIPLES OF SPECTROSCOPY	$160
C-1-9 Clausen :	PRINCIPLES OF INDUSTRIAL CHEMISTRY	$150
oC-1-10 Crompton :	CHEMICAL ANALYSIS OF ADDITIVES IN PLASTICS 2/e	$450
oC-1-11 Callely :	TREATMENT OF INDUSTRIAL EFFLUENTS	$350
ΔC-1-12 Carey :	ADVANCED ORGANIC CHEMISTRY (019)	$460
C-1-13 Coulson :	HÜCKEL THEORY FOR ORGANIC CHEMISTS	$120
S-1-3 Strehlow :	FUNDAMENTALS OF CHEMICAL RELAXATION	$ 65

CHEMISTRY-CHEMICAL ENGINEERING

C-1-14 Cussler : MULTICOMPONENT DIFFUSION	$
C-1-15 Cooper : SPECTROSCOPIC TECHNIQUES FOR ORGANIC CHEMISTS	$190
D-1-1 Drauglis : THE PHYSICAL BASIS FOR HETEROGENEOUS CATALYSIS	$300
D-1-2 Demtroder : LASER SPECTROSCOPY	$ 65
D-1-4 Delmon : PREPARATION OF CATALYSTS 2Vols	$780
△D-1-5 Diehl : N.M.R. BASIC PRINCIPLES & PROGRESS GRUNDLAGEN UND FORTSHRITTE (001)	100
△D-1-6 Drago : PHYSICAL METHOD IN CHEMISTRY 1977 (066)	$350
D-1-7 Delahay : NEW INSTRUMENTAL METHODS IN ELECTROCHEMISTRY	$230
E-1-1 Eliel : STEREOCHEMISTRY OF CARBON COMPOUNDS	$150
E-1-2 Erdey-Gruz : KINETICS OF ELECTRODE PROCESSES	$180
E-1-3 Edwards : INORGANIC REACTION MECHANISMS	$ 90
E-1-4 Emmett : CATALYSIS THEN AND NOW	$250
*E-1-5 Earnshaw : THE CHEMISTRY OF THE TRANSITION ELEMENTS	$ 50
△E-1-6 Eyring : BASIC CHEMICAL KINETICS	$270
△E-1-7 Elias : MACROMOLECULES 2 VOLUME IN ONE (005)	$400
E-1-8 El-Aasser : ADVANCES IN EMULSION POLYMERIZATION AND LATEX TECHNOLOGY	$500
*F-1-1 Hessenden : THE BASIS OF ORGANIC CHEMISTRY**	$180
F-1-2 Fuchs : CARBON-13 NMR—BASED ORGANIC SPECTRAL PROBLEMS (1979)	$150
F-1-3 Freifelder : PRACTICAL CATALYTIC HYDROGENATION	$280
F-1-4 Ferraro : INTRODUCTORY GROUP THEORY	$150
F-1-5 Flygare : MOLECULAR STRUCTURE AND DYNAMICS	$260
▽F-1-6 Finch : POLYVINYL ALCOHOL	$400
G-1-1 Gileadi : INTERFACIAL ELECTROCHEMISTRY	$180
△G-1-2 Gilchrist : ORGANIC REACTIONS AND ORBITAL SYMMETRY (003)	$110
H-1-1 Hirsch : CONCEPTS IN THEORETICAL ORGANIC CHEMISTRY	$100
H-1-2 Huheey : INORGANIC CHEMISTRY (2nd ed)	$300
H-1-3 Huheey : ANSERS TO PROBLEMS IN INORGANIC CHEMISTRY	$ 25
H-1-4 Hightower : THE USES OF HETEROGENEOUS CATALYSIS	$280
H-1-5 Heck : ORGANOTRANSITION METAL CHEMISTRY	$160
H-1-6 Herzberg : SPECTRA OF DIATOMIC MOLECULES	$230
*H-1-8 Harris : SYMMETRY AND SPECTROSCOPY (1978) — AN INTRODUCTION TO VIBRATIONAL AND ELECTRONIC SPECTROSCOPY	$200
*H-1-9 Hammes : PRINCIPLES OF CHEMICAL KINETICS	$100
▽H-1-10 Haslam : IDENTIFICATION AND ANALYSIS OF PLASTICS	$400
△H-1-11 Heftman : CHROMATOGRAPHY 1975 (033)	$360
H-1-12 Himmelblau : FAULT DETECTION AND DIAGNOSIS IN CHEMICAL AND PETROCHEMICAL PROCESSES	$
I-1-1 Ikan : NATURAL PRODUCTS—A LABORATORY GUIDE	$220
I-1-2 Ives : REFERENCE ELECTRODES	$280
J-1-1 Johnson : BASIC LIQUID CHROMATOGRAPHY	$150
J-1-2 Jolly : THE SYNTHESIS AND CHARACTERIZATION OF INORGANIC COMPOUNDS	$220
J-1-3 Jackson : TRANSPORT IN POROUS CATALYSTS	$
J-1-4 Janssen : TWIN SCREW EXTRUSION	$
K-1-1 Kochi : ORGANOMETALLIC MECHANISMS AND CATALYSIS	$250
△K-1-2 Karger : AN INTRODUCTION TO SEPARATION SCIENCE (079)	$250
L-1-1 Laidler : THEORIES OF CHEMICAL REACTION RATES	$ 85
L-1-2 Littlewood : GAS CHROMATOGRAPHY (2nd ed)	$150
L-1-3 Ludewig : POLYESTER FIBRES CHEMISTRY AND TECHNOLOGY	$250
L-1-4 Lowe : QUANTUM CHEMISTRY	$220
*L-1-6 Lambert : ORGANIC STRUCTURAL ANALYSIS	$240
L-1-7 Leyden : ANALYTICAL APPLICATIONS OF NMR	$180

CHEMISTRY-CHEMICAL ENGINEERING

ΔL-1-8 Levine : STUDENT/INSTRUCTOR'S SOLUTION SUPPLEMENT TO
           ACCOMPANY PHYSICAL CHEMISTRY 1978 (059) $120
 L-1-9 Lowenheim : ELECTROPLATING $290
 M-1-1 Mcnair : BASIC GAS CHROMATOGRAPHY $140
 M-1-2 Mcomie : PROTECTIVE GROUPS IN ORGANIC CHEMISTRY $135
 M-1-3 Murphy : FOUNDATIONS OF COLLEGE CHEMISTRY $180
#M-1-4 Darmstadt : MERCK STANDARDS $450
*M-1-5 Mclauchlan : MAGNETIC RESONANCE ....$ 50
 M-1-6 Marshall : BIOPHYSICAL CHEMISTRY (1978) $300
ΔM-1-7 Murrell : VALENCE THEORY (061) $180
ΔM-1-8 Moore : ORGANIC CHEMISTRY: AN OVERVIEW 1978 (036) $250
ΔM-1-9 Murrell : THE CHEMICAL BOND 1978 (053) $150
#ΔN-1-1 Norman : PRINCIPLES OF ORGANIC SYNTHESIS 1978 2/e (068) . $280
 O-1-1 Osipow : SURFACE CHEMISTRY $150
 O-1-2 Orchin : THE IMPORTANCE OF ANTIBONDING ORBITALS ....$ 50
 O-1-3 Overbeek : COLLOID SURFACE CHEMISTRY
               PART 1. SURFACE CHEMISTRY $220
               PART 2. LYOPHOBIC COLLOIDS $220
              #PART 3. ELECTROKINETICS AND MEMBRANE PHENOMENA
              #PART 4. LYOPHILIC COLLOIDS
 O-1-4 Oliv'e : COORDINATION AND CATALYSIS $160
 P-1-1 Pearson : HARD AND SOFT ACIDS AND BASES $250
 P-1-2 Pearson : SYMMETRY RULES FOR CHEMICAL REACTIONS $250
*P-1-3 Pass : IONS IN SOLUTION 3 —— INORGANIC PROPERTIES .....$ 50
 P-1-4 Peterson : HYDROGENATION CATALYSTS $140
 P-1-5 Powell : THE CHEMISTRY OF THE NON-METALS $180
ΔP-1-6 Purcell : INORGANIC CHEMISTRY 1977 (017) $400
 P-1-7 Pickett : ELECTROCHEMICAL REACTOR DESIGN $300
 R-1-1 Roberts : MODERN EXPERIMENTAL ORGANIC CHEMISTRY 3/e
               (1979) $280
 R-1-2 Roberts : AN INTRODUCTION TO THE ANALYSIS OF SPIN-SPIN
               SPLITTING IN HIGH-RESOLUTION NUCLEAR MAGNETIC
               RESONANCE SPECTRA ....$ 60
*R-1-3 Roberts : IONS IN SOLUTION 2 —— AN INTRODUCTION TO
               ELECTROCHEMISTRY ....$ 50
ΔR-1-5 Ray : INORGANIC POLYMERS 1978 (074) $100
ΔR-1-6 Rochow : MODERN DESCRIPTIVE CHEMISTRY 1977 (078) $160
 S-1-1 Stecher : THE MERCK INDEX (9/e) $450
 S-1-2 Sawyer : EXPERIMENTAL ELECTROCHEMISTRY FOR CHEMISTS ...$150
 S-1-3 Strehlow : FUNDAMENTALS OF CHEMICAL RELAXATION ....$ 65
*S-1-4 Stephenson : MATHEMATICAL METHODS FOR SCIENCE STUDENT .$200
*S-1-5 Spedding : AIR POLLUTION ....$ 50
 S-1-6 Stahl : THIN-LAYER CHROMATOGRAPHY $400
 S-1-7 Stone : ADVANCES IN ORGANOMETALLIC CHEMISTRY $170
 S-1-8 Starks : PHASE TRANSFER CATALYSIS $180
#S-1-9 Sittig : HANDBOOK OF CATALYST MANUFACTURE 1978 ....$
 S-1-10 Saunders : THE IDENTIFICATION OF PLASTICS AND RUBBERS ...$ 50
 S-1-11 Sekine : SOLVENT EXTRACTION CHEMISTRY $370
ΔS-1-12 Seyferth : NEW APPLICATIONS OF ORGANOMETALLIC REAGENTS IN
                ORGANIC SYNTHESIS (020) $240
ΔS-1-13 Schlafer : BASIC PRINCIPLES OF LIGAND FIELD THEORY (051) .$240
ΔS-1-14 Sewell : INTRODUCTION TO HIGH PERFORMANCE LIQUID
              CHROMATOGRAPHY (065) $120
 S-1-15 Stritwieser : INTRODUCTION TO ORGANIC CHEMISTRY $560
 S-1-16 Satterfield : HETEROGENEOUS CATALYSES IN PRACTICE $200
 S-1-17 Storch : INDUSTRIAL SEPARATORS FOR GAS CLEANING ......$

CHEMISTRY-CHEMICAL ENGINEEING

S-1-18 Sterbacek : CALCULATION OF PROPERTIES USING CORRESPONDING-$
    STATE METHODS ....
T-1-1 Thomas : INTRODUCTION TO THE PRINCIPLES OF HETEROGENEOUS
    CATALYSIS $150
ΔT-1-2 Ternay : CONTEMPORARY ORGANIC CHEMISTRY 1976 (002) $300
T-1-3 Tinoco : PHYSICAL CHEMISTRY — PRINCIPLES AND APPLICATIONS
    IN BIOLOGICAL SCIENCES $250
V-1-1 Vetter : ELECTROCHEMICAL KINETICS:Theoretical &
    Experimental Aspects $400
V-1-2 Venuto : FLUID CATALYTIC CRACKING WITH ZEOLITE CATALYSTS . $100
#W-1-1 Wilkinson : INTRODUCTION TO PROCESS ECONOMICS $120
W-1-2 Wilkins : INTRODUCTION TO COMPUTER PROGRAMMING FOR CHEMISTS $260
*W-1-3 White : PRINCIPLES OF BIOCHEMISTRY** $200
W-1-4 Watt : COLLECTED READINGS IN INORGANIC CHEMISTRY $130
*W-1-5 Wehrli : INTERPRETATION OF C-13 NUCLEAR MAGNETIC RESONANCE
    SPECTRA $140
*W-1-6 Wormald : DIFFRACTION METHODS .... $ 50
*W-1-7 Warren : DESIGNING ORGANIC SYNTHESES (1978) $110
W-1-8 Wells : THE THIRD DIMENSION IN CHEMISTRY .... $ 60
W-1-9 Williams : POLYMER ENGINEERING .... $
W-1-10 Wakeman : FILTRATION POST-TREATMENT PROCESSES .... $
W-1-11 Walsh : ATOMIC ABSORPTION SPECTROCHEMICAL ANAIYSIS $200
W-1-12:Whitaker:FUDNDAMENTAL PRINCIPLES OF HEAT TRANSFER(1977).. $360
Z-1-1 Zahradnik : ORGANIC QUANTUM CHEMISTRY PROBLEMS $100
▽Z-1-2 Ziabicki : FUNDAMENTALS OF FIBRE FORMATION $350

PHYSICS

*A-2-1 Atkins : QUANTA — A HANDBOOK OF CONCEPTS $200
A-2-2 Allen : ASTROPHYSICAL QUANTITIES $170
A-2-3 Amit : FIELD THEORY THE RENORMALIZATION GROUP AND CRITICAL
    PHENOMENE $170
A-2-4 Abell : EXPLORATION OF THE UNIVERSE 3/e $420
B-2-1 Bjorken : RELATIVISTIC QUANTUM MECHANICS $180
B-2-2 Blatt : THEORETICAL NUCLEAR PHYSICS $300
B-2-3 Balescu : EQUILIBRIUM AND NONEQUILIBRIUM STATISTICAL
    MECHANICS $220
B-2-4 Berry : PRINCIPLES OF COSMOLOGY AND GRAVITATION .... $ 70
B-2-5 Batygin : PROBLEMS IN ELECTRODYNAMICS $180
B-2-6 Bridgman : DIMENSIONAL ANALYSIS .... $ 65
B-2-7 Bailin : WEAK INTERACTIONS $200
B-2-8 Böhm : THE RIGGED HILBERT SPACE AND QUANTUM MECHANICS ... $ 55
B-2-9 Bygrave : ACCELERATOR NUCLEAR PHYSICS $420
    FUNDAMENTAL EXPERIMENTS WITH A VAN DE GRAAFF
    ACCELERATOR
B-2-10 Burns : INTRODUCTION TO GROUP THEORY WITH APPLICATIONS..... $200
D-2-1 Dirac : THE PRINCIPLES OF QUANTUM MECHANICS (4nd ed) $150
D-2-2 Dennery : MATHEMATICS FOR PHYSICISTS $180
E-2-1 Edmonds : ANGULAR MOMENTUM IN QUANTUM MECHANICS .... $ 65
E-2-2 Einstein : THE MEANING OF RELATIVITY .... $ 70
E-2-3 Eadie : STATISTICAL METHODS IN EXPERIMENTAL PHYSICS $200
E-2-4 Economou : GREEN'S FUNCTIONS IN QUANTUM PHYSICS $140
E-2-5 Epifanov : SOLID STATE PHYSICS $160
F-2-1 Feynman : QUANTUM MECHANICS AND PATH INTEGRALS $200
*F-2-2 Fano : PHYSICS OF ATOMS AND MOLECULES — AN INTRODUCTION
    TO THE STRUCTURE OF MATTER $200
F-2-3 Fetter : THEORETICAL MECHANICS OF PARTICLES AND CONTINUA-. $250

PHYSICS

H-2-1 Haken : QUANTUM FIELD THEORY OF SOLIDS	$160
I-2-1 Itzykson : QUANTUM FIELD THEORY	$550
J-2-1 Jones : THEORETICAL SOLID STATE PHYSICS 2.Vols	$520
K-2-2 Kaufmann : THE COSMIC FRONTIERS OF GENERAL RELATIVITY	$180
L-2-1 Landau : MECHANICS AND ELECTRODYNAMICS	$130
L-2-2 Liebey : THE EINSTEIN THEORY OF RELATIVITY	$120
L-2-3 Lurie : PARTICLES AND FIELDY	$180
L-2-4 Lightman : PROBLEM BOOK IN RELATIVITY AND GRAVITATION	$250
L-2-5 Lichtenberg : UNITARY SYMMETRY AND ELEMENTARY PARTICLES	$160
L-2-6 Landau : PHYSICS FOR EVERYONE	
BOOK. 1. PHYSICAL BODIES	$ 85
BOOK. 2. MOLECULES	$ 85
L-2-7 Landau : PHYSICS FOR EVERYONE-MOTION HEAT	$120
M-2-1 Misner : *GRAVITATION	$500
M-2-2 Mayer : STATISTICAL MECHANICS (2nd Ed.)	$160
M-2-3 Mizushima : QUANTUM MECHANICS OF ATOMIC SPECTRA AND ATOMIC STRUCTURE	$130
M-2-4 Migdal : QUALITATIVE METHODS IN QUANTUM THEORY	$170
M-2-5 Mariot : GROUP THEORY AND SOLID STATE PHYSICS	$ 70
M-2-6 Massey : ATOMIC AND MOLECULAR COLLISIONS	$180
M-2-7 Munster : STATISTICAL THERMODYNAMIC 2.Vols.	$680
N-2-1 Nigel : THE KEY TO THE UNIVERSE	$150
N-2-2 Nash : RELATIVISTIC QUANTUM FIELDS	$130
P-2-1 Perl : HIGH ENERGY HADRON PHYSICS	$230
R-2-1 Rosenberg : THE SOLID STATE	$125
S-2-1 Stacey : PHYSICS OF THE EARTH (2nd ed)	$180
S-2-2 Sommerfeld : PARTIAL DIFFERENTIAL EQUATIONS IN PHYSICS	$160
S-2-3 Sequin : CHARGE TRANSFER DEVICES	$135
S-2-4 Schrodinger : WHAT IS LIFE & MIND AND MATTER	$ 65
S-2-5 Scadron : ADVANCED QUANTUM THEORY	$180
T-2-1 Taylor : SPACETIME PHYSICS	$150
T-2-2 Taylor : SOLUTIONS TO THE EXERCISES SPACETIME PHYSICS	$ 30
T-2-3 Talman : SPECIAL FUNCTIONS	$120
U-2-1 Unsoid : THE NEW COSMOS	$180
W-2-1 Wigner : GROUP THEORY	$180
W-2-3 Westenholz : DIFFERENTIAL FORMS IN MATHEMATICAL PHYSICS	$250
W-2-4 Walker : THE FLYING CIRCUS OF PHYSICS WITH ANSWERS	$180
Z-2-1 Ziman : ELEMENTS OF ADVANCED QUANTUM THEORY	$150

ELECTRICAL-ELECTRONIC ENGINEERING

A-3-1 Auld : ACOUSTIC FIELDS AND WAVES IN SOLIDS	$350
B-3-1 Brookner : RADAR TECHNOLOGY	$380
B-3-2 Beynon : CHARGE-COUPLED DEVICES AND THEIR APPLICATIONS	$180
B-3-3 Brandt : ELECTRONIC STRUCTURE OF METALE	$160
D-3-1 Diefenderfer : PRINCIPLES of ELECTRONIC INSTRUMENTATION	$360
E-3-1 Eveleigh : INTRODUCTION TO CONTROL SYSTEMS DESIGN	$300
G-3-1 Garnell : GUIDED EAPON CONTROL SYSTEMS	$200
G-3-2 Gupta : MICROSTRIP LINES AND SLOTLINES	$250
H-3-1 Howe : STRIPLINE CIRCUIT DESIGN	$180
H-3-2 Helszajn : PASSIVE AND ACTIVE MICROWAVE CIRCUITS	$150
H-3-3 Hanlon : A USER'S GUIDE TO VACUUM TECHNOLOGY	$200
H-3-3 Huff : SEMICONDUCTOR SILICON 1981	$600
I-3-1 Itkis : CONTROL SYSTEMS OF VARIABLE STRUCTURE	$140
J-3-2 Johnston : ADAR ELECTRONIC COUNTER — COUNTERMEASURES	$400
J-3-3 Jacoby : ITERATIVE METHODS FOR NONLINEAR OPTIMIZATION PROBLEMS	$180

ELECTRICAL-ELECTRONIC ENGINEERING

K-3-1 Kressel : CHARACTERIZATION OF EPITAXIAL SEMICONDUCTOR FILMS	$110
K-3-2 Kwakernaak : LINEAR OPTIMAL CONTROL SYSTEMS	$250
L-3-1 Louisell : QUANTUM STATISTICAL PROPERTIES OF RADIATION	$200
L-3-2 Laverghetta : MICROWAVE MEASUREMENTS AND TECHNIQUES	$290
M-3-2 Milnes : DEEP IMPURITIES IN SEMICONDUCTORS	$250
M-3-3 Matthews : SURFACE WAVE FILTERS	$230
M-3-4 Melen : CHARGE-COUPLED DEVICES : TECHNOLOGY AND APPLICATIONS	$300
N-3-1 Nathanson : RADAR DESIGN PRINCIPLES	$280
O-3-1 Ott : NOISE REDUCTION TECHNIQUES IN ELECTRONIC SYSTEMS	$130
R-3-1 Roth : VACUUM TECHNOLOGY (1976)	$220
R-3-2 Ray : II-VI COMPOUNDS	$100
R-3-3 Rabiner : SELECTED PAPERS IN DIGITAL SIGNAL PROCESSING	$600
R-3-4 Rosenbrock : COMPUTER-AIDED CONTROL SYSTEM DESIGN	$140
S-3-2 Sharma : SEMICONDUCTOR HETEROJUNCTIONS	$160
S-3-4 Sucher : HANDBOOK OF MICROWAVE MEASUREMENTS	$600
T-3-1 Thomassen : INTRODUCTION TO MICROWAVE FIELDS AND CIRCUITS	$150
W-3-1 Wolf : SILICON SEMICONDUCTOR DATA	$260
W-3-2 Wieder : LABORATORY NOTES ON ELECTRICAL AND GALVANOMAGNETIC MEASUREMENTS	$200
W-3-3 Wiele : PROCESS AND DEVICE MODELING FOR INTEGRATED CIRCUIT DESIGN	$480

NUCLEAR RADIOLOGY

C-4-1 Casarett : RADIATION BIOLOGY	$200
C-4-2 Capp : NUCLEAR RADIATION PHYSICS	$200
E-4-1 Early : TEXTBOOK OF NUCLEAR MEDICINE TECHNOLOGY	$250
F-4-1 Fitzgerald : MATHEMATICAL THEORY OF RADIATION DOSIMETRY	$250
J-4-1 Johns : THE PHYSICS OF RADIOLOGY	$250
K-4-1 Kammash : FUSION REACTOR PHYSICS PRINCIPLES AND TECHNOLOGY	$250
N-4-1 Nicholson : NUCLEAR ELECTRONICS	$200
R-4-1 Rollo : NUCLEAR MEDICINE PHYSICS INSTRUMENTATION AND AGENTS	$400
W-4-1 Wang : RADIOTRACER METHODOLOGY IN THE BIOLOGICAL ENVIRONMENTAL AND PHYSICAL SCIENCES	$200

MATHEMATICS

A-5-1 Abramowitz : HANDBOOK OF MATHEMATICAL FUNCTIONS WITH FORMULAS GRAPHS AND MATHEMATICAL TABLES	$400
ΔB-5-1 Braun : DIFFERENTIAL EQUATIONS AND THEIR APPLICATIONS 1975 (006)	$220
ΔB-5-2 Bhat : ELEMENTS OF APPLIED STOCHASTIC PROCESSES (011)	$220
D-5-3 Dean : ELEMENTS OF ABSTRACT ALGEBRA	$110
G-5-1 Greenleaf : INVARIANT MEANS ON TOPOLOGICAL GROUPS	$50
G-5-2 Goertzel : SOME MATHEMATICAL METHODS OF PHYSICS	$140
H-5-1 Hornbeck : NUMERICAL METHODS 1975 (014)	$200
K-5-1 Karlin : A FIRST COURSE IN STOCHASTIC PROCESSES 2/e	$260
L-5-1 Liptser : STATISTICS OF RANDOM PROCESSES	
Vol. I. GENERAL THEORY	$160
Vol. II. APPLICATIONS	$140
ΔL-5-2 Lin : MATHEMATICS APPLIED TO DETERMINISTIC PROBLEMS IN THE NATURAL SCIENCES 1976 (010)	$220
ΔP-5-1 Powers : BOUNDARY VALUE PROBLEMS (016)	$100
ΔP-5-2 Pennisi : ELEMENTS OF COMPLEX VARIABLES 1976 (045)	$230
ΔS-5-1 Segel : MATHEMATICS APPLIED TO CONTINUUM MECHANICS 1977 (015)	$220